Lasers

Prentice Hall International Series in Optoelectronics

Consultant editors: John Midwinter, University College London, UK
Bernard Weiss, University of Surrey, UK

Fundamentals of Optical Fiber Communications
W. van Etten and J. van der Plaats

Optical Communication Systems (Second Edition)
J. Gowar

Lasers: Theory and Practice
J. F. B. Hawkes and I. D. Latimer

Wavelength Division Multiplexing
J. P. Laude

Optical Fiber Communications: Principles and Practice (Second Edition)
J. M. Senior

Lasers: Principles and Applications
J. Wilson and J. F. B. Hawkes

Optoelectronics: An Introduction (Second Edition)
J. Wilson and J. F. B. Hawkes

Lasers

Theory and Practice

John Hawkes
Ian Latimer

University of Northumbria at Newcastle

Prentice Hall

New York London Toronto Sydney Tokyo Singapore

First published 1995 by
Prentice Hall Europe
Campus 400, Maylands Avenue
Hemel Hempstead
Hertfordshire, HP2 7EZ
A division of
Simon & Schuster International Group

Typeset in 10/12 pt Times
by Mathematical Composition Setters, Salisbury, Wiltshire

Transferred to digital printing 2004

Printed and bound by Antony Rowe Ltd, Eastbourne

Library of Congress Cataloging-in-Publication Data

Hawkes, John.
Lasers : theory and practice / John Hawkes, Ian Latimer.
p. cm. – (Prentice-Hall international series in optoelectronics)
Includes bibliographical references and index.
ISBN 0-13-521485-8
1. Lasers. I. Latimer, Ian. II. Title. III. Series.
QC688.H38 1994
621.36′6–dc20 94-30468
CIP

British Library Cataloguing in Publication Data

A catalogue record for this book is available from the British Library

ISBN 0-13-521493-9

Contents

Preface

The advent of the laser in the second half of the twentieth century has opened up a whole new field of physics which has been taken up and developed with great enthusiasm by many scientists and engineers around the world. Most of the theory required for an understanding of laser action had, in fact, been established before the Second World War and this, together with certain technological advances made during and after the Second World War, led to the first practical realization of the laser in 1960. Since that time the laser has developed rapidly from an interesting application of quantum physics to an extremely useful engineering tool.

The incorporation of laser studies into undergraduate courses is of value for a number of reasons. Not only are the basic concepts of physics and electronics reinforced in an interesting and exciting way, but also students are introduced to important areas of modern technology. The subject, being inherently practical in nature, lends itself to many experimental demonstrations and the widespread availability of personal computers can also be put to good use in the simulation of several laser-based phenomena.

We have written this text with the aim of combining a reasonably rigorous theoretical treatment of lasers together with a discussion of some of the more practical aspects of laser technology. It is perhaps unfortunate that in a text of this size little space can be found for the many and varied applications of lasers. The book is the outcome of many years' teaching lasers at Newcastle across a wide spectrum of courses, and much of the data we have used to illustrate the theory have been derived from the results of student laboratory experiments. As in any text of this nature, it is inevitable that the research interests of the authors are emphasized to some extent. Here this manifests itself in the discussion of the frequency stability of gas lasers in Chapter 6.

The book is intended to be suitable for students on the final year of a UK honours degree/M.Sc. The first three chapters deal with the basic theory required for an understanding of laser action. Inevitably, this involves some quantum mechanics which may cause difficulties for students who are not too familiar with it. It is possible, however, on a first reading to omit a number of the sections in Chapters 1 and 3 without unduly impairing an understanding of the laser physics developed in later chapters.

Acknowledgements

We are indebted to the many students and colleagues who have, over the course of many years, helped us in our understanding of the subject and provided us with much useful experimental data. In particular, we would like to thank Dr Stephen Spoor for providing data on the iodine-stabilized helium–neon laser, the Pound–Drever stabilization scheme and on the stabilization of semiconductor lasers. Finally, a special word of thanks to our wives Carolyn and Jean for their patience and understanding during the writing of this text.

List of symbols

Wherever possible, we have tried to use the commonly accepted symbols for the various physical parameters needed. Inevitably, a number of symbols have duplicate meanings. The following list does not include every variant formed by adding a suffix, nor all cases where a symbol is used as a measure of physical distance. A symbol in bold type indicates that it represents a vector quantity.

a	aperture radius, fibre radius, lattice periodicity, Bohr radius (a_0)
A	area, spontaneous transition rate (A_{21}), mirror power absorption, element of *ABCD* matrix of optical system, vector potential (**A**)
b	mirror curvature (concave positive)
B	Einstein coefficient (B_{12}, B_{21}), element of *ABCD* matrix of optical system
$\mathscr{B}$	magnetic flux density
c	speed of light in vacuum
C	element of *ABCD* matrix of optical system, capacitance
d	cavity length, mode volume thickness
D	diameter, diffusion coefficient of electrons and holes (D_e, D_h), element of *ABCD* matrix of optical system, Density of states ($D(E)$), dipole moment (**D**)
e	electronic charge
E	energy, band gap (E_g), Fermi level (E_F), quasi-Fermi level for conduction and valence bands (E_{Fc}, E_{Fv})
$\mathscr{E}$	electric field, peak amplitude of electric field for Gaussian beam ($\mathscr{E}_0$)
$\mathbb{E}$	field component parallel to dielectric boundary
f	modulation frequency, focal length of lens, characteristic length (f_0)
F	Fermi–Dirac distribution ($F(E)$), F number of lens ($F^{\#}$)
$\mathscr{F}$	power-broadening factor, finesse, reflectivity finesse ($\mathscr{F}_R$)
g	cavity g parameter, degeneracy, lineshape function (area normalized: $g(\nu)$, height normalized: $\bar{g}(\nu)$)
G	large signal power gain, small signal power gain (G_0)
h	Planck's constant
$\hbar$	$=h/2\pi$
H	Hamiltonian operator ($\hat{H}$), Hermite polynomial of order m ($H_m(\xi)$)
$\mathscr{H}$	magnetic field
$\mathbb{H}$	magnetic field component parallel to dielectric boundary
i	electric current
I	irradiance, saturation irradiance (I_s)
j	$\sqrt{-1}$

k wave vector ($=2\pi/\lambda$), Boltzmann's constant

l angular momentum, length of gain medium, azimuthal mode number for Laguerre–Gaussian mode, coherence length (l_c)

L total angular momentum operator ($\hat{L}$), diffusion length of electrons and holes (L_e, L_h), perimeter of a ring cavity, total intracavity loss, Laguerre polynomial of order l, p and argument u. ($L_p^l(u)$), Lorentzian lineshape function ($L(\omega)$), effective cavity length (L_c)

m mass, electron mass (m_e), reduced mass (m_r), effective mass (m_e^*, m_h^*), magnetic quantum number, mode number

M mass, times diffraction limited number (M^2)

$\mathcal{M}$ general matrix element ($\mathcal{M}_{12}$)

n principal quantum number, mode number, photon density (n_p), neutral atom density, electron density (n_e), intrinsic carrier concentration (n_i), ion density (n_i), refractive index, real and imaginary parts of refractive index (n_r, n_i), ordinary and extraordinary refractive indices (n_0, n_e)

$\mathfrak{n}$ normalized photon density

N atomic population density (N_0), population inversion density, number of cavity roundtrips for a re-entrant ray, number of photons in cavity mode (N_p), Fresnel number (N_F), threshold inversion density (N_{th}), effective density of states in conduction/valence band (N_c, N_v), Avogadro's number (N_A)

$\mathcal{N}$ normalized population densities, normalized population inversion

NA numerical aperture

O optical field at frequency 2ω ($O_{2\omega}(t)$)

p hole concentration, momentum, radial mode number for Laguerre–Gaussian modes

P power, dipole moment, polarization, transition probability ($P_{12}(t)$), pressure, velocity probability distribution ($P(v_z)$), longitudinal phase factor ($P(z)$)

$\mathcal{P}$ transition probability ($\mathcal{P}_{12}(\omega)$), polarization

q longitudinal mode number, complex beam parameter $q(z)$

r radial distance, linear electro-optic coefficient, field reflectivity of a mirror, normalized pumping rate

R mirror power reflectivity, pumping rate, threshold pumping rate (R_t), radius of curvature of wavefront ($R(z)$), total reflectance of a Fabry–Perot interferometer or cavity (R_{FP}), electrical resistance

$\mathcal{R}$ matrix element of $\mathbf{r}$ ($\mathcal{R}_{12}$)

s number of cavity round trips, spin operator ($\hat{s}$)

t time, mirror field transmittance, thickness of thin film

T temperature, electron temperature (T_e), ion temperature (T_i), mirror power transmittance, total transmittance of a Fabry–Perot interferometer or cavity (T_{FP})

v velocity

V voltage, half-wave voltage (V_π), potential energy, volume

$\mathcal{V}$ electric dipole operator matrix element ($\mathcal{V}_{21}$)

w spot size of a Gaussian beam ($w(z)$), waist size of a Gaussian beam (w_0), waist size at locations 1, 2 (w_{01}, w_{02})

W power, spot size of a multimode beam ($W(z)$), waist size of a multimode beam (W_0)

x coordinate distance

X matrix element of x (X_{ij})

y coordinate distance

Y admittance of a thin film

$\mathcal{Y}$ admittance of free space

z coordinate distance, Rayleigh range (z_R), Rayleigh range in tangential and sagittal planes (z_{RT}, z_{RS})

Greek letters

α absorption coefficient, angle

β temperature coefficient of refractive index (β_n)

γ mutual coherence function (γ_{12}), saturated gain coefficient ($\gamma(\omega)$), unsaturated gain coefficient ($\gamma_0(\omega)$), threshold gain coefficient ($\gamma_{th}(\omega)$)

Γ mode confinement factor (Γ_m), active mode locking pulsewidth parameter

δ phase angle

$\Delta\nu$ free spectral range or longitudinal mode spacing, homogeneous linewidth ($\Delta\nu_h$), hole width in gain curve ($\Delta\nu_H$), resolution of Fabry–Perot interferometer ($\Delta\nu_{1/2}$); linewidth ($\Delta\nu_L$)

ε relative permittivity of medium (ε_r), permittivity of free space (ε_0)

η admittance ($=\mathbb{E}/\mathbb{H}$), admittance for s and p polarization (η_s, η_p), second harmonic generation efficiency (η_{SHG})

θ angle, the Brewster angle (θ_B), angular divergence of Gaussian beam, geometric constant of cavity

Θ angular divergence of multimode beam

κ spin eigenfunctions

λ wavelength of light in medium, (in vacuum $=\lambda_0$), Debye length (λ_D)

Λ wavelength (nonoptical)

μ Bohr magneton (μ_B), z component of magnetic moment (μ_z)

ν optical frequency (Hz)

ρ energy radiation density at angular frequency ω per unit frequency interval ($\rho(\omega)$), energy radiation density at angular frequency ω (ρ_ω)

$\sigma(\omega)$ stimulated emission cross-section

τ^{se} lifetime, cavity lifetime (τ_c), cavity roundtrip time (τ_{RT})

ϕ angle, quantum yield, roundtrip phase change in a cavity or thin film, phase angle

ψ wave function, phase change on reflection from a cavity, transverse field distribution of TEM mode ($\psi(x, y, z)$) (subscript 0 indicates TEM_{00} field)

ω angular frequency

1

Energy levels in atoms and molecules

1.1 Introduction

The development of the laser has been one of the great triumphs of science in the twentieth century. The foundations were laid by Einstein in 1917 [Ref. 1], who pointed out that the equation proposed by Planck to describe the spectral distribution of light emitted from a black body could be derived quite simply by assuming the existence of a hitherto-unknown type of light-emission process which has since become known as *stimulated emission*. For something like 30 years after this, however, the concept of stimulated emission was only used in theoretical discussions and rarely had any relevance for experimental work.

The idea that a system of molecules or atoms could give rise to the amplification of a beam of light follows directly from Einstein's work, but equally well it is easy to show that amplification can only be obtained under a condition known as *population inversion*. However, systems in thermal equilibrium cannot exhibit population inversion, and it took some considerable time before physicists overcame their instinctive mistrust of nonequilibrium situations. The situation changed after the Second World War when microwave sources, developed for use in radar systems, opened up a whole new wavelength region for spectroscopic studies. One of the areas of interest was the study of *magnetic resonance*, where transitions between energy levels involving the magnetic moment of the nucleus were investigated. It was observed that the nuclear magnetic system could take a considerable time (i.e. up to several minutes) to respond to temperature changes of the lattice. It was then realized that nonequilibrium situations could be prepared and studied much more easily than had previously been supposed.

Population inversion between nuclear magnetic energy levels was first unambiguously demonstrated by Purcell and Pound in 1951 [Ref. 2] and after this date the pace began to quicken. The first public description of the idea of amplification of (microwave) radiation due to stimulated emission was given by Weber in 1952 and the first amplifier using stimulated emission (within the ammonia molecule) was made in 1953. The device was called a *MASER* (*M*icrowave *A*mplification by the *S*timulated *E*mission of *R*adiation). Maser operation was soon obtained in a wide variety of different systems including ruby (that is, sapphire containing small amounts of chromium). The push was then on to extend the maser principle to include wavelengths in the optical region (i.e. to make an *optical maser*). This was achieved in 1960 by T. H. Maiman at the Hughes Research Laboratory, using ruby as the active material [Ref. 3]. Thereafter there followed a veritable explosion of development which produced a huge number of diverse lasers in many different kinds of media (the term *LASER*, *L*ight *A*mplification by the *S*timulated *E*mission of *R*adiation, soon replacing the term 'optical maser'). For those interested a detailed coverage of the history of the development of the laser is contained in Ref. 4.

Any understanding of lasers must begin with a treatment of the interaction between light and matter. The most accurate theoretical technique available for this is quantum electrodynamics, a discipline with which few undergraduates are familiar (let alone authors of laser textbooks!). Fortunately it is possible to treat the interaction with reasonable accuracy by 'semi-classical' techniques, that is, by using a judicious mixture of classical and quantum physics. Even so, there is quite a daunting amount of theory to be covered. Students who find quantum mechanics difficult (or even impossible!) need not despair since they may omit sections of the present chapter and Chapter 3 at a first reading without being disadvantaged in the remainder of the book. It is suggested that in the present chapter Sections 1.2.2 onwards need not be followed in detail, although students should make sure they have a reasonably good qualitative idea of the origins of the energy levels in atoms, molecules and semiconductors.

Before launching into detailed theory it may be useful to give a brief introduction to the basic ideas behind laser action. As we shall see later in this chapter, atoms (and molecules) can exist in a variety of different energy states (called *stationary states*). In the presence of an electromagnetic field there is the possibility that the atom can change between one energy state and another. Thus atoms can change from a lower energy state to a higher one provided that at the same time a photon (i.e. quantized packet of radiation energy) is *absorbed*.

The reverse process can also take place, that is, an atom can change from a higher energy state to a lower one with the emission of a photon. Assuming that the law of energy conservation holds for these processes, the photon energy must equal the energy change undergone by the atom. However, as will be seen in Chapter 2, the energy of a photon (E_p) is related to the frequency of the radiation involved (ν) by the equation

$$E_p = h\nu \tag{1.1}$$

where h (*Planck's constant*) has the value 6.626×10^{-34} Js. If an atom changes its energy from E_2 to E_1 ($E_2 > E_1$), then we can write

$$h\nu = E_2 - E_1 \tag{1.2}$$

Thus if a group of atoms is excited into a variety of upper energy states then the emission spectra should consist of a series of single frequencies each corresponding to a particular transition between pairs of energy levels. When observed in a spectrometer each single frequency shows up as a line, and is thus usually referred to as a *spectral line*. In fact, as we shall see in Chapter 3, all spectral 'lines' have a finite frequency width but the term 'line' is still used. Figure 1.1 shows part of the emission spectrum resulting from an electrical discharge in a mixture of helium and neon.

An important feature of light emission from an atom is that there are in general two different processes at work which are termed *spontaneous* and *stimulated* emission. As its name implies, the former does not require the presence of radiation and occurs spontaneously whereas the latter is found only in the presence of radiation. Stimulated emission can occur when a photon passes near to an atom which is in an excited state and where the photon energy is equal to the energy difference between the two levels. One very important feature is that the photon emitted in this way travels in exactly the same direction as the stimulating photon.

Let us suppose we have a collection of atoms and that we assume, for simplicity, that each atom has only two energy states, E_1 and E_2. The numbers of atoms in these energy states is denoted by N_1 and N_2 respectively. If a beam of radiation of frequency ν_{12}, where $\nu_{12} = (E_2 - E_1)/h$, is passed through this collection of atoms then both absorption and stimulated emission processes take place.

It turns out that the probability of a photon being absorbed, when it encounters an atom in the lower energy state, is the same as the probability of the photon causing a stimulated emission process when it encounters an atom in the upper energy state.[1] Thus as the beam passes through the group of atoms photons will be

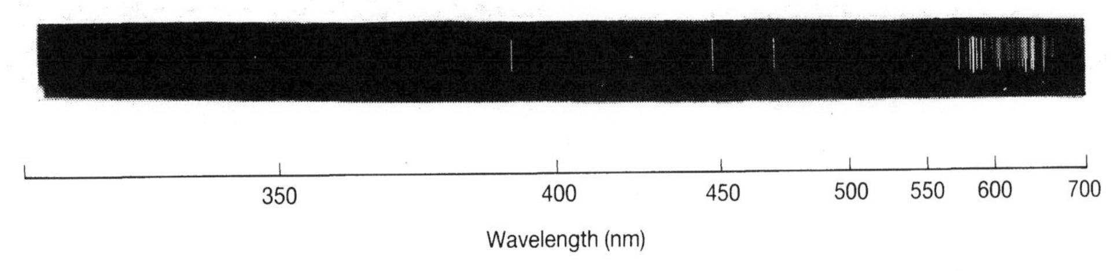

Figure 1.1 Part of an emission spectrum of neon. Each 'line' results from a strong emission which takes place over a narrow range of wavelengths and corresponds to a particular transition between the electron energy levels of neon.

lost, through absorption, at a rate which is proportional to N_1 and will be generated, through stimulated emission, at a rate proportional to N_2. Provided photons are not lost through any other processes, the numbers of photons, n, will change in time according to

$$\frac{dn}{dt} = C(N_2 - N_1) \tag{1.3}$$

where C is some constant.

If $N_2 > N_1$ then the photon numbers will grow with time, that is, the beam irradiance will be increased or amplified as it passes through the collection of atoms. This is the process of *L*ight *A*mplification by the *S*timulated *E*mission of *R*adiation (or *LASER* action) which occurs in all lasers. The requirement that N_2 be larger than N_1 is the condition referred to as population inversion. Unfortunately, when the atoms are in thermal equilibrium N_2 will always be less than N_1 and special techniques have to be employed to achieve population inversion. These inevitably involve the input of external energy into the laser system (a process often described by the term 'pumping').

To get a general 'feel' at this stage as to how a laser operates it is useful to consider a relatively simple example, namely the ruby laser. The lasing medium is a rod of aluminium oxide (Al_2O_3) doped with about 0.05 per cent by weight of chromium. Chromium ions, Cr^{3+}, replace aluminium ions on the lattice and provide the energy levels for the lasing transitions. For the present purposes we need only consider three of the energy levels of the Cr^{3+} ion of which the lowest is the ground state (Figure 1.2). We will refer to these levels in increasing order of energy by '0', '1' and '2'. The two higher levels are sufficiently far above the ground state so that, in thermal equilibrium, their populations are negligible compared with that of the ground state.

Pumping is provided by illuminating the rod with the output from a high-power flash tube. This provides an intense pulse of radiation lasting for several milliseconds. The absorption process $0 \Rightarrow 2$ is relatively strong and a large number of atoms are raised into level 2. Atoms in level 2 then make a rapid (nonradiative) transition to level 1, which has a comparatively long lifetime ($\cong$ 3 ms). Thus atoms tend to 'pile up' in level 1 and, given sufficiently intense pumping, a population

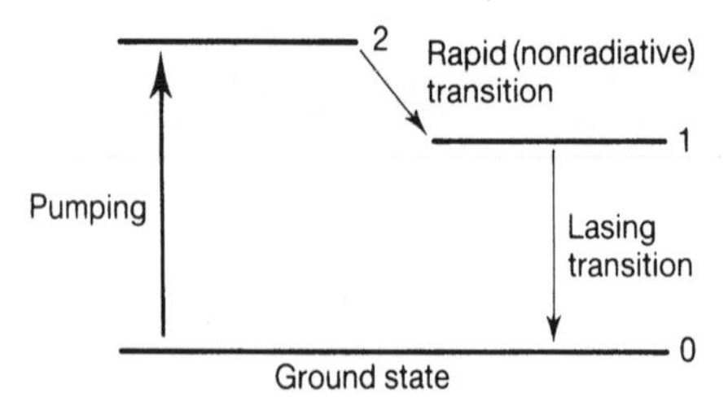

Figure 1.2 Schematic diagram of the lowest energy levels of the Cr^{3+} ion in Al_2O_3 showing the pumping and laser transitions.

inversion can subsequently be built up between levels 1 and 0. Any light of the correct frequency (i.e. resulting from spontaneous emissions between levels 1 and 0) which is travelling down the axis of the rod can then be amplified.

Unless the pumping is exceptionally intense, however, the beam irradiance will not be particularly strong after just one transit of the lasing medium. To be able to make use of and study the special properties of laser radiation the output often needs to be increased. One way of achieving this is to make the gain medium appear longer than it is by the simple expedient of placing reflecting mirrors at either end (Figure 1.3). In the first laser the mirrors were formed by evaporating silver on the plane ends of the ruby rod. The mirrors and the space between constitute what is called the *laser cavity*. The light beam is then reflected back and forth through the medium within the cavity. On each pass the irradiance will increase. However, as the beam irradiance increases, the gain will tend to decrease, since when gain takes place the population inversion is reduced. In fact, the irradiance saturates when the losses per transit (that are inevitably present) exactly match the gain per transit.

Obviously if both mirrors are 100 per cent reflecting at the lasing wavelength then no radiation will actually emerge from the laser cavity, so that it is usual to have only one very highly reflecting mirror while the other (with somewhat less than 100 per cent reflectance, say 95 per cent) allows light to emerge. As might be expected, the earliest lasers used plane mirrors, but, as we shall see in later chapters, there are a number of advantages in using curved mirrors.

The mirrors provide the system with *optical feedback* and it is possible to regard the laser as the optical equivalent of an electronic oscillator (which itself is basically an amplifier provided with the correct amount of positive feedback). Thus a laser is somewhat more than just an optical amplifier and is more accurately termed a *laser oscillator*. The influence of the mirrors on the behaviour of the laser is considerably more subtle than might at first be expected on a 'simple reflection' basis. For example, if we imagine a 'plane wave'-like wavefront propagating backwards and forwards through the cavity then a requirement that must be imposed for 'stable' behaviour is that the phase of the wavefront at a particular point in the cavity must always be the same (to within an integer multiple of 2π). This places quite severe restrictions on the form of the wavefront and also on the frequencies of the radiation that can propagate.

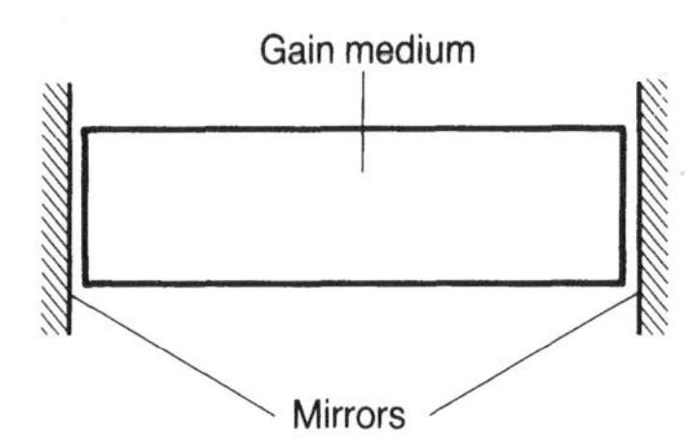

Figure 1.3 The basic elements of a laser. A rod-shaped gain medium has mirrors placed at either end to provide optical feedback.

So far, we have not mentioned the origin of the initial beam. In fact, it originates in a photon derived from *spontaneous emission* that just happens to be emitted along the axis of the laser shortly after population inversion has been achieved.

This brief discussion highlights some of the important areas that need to be addressed when trying to achieve an understanding of lasers. First, there is the basic energy level structures of atoms and molecules themselves and these are dealt with in the remainder of this chapter. Next, we need to understand some of the basic properties of light and its interactions with matter, topics covered in Chapters 2 and 3, respectively. As mentioned above, the laser cavity itself exercises an important influence on the characteristics of laser radiation and, accordingly, Chapters 4 and 5 contain a detailed analysis of the cavity properties. Since mirrors play such a vital role, a section on their technology is also included. Chapter 6 discusses the general properties of the radiation emitted from lasers while Chapter 7 describes the various types of laser that are most common. Finally, Chapter 8 discusses various techniques used to modify the characteristics of laser radiation.

1.2 The hydrogen atom

We start our investigation by looking at the simplest atom of all, the hydrogen atom. This consists of a single electron moving in the potential of a single proton. Before we consider a quantum-mechanical treatment, however, it is useful to review the model proposed by Niels Bohr in 1913, which was the first to give a satisfactory explanation of the line spectra of hydrogen.

1.2.1 The Bohr model

Bohr based his model on the work of Rutherford, who had shown that atoms consist of a heavy, positively charged nucleus surrounded by a number of negative electrons. To explain why the electrons did not collapse onto the nucleus it was suggested that they orbited the nucleus in circular orbits in much the same way that planets orbit the sun.

If the radius of the electron orbit is r and the electron velocity v, then the acceleration towards the centre is v^2/r. The electrostatic force acting on the electron is $e^2/(4\pi\varepsilon_0 r^2)$, where ε_0 is the permittivity of free space, and so, from Newton's second law of motion, we must have

$$\frac{m_e v^2}{r} = \frac{e^2}{4\pi\varepsilon_0 r^2} \tag{1.4}$$

where m_e is the electron mass. The total energy of the electron (kinetic + potential[2]) is readily calculated to be (Problem 1.1)

$$E = -\frac{e^2}{8\pi\varepsilon_0 r} \tag{1.5}$$

So far, provided Eq. (1.4) is satisfied, there are no restrictions on the values for the electron orbit radius (r) or velocity (v). However, Bohr suggested that the electron angular momentum (i.e. $m_e vr$) had to take on multiple values of the quantity $h/2\pi$ (or $\hbar$). Thus

$$m_e vr = nh/(2\pi) \tag{1.6}$$

where n is a positive integer. v may now be eliminated from Eqs (1.3) and (1.5) to yield

$$r_n = \frac{n^2 h^2 \varepsilon_0}{e^2 \pi m_e} \tag{1.7}$$

that is,

$$r_n = n^2 a_0$$

where

$$a_0 = h^2 \varepsilon_0 / (e^2 \pi m_e) \tag{1.8}$$

a_0 is called the *Bohr radius*.

Thus the orbital radii are restricted to be multiples (i.e. 1, 4, 9, 16, etc.) of the Bohr radius and this in turn implies that the energy is also restricted. Substituting for r_n from Eq. (1.7) in Eq. (1.5) yields

$$E_n = -\frac{1}{n^2}\left[\frac{e^4 m_e}{8\varepsilon_0^2 h^2}\right] \tag{1.9}$$

Example 1.1 Line emission in hydrogen

We may use Eq. (1.9) in conjunction with Eq. (1.2) to calculate the wavelengths of some of the emission (or absorption) lines in hydrogen. By direct evaluation we have that

$$\frac{e^4 m_e}{8\varepsilon_0^2 h^2} = 2.18 \times 10^{-18}\ \text{J}$$

Thus a transition from the level $n = 5$ to the level $n = 2$, for example, corresponds to a change in energy of

$$2.18 \times 10^{-18}\left[\frac{1}{4} - \frac{1}{25}\right]$$

or 4.57×10^{-19} J. The corresponding wavelength is then given by

$$\frac{hc}{\lambda} = 4.57 \times 10^{-19}\ \text{J}$$

whence $\lambda = 0.433 \times 10^{-6}$ m, or 0.43 μm (i.e. violet radiation).

In our derivation of Eq. (1.9) we have ignored the fact that, because of the finite mass of the nucleus, the electron and nucleus will in fact revolve round their common centre of mass. It is easy to show (see Problem 1.2) that this may be accounted for by replacing the electron mass (m_e) in Eq. (1.9) by the *reduced mass* m_r given by

$$\frac{1}{m_r} = \frac{1}{m_e} + \frac{1}{m_p} \qquad (1.10)$$

One of the problems with this model is that because the electron is moving in a circle it is accelerating. According to classical theory, an accelerating charge radiates electromagnetic energy, and the electron should thus lose energy and rapidly spiral into the nucleus. Bohr had to assume, on a rather *ad hoc* basis, that somehow this was not allowed to happen, and that the electron orbits were stable.

The energy levels predicted by Eq. (1.9) are illustrated in Figure 1.4. When the emission spectra of a hydrogen discharge is examined the frequencies of the lines correspond with a high degree of accuracy to transitions between these energy levels. Figure 1.5 shows some of these transitions.

Unfortunately, the Bohr model can only be used where one electron is circulating the nucleus. Where more than one electron is involved, then the interactions between the electrons make the problem impossibly complicated. Furthermore, when the hydrogen spectrum is examined under high resolution, a *fine structure* is revealed, that is, lines which are supposed to represent transitions between single energy levels split into two or more lines close together, indicating that the energy level structure

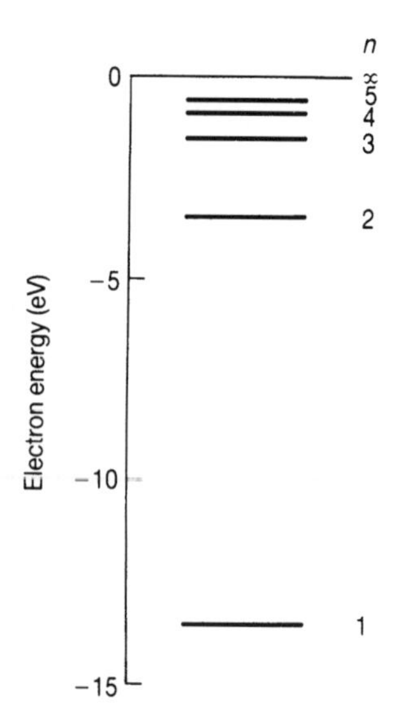

Figure 1.4 Electron energy levels in hydrogen as given by the Bohr model.

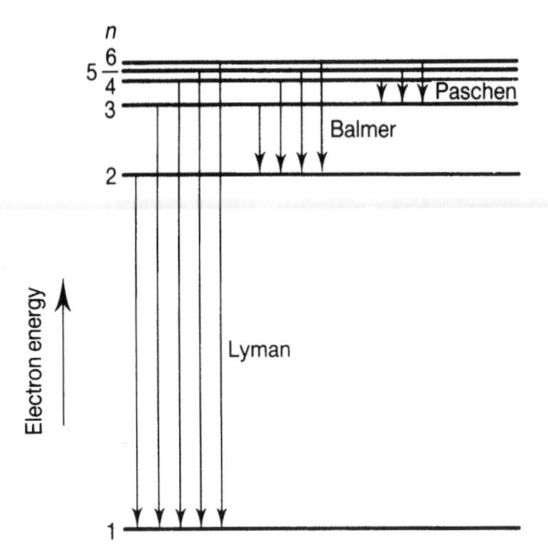

Figure 1.5 Some of the possible transitions between the electron energy levels of the hydrogen atom.

must be more complicated than that given by the Bohr model. These problems cannot be dealt with properly without recourse to quantum mechanics.

1.2.2 A quantum-mechanical treatment of the hydrogen atom

In quantum mechanics[3] the measurable quantities associated with a particular physical system are represented by linear Hermitian *operators*. A measurement of a particular quantity will then yield a value which is an *eigenvalue* of the corresponding operator. Thus if the operator is $\hat{O}$ then the possible eigenvalues (O) are found by seeking functions ψ that satisfy the *eigenvalue equation*:

$$\hat{O}\psi = O\psi \tag{1.11}$$

We are interested here in the possible energy states of an atomic system and accordingly we need to solve the energy eigenvalue equation (i.e. Eq. (A1.18)):

$$\hat{H}\psi = E\psi \tag{1.12}$$

where $\hat{H}$, the energy operator, is known as the *Hamiltonian*[4] operator, and E is the energy of the system.

The Hamiltonian operator is obtained by writing down the classical expression for the energy of the system in terms of spatial coordinates and momenta and then by replacing the momenta by the equivalent operators so that, for example p_x is replaced by the operator

$$\hat{p}_x = -\mathrm{j}\hbar \frac{\partial}{\partial x}$$

where $\mathrm{j} = \sqrt{-1}$.

If we consider a single particle of mass m and momentum p moving in a potential $V(\mathbf{r})$, then the energy (H) can be written

$$H = \frac{p^2}{2m} + V(\mathbf{r})$$

Since we can write

$$p^2 = p_x^2 + p_y^2 + p_z^2$$

where p_x, p_y and p_z are the x, y and z components, respectively, of the momentum, we then have

$$\hat{p}^2 = -\hbar^2\left[\frac{\partial^2}{\partial x^2} + \frac{\partial^2}{\partial y^2} + \frac{\partial^2}{\partial z^2}\right]$$

Introducing the operator ∇^2 where

$$\nabla^2 = \frac{\partial^2}{\partial x^2} + \frac{\partial^2}{\partial y^2} + \frac{\partial^2}{\partial z^2} \tag{1.13}$$

we can write

$$\hat{H} = -\frac{\hbar^2}{2m}\nabla^2 + V(\mathbf{r})$$

The eigenvalue equation (Eq. (1.12)) then becomes

$$-\frac{\hbar^2}{2m}\nabla^2\psi(\mathbf{r}) + (V(\mathbf{r}) - E)\psi(\mathbf{r}) = 0 \tag{1.14}$$

where we have made explicit the functional dependence of the eigenfunction on spatial coordinates. This equation is often known as the (time-independent) Schrödinger equation, and the resulting eigenfunctions (i.e. ψ) as *wavefunctions*[5]. The wavefunctions themselves have a direct physical interpretation. Thus the quantity

$$\psi^*(x, y, z)\psi(x, y, z)\,\mathrm{d}x\,\mathrm{d}y\,\mathrm{d}z$$

where ψ^* represents the complex conjugate of ψ, is proportional to the *probability* that the particle will be found within a small volume element $\mathrm{d}x\,\mathrm{d}y\,\mathrm{d}z$ of space at the point x, y, z.

We now turn to the problem of the hydrogen atom. For simplicity, we regard the centre of mass of the proton as fixed and take the origin of our coordinate system to be at the proton. The potential energy $V(\mathbf{r})$ of the electron is given by

$$V(\mathbf{r}) = -\frac{e^2}{4\pi\varepsilon_0 r} \tag{1.15}$$

The Hamiltonian operator is then

$$\hat{H}_c = -\frac{\hbar^2}{2m_e}\nabla^2 - \frac{e^2}{4\pi\varepsilon_0 r} \tag{1.16}$$

Because the potential is spherically symmetric then the operator ∇^2 is most conveniently expressed in spherical polar coordinates. Thus

$$\nabla^2 = \frac{1}{r^2}\left[\frac{\partial}{\partial r}\left(r^2\frac{\partial}{\partial r}\right) + \frac{1}{\sin\theta}\frac{\partial}{\partial\theta}\left(\sin\theta\frac{\partial}{\partial\theta}\right) + \frac{1}{\sin^2\theta}\frac{\partial^2}{\partial\phi^2}\right] \tag{1.17}$$

The solution of Eq. (1.13) under these circumstances is dealt with in standard texts on quantum mechanics (see, for example, Ref. 5). It is usual to assume that we can write

$$\psi(r, \theta, \phi) = R(r)\Theta(\theta)\Phi(\phi)$$

a technique known as *separation of the variables.*

The Schrödinger equation can be split up into three separate (but linked) equations which may be solved separately. In writing down the solutions it is customary to take the product of the two angular functions together and call them *spherical harmonics*, $Y_{l,m}(\theta, \phi)$. The first few spherical harmonics are tabulated in Appendix 2.

That portion of the Schrödinger equation which involves the spherical harmonics can be written in the form

$$\hat{\mathbf{L}}^2 Y_{l,m}(\theta, \phi) = l(l+1)\hbar^2 Y_{l,m}(\theta, \phi) \tag{1.18}$$

where $\hat{\mathbf{L}}^2$ is an operator representing the sum of the squares of the x, y and z angular momenta (i.e. $\hat{\mathbf{L}}^2 = \hat{l}_x^2 + \hat{l}_y^2 + \hat{l}_z^2$). Thus Eq. (1.18) can be regarded as an eigenvalue equation showing that the eigenvalues of $\hat{\mathbf{L}}^2$ are given by $l(l+1)\hbar^2$. The total angular momentum of the system can then be written $(l(l+1))^{1/2}\hbar$. l is known as the *angular momentum quantum number.* All positive values of l are allowed up to a maximum of $n-1$, where n is another quantum number involved in the radial equation.

It can also be shown (Problem 1.4) that the operator $\hat{\mathbf{L}}^2$ commutes with the operators $\hat{l}_x$, $\hat{l}_y$ and $\hat{l}_z$, but that the latter three do *not* commute with each other. The significance of this (see Appendix 1) is that it should then be possible to construct eigenfunctions which are simultaneously eigenfunctions of $\hat{\mathbf{L}}^2$ and *one* of the components of $\hat{l}$ (conventionally taken to be l_z) but not all three. In spherical polar coordinates $\hat{l}_z$ is represented by the operator $-j\hbar(\partial/\partial\phi)$, and since the spherical harmonics have a ϕ dependence given by $\exp(jm\phi)$, we then have

$$\hat{l}_z Y_{l,m}(\theta,\phi) = m\hbar Y_{l,m}(\theta,\phi) \tag{1.19}$$

Thus the value of the z component of the angular momentum is $m\hbar$. In fact, the z designation here has no particular significance. Because the system is spherically symmetrical any particular axis will do. It is often useful to imagine an axis (referred to as the z axis) to be picked out by the presence of, say, a (vanishingly small) magnetic field. A finite magnetic field would, in fact, cause states with different m values to have different energies. Because of this m is known as the *magnetic quantum number*. m can take on all the integer values between -1 and $+1$.

The solutions of the radial equation, $R_{n,l}(r)$, depend on the two quantum numbers n and l mentioned above, and the first few are tabulated in Appendix 2. Figure 1.6 shows the variation of $R_{nl}(r)$ with r for $n = 1$ and 2.

When the wavefunction $\psi(r,\theta,\phi)$ $(=R_{nl}(r)Y_{l,m}(\theta,\phi))$ is used in Eq. (1.12), the energy eigenvalues are given by

$$E_n = -\frac{e^4 m_e}{2(4\pi\varepsilon_0\hbar^2)n^2} \tag{1.20}$$

an expression which is identical to that given by the simple Bohr model (i.e. Eq. (1.9)).

Thus the energy is determined solely by the quantum number n, which is called the *principal quantum number*. Since for each value of n, l can vary from 0 to $n-1$ and m from -1 to $+1$, there will be a number of different wavefunctions which

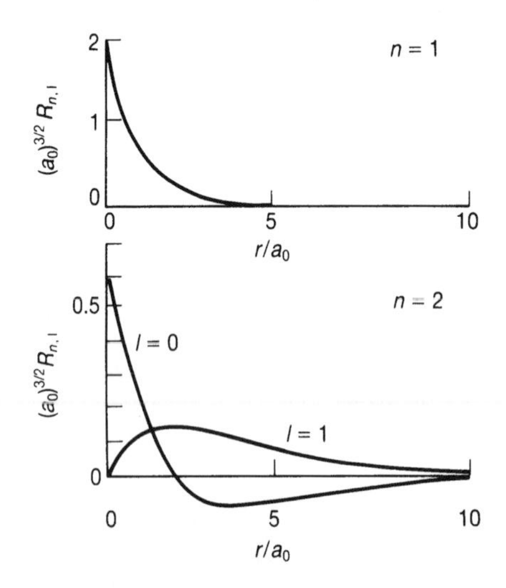

Figure 1.6 Variation in the radial part of the electron wavefunction (i.e. $R(r)$) of the hydrogen atom for the $n = 1$ and $n = 2$ wave functions.

have the same value of the energy. This number is known as the *degeneracy* of the level. In the present instance the degeneracy of each energy level is given by $\Sigma_1^{n-1}(2l+1)$ or n^2.

1.2.3 Spin–orbit coupling

We turn now to the problem of the fine structure which is observed in the hydrogen spectra but which is not predicted by Eq. (1.20). In 1925 Uhlenbeck and Gouldsmit showed that the splittings could be obtained by assuming that the electrons had both an intrinsic angular momentum (called *spin*) and a magnetic moment. The theoretical explanation of the electron spin and magnetic moment came when Dirac sought to derive a relativistically correct Hamiltonian operator. This resulted in an equation whose solution predicted particles (i.e. electrons) with a total intrinsic angular momentum of $\hbar\sqrt{3}/2$ and a magnetic moment of $e\hbar/(2m_e)$. The latter quantity is known as the *Bohr magneton* (see Problem 1.9) and designated by μ_B.

As in the case of the angular momentum we can construct simultaneous eigenfunctions of the spin operators $\hat{s}^2$ and $\hat{s}_z$. If we write these eigenfunctions as $\varkappa$, then

$$\hat{s}^2\varkappa = s(s+1)\hbar^2\varkappa \tag{1.21}$$

and

$$\hat{s}_z\varkappa = m_s\hbar\varkappa \tag{1.22}$$

where $s = 1/2$ and m_s can take the values $\pm 1/2$. For convenience, we adopt the notation that $\varkappa_+$ and $\varkappa_-$ represent spin eigenfunctions with $m_s = 1/2$ and $-1/2$, respectively.

The component of magnetic moment along any particular axis (z, say) is related to the corresponding value of the spin component by

$$\mu_z = -(2\mu_B/\hbar)m_s \tag{1.23}$$

The presence of an electronic magnetic moment implies that there will be additional energy terms whenever magnetic fields are present. The fact that the electron is itself circulating a charged nucleus produces such an interaction. This is most easily appreciated by imagining ourselves at rest with regard to the electron. On the simple Bohr model the nucleus will then appear to rotate round the electron. Because the nucleus is charged the electron is effectively at the centre of a current loop. The magnetic field from the current loop will then interact with the magnetic moment of the electron (see Problem 1.10). Because of its origin this is referred to as the *spin–orbit* interaction. A proper treatment of the effect involves relativistic corrections and the resulting expression for the spin–orbit Hamiltonian can be written:

$$\hat{H}_{so} = \zeta(r)\hat{\mathbf{s}}.\hat{\mathbf{l}} \tag{1.24}$$

where

$$\hat{\mathbf{s}}.\hat{\mathbf{l}} = \hat{s}_x\hat{l}_x + \hat{s}_y\hat{l}_y + s_z\hat{l}_z$$

and

$$\zeta(r) = \frac{1}{2m^2c^2r}\frac{\mathrm{d}V}{\mathrm{d}r}$$

V being the potential in which the electron moves. The total Hamiltonian for the hydrogen atom can now be written

$$\hat{H} = \hat{H}_c + \hat{H}_{so}$$

Rather than solving the Schrödinger equation all over again for the new Hamiltonian, it is convenient to use perturbation theory since the effects of the spin–orbit term are small compared with the Coulomb term (see Appendix 1). The energy shifts from the values given by Eq. (1.20) can then be written (Eq. (A1.25))

$$\Delta E_{so} = \langle \psi | \hat{H}_{so} | \psi \rangle \tag{1.25}$$

where ψ represents the unperturbed wavefunction.

Unfortunately, we are dealing here with a set of degenerate wavefunctions so the procedure is not entirely straightforward (see Ref. 6, for example). To further complicate the picture two additions to the Hamiltonian are required: first, a term representing a relativistic correction and second, the Darwin term which only applies to states where $l = 0$. When all these are taken into account the calculated positions of the energy levels agree very well with experiment. It should be mentioned, however, that when the lines are observed under even higher resolution, further splitting (the *hyperfine* splittings) can be observed. These are attributable to several other (much smaller) interactions arising from such things as the finite nuclear size and the nuclear magnetic moment. We need not pursue these further.

1.3 The many-electron atom

1.3.1 Exchange parity and the Pauli exclusion principle

As soon as there is more than one electron surrounding the nucleus the problem is complicated by the presence of an electrostatic repulsion between the electrons. The force acting on one particular electron then depends on the positions of all the other electrons. But then we do not know where *they* are until we have solved the first problem! An exact solution is not possible and we have to resort to techniques involving successive approximations. Before we tackle this, however, there is another important matter to be discussed which arises because electrons are identical particles. Our labelling of any one electron is therefore completely arbitrary, there can be no physical difference between wavefunction $\psi(\mathbf{r}_1, \mathbf{r}_2)$ and $\psi(\mathbf{r}_2, \mathbf{r}_1)$. To examine the consequences of this we introduce an operator $\hat{P}$, the *exchange* operator, where

$$\hat{P}_{12}\psi(\mathbf{r}_1, \mathbf{r}_2) = \psi(\mathbf{r}_2, \mathbf{r}_1) \tag{1.26}$$

If the eigenvalues of the operator are p, that is, if

$$\hat{P}_{12}\psi(\mathbf{r}_1, \mathbf{r}_2) = p\psi(\mathbf{r}_1, \mathbf{r}_2)$$

then

$$\begin{aligned}\hat{P}_{12}(\hat{P}_{12}\psi(\mathbf{r}_1, \mathbf{r}_2)) &= p^2\psi(\mathbf{r}_1, \mathbf{r}_2)\\ &= \psi(\mathbf{r}_1, \mathbf{r}_2)\end{aligned}$$

Thus $p^2 = 1$, and so $p = \pm 1$.

Wavefunctions with $p = 1$ are said to be *symmetric* and have *even exchange parity* while those with $p = -1$ are *antisymmetric* and have *odd exchange parity*. The particles of nature can be divided into two groups, *fermions*, which have half integer spin, and *bosons*, which have integer spin. Fermions (e.g. electrons, protons, neutrons, etc.) are only found in odd exchange parity states, while bosons (e.g. pions, kaons, photons, etc.) are only found in even exchange parity states.

For simplicity, we consider first a two-electron atom and ignore the effects of inter-electron interactions. If one electron is described by a (normalized) spatial wavefunction ψ_a and the other by ψ_b and both spins are $+1/2$, then a suitable antisymmetric wavefunction (see Problem 1.11) can be written

$$\begin{aligned}\psi_{12} &= \frac{1}{\sqrt{2}}(\psi_a(\mathbf{r}_1)\psi_b(\mathbf{r}_2) - \psi_b(\mathbf{r}_1)\psi_a(\mathbf{r}_2))\chi_+^{(1)}\chi_+^{(2)}\\ &= \frac{1}{\sqrt{2}}\begin{vmatrix}\psi_a(\mathbf{r}_1)\chi_+^{(1)} & \psi_a(\mathbf{r}_2)\chi_+^{(2)}\\ \psi_b(\mathbf{r}_1)\chi_+^{(2)} & \psi_b(\mathbf{r}_2)\chi_+^{(2)}\end{vmatrix}\end{aligned} \tag{1.27}$$

We may also note that if the functions $\psi_a\chi$ and $\psi_b\chi$ are identical then the total antisymmetric wavefunction will vanish. This is a result known as the *Pauli exclusion principle*. It is sometimes stated in the form that no two electrons can share the same set of quantum numbers.

1.3.2 The central field approximation

Our first approximation for a multi-electron atom involves neglecting any interactions between electrons. The only remaining interactions are then those between the individual electrons and the Z protons on the nucleus. The wavefunctions and charge distributions will be as for the hydrogen atom except for the replacement of the nuclear charge of e by Ze. Bearing in mind the Pauli exclusion principle, the lowest energy state of an atom is then given by allocating the Z electrons one by one to the lowest available hydrogen-like wavefunctions.

Electrons with the same values of the principal quantum number n are said to be in the same *shell* while those with the same value of n and l are said to be in the same *subshell*. The l value of an electron is referred to by using a letter as given by the following:[6]

l value:	0	1	2	3	4	5
letter:	s	p	d	f	g	h

A complete listing of the electron n and l values in a particular atom is called the *electron configuration*. For example, the electron configuration for sodium, which has 11 electrons, is written:

$$\text{Na: } 1s^2 2s^2 2p^6 3s^1$$

Sodium thus has two electrons in the 1s subshell (i.e. with $n = 1$ and $l = 0$), two electrons in the 2s subshell, six electrons in the 2p subshell and one electron in the 3s subshell. At the present level of approximation all the electrons in the same shell are degenerate. This degeneracy is lifted when the electrostatic interaction between electrons is considered and is usually carried out by using the so-called *central field approximation*. In this it is assumed that each electron moves in a symmetric central potential which can be written

$$V(r) = -\frac{Z(r)e^2}{4\pi\varepsilon_0 r} \tag{1.28}$$

This is the same as the simple Coulomb potential except for the function $Z(r)$, which represents the effective charge as seen by an electron when at a distance r from the nucleus. When an electron is very close to the nucleus, with little probability of any other electron being between it and the nucleus, then we would expect $Z(r) \cong Z$. At very large distances from the nucleus, when the nuclear charge of Ze will be almost completely screened by the remaining $Z - 1$ electrons, we would expect $Z(r) \cong 1$.

One way of determining $Z(r)$ is to assume that the electron wavefunctions are the same as those of the hydrogen atom (i.e. as indicated in the electron configuration). $Z(r)$ is then evaluated by calculating the probability that each electron is inside a sphere of radius r, and then assuming that the resulting charge distribution inside the sphere is symmetrical. Using the potential of Eq. (1.28) a new set of wavefunctions is then calculated. This new set can then in turn be used for a more accurate determination of $Z(r)$. This iterative process is repeated until there is no significant change in the wavefunctions on each iteration. The technique is called the *Hartree self-consistent field* method. A more accurate version of the same method (the *Hartree–Fock* method) results when determinental product states are used (as in Eq. (1.27)) as the starting wavefunctions to determine $Z(r)$, although the calculations are much more lengthy.

The resulting set of single-particle energy levels are similar to those of the hydrogen atom, except that they now depend on l as well as n. This is *not* a fine structure-splitting effect, but arises because states with different l values have different radial probability densities. Thus *s* states give rise to higher electron concentrations near the nucleus than do *p* states. They are therefore not as well screened from the nucleus and thus have a lower energy. The assumption of spherical symmetry for the charge distribution is not as restricting as it might sound. Because of the properties of spherical harmonic functions, filled subshells are always spherically symmetric, so that the only lack of spherical symmetry arises from the electrons in incomplete subshells. These usually involve only the outermost electrons.

1.3.3 The Coulomb and spin-orbit interactions

The previous section showed how the ground state (i.e. lowest energy state) of a multi-electron atom could be determined. Higher energy levels of the atom result if electrons are excited into higher states. Laser transitions usually only involve relatively low-lying energy states resulting from the excitation of electrons which are not in closed (i.e. full) shells and subshells. Thus if we take the Nd^{3+} ion it has the electron configuration

$$1s^2 2s^2 2p^6 3s^2 3p^6 3d^{10} 4s^2 4p^6 4d^{10} 5s^2 5p^6 4f^3$$

As far as the low-lying energy levels are concerned, we need only consider the three 4f electrons which lie outside closed subshells. At our present level of approximation the resulting determinental product states are highly degenerate (there are in fact some 364 of them, see Problem 1.12). This degeneracy is partially lifted when the two next most important interactions are considered. These are the inter-electron Coulomb interaction, which is not spherically symmetric, and cannot therefore be adequately dealt with using the central field approximation, and the spin–orbit interaction. We may write these additional contributions to the Hamiltonian as

$$\hat{H}_c + \hat{H}_{so} = \sum_{\substack{i,j \\ i \neq j}} \frac{e^2}{4\pi\varepsilon_0 r_{ij}} + \sum \zeta(r_i)\hat{\mathbf{s}}_i.\hat{\mathbf{l}}_i \tag{1.29}$$

The way in which we now proceed depends on the relative magnitudes of the splittings produced by these terms. If the effects of the Coulomb term dominate, as is the case in the lighter elements, then the *LS* (or *Russell–Saunders*) coupling approximation is used. If the reverse is the case, then the *j–j* coupling approximation is used.

1.3.3.1 LS coupling

In the *LS* coupling scheme we first determine the energy splitting arising from the inter-electron Coulomb term $\hat{H}_c$. The effect is to form states called *terms* which are described in terms of their total angular (L) and spin (S) momentum values. If, for example, we have two electrons with l values of l_1 and l_2, the possible total angular momentum quantum numbers (i.e. L values) are integers with values which lie between $|l_1 + l_2|$ and $|l_1 - l_2|$. Similarly, S can take on integer values between $|s_1 + s_2|$ and $|s_1 - s_2|$.

The L values of the terms are denoted using a variant of the spectroscopic notation, that is, L values of 1, 2, 3, 4, ... are represented by the capital letters S, P, D, F..., respectively. The value of S is not indicated directly, instead the value of $2S + 1$ is quoted (a quantity called the *multiplicity*). For example, the term ^{3}F represents a state with $L = 4$ and $S = 1$. Each term has a degeneracy of $(2S + 1)(2L + 1)$. Example 1.2 illustrates these ideas by considering the terms resulting from a 3p and 3d electron.

Example 1.2 Term formation (3p + 3d electron)

With a 3p and a 3d electron we have $l_1 = 1$ and $l_2 = 2$, and the possible L values are then 3, 2 and 1. Similarly, the possible S values from the two spins of 1/2 are 1 and 0. Altogether six terms may be formed, namely,

$$^3\mathrm{F},\ ^3\mathrm{D},\ ^3\mathrm{P},\ ^1\mathrm{F},\ ^1\mathrm{D},\ ^1\mathrm{P}$$

The total degeneracy of all the above states will then be

$$3 \times 7 + 3 \times 5 + 3 \times 3 + 1 \times 7 + 1 \times 5 + 1 \times 3 = 60$$

As far as the initial 3p and 3d electron states are concerned these have individual degeneracies of 2×3 (i.e. 6) and 2×5 (i.e. 10), respectively, again giving rise to a total degeneracy of 6×10 (i.e. 60).

When two or more of the electrons have the same n and l quantum numbers (i.e. are *equivalent*) then the Pauli exclusion principle acts to rule out some of the terms that would otherwise be present. For example, if we had a $2p^2$ configuration, then only the terms ^{3}P, ^{1}D and ^{1}S are allowed (if the p electrons were nonequivalent another three terms, ^{3}D, ^{3}S and ^{1}P would be present).

The reader should note that the general treatment of any number of equivalent electrons is not a completely trivial problem, and is best undertaken using group theory. Even the $2p^2$ configuration requires some careful analysis [Ref. 7].

Using first-order perturbation theory the changes in the term energies when the Coulomb interaction is introduced are given by

$$\Delta E = \left\langle LSM_LM_S \left| \frac{e^2}{4\pi\varepsilon_0 r_{12}} \right| LSM_LM_S \right\rangle$$

An explicit evaluation of this matrix element requires that the states[7] $|LSM_LM_S\rangle$ be expressed in terms of the determinental product states, which have, in principle, a known dependence on the spatial coordinates (Ref. 8 contains further details). When this is done two general trends are observed:

1. Terms with the same value of spin multiplicity tend to be grouped together, those groups with the highest values of S having the lower energies.
2. Within the same spin multiplicity group terms are ordered according to their L values, those with the highest L values having the lowest energies.

Both these 'tendencies'[8] (they are by no means always strictly obeyed) arise from the nature of the inter-electron Coulomb force. The spin dependency may seem rather puzzling at first since we are not dealing here with a spin-dependent force. However, it is one of the subtler outcomes of the requirement that the wavefunction be antisymmetric with respect to particle exchange. Electrons with the same spin cannot have the same space-dependent wavefunctions, and they will thus tend to stay away from each other (we could say rather loosely that they 'repel' each other). Electrons with different spins, on the other hand, *can* have identical spatial wavefunctions and so will tend to be closer together (i.e. the particles 'attract' each other). These apparent repulsive and attractive 'forces' are sometimes called *exchange forces*. When the inter-electron Coulomb force is introduced the state in which the electrons have the greatest separations will have the lowest energies. As we have just seen, this will tend to be the states with the largest S values.

With regard to the L value ordering, if the electrons are seeking to maximize their separation, then they could do so by being arranged equidistantly and symmetrically about the nucleus and by then all rotating in the same direction. This would involve the electrons all having the same l values and thus give rise to the largest L value. Hence the association of the largest L value with the lowest energy.

The second interaction we have to consider in this section is the spin–orbit interaction which can be written [Ref. 9] as

$$\hat{H}_{so} = \zeta(L, S)\hat{\mathbf{L}}.\hat{\mathbf{S}} \tag{1.30}$$

The effect of this operator on the terms depends not only on the value of L and S but also on the values of another quantum number J, which represents the total angular momentum. J can take on integer (if S is an integer) or half-integer (if S is a half-integer) values from a maximum of $L + S$ down to $|L - S|$. So that, for example, the ^{3}F term gives rise to states with $J = 2$, 3 and 4. The J value of a particular state is denoted by a suffix which directly indicates the J value. Thus in the case of the ^{3}F term we obtain the states (or *multiplets* as they are often called) $^3F_2$, $^3F_3$ and $^3F_4$. The multiplets have a remaining degeneracy of $2J + 1$.

The energy shifts of the multiplets from their term values can be shown to be given by

$$\Delta E = \tfrac{1}{2}\zeta(L, S)[J(J+1) - L(L+1) - S(S+1)] \tag{1.31}$$

Thus states with the same L and S values will be split according to the value of J. Provided $L > S$ the number of different J states is just $2S + 1$ and so the term will split into $2S + 1$ multiplets. This explains why the value of $2S + 1$ is quoted when designating the term, and is also why terms with $S = 0$ and $S = 1$ are referred to as 'singlets' and 'triplets', respectively. Assuming that $\zeta(L, S)$ is positive, states with the lowest values of J will lie lowest. The effects of successively applying the inter-electronic Coulomb interaction and spin–orbit coupling on a 3p3d electron configuration is shown schematically in Figure 1.7.

One of the most important solid-state lasers is based on neodymium-doped yttrium aluminium garnet ($Y_3Al_5O_{12}$), or Nd : YAG. The Nd^{3+} ions have four 4f electrons outside closed subshells. Figure 1.8 shows the effects of electrostatic and spin–orbit interactions on the $4f^3$ configuration.

1.3.3.2 j–j coupling

As the atomic number of an atom increases there is a general tendency for the inter-electron Coulomb interaction to become smaller and for the spin–orbit interaction to increase, so that in atoms such as lead, for example, we must consider the effects of the spin–orbit interaction before we deal with the inter-electron Coulomb interaction. This is a scheme known as *j–j coupling*. (There are a number of instances, for example in neon, where both interactions producc splitting of the same magnitude. In these cases neither *LS* or *j–j* coupling is valid.)

In *j–j* coupling the individual electrons in the configuration are all subject to spin–orbit coupling and will give rise to states with different j_i values. The lowest level is expected to be that where all the electrons have their smallest j_i values $(= l_i - 1/2)$. The resulting levels may be described using the notation $(j_1, j_2 \ldots)$:

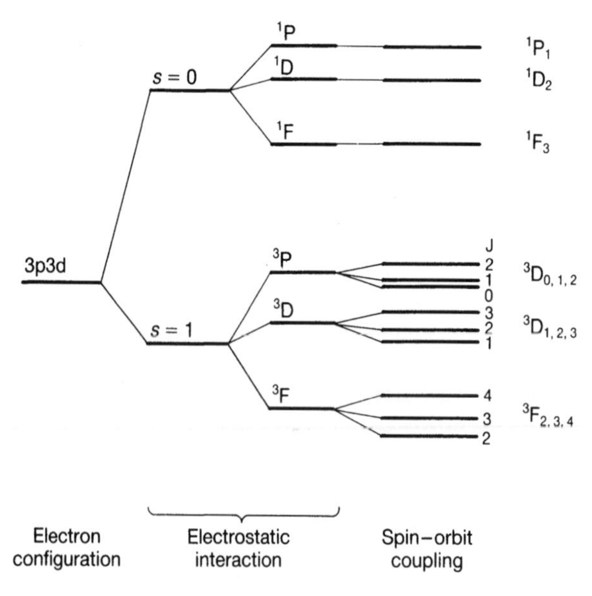

Figure 1.7 Illustration of the effects of inter-electron electrostatic interactions and spin–orbit coupling on the 3p3d electron configuration.

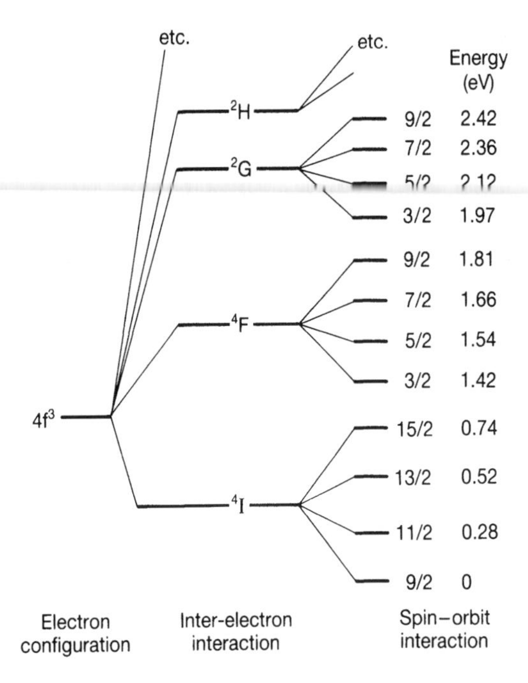

Figure 1.8 Origins of the low-lying energy levels of Nd^{3+}.

The inter-electron Coulomb interaction then splits these states into a number of substates characterized by the possible values of J (these will have integer spacings between $j_1 + j_2$ and $|j_1 - j_2|$). The final states are thus designated by the notation $(j_1, j_2 \ldots)_J$.

1.3.4 The crystal field interaction

So far, we have been considering the energy levels of isolated atoms. A number of interesting laser materials involve ions which are present as substitutional impurities in crystalline solids (obvious examples being ruby and Nd : YAG). The interaction of the ion with the lattice is called the *crystal field* interaction. One obvious origin for the interaction is the Coulomb force between the host ionic charges and the electrons on the impurity ion (this in fact is not the only interaction but we need not consider others here). An important feature of the crystal field is that it will have the same symmetry as that of the lattice. Indeed, part of the notation used to label the resulting states derives from the symbols used in group theory to deal with the same type of symmetry.

The effect of the crystal field depends markedly on how well shielded are the electrons outside filled subshells. In the case of Nd^{3+}, for example (a 'rare earth ion'), the electron configuration is $4f^3$. Now in fact these are not the 'outermost' electrons since the 5s and 5p subshells are filled and these electrons tend to screen the 4f electrons from any external electrostatic field. Consequently the effect of the crystal field is smaller than that of the spin–orbit coupling, and can be treated as a perturbation on the *LSJ* multiplets. As might be expected, the crystal field splits the multiplets into further states, the number of resulting states depending on the symmetry of the crystal field. For fields of low symmetry each multiplet with half-integer J is split into, at most, $J + 1/2$ crystal field levels. Since the total degeneracy of a J multiplet is $2J + 1$ the crystal field states are still at least twofold degenerate (the remaining degeneracy can be removed by applying a magnetic field).

A feature of the crystal field levels is that they are not as 'narrow' as the isolated atom energy levels. Figure 1.9 shows the absorption spectrum of a piece of Nd^{3+} : YAG and the broad nature of the crystal field levels is apparent. This broadening has two main origins. First, the strength of the crystal field will vary slightly throughout the material due to the presence of inhomogeneities and second, the constant thermal motion of the lattice will continually 'modulate' the energy levels of any one particular ion and smear them out.

In the case of Cr^{3+} (a transition metal ion) there are three electrons in a 3d subshell. In contrast to Nd^{3+}, these electrons are not at all well shielded from the crystal field and in consequence the crystal field interaction is much larger than the spin–orbit interaction. Thus in Cr^{3+} we must consider the effects of the crystal field interaction before those of spin–orbit coupling. As in the case of Nd^{3+}, the degree of splitting depends on the symmetry of the lattice, and so it is difficult to generalize without recourse to group-theoretical arguments. The term energy states are split into substates which are labelled with a letter relating to the symmetry of the crystal field (*not* to the angular momentum). The spin multiplicity is still quoted in the usual way. Thus in the $3d^3$ configuration of Cr^{3+} : Al_2O_3 the ^4F term (which is the ground state) is split into three crystal field levels $^4A_2$, $^4T_2$ and $^4T_1$ (A_2, T_2 and T_1 designate different representations of the octahedral group; the suffices are *not* related to any J values). Figure 1.10 shows the successive splittings of the $3d^3$ configuration.

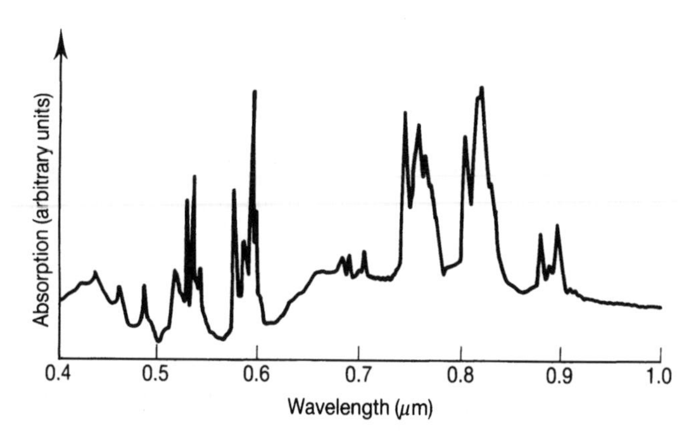

Figure 1.9 Optical absorption spectrum of Nd:YAG.

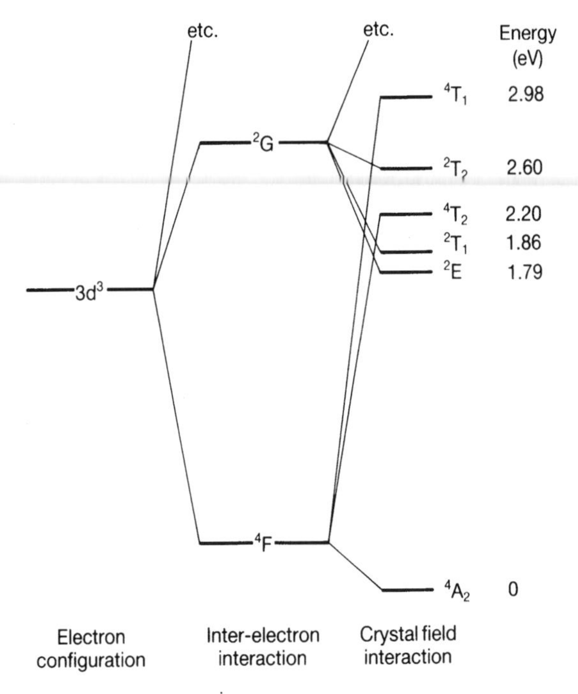

Figure 1.10 The origins of the low-lying electronic energy levels in Cr^{3+}.

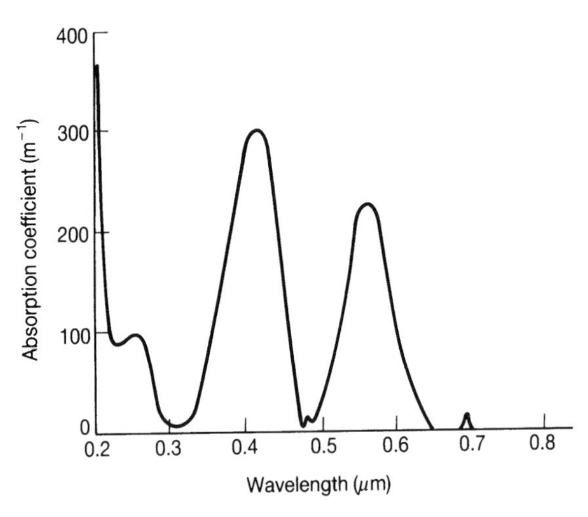

Figure 1.11 The room temperature absorption spectrum of ruby which has a Cr^{3+} ion concentration of 2×10^{25} m^{-3}. The absorption spectrum is in fact dependent on the direction of polarization of the light relative to the crystallographic axes. For simplicity, the curve here has been drawn assuming unpolarized light.

In theory we should then go on to consider the effects of spin–orbit coupling. However, the crystal field blurring effects noted in the case of Nd^{3+} : YAG are so large here that they mask any spin–orbit effects. An indication of the relatively large line broadening is given in Figure 1.11, which shows the absorption spectrum of ruby at room temperature.

1.4 Molecular spectra

In the previous sections of this chapter we have considered the energy levels of a single atom (or ion). Unfortunately, in the real world atoms interact, and, because of the interaction, can form states involving atomic aggregates (for example, molecules) which have lower overall energies than would the same number of separate atoms. Not surprisingly, chemistry was one of the first scientific disciplines to study and classify these interactions, although a full understanding of their nature had to await the arrival of quantum mechanics.

The binding forces are usually described in terms of a *chemical bond*. The two main types of bond are called *ionic* and *covalent*. The easiest of these to understand is the ionic bond, which occurs, for example, in sodium chloride (NaCl). Here sodium has filled subshells apart from a single 3s electron, while chlorine has a $3p^5$ outer subshell. Thus if an electron is transferred from the sodium to the chlorine then both atoms will attain closed subshell status. This process will require energy, but since the closed subshell is a particularly stable (i.e. low-energy) configuration the energy required is more than offset by the resulting loss in electrostatic energy since the atoms have become electrically charged (i.e. have become ions).

Covalent bonds are rather more difficult to describe and are best dealt with by analyzing the simplest possible type of molecule consisting of a single electron and two (separated) protons (the hydrogen molecule ion).

1.4.1 The hydrogen molecule ion

We consider here the problem of a single electron which is shared between two protons. The protons are much heavier than the electron and will therefore respond much more 'slowly' to any forces that are present. It is therefore possible to simplify the problem by adopting the *adiabatic* approximation. This assumes that the electron will be always be able to reach an equilibrium state before the relative positions of the protons can change significantly. The problem then divides into two parts. First, we seek the electron wavefunctions (and corresponding energy levels) when protons have a fixed separation. Then the motions of the protons are dealt with by using as a potential function the electrostatic repulsion between them combined with the instantaneous total electronic energy. The former analysis is at the heart of chemical bonding, while the latter gives rise to the idea of rotational and vibrational states.

We make no attempt here to solve the problem of an electron moving in the field of two fixed protons. However, it is possible to make some progress by examining the form the wavefunctions are expected to take. The origin of the coordinate system is taken to be a point midway between the protons, the line joining the protons being the x axis (Figure 1.12). If the proton spacing is r, then the Hamiltonian for the electronic motion can be written

$$\hat{H}_e = \frac{\hat{p}^2}{2m} - \frac{e^2}{4\pi\varepsilon_0}\{[(x-r/2)^2+y^2+z^2]^{-1/2} + [(x+r/2)^2+y^2+z^2]^{-1/2}\} \tag{1.32}$$

We may introduce here the *parity* operator, $\hat{P}_x$ (note this is *not* the same as the exchange parity operator of Section 2.1). The parity operator has the effect of replacing x by $-x$. If the eigenvalue of $\hat{P}_x$ is P_x, then

$$\begin{aligned}\hat{P}_x^2\psi(x,y,z) &= P_x^2\psi(x,y,z)\\ &= \psi(x,y,z)\end{aligned}$$

Thus (as in the case of exchange parity) $P_x = \pm 1$.

We now consider the situation when the protons are separated by relatively large distances. It seems reasonable that solutions to the Schrödinger equation can be solved by allowing the electron to inhabit hydrogen-like states centred round one of the protons. We let $\psi_n(1)$ and $\psi_n(2)$ represent the (normalized) hydrogen-like wavefunctions when the electron is orbiting protons 1 and 2, respectively. Unfortunately, neither $\psi_n(1)$ or $\psi_n(2)$ are eigenfunctions of the operator $\hat{P}_x$. However, as may readily be verified (see Problem 1.15), the functions

$$\begin{aligned}\psi_n^s &= 2^{-1/2}[\psi_n(1)+\psi_n(2)]\\ \psi_n^a &= 2^{-1/2}[\psi_n(1)-\psi_n(2)]\end{aligned} \tag{1.33}$$

are eigenfunctions of $\hat{P}_x$, with eigenvalues of $+1$ and -1, respectively. Figure 1.13(a) is a sketch of the symmetric and antisymmetric functions ψ_n^s and ψ_n^a.

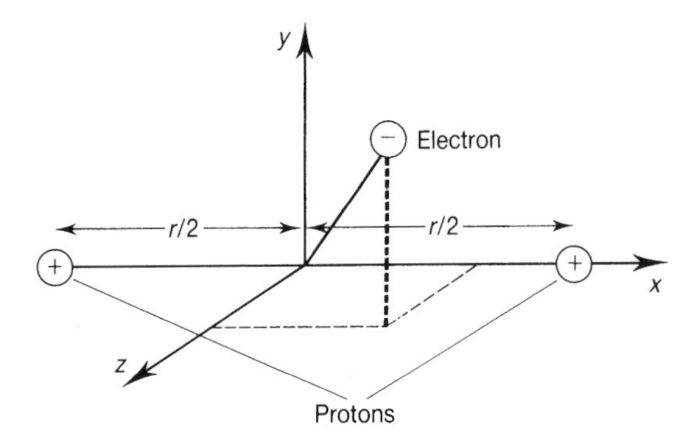

Figure 1.12 Coordinates used in the analysis of the energy levels of the hydrogen molecular ion.

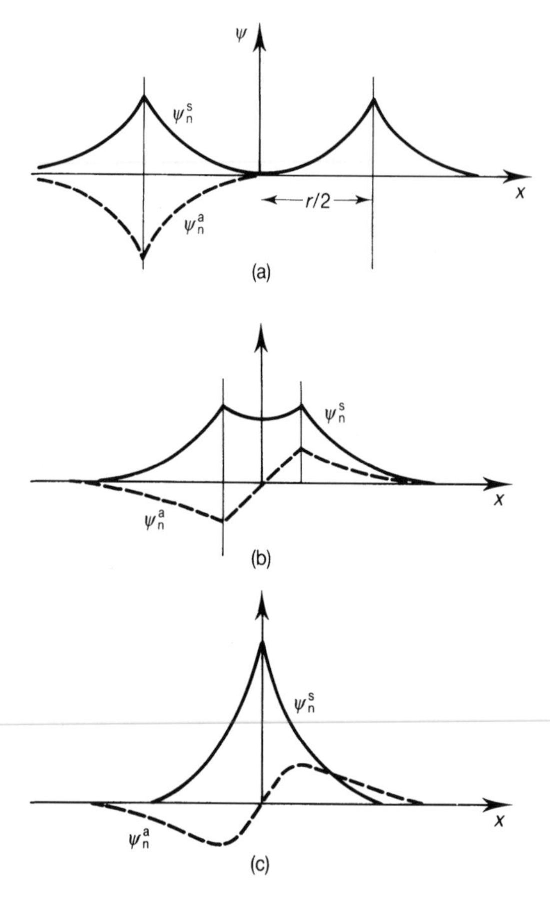

Figure 1.13 Behaviour of the lowest energy symmetric and antisymmetric wavefunctions of the hydrogen molecular ion.

Although these functions are obviously degenerate when the proton separation is large, this is not so when the protons move closer together. Figures 1.13(b) and 1.13(c) show how these two wavefunctions might be expected to change as r is decreased. When $r = 0$ we effectively have a singly ionized helium atom, that is, the wavefunctions must be the same as for hydrogen but with the nuclear charge now $2e$ instead of e. A comparison between the radial hydrogenic wavefunctions of Figure 1.6 and Figure 1.13(c) shows that the state ψ^s_n $(r = \infty)$ tends to the state ψ_{100} of ionized helium (when $r = 0$) while ψ^a_n tends to the state ψ_{210}. Figure 1.14 shows the energy changes of the electron wavefunctions as a function of r. (Note that the $n = 2$ state of ionized helium has exactly the same energy as the $n = 1$ state of hydrogen, which explains the almost constant energy of the antisymmetric state.)

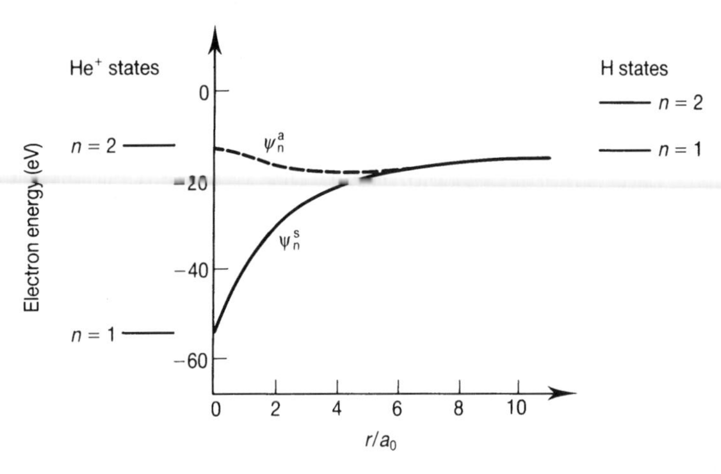

Figure 1.14 Electron energy of the symmetric and antisymmetric wavefunctions of the hydrogen molecular ion as a function of ionic separation.

To obtain the total energy of the molecule we must add in the electrostatic energy $(= e^2/(4\pi\varepsilon_0 r))$ between the protons. The total energy for the symmetric and antisymmetric states as a function of r is shown in Figure 1.15. There is a minimum in the energy of the symmetrical wavefunction, which represents the equilibrium separation of the protons in ionized helium. The depth of the minimum below the $r = \infty$ energy represents the binding energy of the molecule. The binding results from the lower energy exhibited when the electron is shared between the two nuclei.

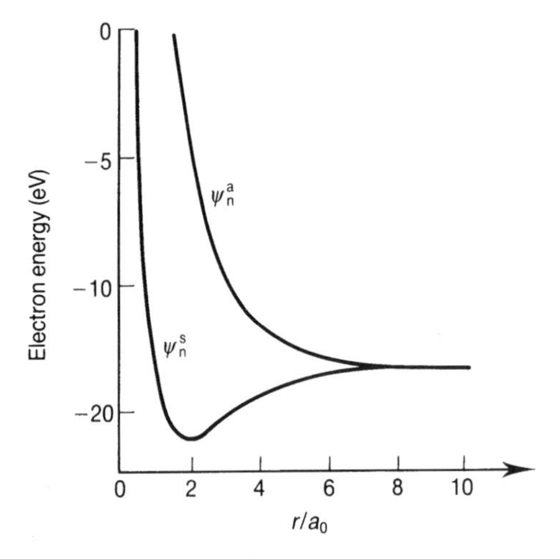

Figure 1.15 Total energy for the symmetric and antisymmetric electronic states of the hydrogen molecular ion.

1.4.2 The hydrogen molecule

The next simplest molecule is the neutral hydrogen molecule. This in fact has several features in common with the ionized hydrogen molecule. The wavefunctions must again have even or odd parity with respect to x so that, for large proton separations, they can be expressed as symmetric or antisymmetric combinations of the wavefunctions for either two neutral hydrogen atoms or a negative and positive hydrogen ion. The behaviour of the electronic energy of the ground state with r is similar to that of Figure 1.14. Now, however, the energy at large r is approximately equal to twice the ground state energy of the hydrogen atom, while when $r = 0$, it is equal to the ground state energy of the helium atom. The total ground state energy of the molecule shows a minimum similar to that in Figure 1.15. The presence of two electrons, however, leads to some additional features. For example, the wavefunctions (including spin) must be antisymmetric with regard to electron exchange. This requirement leads to effects similar to those discussed in Section 1.3.3.1, that is, to a tendency for parallel spins to repel and for anti-parallel spins to attract. The ground state of the helium atom then has an electron wavefunction which has even spatial parity and zero total spin. This is a situation characteristic of the covalent bond, where two electrons are shared (or 'exchanged') between two similar atoms giving rise to eigenfunctions which are symmetric in the space coordinates but antisymmetric in the spins.

1.4.3 Electronic molecular states

For the simple diatomic molecule there is an obvious axis of symmetry for the system which joins the two nuclei (i.e. the z axis). This cylindrical symmetry of the diatomic molecule implies that only the component of angular momentum along the z axis is now a 'good' quantum number. Thus the single-electron angular momentum components are designated by a quantum number λ which takes on integer values, the actual value being designated by a Greek letter according to the scheme

λ value	0	1	2	3
designation	σ	π	δ	ϕ

The actual axial angular momentum for a particular value of λ is then $\pm\lambda\hbar$ (i.e. the state is twofold degenerate).

When more than one electron is present the total axial component of momentum (Λ) is formed by addition of the individual λ values (bearing in mind that each can take both a negative and a positive value). Thus

$$\Lambda = \sum_i \lambda_i \tag{1.34}$$

The numbers are again labelled using Greek letters (this time capitals), thus:

Λ value	0	1	2	3
designation	Σ	Π	Δ	Φ

States still have a twofold degeneracy. This remains true even when the residual interactions are considered except for the Σ state, which splits into Σ^+ and Σ^- states.

Provided spin–orbit coupling is not too large, the total spin S remains a good quantum number and can be included in the orbital designation. Thus from two electrons in nonequivalent π and δ orbitals, states (or terms) can be formed with $\Lambda = 1$ and 2 in combination with triplet and singlet states. That is, we have $^1\Sigma$, $^1\Delta$, $^3\Sigma$ and $^3\Delta$ states. If two electrons are in the same orbital then the Pauli exclusion principle acts to exclude some terms. In the above, for example, the term $^3\Delta$ would be forbidden. Spin–orbit coupling then splits the terms into states which can be labelled by the total value of the angular momentum (spin plus orbital) about the inter-nuclear axis. This angular momentum can be written as $\Omega\hbar$, where Ω takes on values an integer step apart between $|\Lambda + S|$ and $|\Lambda - S|$.

1.4.4 Vibrational and rotational states

In discussing the molecular states for diatomic molecules we have so far neglected any motion of the nuclei. The two most important motions are vibrations along the line joining the nuclei and rotations about any axis perpendicular to this line. When the molecule is in its lowest energy state we expect the total energy to be at a minimum (i.e. see Figure 1.15), and hence as far as small variations in separation (r) are concerned, the potential is parabolic in r. This potential is, in fact, that of the *harmonic oscillator*. This is a problem of great importance in quantum mechanics, but unfortunately we have not enough space to deal with the problem here (see, for example Ref. 9). If the potential variation with r is written as

$$V(r) = V_0 + \tfrac{1}{2}kx^2 \tag{1.35}$$

then the vibrational energy eigenvalues are given by

$$E_n = (n + \tfrac{1}{2})h\nu_0 \tag{1.36}$$

where

$$\nu_0 = \frac{1}{2\pi}\left(\frac{k}{m_r}\right)^{1/2} \tag{1.37}$$

and m_r is the reduced mass of the two atoms (i.e. see Eq. (1.10)), that is,

$$m_r = \frac{m_1 m_2}{m_1 + m_2} \tag{1.38}$$

The vibrational energy levels are thus expected to be equally spaced, with typical spacings (see Example 1.3) of a few tenths of an electron volt. This is much less than the typical level spacing for the electronic energy levels which are an order of magnitude or so larger. In practice, because the potential differs from the ideal of Eq. (1.35), when the atoms are far away from their equilibrium positions the higher vibrational energy levels tend to become closer together (rather than being equally spaced).

The other type of motion we need to consider is rotation of the atoms about their common centre of mass. They have a total moment of inertia, I, of $m_r r^2$ (see Problem 1.16) so that the rotational energy of the molecule can be written

$$E_r = R^2/(2I) \tag{1.39}$$

where R is the rotational angular momentum. If the problem is dealt with quantum mechanically the operator corresponding to R^2 has eigenvalues of $K(K+1)\hbar$, where K is a positive integer. The rotational energy levels can thus be written

$$E_K = \frac{K(K+1)\hbar^2}{2I} \tag{1.40}$$

The spacing between adjacent levels is then

$$E_K - E_{K-1} = K\hbar^2/I \tag{1.41}$$

For the smaller values of K the energy spacings of the rotational levels are some one or two orders of magnitude smaller than the vibrational level spacings (see Example 1.3).

Example 1.3 Vibration and rotational energy levels in HCl

In the ground state of the HCl molecule the equilibrium atomic separation is 0.129 nm and the force constant (i.e. k value) is 480 N m^{-1}. The mass of a hydrogen atom is 1.67×10^{-27} kg while that of a chlorine atom is 5.89×10^{-26} kg. Thus the reduced mass m_r is given by

$$m_r = \frac{1.67 \times 10^{-27} \times 5.89 \times 10^{-26}}{1.67 \times 10^{-27} + 5.89 \times 10^{-26}}$$
$$= 1.62 \times 10^{-27} \text{ kg}$$

The vibrational energy level separation is $h\nu_0$, where ν_0 is given by Eq. (1.37):

$$\nu_0 = \frac{1}{2\pi}\left(\frac{480}{1.62 \times 10^{-27}}\right)^{1/2}$$
$$= 8.66 \times 10^{13} \text{ Hz}$$

The vibrational level spacing, ΔE_v (in eV), is then

$$\Delta E_v = h\nu_0/e$$
$$= 6.626 \times 10^{-34} \times 8.66 \times 10^{13}/1.6 \times 10^{-19}$$
$$= 0.36 \text{ eV}$$

To calculate the positions of the rotational energy levels (Eq. (1.40)) we need a value for $\hbar^2/I$. The moment of inertia, I, is given by

$$\begin{aligned} I &= m_r r_0^2 \\ &= 1.62 \times 10^{-27} \times (0.129 \times 10^{-9})^2 \text{ kg m}^2 \\ &= 2.7 \times 10^{-47} \text{ kg m}^2 \end{aligned}$$

Therefore

$$\begin{aligned} \hbar^2/I &= (1.055 \times 10^{-34})^2/2.7 \times 10^{-47} \\ &= 4.12 \times 10^{-22} \text{ J or } 2.6 \times 10^{-3} \text{ eV} \end{aligned}$$

Thus the first excited rotational level ($K = 1$) will be 2.6×10^{-3} eV above the ground state.

The wavefunctions associated with the rotational motion are, perhaps not surprisingly, the same as the angular part of the hydrogen atom wavefunctions with l and m replaced by J and M_J (M_J having integer values from $-J$ to $+J$). The parity of such states is even (symmetric) when J is even and odd (antisymmetric) when J

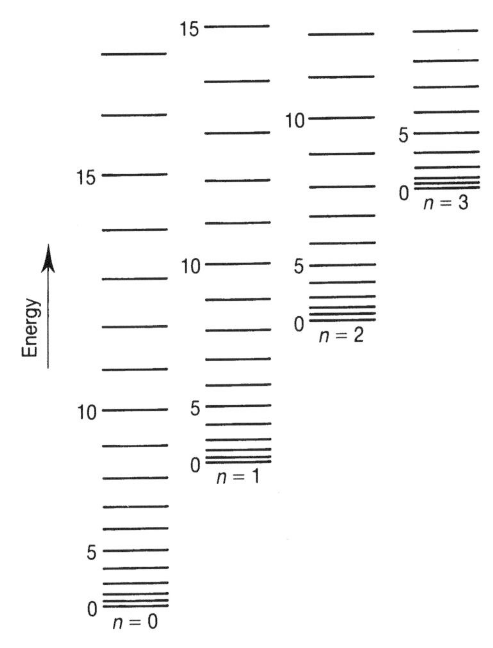

Figure 1.16 Schematic diagram of the vibrational and rotational energy levels resulting from a single electronic term. The levels have been grouped, for convenience, according to their differing vibrational origins. In practice the levels would be superimposed on each other.

is odd. If a molecule is symmetric (i.e. H_2, O_2, etc.) then the odd parity wavefunctions are not allowed.

To a reasonably good approximation the wavefunction of the whole molecule is given by forming antisymmetric products of the electronic, vibrational and rotational wavefunctions. The energy of the wavefunction (E_T) is then just the sum of the individual energies, i.e.

$$E_T = E_e + E_n + E_J$$

where E_e is the energy of the electronic wavefunction and E_n and E_J are given by Eqs (1.44) and (1.48) below.

A schematic representation of the molecular energy levels for a particular electronic term is shown in Figure 1.16.

1.5 Energy levels in semiconductors

In semiconductors we are dealing with the situation where the constituent atoms are sufficiently close enough for the outermost electrons to interact strongly. In a sense, the whole piece of material forms a 'giant molecule'. There are two fundamental methods of analyzing the situation which in fact start from opposing viewpoints. These are the *tight binding approximation* and the *nearly free electron model.*

1.5.1 The tight binding approximation

In this approach the electrons are assumed to inhabit atomic-like energy levels which are then perturbed due to the interactions with electrons on neighbouring atoms. We will not investigate this model in any mathematical detail. However, it is easy to see qualitatively what happens by supposing that the atomic spacing can be altered at will. If the spacing is increased sufficiently so that the interactions between neighbouring atoms are vanishingly small then the energy level structure must closely resemble that of a single atom (we assume a monatomic material for simplicity). Each single 'atomic' level in the system will, of course, be highly degenerate (at least equal to the number of atoms present).

If, however, the atomic spacing is decreased the inter-atomic interactions will assume an increasing importance. It should be no surprise that the effect of the interaction is to split the degenerate levels with the splitting increasing as the atoms get close together. Thus each original energy level becomes a *band* of energy levels where each band contains a *very* large number (approximately equal to the number of atoms present) of individual levels. We expect the width of the band to increase as the atomic spacing decreases as illustrated in Figure 1.17. There is even the possibility that the bands will overlap. In typical semiconductors, however, the equilibrium atomic spacing results in a sequence of *energy bands* with *energy gaps* in between.

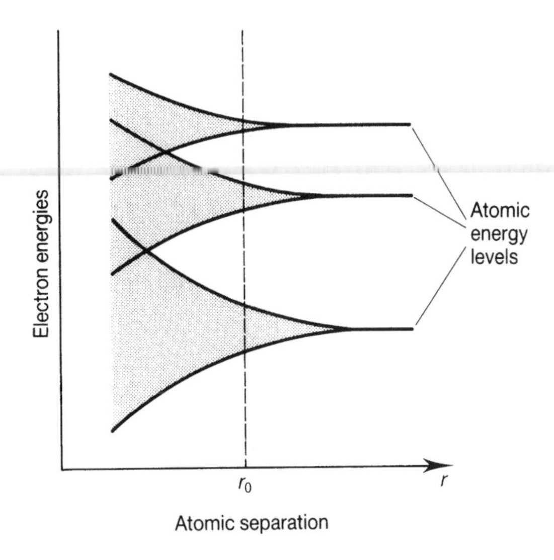

Figure 1.17 A schematic illustration of the origin of the energy bands in solids. The value r_0 represents a typical value for the atomic separation.

Not all the bands will be occupied since not all the original atomic-like energy levels are occupied. At low temperatures the isolated atoms will be in their ground states. This follows through into the band picture; all bands up to a particular one are full and are empty thereafter. The most important features as far as we are concerned are the highest occupied band (the *valence band*) and the first empty (at $T = 0$ K) band (the *conduction band*). The energy difference between the top of the valence band and the bottom of the conduction band is called the *energy gap* (Figure 1.18).

1.5.2 The nearly free electron model

For a more quantitative treatment of the semiconductor band structure we turn to the alternative theoretical treatment, the *nearly free electron model*. Here it is assumed that when the semiconductor material is formed a certain number of electrons which were originally on each atom (this number is called the *valency*) can 'break loose' and form an electron 'gas', the electrons being free to move about anywhere within the confines of the solid. To solve the Schrödinger equation (i.e. Eq. (1.14)) for these electrons we need to know the potential in which the electrons move. When an electron is close to an atom it will experience a screened nuclear potential which will be similar to that seen by the outermost electrons on the isolated atom (i.e. Eq. (1.28)). In addition, there will be an inter-electron interaction, which poses a potentially difficult problem. Fortunately, it can be dealt with by modifying the screening function in the nuclear potential. At the material boundaries the

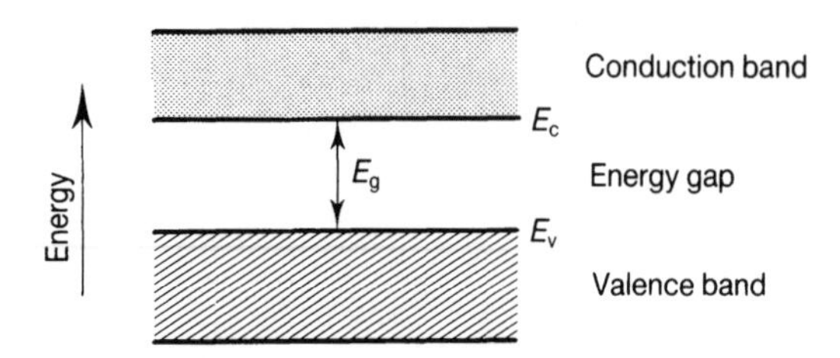

Figure 1.18 The energy band structure at the top of the occupied energy levels at 0 K. An energy gap (E_g) separates the valence band (full) with the conduction band (empty).

electrons must experience an additional potential that serves to keep the electrons within the solid. This is usually modelled by a finite potential step. An illustration of the resulting free-electron potential in a one-dimensional solid is given in Figure 1.19.

It is not possible to solve the Schrödinger equation explicitly with this form of potential, and in any case the solutions give little physical insight into their significance. We start therefore by making some rather drastic simplification. Initially, we neglect the electron–ion interaction (the *free electron model*) and suppose that the potential step at the boundary of the solid is an infinite one. Thus for a one-dimensional solid of length L the simplified potential is given by

$$V = 0: \quad 0 \leqslant x < L$$
$$V = \infty: \quad x < 0, \ x \geqslant L$$

This potential is often referred to as an *infinitely deep potential well*. The Hamiltonian operator for $0 \leqslant x < L$ then becomes

$$\hat{H} = \hat{V} + \hat{p}^2/(2m)$$
$$= -\frac{\hbar^2}{2m}\frac{\partial^2}{\partial x^2}$$

Thus the Schrödinger equation becomes

$$-\frac{\hbar^2}{2m}\frac{\partial^2}{\partial x^2}\psi(x) = E\psi(x) \tag{1.42}$$

A number of different functions will solve this equation, but it is convenient in the present instance to assume a solution of the form

$$\psi(x) = A\exp(\mathrm{j}kx) \tag{1.43}$$

For Eq. (1.43) to be a solution of Eq. (1.42) we must have

$$E = \frac{\hbar^2 k^2}{2m} \tag{1.44}$$

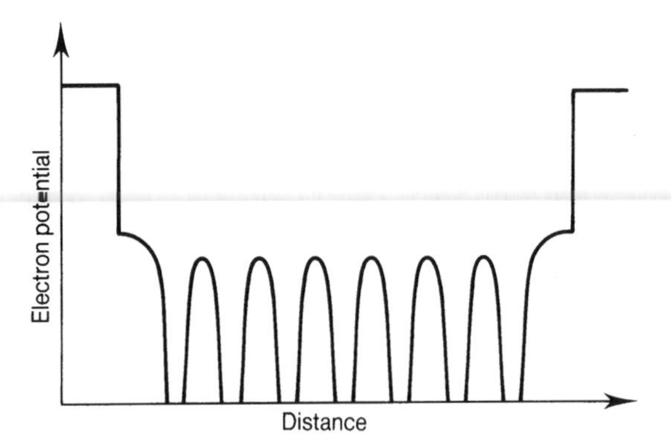

Figure 1.19 Schematic diagram of the energy of an electron within a one-dimensional solid.

We must also decide on the boundary conditions, that is, the behaviour of $\psi(x)$ at the edges of the potential well. Strictly speaking, these should be that $\psi(x)=0$ at both boundaries. This condition presents some difficulties when applied to an irregular-shaped three-dimensional solid, since different shapes imply different boundary conditions and hence different wavefunctions. Yet we are only really interested in the bulk material properties and not in any surface effects. One way of obviating this problem is to adopt *periodic boundary conditions*. These ensure in a sense that an electron never encounters any boundaries and yet never leaves the solid. In one dimension the condition is that:

$$\psi(x+L)=\psi(x) \tag{1.45}$$

When applied to the wavefunction of Eq. (1.43) we require that

$$\exp(\mathrm{j}k(x+L))=\exp(\mathrm{j}kx)$$

which implies that

$$k=\frac{2\pi}{L}n \tag{1.46}$$

where n is an integer (note that n here is *not* the same as the principal quantum number of the hydrogen atom). We see that k is quantized in units of $2\pi/L$, and the energy levels are then given by

$$E_n=\frac{2\pi^2\hbar^2}{mL^2}n^2 \tag{1.47}$$

For a three-dimensional infinitely deep potential well of side length L it is easily verified (see Problem 1.18) that a suitable wavefunction is

$$\psi(x,y,z)=A\exp(\mathrm{j}(k_xx+k_yy+k_zz))$$

where

$$k_x = \frac{2\pi}{L} n_x, \; k_y = \frac{2\pi}{L} n_y \text{ and } k_z = \frac{2\pi}{L} n_z$$

n_x, n_y and n_z being integers, and provided that the electron energy is given by

$$E_{k_x k_y k_z} = \frac{\hbar^2}{2m} (k_x^2 + k_y^2 + k_z^2) \tag{1.48a}$$

or, alternatively,

$$E_{n_x n_y n_z} = \frac{2\pi^2 \hbar^2}{mL^2} (n_x^2 + n_y^2 + n_z^2) \tag{1.48b}$$

We must now try to include the effects of the free electron–ion interaction. One thing we can be certain about is that the potential will have exactly the same repeating pattern as the lattice itself, and because of this it is referred to as a *periodic potential*. In the early days of solid-state physics the Schrödinger equation was solved for a number of simple periodic potentials. It was found that they all gave fairly similar results which only differed in 'fine' detail. One of these was the *Krönig–Penney* potential [Ref. 10], which is shown in Figure 1.20.

Although this is one of the simplest potentials the solution of the Schrödinger equation is still too long to reproduce here in detail (see, for example, Ref. 11 for a full treatment) and we must content ourselves with a discussion of the results. It turns out that it is convenient to consider the case where the potential 'barrier' height (V_0) becomes very small while its width (b) becomes very small, so that the product $V_0 b$ remains finite and, accordingly, we introduce a parameter P where

$$P = \lim_{\substack{V_0 \Rightarrow \infty \\ b_0 \Rightarrow \infty}} \left(\frac{mabV_0}{\hbar^2} \right) \tag{1.49}$$

The condition for a solution to the Schrödinger equation to exist is then found to be:

$$P \frac{\sin(\alpha a)}{\alpha a} = \cos(\alpha a) = \cos(ka) \tag{1.50}$$

where

$$\alpha^2 = 2mE/\hbar^2 \tag{1.51}$$

Equation (1.50) is actually a relationship between E and the parameter k which is the equivalent of Eq. (1.48a) in this new situation. This equation can only be solved when the left-hand side ($F(\alpha a)$, say) lies between ± 1. Figure 1.21 shows a plot of $F(\alpha a)$ as a function of αa. We see that provided $P > 0$ there are a number of ranges of αa (i.e. energy) when we cannot solve Eq. (1.50). These represent energy gaps, whereas the regions where we have solutions represent energy bands.

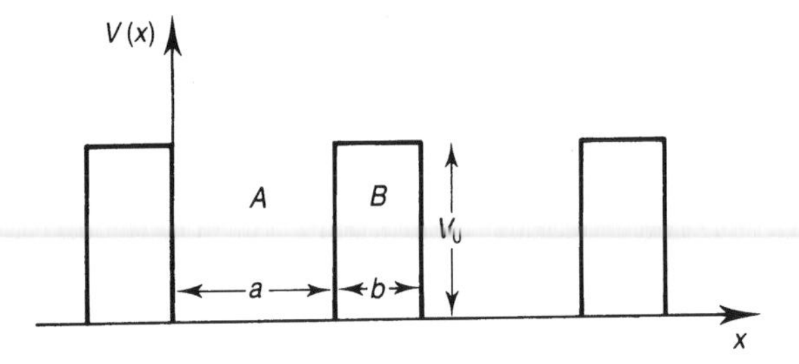

Figure 1.20 Form of the potential used in the Krönig–Penney model.

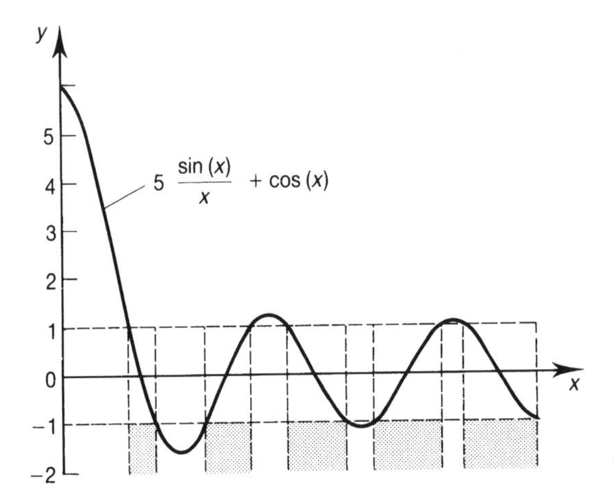

Figure 1.21 Plot of the function $P(\sin(x))/x + \cos(x)$ with $P = 5$. The shaded regions along the x axis represent regions where the function lies between +1 and −1, and hence correspond to energy bands.

The variation of E with k as given by Eq. (1.50), is shown in Figure 1.22. We see that the general trend of the curve follows that of the free electron parabola (i.e. Eq. (1.44)). However, when k is a multiple of π/a there is a discontinuity (the size of the discontinuity being the energy gap). If P (effectively the 'strength' of the potential) is small then the curve is very close to the free electron parabola and the energy gaps are small. As P increases, the curve tends towards a series of nearly horizontal line segments, the energy gaps increase and the band widths decrease.

There are a number of different ways of presenting the E-k relationship which result from the fact that k is not unique. This is evident from an inspection of Eq. (1.50). The right-hand side of this equation is $\cos(ka)$, so that its value will not change if k is replaced by $k + 2n\pi/a$ (n an integer). Thus k need only be defined within a range of $2\pi/a$. In the *reduced zone scheme* the range of k is restricted such that $-\pi/a < k \leqslant \pi/a$. The resulting E-k relationship is shown in Figure 1.23. It

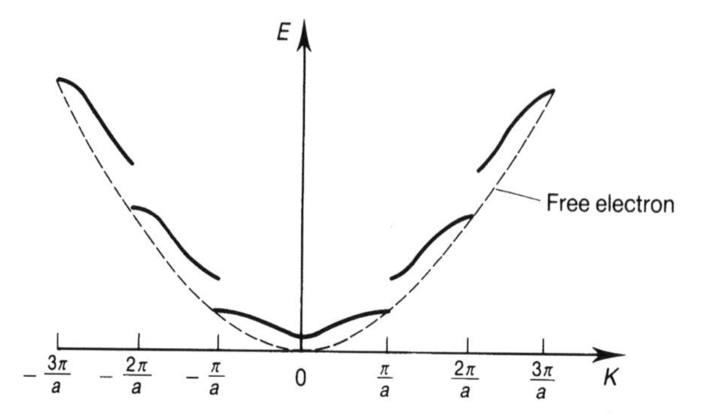

Figure 1.22 The behaviour of the electron energy with the wavevector k for an electron moving in a Krönig–Penney type of potential. Qualitatively similar results are obtained for other periodic potentials. Also shown is the E-k relationship for a free electron (that is, when no periodic potential is present).

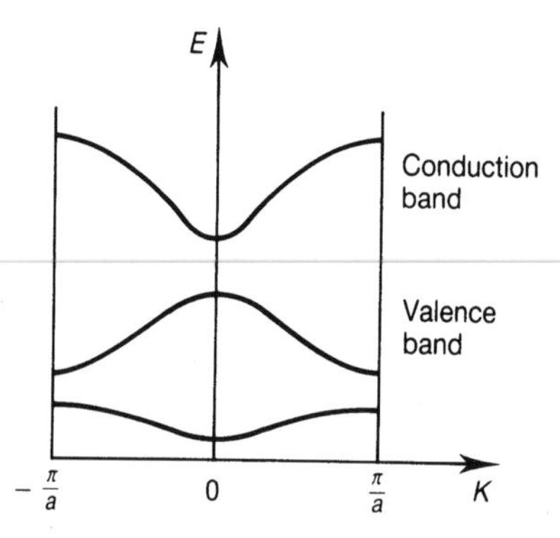

Figure 1.23 The E-k relationship (i.e. *band structure*) for an electron moving in a periodic potential shown in the reduced zone scheme.

must not be forgotten that the k values are still restricted by the periodic boundary conditions to take values which are integer multiples of $2\pi/L$ (i.e. Eq. (1.46)), and it is of interest to ask how many different k states there are within each band. This is easily answered when we remember that the one-dimensional solid considered here has a lattice spacing of a and a total length of L. If the total number of atoms in the solid is N, then

$$L = (N-1)a \tag{1.52}$$

so that

$$\frac{2\pi}{a} = (N-1)\frac{2\pi}{L} \tag{1.53}$$

The number of different k states within a band is thus *exactly* equal to the number of atoms in the solid. Because of the Pauli exclusion principle (and allowing for spin), only two electrons can have the same value of k. If the atomic valency is z, there are Nz 'free' electrons altogether, and these will occupy $Nz/2$ different k states. The number of bands required is just $z/2$. This is a simple and quite far-reaching result. It predicts that, at $T = 0$, odd valency monatomic solids will always have one half-filled band, while those of even valency will have completely filled bands. Whether or not a band is completely filled has a direct bearing on the electrical conduction properties of the material.

In a half-filled band electrons can easily change their state in response to an applied electric field and conduction is therefore readily possible. Half-filled bands then are associated with high conductivity, that is, *metallic* behaviour. A full band, on the other hand, acts as a perfect insulator. This may seem surprising at first, but if conduction is to take place then some of the electrons must change their state, and within a full band this is impossible since all possible states are occupied. At temperatures above absolute zero, however, a few electrons in the highest filled band (i.e. the valency band) will be be thermally excited into the next empty band (the conduction band) where they can take part in electrical conduction. The conductivity resulting from this excitation is, however, very temperature dependent and much smaller than in metals, hence the nomenclature *semi*conductor.

In quadrivalent materials such as silicon, germanium or gallium arsenide[9] we would therefore expect the *E-k* diagrams (or band *structure* diagrams, as they are also called) to be as indicated in Figure 1.23. The two lower bands should be completely occupied by electrons while bands above should be empty. If we compare this diagram with the results of much more sophisticated calculations (Figure 1.24) they seem to have little in common. However, they are not as different as might at first appear. Figure 1.24 deals with a three-dimensional material whereas Figure 1.23

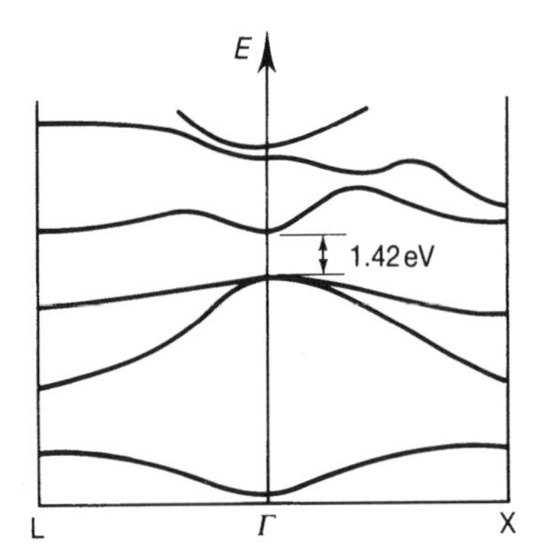

Figure 1.24 The band structure for GaAs. L, X and Γ represent particular points within the first Brillouin zone. Γ is at the origin while X and L lie on the boundary of the Brillouin zone (i.e. along the [100] and [111] directions, respectively).

is for a one-dimensional situation. Since the behaviour of E with k for positive k is the same as for negative k, the latter part of the one-dimensional diagram is redundant. In the three-dimensional case we need to show how E varies with k along an 'infinite' number of k directions. Obviously, this is impossible. Fortunately the crystalline symmetry of the materials means that quite a number of directions are equivalent and it is often sufficient to show how E varies with k along just two (sometimes more) principal crystallographic directions. Because of the redundancy of negative k values, these can both be drawn on the same diagram. This then explains the asymmetric nature of Figure 1.24 compared with Figure 1.23.

A surprising feature seen in the diagram for silicon is that there appear to be three occupied valency bands rather than the two expected. In fact there *are* only two principal bands but one of these is split into two by spin–orbit coupling. When a free electron is close to an atomic nucleus the forces it encounters will be similar to those seen by the bound electrons, so that the appearance of effects due to spin–orbit coupling should not come as too much of a surprise.

Another obvious difference between the predictions of the simple model and those of more sophisticated calculations is that the former predicts that the minimum of the conduction band and the maximum of the valence band both occur at the same value of k (i.e. $k = 0$). When this is true (as, for example, in GaAs) the semiconductor is known as a *direct* band gap semiconductor. If, on the other hand, they do not (as, for example, in Si and Ge), then the semiconductor is known as an *indirect* band gap semiconductor.

1.5.3 The density of states function

In situations where a large number of electron energy states are spread out over an appreciable range of energy the concept of a *density of states* becomes very useful. Formally, we define the density of states function $D(E)$ by requiring that $D(E)\,\mathrm{d}E$ be equal to the number of electron states per unit volume of material between energy E and $E+\mathrm{d}E$. We now evaluate $D(E)$ for the free electron model.

We recall that in three dimensions each electron state is described by three k quantum numbers k_x, k_y and k_z (see the discussion just after Eq. (1.47)). It is convenient to represent each triple of k values as a point in a cartesian 'space' (*k-space*) whose axes represent k_x, k_y and k_z. Since each k value must be a multiple of $2\pi/L$, allowed states are represented by points on a cubic lattice in k space with lattice spacing $2\pi/L$. The number of allowed points per unit volume of k space (the 'density') is then $1/(2\pi/L)^3$ and hence the density of electron states is $2/(2\pi/L)^3$.

We proceed by evaluating the number of states lying within a thin shell of k space of radius k and thickness $\mathrm{d}k$. The volume of k space involved is $4\pi k^2\,\mathrm{d}k$ and hence the number of states is just the volume involved multiplied by the state density, that is, $2k^3L^3\,\mathrm{d}k/\pi^2$. Introducing $D(k)\,\mathrm{d}k$, which is the density of states as far as k is concerned, we then have

$$D(k)\,\mathrm{d}k = \frac{2k^2L^3\,\mathrm{d}k}{\pi^2L^3}$$

$$= \frac{2k^2}{\pi^2}\,\mathrm{d}k \tag{1.54}$$

We can now make use of the relationship between k and E (i.e. Eq. (1.44)) to obtain

$$D(E) = \frac{4(2m)^{3/2}L^3}{\pi^2\hbar^3}\,E^{1/2} \tag{1.55}$$

Now in a semiconductor relatively few electrons are excited into the conduction band and hence only states at the very bottom of the conduction band are ever usually occupied. Thus the electrons will always be close to a minimum in the E-k curve and so, as far as these electrons are concerned, the curve will resemble a parabola. In this situation we can write

$$E = E_c + Ak^2 \tag{1.56}$$

where E_c is the energy of the bottom of the conduction band.

When the motions of electrons in the conduction band is examined it is found (see, for example, Ref. 12) that we can ignore any internal forces and use Newton's law of motion *provided* we replace the normal electron mass by an *effective mass*, m_e^*, where (in one dimension)

$$m_e^* = \hbar^2\left(\frac{\mathrm{d}^2E}{\mathrm{d}k^2}\right)^{-1} \tag{1.57}$$

The quadratic relationship of Eq. (1.56) thus gives rise to an effective mass of

$$m_e^* = \hbar^2/(2A)$$

Provided an electron does not move too far from the minimum of the conduction band it will always have a constant effective mass (i.e. independent of k). Writing the E-k relationship of Eq. (1.44) in terms of the effective mass we then have

$$E = E_c + \frac{\hbar^2}{2m_e^*}k^2 \tag{1.58}$$

The density of states function for states at the bottom of the band can then be written

$$D(E) = \frac{4(2m_e^*)^{3/2}}{\pi^2\hbar^3}\,(E - E_c)^{1/2} \tag{1.59}$$

1.5.4 Holes

When electrons are excited by some means from the valence band to the conduction band the valence band will no longer be completely full and will therefore be able to contribute to the material conductivity. It may be shown (see, for example, Ref. 13) that this contribution may be calculated by assuming that the valence band is empty except for the presence of a number of positively charged particles which are called *holes*. The number of holes is equal to the number of empty states in the band. The effective mass of a hole m_h^* is $-m_e^*$, where m_e^* is the effective mass of the missing electron (this in fact results in holes having a positive effective mass). Because of their positive charge, the energy of a hole on a conventional electron E-k diagram increases as electron energy decreases.

What was said in the previous paragraph about electrons can also be applied to holes, so that the E-k relation for holes becomes

$$E = E_v - \frac{\hbar^2}{2m_h^*} k^2$$

and the hole density of states can be written

$$D(E) = \frac{4(2m_h^*)^{3/2}}{\pi^2 \hbar^3} (E_v - E)^{1/2} \qquad (1.60)$$

1.5.5 Extrinsic semiconductors

So far, we have discussed 'pure' semiconductor materials. However, one of the most interesting properties of semiconductors is that their electrical properties can be considerably altered by introducing certain impurities (*dopants*) whose valency differs from that of the host. The pure material is described as *intrinsic*, and the doped *extrinsic*.

Suppose that a pentavalent impurity (e.g. phosphorus) is introduced into a four-valent semiconductor. (The term 'valence' is used here in its chemical bonding sense.) The bonding requirements of the host material can be fully satisfied by involving just four of the valency electrons so that one 'redundant' electron is left over. This is very loosely bound to the atom and is easily excited into the conduction band. Once in the conduction band, it can take part in conduction just like any other electron in the band.

Because the impurity atom 'donates' an electron to the conduction band, the impurity is called a *donor* atom. When the electrons are bound to the donor atom their energy levels lie just below the bottom of the conduction band (Figure 1.25). In silicon, for example, the donor levels lie some 0.04 eV below the conduction band, the exact value depending on the type of donor. (We cannot show these levels on an E-k diagram because k is not a 'good' quantum number for an electron on the donor atom). Doping with donors thus increases the number of electrons in the conduction band.

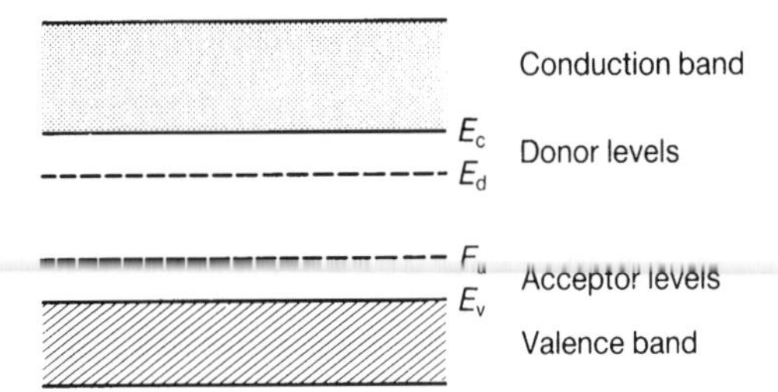

Figure 1.25 Position of donor and acceptor energy levels in a semiconductor. The levels arise because of doping with impurities having a valence of one more or less than that of the host.

When an impurity with a valency of one less than the host material is added, empty states are formed just above the top of the valence band into which electrons can be excited (i.e. they 'accept' electrons, and hence the name *acceptors*). When an electron is excited from the valence band into an acceptor state a hole is formed. Thus doping with acceptors increases the number of holes in the valence band.

Problems

1.1 Show that the total energy of an electron moving in a Bohr-type orbit round a fixed positive charge $(+e)$ is given by

$$E = -\frac{e^2}{8\pi\varepsilon_0 r}$$

where the zero of potential energy is at $r = \infty$.

1.2 Show that when the finite mass of the nucleus is taken into account, Eq. (1.9) may still be used to describe the energy of the system provided the electron mass is replaced by the reduced mass, m_r, where

$$\frac{1}{m_r} = \frac{1}{m_e} + \frac{1}{m_p}$$

Determine the effect this correction has on the emission wavelength of the transition considered in Example 1.1.

1.3 Using the simple Bohr model calculate the second ionization potential of helium and the third ionization potential of lithium (the experimental values are 54.4 eV and 122.4 eV, respectively).

1.4 Show that the operator representing the sum of the squares of x, y and z angular momenta, $\hat{L}^2$ (where $\hat{L}^2 = \hat{l}_x^2 + \hat{l}_y^2 + \hat{l}_z^2$) is given by

$$-\hbar^2\left[\frac{1}{\sin(\theta)}\frac{\partial}{\partial\theta}\left(\sin(\theta)\frac{\partial}{\partial\theta}\right) + \frac{1}{\sin^2(\theta)}\frac{\partial^2}{\partial\phi^2}\right]$$

Show also that the operator $\hat{L}^2$ commutes with the operators $\hat{l}_x$, $\hat{l}_y$ and $\hat{l}_z$, but that the latter three operators do *not* commute with each other.

1.5 Illustrate the dependence of the hydrogenic wavefunctions on angle by plotting the function $|Y_{l,m}(\theta,\phi)|^2$ on a polar diagram for $l = 0, 1, 2$. Appendix 2 lists the appropriate functions.

1.6 Calculate the probability that an electron in the ground state of the hydrogen atom will be found at a distance greater than $2a_0$ from the nucleus.

1.7 Show that, for the $n = 2$, $l = 1$ in hydrogen, the electron probability distribution reaches a maximum in the radial direction when $r = 2a_0$.

1.8 Consider the six 2p states in hydrogen and show that the sum of $|\psi|^2$ for all the six states together is a function of r only.

1.9 The magnetic moment of a plane current-carrying loop may be defined as the product of the current flowing in the loop multiplied by the area of the loop. An electron circulating in a Bohr-type orbit can be regarded as constituting such a current loop. Assuming that the charge and mass are uniformly distributed round the (circular) orbit, show that the magnetic moment of the electron is then

$$\mathbf{M} = -\frac{1}{2}\frac{e}{m_e}\mathbf{l}$$

where $\mathbf{l}$ is the angular momentum of the electron. In the case of spin angular momentum, however, it should be noted that the proportionality factor is *twice* that obtained above, which then gives the Bohr magneton to be $e\hbar/(2m_e)$.

1.10 Derive an expression for the spin–orbit coupling interaction between an electron and the nucleus by assuming that the nucleus rotates round the electron thus forming a current loop. The magnetic moment of the electron will then interact with the magnetic field generated by the current loop.

1.11 Verify that a suitable antisymmetric wavefunction for two noninteracting electrons whose spatial wavefunctions (normalized) are ψ_a and ψ_b is given by

$$\psi_{12} = \frac{1}{\sqrt{2}}\begin{vmatrix} \psi_a(\mathbf{r}_1)\chi^{(1)}_{1/2}\psi_a(\mathbf{r}_2)\chi^{(2)}_{1/2} \\ \psi_b(\mathbf{r}_1)\chi^{(1)}_{1/2}\psi_b(\mathbf{r}_2)\chi^{(2)}_{1/2} \end{vmatrix}$$

1.12 The ion Nd^{3+} has three 4f electrons lying outside closed shells. Show that in the central field approximation the determinental product states are 364thfold degenerate.

1.13 The element arsenic has 32 electrons. Write down the electron configuration and hence, using Hundt's rules (i.e. see Section 1.3.3.1), deduce that the ground state is $^4S_{3/2}$.

1.14 Use the energy levels of Figure 1.8 to identify the transitions involved in the absorption spectrum shown in Figure 1.9.

1.15 A single electron is moving in the field due to two protons. If $\psi_n(1)$ and $\psi_n(2)$ represent the (normalized) hydrogen-like wavefunctions when the electron is orbiting protons 1 and 2 respectively, verify that the functions

$$\psi_n^s = 2^{-1/2}\,[\psi_n(1) + \psi_n(2)]$$

$$\psi_n^a = 2^{-1/2}\,[\psi_n(1) - \psi_n(2)]$$

are eigenfunctions of the exchange operator $\hat{P}_x$, with eigenvalues of +1 and −1, respectively.

1.16 Two atoms of different mass and a constant distance r apart rotate about their common centre of gravity. Show that the total moment of inertia is given by

$$I = m_r r^2$$

where m_r is the reduced mass of the two atoms.

1.17 Comment on the occupancy at room temperature of the vibrational and rotational levels of HCl.

1.18 An electron moves in a three-dimensional infinitely deep cubical potential well of side length L. Verify that a suitable energy eigenfunction is

$$\psi(x, y, z) = A \exp(\mathrm{j}(k_x x + k_y y + k_z z))$$

where the electron energy is given by

$$E = \frac{\hbar^2}{2m}(k_x^2 + k_y^2 + k_z^2)$$

and that the periodic boundary requirements yield

$$k_x = \frac{2\pi}{L} n_x, \; k_y = \frac{2\pi}{L} n_y \text{ and } k_z = \frac{2\pi}{L} n_z$$

where n_x. n_y and n_z are all integers.

1.19 Given that the E-k relation for an electron in the conduction band of a semiconductor can be written

$$E = A - B\cos(ka)$$

determine how the effective mass depends on k and sketch a graph of the relationship.

1.20 Show that the density of states for an electron moving in a two-dimensional potential well is independent of energy and given by

$$D(E) = 4\pi m/h^2$$

Notes

1. Technically this is only true if the so-called *degeneracies* of the two energy states are the same. However this point is of no real consequence here.
2. The electron is assumed to have zero potential energy when it is infinitely far removed from the nucleus.
3. Appendix 1 contains a brief discussion of the relevant parts of quantum mechanics.
4. The Hamiltonian operator is named after the distinguished Irish mathematician Sir William Hamilton, who in the nineteenth century succeeded in writing Newton's laws of motion so that they were independent of the coordinate system used. The formalism used by Hamilton has a close correspondence with that of quantum mechanics.
5. The term 'wavefunction' is somewhat misleading since although some solutions of the Schrödinger equation are indeed 'wavelike' others are not. Nevertheless usage of the term is now firmly established.

6. The use of the letters s, p, d dates from the early days of spectroscopy when certain characteristic spectral lines were designated according to their appearance, i.e. strong, principal and diffuse. It was only later that they were shown to involve the $l = 0$, 1 and 2 states, but by then the notation had become well entrenched. The letter ordering after d is alphabetical except that e is omitted because of its association with the charge on the electron.
7. The states used here involve the quantum numbers M_L and M_S which are the components of total angular and spin momenta along the z axis.
8. These are a more general form of the so-called Hundt's rule, which applies to equivalent electrons in LS coupling. This states that the ground term will be the one with the largest value of S, and if there are more than one of these then the one with the largest value of L will have the lowest energy.
9. In gallium arsenide, gallium has a valency of three while arsenic has a valency of five, thus giving an 'average' effective valency of four.

References

[1] A. Einstein, *Physikalilische Zeitschrift*, **18**, 121, 1917. A translation appears in D. ter Haar, *The Old Quantum Theory*, Pergamon, Oxford and New York, 1967.

[2] E. M. Purcell and R. V. Pound, 'A nuclear spin system at negative temperature', *Phys. Rev.* **81**, 279–80, 1951.

[3] T. H. Maiman, 'Stimulated optical radiation in ruby', *Nature*, **187**, 493–4, 1960.

[4] M. Bertolotti, *Masers and Lasers: An Historic Approach*, Adam Hilger, Bristol, 1983: J. L. Bromberg, *The Laser in America, 1950–1970*, MIT Press, Boston, MA, 1971: J. Hecht, *Laser Pioneers*, Academic Press, New York, 1992.

[5] R. Eisberg and R. Resnick, *Quantum Physics of Atoms, Molecules, Solids, Nuclei and Particles*, Wiley, Chichester, 1974, Chapter 7.

[6] J. M. Cassels, *Basic Quantum Mechanics*, McGraw-Hill, London, 1970, Chapter 5, Sections 15 and 16.

[7] G. K. Woodgate, *Elementary Atomic Structure*, 2nd edn, McGraw-Hill, London, 1980, Section 7.2.

[8] G. K. Woodgate, *Elementary Atomic Structure*, 2nd edn, McGraw-Hill, London, 1980, Section 7.3

[9] J. M. Cassels, *Basic Quantum Mechanics*, McGraw-Hill, London, 1970, Section 3.8.

[10] R. de L. Krönig and W. G. Penney, 'Quantum mechanics of electrons in crystal lattices', *Proc. Roy. Soc. (Lond.)*, **A. 130**, 499–513, 1931.

[11] C. Kittel, *Introduction to Solid State Physics*, 5th edn, Wiley, New York, 1976, p. 191.

[12] J. S. Blakemore, *Solid State Physics*, 2nd edn, W. B. Saunders, Philadelphia, 1974, p. 237.

[13] H. P. Myers, *Introductory Solid State Physics*, Taylor and Francis, London, 1990, Section 9.6.

2

Light and its properties

In the previous chapter we saw that atoms have well-defined energy states and that transitions between them can involve the absorption and emission of light. In this chapter we are concerned with those basic properties of light (indeed, of electromagnetic radiation in general) which will enable us to understand the emission and absorption processes in more detail. Since this is essentially a quantum process, it would seem necessary to consider the quantum theory of light. However, this would be an almost impossible undertaking in a book of this size and level. Fortunately, when it comes to calculating the transition rates for the absorption and emission processes we can obtain the correct quantum-mechanical results by adopting a semi-classical viewpoint. That is, we treat the atom quantum mechanically but deal with the radiation as an oscillating electric field which acts on the atom. Of course, the quantum nature of light cannot be forgotten. When light exchanges energy with matter, for example, the energy change must always be a multiple of $h\nu$. Sometimes we will treat the light in a laser cavity as an electromagnetic wave and at others as a collection of photons. Having to adopt such a schizophrenic approach is perhaps not so intellectually satisfying as adopting a wholly quantum model, but it does ensure that the level of mathematics remains within reasonable bounds!

We may start therefore with a review of the wave (or 'classical') view of light.

2.1 The classical wave view of light

The appearance of Maxwell's equations in 1864 represents one of the high points of classical physics. They summarized the results of many years of investigation into electricity and magnetism, and predicted the existence of electromagnetic waves with a (theoretically) infinite range of wavelengths. At the time scientists were only aware of visible radiation where the wavelength extends from about 0.4 μm to 0.7 μm, but shortly afterwards the known wavelength range was extended. In 1887, for example, Hertz succeeded in generating non-visible electromagnetic waves with wavelengths of the order of 10 m.

Using vector calculus, Maxwell's equations for the *electric field* $\mathscr{E}$ and *magnetic field* $\mathscr{B}$ can be written (see, for example, Ref. 1):

$$\nabla . \mathscr{E} = \frac{\rho}{\varepsilon_0} \tag{2.1}$$

$$\nabla . \mathscr{B} = 0 \tag{2.2}$$

$$\nabla \times \mathscr{E} = -\frac{\partial \mathscr{B}}{\partial t} \tag{2.3}$$

$$\nabla \times \mathscr{B} = \mu_0 \mathbf{J} + \varepsilon_0 \mu_0 \frac{\partial \mathscr{E}}{\partial t} \tag{2.4}$$

where ρ and $\mathbf{J}$ are the electrical charge and the electric current densities, respectively. The constants μ_0 (the *permeability of vacuum*) and ε_0 (the *permittivity of vacuum*) have the values $4\pi \times 10^{-7}$ H m^{-1} and 8.854×10^{-12} F m^{-1}, respectively.

For simplicity we start with the situation in a vacuum. Here there can be no electric charges or currents, so that both ρ and $\mathbf{J}$ are zero and Maxwell's equations reduce to:

$$\nabla . \mathscr{E} = 0 \tag{2.5}$$

$$\nabla . \mathscr{B} = 0 \tag{2.6}$$

$$\nabla \times \mathscr{E} = -\frac{\partial \mathscr{B}}{\partial t} \tag{2.7}$$

$$\nabla \times \mathscr{B} = \varepsilon_0 \mu_0 \frac{\partial \mathscr{E}}{\partial t} \tag{2.8}$$

From these equations it is possible to derive two more equations, each one involving either only $\mathscr{E}$ or $\mathscr{B}$. Thus if we take the curl of Eq. (2.7) we have

$$\nabla \times \nabla \times \mathscr{E} = -\nabla \times \frac{\partial \mathscr{B}}{\partial t} = -\frac{\partial (\nabla \times \mathscr{B})}{\partial t} = -\varepsilon_0 \mu_0 \frac{\partial^2 \mathscr{E}}{\partial t^2} \qquad \text{(from Eq. (2.8))}$$

Using the vector identity (see Problem 2.1)

$$\nabla \times \nabla \times \mathscr{E} = \nabla(\nabla . \mathscr{E}) - \nabla^2 \mathscr{E}$$

together with Eq. (2.5) we obtain

$$\nabla^2 \mathscr{E} = \varepsilon_0 \mu_0 \frac{\partial^2 \mathscr{E}}{\partial t^2} \tag{2.9}$$

Similarly, it may be shown that

$$\nabla^2 \mathscr{B} = \varepsilon_0 \mu_0 \frac{\partial^2 \mathscr{B}}{\partial t^2} \tag{2.10}$$

Each of Eqs (2.9) and (2.10) corresponds to three equations, each involving one of the three components of the corresponding vectors $\mathscr{E}$ and $\mathscr{B}$. If we designate a general component by ψ, then we can write

$$\nabla^2 \psi = \varepsilon_0 \mu_0 \frac{\partial^2 \psi}{\partial t^2} \tag{2.11}$$

where

$$\nabla^2 = \frac{\partial^2}{\partial x^2} + \frac{\partial^2}{\partial y^2} + \frac{\partial^2}{\partial z^2}$$

As may easily be verified (see Problem 2.2) Eq. (2.11) is satisfied by

$$\psi = f\left(t \pm \frac{1}{c}\,\mathbf{n.r}\right) \tag{2.12}$$

where f is any function whose second derivative exists and where

$$c = \frac{1}{\sqrt{\varepsilon_0 \mu_0}} \tag{2.13}$$

$$\mathbf{r} = x\mathbf{a} + y\mathbf{b} + z\mathbf{c}$$

and

$$\mathbf{n} \text{ (a unit vector)} = n_x\mathbf{a} + n_y\mathbf{b} + n_z\mathbf{c}$$

a, **b** and **c** being unit vectors along the x, y and z axes, respectively.

As time passes the shape of the function f remains unchanged but moves along the direction defined by **n** with a speed given by the value of c. When the '$-$' sign is used the motion has the same direction as the vector **n**, while the '$+$' sign results in motion in the opposite direction. The constant c is called the speed of light in a vacuum, and can either be calculated from the constants ε_0 and μ_0 as in Eq. (2.13) or determined experimentally. Whichever method is adopted, the same value (to within experimental error) is obtained. Since there is no dependence on frequency in Eq. (2.13), *all* electromagnetic waves, whatever their origin (i.e. radiowaves, microwaves, lightwaves, etc.), travel in a vacuum with exactly the same speed: the 'speed of light'.

2.1.1 Plane waves

To simplify the discussion further we assume that we are dealing with *plane waves*. That is, waves where, at any particular instant in time, the field components are the same over a plane which is perpendicular to the direction of propagation. We take this plane to be the xy plane and the direction of propagation to be the z axis. Equation (2.12) then becomes

$$\psi = f\left(t - \frac{z}{c}\right) \tag{2.14}$$

In addition, it is convenient to assume a purely oscillatory waveform (that is, a cosine or sine-like curve), since by Fourier analysis all waveforms that are likely to be met with in practice can be expressed in terms of sums over different oscillation frequencies or integrations over continuous frequency distributions (or some combination of the two). We may then introduce the *wavelength* λ and the *frequency* ν of the wave, and since the product of the frequency and wavelength of a wave is its (phase) velocity, we have $\nu\lambda = c$.

Instead of frequency and wavelength it is sometimes convenient to use the *angular frequency* ω $(=2\pi\nu)$ and the *wavevector* k $(=2\pi/\lambda)$. Using these variables and assuming a cosine functional form for ψ, gives

$$\psi = A\cos(\omega t - kz) \tag{2.15a}$$

Sometimes, instead of the cosine function, it is convenient to express ψ, in terms of complex exponentials, thus

$$\psi = A\mathbb{R}\{\exp(\mathrm{j}(\omega t - kz))\} \tag{2.15b}$$

where $\mathbb{R}$ indicates that the real part of the expression following must be taken. When using this latter form it is customary to omit the operator $\mathbb{R}$.

A number of interesting deductions can be made for the plane waves (see Problem 2.3). First, $\mathscr{E}$ and $\mathscr{B}$ are perpendicular to each other and to the direction of propagation (see Figure 2.1). Second, the magnitudes of $\mathscr{E}$ and $\mathscr{B}$ are related by

$$|\mathscr{B}| = \frac{|\mathscr{E}|}{c} \tag{2.16}$$

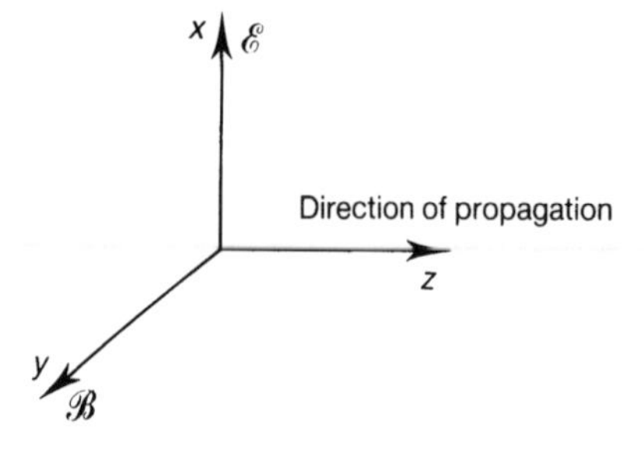

Figure 2.1 For a plane electromagnetic wave in a vacuum the direction of propagation, and the electric field ($\mathscr{E}$) and magnetic field ($\mathscr{B}$) vectors are all at right angles.

2.1.2 Polarization

A plane wave in which the direction of $\mathscr{E}$ remains fixed along a particular axis is said to be *linearly* (or *plane*) *polarized*. Consider, however, a plane wave composed of electric fields along the x and y axes which have a phase difference ϕ so that

$$\mathscr{E}_x = \mathscr{E}_{0x} \cos(\omega t - kz)$$

and

$$\mathscr{E}_y = \mathscr{E}_{0y} \cos(\omega t - kz + \phi)$$

If $\phi = \pi/2$ and if $\mathscr{E}_{0x} = \mathscr{E}_{0y} = \mathscr{E}_0$, then the total electric field can be written

$$\mathscr{E} = \mathscr{E}_0(\mathbf{a} \cos(\omega t - kz) - \mathbf{b} \sin(\omega t - kz)) \tag{2.17}$$

At a fixed point z, the total electric field remains constant but its direction rotates. The tip of the field vector describes a circle (Figure 2.2(a)), and the wave is said to be *circularly polarized*. If the two amplitudes along the x and y axes are not equal then the tip of the resultant vector will describe an ellipse (Figure 2.2(b)) resulting in an *elliptically polarized* wave.

The sense in which the field rotates determines whether the polarization is *left-handed* or *right-handed*. In the above example the vector described by Eq. (2.17) rotates anti-clockwise with time for a fixed observer facing into the beam (i.e. along the z axis). The polarization is then defined as being left-handed.

2.1.3 Propagation in a dielectric medium

So far, we have only dealt with waves which are travelling in a vacuum. In most instances we encounter electromagnetic waves travelling in a medium even if that medium is only air. We assume that the medium is nonmagnetic and nonconducting. The difference between this situation and that of a vacuum is that the medium can

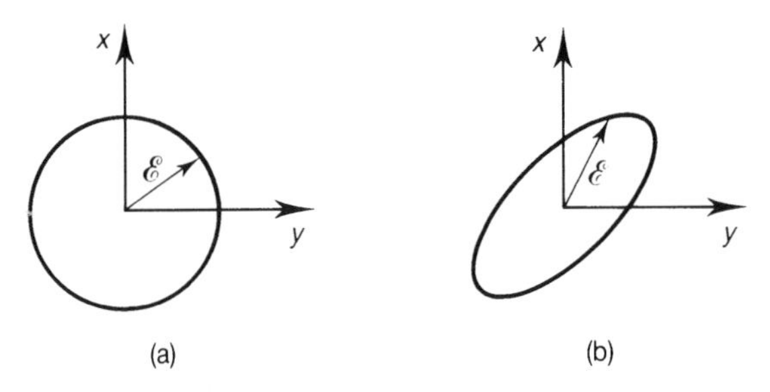

Figure 2.2 Behaviour of the electric field vector ($\mathscr{E}$) in (a) circularly polarized and (b) elliptically polarized light.

become *polarized*. The *dipole moment*, $\mathbf{p}$, of an atom or molecule may be defined by

$$\mathbf{p} = \int \rho(\mathbf{r})\mathbf{r}\,\mathrm{d}\tau \tag{2.18}$$

where $\rho(\mathbf{r})$ is the charge density at the point $\mathbf{r}$ and $\mathrm{d}\tau$ a small element of volume, the integral is carried out over all space. When no electric field is applied the 'centres of mass' of the positive and negative charge distributions in isolated atoms are usually coincident and so the atom has no dipole moment. However, when an electric field is applied the electrons and protons experience forces in opposite directions and so the centres of gravity of the positive and negative charge tend to move apart and the atom then acquires a dipole moment. The same ideas can be extended to cover molecules, although some molecules have a dipole moment in the absence of any external field and are known as *polar* materials.

If we have N atoms or molecules (nonpolar) per unit volume, each with a dipole moment $\mathbf{p}$, then the polarization is given by

$$\mathscr{P} = N\mathbf{p} \tag{2.19}$$

For relatively small electric fields the relationship between the dipole moment per unit volume and the applied field $\mathscr{E}$ can usually be taken as linear and so we write[1]

$$\mathscr{P} = \chi\varepsilon_0\mathscr{E} \tag{2.20}$$

where χ is known as the *polarizability* of the medium. In general, χ will depend on the frequency of the applied electric field. A simple form for $\chi(\omega)$ can be obtained from a model in which it is assumed that any displacement of the electrons in an atom/molecule results in a restoring force which is proportional to the displacement. When an oscillating field is present the electron will then undergo forced harmonic oscillation. Because it is a charged particle the resulting acceleration and deceleration will cause it to radiate energy. This energy loss may be allowed for by introducing a damping term which is proportional to the electron velocity. The equation of motion of the electron in the presence of an oscillating field $\mathscr{E}_x$ along the x axis can be written

$$-e\mathscr{E}_x = m\left(\frac{\mathrm{d}^2x}{\mathrm{d}t^2} + 2\gamma\frac{\mathrm{d}x}{\mathrm{d}t} + \omega_0^2 x\right) \tag{2.21}$$

Here 2γ is the damping constant and ω_0 the 'natural oscillation frequency' of the undamped oscillation (the reason for using 2γ rather than γ will become clear later). We write the electric field oscillation in complex form, that is, $\mathscr{E}_x = \mathscr{E}_0 \exp(\mathrm{j}\omega t)$, and assume that the electron displacement x can similarly be written $x = x_0 \exp(\mathrm{j}\omega t)$. Substitution of these expressions into Eq. (2.21) gives

$$x = \frac{-e}{m(\omega_0^2 - \omega^2 + 2\mathrm{j}\gamma\omega)}\mathscr{E}_x$$

Thus the polarization of each atom/molecule can be written

$$p = \frac{e^2}{m(\omega_0^2 - \omega^2 + 2j\gamma\omega)} \tag{2.22}$$

From Eqs (2.19), (2.20) and (2.22) we then know that[2]

$$\varkappa(\omega) = \frac{Ne^2}{\varepsilon_0 m(\omega_0^2 - \omega^2 + 2j\gamma\omega)} \tag{2.23}$$

The *electric displacement* $\mathscr{D}$ of a medium is defined by

$$\mathscr{D} = \varepsilon_0 \mathscr{E} + \mathscr{P} \tag{2.24}$$

so that in a linear medium

$$\mathscr{D} = \varepsilon_0(1 + \varkappa)\mathscr{E}$$

or

$$\mathscr{D} = \varepsilon_0 \varepsilon_r \mathscr{E} \tag{2.25}$$

where ε_r, the *dielectric constant* (or *relative permittivity*), is given by

$$\varepsilon_r = 1 + \varkappa \tag{2.26}$$

If the medium is also magnetic it is useful to introduce the *magnetizing field* $\mathscr{H}$, defined by

$$\mathscr{H} = \frac{\mathscr{B}}{\mu_0} - \mathscr{M} \tag{2.27}$$

where $\mathscr{M}$ is the magnetization of the medium. An alternative relation between $\mathscr{B}$ and $\mathscr{H}$ (similar to that between $\mathscr{E}$ and $\mathscr{D}$ in Eq. (2.25)) is given by introducing the *relative permeability*, μ_r, where

$$\mathscr{B} = \mu_r \mu_0 \mathscr{H} \tag{2.28}$$

Fortunately, all the laser media we have to deal with later are nonmagnetic, so that we may assume $\mu_r = 1$.

We now need to consider Maxwell's equations within an isotropic linear dielectric medium. Equations (2.5), (2.6) and (2.7) remain intact, but Eq. (2.8) must be replaced by

$$\nabla \times \mathscr{B} = \mu_0 \frac{\partial \mathscr{D}}{\partial t}$$

or, using Eq. (2.25),

$$\nabla \times \mathscr{B} = \varepsilon_0 \mu_0 \varepsilon \frac{\partial \mathscr{E}}{\partial t} \tag{2.29}$$

It is then a simple matter to repeat the steps leading to Eq. (2.11), with the result that

$$\nabla^2\psi = \varepsilon_0\mu_0\varepsilon_r \frac{\partial^2\psi}{\partial t^2} \tag{2.30}$$

Thus we again predict the existence of electromagnetic waves, the only difference being that the phase velocity v_p of the wave ($=\omega/k$) is now given by

$$v_p = 1/(\varepsilon_0\mu_0\varepsilon_r)^{1/2}$$

or

$$v_p = \frac{c}{\sqrt{\varepsilon_r}}$$

In dealing with light it is customary to define the ratio of the phase velocity in a vacuum to that in a medium as the *refractive index*, n. Thus

$$n = \sqrt{\varepsilon_r} \tag{2.31}$$

Thus for a plane wave travelling in a dielectric medium along the z axis with its electric field vector along the x axis we may write

$$\mathscr{E}_x = \mathscr{E}_0 \exp\left(\mathrm{j}\omega\left(t - \frac{nz}{c}\right)\right) \tag{2.32}$$

The counterpart of Eq. (2.16) now becomes

$$|\mathscr{B}| = \frac{|\mathscr{E}|}{v_p} \tag{2.33}$$

2.1.4 Energy flow

As electromagnetic waves travel along they carry energy with them, and the amount of energy was first determined by Poynting in 1884. The energy flow rate through any surface due to the presence of an electromagnetic field is given by evaluating the integral $\int \mathscr{S}.\,d\mathbf{A}$, where $\mathscr{S}$, the *Poynting vector*, has the value $\mathscr{E} \times \mathscr{H}$ [Ref. 2]. For the plane wave considered above, the instantaneous flow of energy through a unit area perpendicular to the direction of propagation is given by

$$\mathscr{E}_0 \cos(\omega t - kz) \frac{\mathscr{B}_0}{\mu_0} \cos(\omega t - kz)$$

or

$$\frac{1}{c\mu_0} \mathscr{E}_0^2 \cos^2(\omega t - kz)$$

Averaged over a time that is long compared with $2\pi/\omega$, this expression gives an irradiance[3] I of

$$I = \frac{\varepsilon_0 c}{2} \mathscr{E}_0^2 \tag{2.34}$$

Another useful concept is that of energy density. If we again imagine a unit area perpendicular to the direction of propagation then in one second a 'column' of radiation of length c will have passed through the surface (Figure 2.3). The time-averaged energy density, ρ, in the column is thus the irradiance on the surface divided by c. Thus

$$\rho = \frac{\varepsilon_0}{2} \mathscr{E}_0^2 \tag{2.35}$$

2.1.5 Refractive index variation with frequency

We must take care when using Eq. (2.31) that the refractive index and the relative permittivity are measured at the same frequency. Normally when we refer to the refractive index we are concerned with optical frequencies, that is, frequencies of the order of 10^{14} Hz. Whereas ε_r is usually determined from electrical measurements at much lower frequencies. It is useful therefore at this stage to examine the variation in refractive index with frequency that can be expected. We make use of the the electron oscillator model that was described in the previous section, combining Eqs (2.23), (2.26) and (2.31):

$$n^2 - 1 = \frac{Ne^2}{\varepsilon_0 m(\omega_0^2 - \omega^2 + 2\mathrm{j}\gamma\omega)} \tag{2.36}$$

This equation implies that we have a *complex* refractive index. Before dismissing this result out of hand we examine the consequences by putting

$$n = n_r + \mathrm{j}n_i \tag{2.37}$$

Taking Eq. (2.32) for the the variation of electric field in a plane wave, and inserting the complex refractive index of Eq. (2.37) we obtain

$$\mathscr{E}_x = \mathscr{E}_0 \exp\left[\mathrm{j}\omega\left(t - \frac{n_r z}{c}\right)\right]\exp\left(-\frac{n_i\omega}{c}z\right)$$

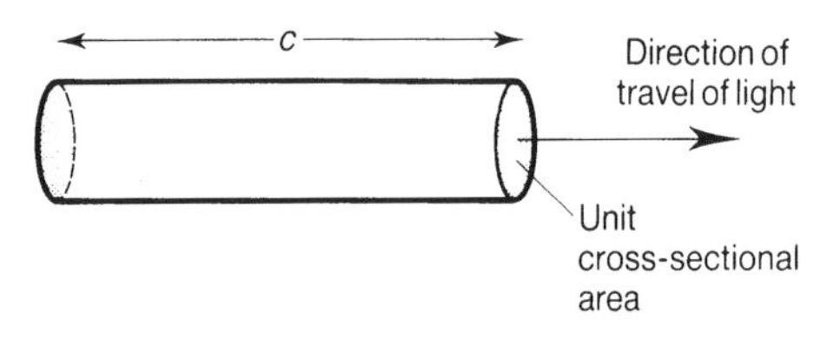

Figure 2.3 Diagram used to relate radiation energy density to irradiance.

This represents a wave whose amplitude decays with increasing z.

Since the irradiance of the wave, $I(z)$, is proportional to $\mathscr{E}_x^2$ the irradiance decay factor with z is given by

$$\exp\left(-\frac{2n_i\omega}{c}z\right)$$

Such processes are often described in terms of an *absorption coefficient*, α, defined such that

$$I(z) = I(0)\exp(-\alpha z) \tag{2.38}$$

We can then write

$$\alpha = \frac{2n_i\omega}{c} \tag{2.39}$$

Thus the imaginary part of the refractive index may be associated with absorption in the material. This is not unreasonable since we note that from Eq. (2.36) n is only imaginary when the damping (or energy loss) term γ is present. Now Eq. (2.36) can be written:

$$n^2 - 1 = \frac{Ne^2(\omega_0^2 - \omega^2 + 2j\gamma\omega)}{\varepsilon_0 m((\omega_0^2 - \omega^2)^2 + 4\gamma^2\omega^2)} \tag{2.40}$$

By squaring both sides of Eq. (2.37) we have

$$n^2 = n_r^2 + 2jn_r n_i - n_i^2 \tag{2.41}$$

If we are dealing with materials which are relatively weakly absorbing, that is, if $n_r \gg n_i$, then

$$n^2 \approx n_r^2 + 2jn_r n_i \tag{2.42}$$

Substituting for n from Eq. (2.42) into Eq. (2.40) and equating the real and imaginary parts yields

$$n_r^2 = 1 + \frac{Ne^2}{\varepsilon_0 m}\left(\frac{\omega_0^2 - \omega^2}{(\omega_0^2 - \omega^2)^2 + 4\gamma^2\omega^2}\right) \tag{2.43}$$

and

$$2n_i n_r = \frac{Ne^2}{\varepsilon_0 m}\left(\frac{2\gamma\omega}{(\omega_0^2 - \omega^2)^2 + 4\gamma^2\omega^2}\right) \tag{2.44}$$

For the dilute gas situation we are discussing here we know that $n_r \approx 1$, so that (from Eq. (2.44)) n_i will reach a maximum value when $\omega \approx \omega_0$, and will fall to half its maximum value when $\omega_0^2 - \omega^2 \approx 2\gamma\omega$.

It is convenient to simplify Eqs (2.43) and (2.44) at this juncture. Normally we will only be interested in refractive index values close enough to the peak so that we can replace $\omega + \omega_0$ by 2ω. Equation (2.43) can then then be rewritten as

$$n_r^2 = 1 + \frac{Ne^2}{2m\varepsilon_0\omega_0}\left(\frac{\omega_0 - \omega}{(\omega_0 - \omega)^2 + \gamma^2}\right) \tag{2.45}$$

Making the additional assumption that $n_r \approx 1$ we have

$$n_r = 1 + \frac{Ne^2}{4m\varepsilon_0\omega_0}\left(\frac{\omega_0 - \omega}{(\omega_0 - \omega)^2 + \gamma^2}\right) \tag{2.46}$$

Similarly Eq. (2.44) becomes

$$n_i = \frac{Ne^2}{4\varepsilon_0 m\omega_0}\left(\frac{\gamma}{(\omega_0 - \omega)^2 + \gamma^2}\right) \tag{2.47}$$

Both n_r and n_i are plotted as functions of ω about the value $\omega = \omega_0$ in Figure 2.4. We know from Eq. (2.39) that the absorption coefficient is directly proportional to the value of n_i, so that the peak in the curve of n_i represents an *absorption line* in the spectrum.

The variation in n_i close to the absorption line is dominated by the term in square brackets (we may ignore the effects of ω in the denominator of the multiplying expression so long as we remain close to ω_0). This type of functional variation with ω is one that we will meet on a number of future occasions and so it is convenient to introduce here the *Lorentzian function* $L(\omega - \omega_0)$ shown in Figure 2.5 where

$$L(\omega - \omega_0) = \frac{\gamma/\pi}{(\omega - \omega_0)^2 + \gamma^2} \tag{2.48}$$

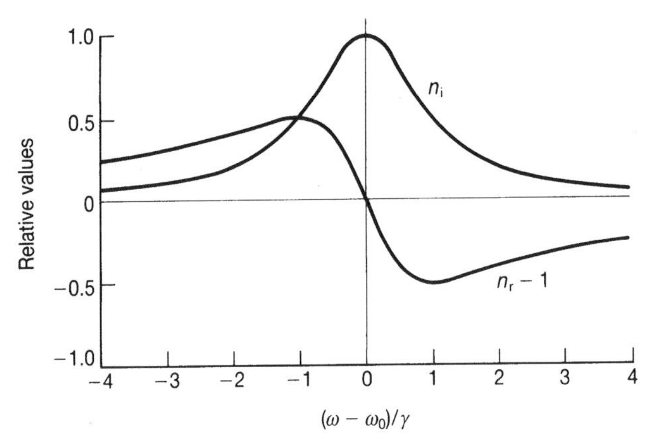

Figure 2.4 Variation of the real (n_r) and imaginary (n_i) parts of the refractive index as a function of ω as predicted by Eqs (2.46) and (2.47).

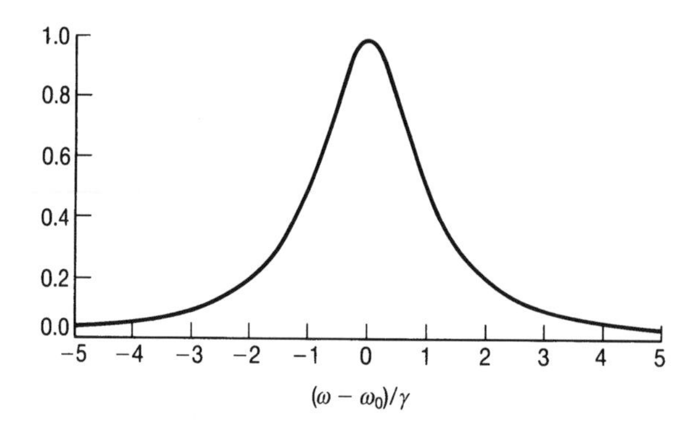

Figure 2.5 The Lorentzian function $L(\omega - \omega_0)$.

The factor π ensures that the function is normalized, that is, that[4]

$$\int_{-\infty}^{\infty} L(x)\, \mathrm{d}x = 1$$

The *linewidth* of an absorption line may be defined as the frequency range over which the absorption exceeds one half of its peak value[5] (i.e. the *full-width half-maxium* or FWHM linewidth). For a Lorentzian type curve the linewidth is 2γ and so an alternative way of writing the normalized Lorentzian is

$$L(\omega - \omega_0) = \frac{1}{2\pi} \frac{\Delta\omega_L}{(\omega - \omega_0)^2 + (\Delta\omega_L/2)^2} \tag{2.49}$$

where $\Delta\omega_L$ is the linewidth $(=2\gamma)$.

In the absorption spectrum of a typical molecular gas there are absorption peaks in several different frequency regions corresponding to different resonant absorption processes. At very small values of ω, that is, where $\omega \ll$ the smallest ω_0, n_r reaches its largest (nonresonant) value. As the frequency is increased the refractive index passes through a series of resonances corresponding to molecular rotations, molecular vibrations, atomic spectra etc. as shown schematically in Figure 2.6. Note that on passing through a resonance the overall value of n_r decreases. Between resonances n_r increases with increasing ω, a situation called *normal dispersion*. Within a frequency range of γ about the centre of a resonance, however, we have *anomalous dispersion*, where the reverse is true.

At very high frequencies, where $\omega \gg$ the largest value of ω_0, the second term on the right-hand side of Eq. (2.45) will become negative, leading to a value of the refractive index which is less than unity. Thus for X-rays and γ-rays we would expect refractive indices slightly less than unity and this is indeed found to be so, even for dense media such as metals.

The variation of refractive index with frequency (i.e. *dispersion*) also means that we have to be careful about what we mean by the speed of light in a medium. In the previous section we introduced the phase velocity, v_p, which is the velocity of

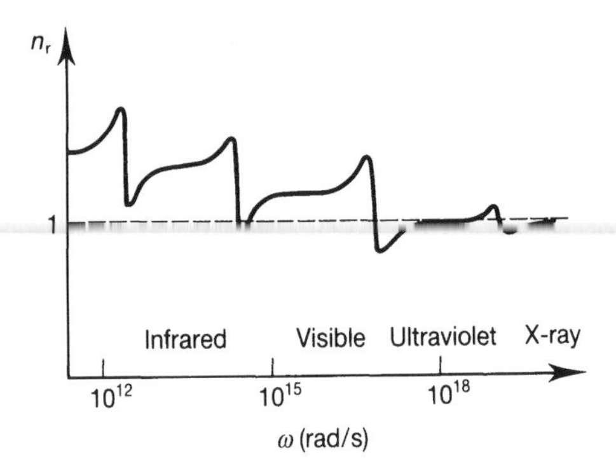

Figure 2.6 Schematic diagram of the variation of the refractive index across the electromagnetic spectrum.

a 'perfect' sine wave. Unfortunately, such a wave can never exist in the real world since it would have to extend from $-\infty$ to $+\infty$ with a constant amplitude for all time. Any real wave must have a finite spatial extent and would probably also have a varying amplitude. However, any such 'real' wave can be shown by Fourier analysis to be equivalent to a 'band' or *group* of perfect sinewaves. Because of dispersion, each single wavelength in the group will travel with a slightly different velocity (i.e. $c/n(\omega)$) and the question now arises as to what is the effective velocity (v_g) of the group. It turns out (see Ref. 3, and Problem 2.4) that

$$v_g = \frac{\partial \omega}{\partial k} \tag{2.50}$$

Using the relations $\omega = 2\pi f$, $k = 2\pi/\lambda$ and $f\lambda = c/n(\lambda)$ Eq. (2.50) can be also written as

$$v_g = \frac{c}{n}\left(1 + \lambda \frac{\partial n}{\partial \lambda}\right) \tag{2.51}$$

Defining an *effective refractive index* n_{eff}, where $v_g = c/n_{\text{eff}}$,

$$n_{\text{eff}} = \frac{n}{1 + \lambda(\partial n/\partial \lambda)} \tag{2.52}$$

Since $\lambda(\partial n/\partial \lambda)$ is usually much smaller than unity

$$n_{\text{eff}} \approx n\left(1 - \lambda \frac{\partial n}{\partial \lambda}\right) \tag{2.53}$$

2.1.6 The Fresnel equations

In Section 2.1.3 we considered the propagation of a plane electromagnetic wave through a dielectric medium. Here we discuss what happens when a wave encounters a boundary between two such media. Because the material is covered in most optics textbooks (see, for example, Ref. 4) the results will not be derived in detail.

The boundary conditions are that the tangential components of $\mathscr{E}$ and $\mathscr{H}$ and the normal components of $\mathscr{D}$ and $\mathscr{B}$ must be continuous across the interface. We consider a plane wave incident at an angle θ_i to an interface between two dielectric media. The interface is taken to be the x-y plane, whereas the direction of propagation of the incident wave lies in the y-z plane. Since the wave is no longer propagating along one of the axes, we generalize Eq. (2.15b) to read

$$\mathscr{E}_i = \mathscr{E}_{0i} \exp(j(\omega_i t - \mathbf{k}_i.\mathbf{r}) \tag{2.54}$$

Here $\mathbf{k}_i$ is a vector (thus $\mathbf{k}$ is now truly a wave*vector*) which has a magnitude of $2\pi n/\lambda_0$ and which points along the direction of propagation of the wave. The reflected and transmitted waves propagate at angles of θ_r and θ_t and their electric fields are written $\mathscr{E}_r$ and $\mathscr{E}_t$ respectively where

$$\mathscr{E}_r = \mathscr{E}_{0r} \exp(j(\omega_r t - \mathbf{k}_r.\mathbf{r}))$$
$$\mathscr{E}_t = \mathscr{E}_{0t} \exp(j(\omega_t t - \mathbf{k}_t.\mathbf{r}))$$

Using the above boundary conditions the following results are readily obtained (see Problem 2.5)

1. All three k vectors lie in the same plane (the *plane of incidence*).
2. The angle of reflection is equal to the angle of incidence ($\theta_i = \theta_r$).
3. The angle of incidence (θ_i) and the angle of the transmitted wave are related by Snell's law:

$$\frac{\sin(\theta_i)}{\sin(\theta_t)} = \frac{n_2}{n_1} \tag{2.55}$$

We now consider how much of the initial wave is reflected and how much transmitted at the interface. There are two different situations to be dealt with when the incident electric field vector is (1) perpendicular to the plane of incidence and (2) in the plane of incidence. Figure 2.7 shows the directions of $\mathscr{E}$ and $\mathscr{H}$ in these two

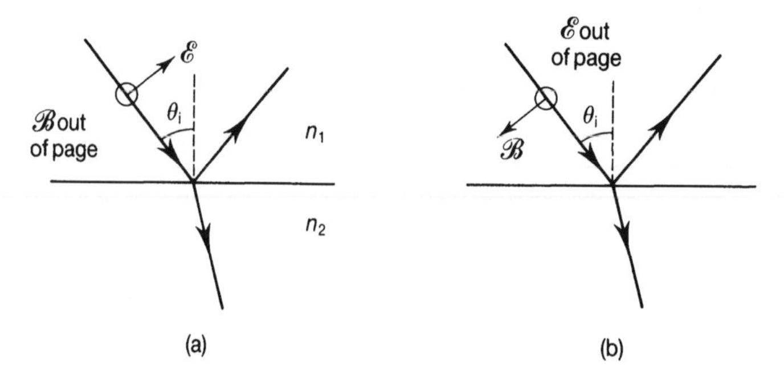

Figure 2.7 Illustration of (a) TM and (b) TE polarization when light is incident on an interface.

situations. In the first of these the electric field is always transverse to the direction of propagation, and is thus sometimes referred to as TE (i.e. *T*ransverse *E*lectric) polarization. Similarly, the second is referred to as TM (*T*ransverse *M*agnetic) polarization. An alternative notation is to use the symbols s and p, respectively. Once we have determined what happens in both of these polarization situations then any arbitrary state of polarization is readily dealt with.

We consider first the case where the electric field is perpendicular to the plane of incidence (i.e. TE polarization). The relative magnitudes of the reflected and transmitted fields may be deduced from the requirements that the tangential electric fields (i.e. y direction) must be equal on either sides of the interface, as must the tangential components of magnetic field (i.e. y direction). The result is (see Problem 2.6) that

$$\mathscr{E}_{0\mathrm{r}}^{\perp} = \frac{n_1 \cos(\theta_\mathrm{i}) - n_2 \cos(\theta_\mathrm{t})}{n_1 \cos(\theta_\mathrm{i}) + n_2 \cos(\theta_\mathrm{t})} \mathscr{E}_{0\mathrm{i}}^{\perp} \tag{2.56}$$

and

$$\mathscr{E}_{0\mathrm{t}}^{\perp} = \frac{2n_1 \cos(\theta_\mathrm{i})}{n_1 \cos(\theta_\mathrm{i}) + n_2 \cos(\theta_\mathrm{t})} \mathscr{E}_{0\mathrm{i}}^{\perp} \tag{2.57}$$

The $\perp$ suffix is introduced to indicate that these equations apply when the electric field is perpendicular to the plane of incidence. When the electric field is in the plane of incidence (i.e. TM polarization) we have

$$\mathscr{E}_{0\mathrm{r}}^{\parallel} = \frac{n_1 \cos(\theta_\mathrm{t}) - n_2 \cos(\theta_\mathrm{i})}{n_1 \cos(\theta_\mathrm{t}) + n_2 \cos(\theta_\mathrm{i})} \mathscr{E}_{0\mathrm{i}}^{\parallel} \tag{2.58}$$

and

$$\mathscr{E}_{0\mathrm{t}}^{\parallel} = \frac{2n_1 \cos(\theta_\mathrm{i})}{n_1 \cos(\theta_\mathrm{t}) + n_2 \cos(\theta_\mathrm{i})} \mathscr{E}_{0\mathrm{i}}^{\parallel} \tag{2.59}$$

Equations (2.56) to (2.59) are known as *Fresnel's equations* and relate the incident, reflected and transmitted fields. To determine the reflected and transmitted powers, we need consider the square of the electric field amplitude (see the discussion in Section 2.1.3). The reflection coefficient, R, can be thus be written

$$R = \left(\frac{\mathscr{E}_{0\mathrm{r}}}{\mathscr{E}_{0\mathrm{i}}}\right)^2 \tag{2.60}$$

Since energy must be conserved at the interface, the transmission coefficient T is related to R by

$$R + T = 1 \tag{2.61}$$

The behaviour of the reflection coefficient with angle of incidence when light moves from a less dense to a more dense optical medium is illustrated in Figure 2.8. With the magnetic field in the plane of incidence, the reflection coefficient increases monotonically to a value of unity at $\theta = 90^\circ$. When the electric field is in the plane of incidence R declines to zero and then increases to the value of unity. Thus for

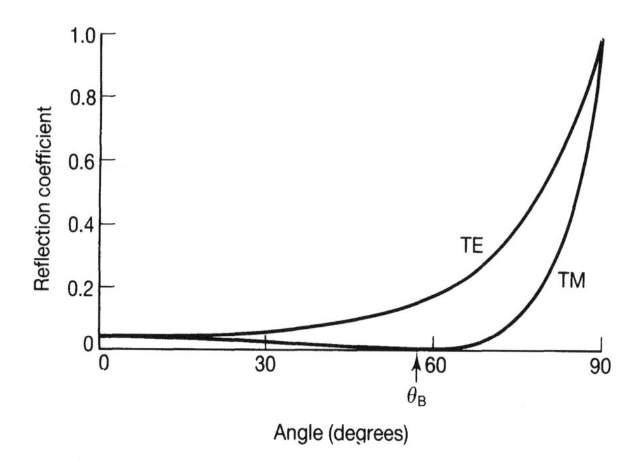

Figure 2.8 Reflectance as a function of angle of incidence for TE and TM polarized radiation when light is incident on to a medium of refractive index 1.5 from one of refractive index unity.

the latter situation there is an incident angle, called the *Brewster angle*, θ_B, at which there is no reflection. From Eq. (2.58) we can see that this occurs when

$$n_1 \cos(\theta_t) = n_2 \cos(\theta_i)$$

When combined with Snell's law this condition can be written (see Problem 2.7) as

$$\tan(\theta_B) = \frac{n_2}{n_1} \tag{2.62}$$

It is interesting to note that at the Brewster angle the reflected and transmitted beams are at right-angles.

When light travels from an optically more dense to a less dense medium another interesting and important phenomenon can occur, that of *total internal reflection*. Snell's law gives

$$n_1 \sin(\theta_1) = n_2 \sin(\theta_2)$$

When $n_1 > n_2$ then $\theta_2 > \theta_1$, so that if θ_1 is increased from 0° a value of θ_1 will be reached when $\theta_2 = 90^\circ$. This particular value of θ_1 is known as the *critical angle*, θ_c, where

$$\sin(\theta_c) = n_2/n_1 \tag{2.63}$$

It may be shown (see Problem 2.9) that the transmission coefficient becomes zero at the critical angle while the reflection coefficient becomes unity. If $\theta_1 > \theta_c$ then the reflection coefficient remains at unity.

Another, less obviously important, result is that when total internal reflection takes place there is a phase shift between the incident and reflected beams:

the phase of the reflected beam leads that of the incident beam). The phase shift depends both on the angle of incidence and on the polarization of the incident radiation. For TE polarization it can be written as $\phi_{TE}(\theta)$ where

$$\tan(\phi_{TE}(\theta)/2) = \left[\sin^2\theta - \left(\frac{n_2}{n_1}\right)^{1/2}\right] \Big/ \cos(\theta) \tag{2.64}$$

while for TM polarization

$$\tan(\phi_{TM}(\theta)/2) = \left(\frac{n_1}{n_2}\right)^2 \tan(\phi_{TE}(\theta)/2) \tag{2.65}$$

An example of the variation of the TE and TM phase shifts with angle is shown in Figure 2.9.

2.1.7 Propagation of light in dielectric waveguides

Semiconductor lasers are fabricated from thin ($\approx \mu$m) layers of materials which act as waveguides. To appreciate some of the properties of these lasers we need to examine the behaviour of light within such slab-like structures. We take the simplest waveguide possible where a layer of dielectric material of thickness d and refractive index n_1 (the *core*) is sandwiched between 'infinitely thick' layers of refractive index n_2 (the *cladding* layers) where $n_1 > n_2$. A ray within the core which is incident on the core–cladding interface at an angle greater than the critical angle will suffer total internal reflection (see discussion at end of Section 2.1.6) and can therefore travel along the slab with the zigzag path shown in Figure 2.10.

In this simple analysis *any* ray with an internal angle θ,[6] where $\theta_c < \theta < 90°$, will be able to travel along the waveguide. However, considerations of the wave-like nature of the ray show that only certain angles are possible between these limits,

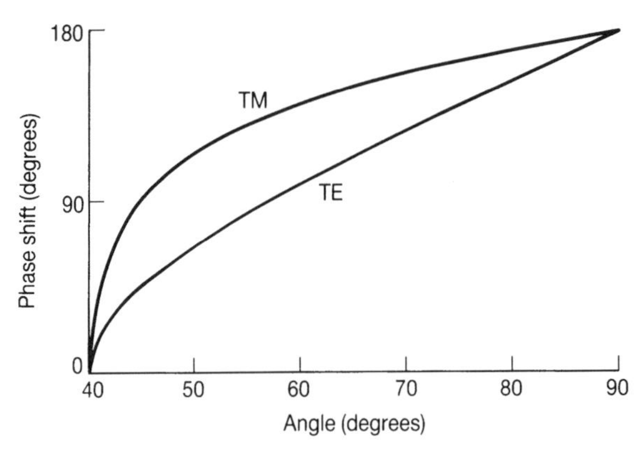

Figure 2.9 Phase shifts for TE and TM polarized radiation when undergoing total internal reflection at an interface with $n_1 = 1.53$ and $n_2 = 1$.

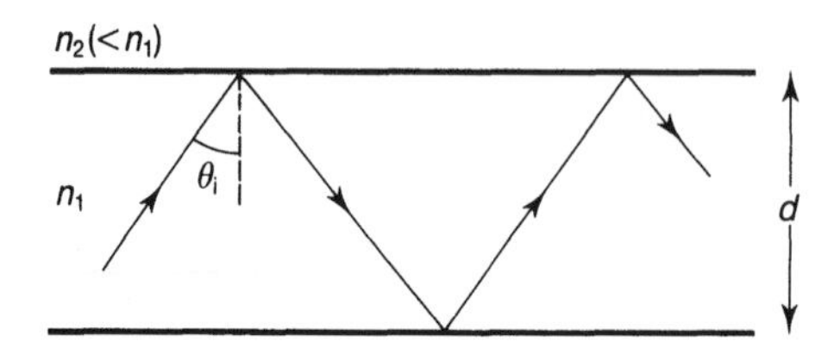

Figure 2.10 Zigzag path of a light ray down a planar dielectric waveguide that results when the angle of incidence at the boundary, θ_i, is greater than the critical angle θ_c.

otherwise destructive interference takes place preventing propagation. The result (see Appendix 3) is that the allowed values of θ must satisfy the equation

$$\frac{2\pi \, d n_1 \cos(\theta)}{\lambda_0} - \phi(\theta) = \pi m \tag{2.66}$$

where m is an integer and $\phi(\theta)$ is the phase change on reflection. Given the condition that θ must lie between θ_c and 90° then there is an upper limit to the value that m can take in Eq. (2.66) (see Appendix 3). For each possible value of m there will be two allowed polarization directions leading to both TE_m and TM_m modes. The variations in electric field across the guide for these modes have a cosine-like behaviour and are proportional to the function

$$\cos\left[\frac{\pi m}{2} - \frac{y}{d}(\pi m + \phi(\theta_m))\right] \tag{2.67}$$

where y is the distance across the guide measured from the centre of the guide.

A rather surprising yet important outcome of the analysis is that the mode field actually extends into the cladding, where it declines exponentially with distance away from the guide. Figure 2.11 illustrates a typical field variation across the guide and cladding for the case of $m = 2$. The rate of the exponential decline in the cladding depends on how close θ_m is to the value of θ at which the mode ceases to propagate (this is referred to as mode *cut-off*). As the mode approaches cut-off the exponential decline becomes increasingly less rapid, so that a greater fraction of the mode power is within the cladding.[7] At cut-off all the mode power moves into the cladding and the mode is no longer a guided mode.

The number of allowed modes for a particular waveguide depends on a parameter V (known as the *normalized film thickness* or *normalized frequency*) where

$$V = \frac{\pi d}{\lambda_0}(n_1^2 - n_2^2)^{1/2} \tag{2.68}$$

If $V \gg 1$ then the number of modes is *approximately* equal to $2V/\pi$, while when $V < \pi/2$ only one mode can propagate. A waveguide where only a single mode can propagate is called a *single-mode* waveguide.[8] The intermediate situation is such that no explicit formula can be given for the number of modes as a function of V.

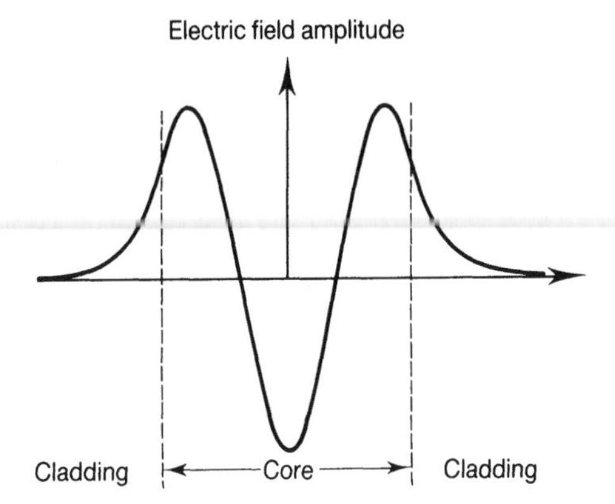

Figure 2.11 Variation in the transverse electric field amplitude of the $m=2$ mode as a function of distance across a planar dielectric waveguide of thickness 2 μm, where $n_1 = 1.5$, $n_2 = 1$ and $\lambda_0 = 1.2$ μm.

Although in theory the lowest-order mode can propagate whatever the value of V, we must bear in mind that if the guide thickness (d) is made very small or if the refractive indices n_1 and n_2 are very close to each other, then the mode field may extend for appreciable distances into the cladding.

Apart from the semiconductor laser, another recently developed laser type in which dielectric waveguide theory is important is the fiber laser. The waveguide in this case has a circular geometry, and is illustrated in Figure 2.12. If we imagine a ray travelling down the fiber core which passes through the centre of the fiber (a *meridional ray*) then the situation is very similar to that shown in Figure 2.10. However, it is also possible to have rays which 'corkscrew' their way down the fiber in a helical-type path without passing through the centre (*skew rays*). In truth, simple ray theory is not really adequate to deal with the situation properly and recourse must be made to solving Maxwell's equations. As might be expected, this involves some quite unpleasant mathematics [Ref. 5] and we must content ourselves with a brief outline of the results.

As with the planar dielectric guide, the circular wave guide can support a number of modes, each of which corresponds to a particular pattern of field distribution

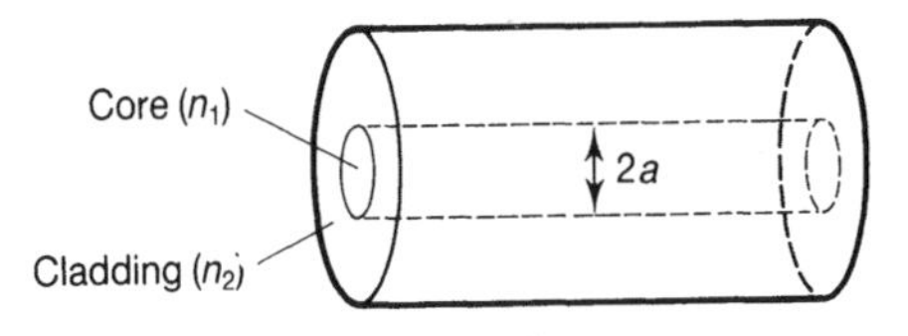

Figure 2.12 Structure of a circular optic waveguide, a core of radius a and refractive index n_1 is surrounded by an annulus of material of refractive index n_2.

across the guide. The radial field distributions are now governed by Bessel functions rather than by cosine functions. Again, fields extend into the cladding where they die away with distance.[9] There are now four different types of mode which are designated TE_{mn}, TM_{mn}, EH_{mn} and HE_{mn}. We note that two integers, m and n, are needed to specify the mode fully rather than one as in the case of the planar waveguide. Again the symbols TE and TM indicate modes where there are no components of electric and magnetic field, respectively, along the direction of propagation. The designations EH and HE indicate modes where there are components of both types of field along the propagation direction (these modes originate in skew rays).

As in the case of the planar waveguide, the number of modes in a circular waveguide depends on a parameter V, now given by

$$V = \frac{2\pi a}{\lambda_0}(n_1^2 - n_2^2)^{1/2} \tag{2.69}$$

$$= \frac{2\pi a}{\lambda_0}(NA)$$

where

$$NA = (n_1^2 - n_2^2)^{1/2} \tag{2.70}$$

here a is the fiber core radius and the quantity NA is known as the *Numerical Aperture* of the fiber.

Provided that $V \gg 1$ the number of modes in the waveguide is given, approximately, by $V^2/2$. Of more importance to us is the fact that the fiber will only support a single mode[10] (the HE_{11} mode) when:

$$V < 2.405 \tag{2.71}$$

since fiber lasers almost inevitably use single-mode fiber. In terms of the radius of the fiber Eq. (2.71) is equivalent to

$$a < \frac{2.405}{2\pi(NA)}\lambda_0 \tag{2.72}$$

Because of Eq. (2.72) the diameters of single-mode fibers tend to be small, say between 5 and 8 μm (see Example 2.1).

Example 2.1 Single-mode fiber diameters

We consider a fiber where $n_1 = 1.46$ and $n_2 = 1.455$. The value of the numerical aperture is given by

$$NA = (1.46^2 - 1.455^2)^{1/2} = 0.12$$

From Eq. (2.72) the fiber radius must be such that

$$a < \frac{2.405}{2\pi\ 0.12}\lambda_0$$

i.e. $a < 3.2\lambda_0$

Thus if the wavelength used is 1.3 μm, then the fiber diameter would have to be less than 8.3 μm.

Another parameter often quoted in connection with single-mode fiber is the *cut-off wavelength*, λ_c. This is defined by

$$\lambda_c = \frac{2\pi(NA)a}{2.405} \tag{2.73}$$

If $\lambda_0 > \lambda_c$ the fiber will be single mode, and multimode if $\lambda_0 < \lambda_c$.

High-quality fibers are usually based on silica, which in its pure state exhibits a high degree of transparency between about 0.5 μm and 1.6 μm. Additional losses can be caused by bends in the fiber ('bending loss') although such losses do not become really significant until the bending radius is less than a few millimetres. Obviously, the flexibility of the fiber is a very useful property. However, with diameters greater than about 600 μm the fiber becomes increasingly inflexible and brittle. Because of this, most silica-based fibers are limited in their overall diameters to a few hundred microns.

The necessary changes in refractive index to obtain both a core and a cladding can be made by doping with suitable impurities (for example, doping with germanium lowers the refractive index), although the changes are limited to a few per cent which means that numerical apertures are never very large (typically, they range from 0.1 to 0.2).

The distribution of energy across a single-mode fiber is circularly symmetric and, to a reasonably good degree of approximation when $1.2 < V < 2.4$, the irradiance distribution can often be regarded as *Gaussian*, that is,

$$I(r) = I(0)\exp(-2r^2/w^2) \tag{2.74}$$

The parameter w is referred to as the *mode field radius* or the *spot size*.

Although in theory the lowest-order mode (HE_{11}) is always able to propagate no matter what value of V, as V is made smaller more and more of the mode energy moves into the cladding. The mode field radius can be expressed as a function of $V (1.2 < V < 2.4)$ as follows [Ref. 6]:

$$w/a = 0.65 + 1.619\ V^{-3/2} + 2.879\ V^{-6} \tag{2.75}$$

The variation of irradiance across the core of a fiber where $V = 1.6$ as given by Eqs (2.74) and (2.75) is shown in Figure 2.13. Extension of the mode field too far into the cladding can cause unwelcome problems, principally that the fiber becomes

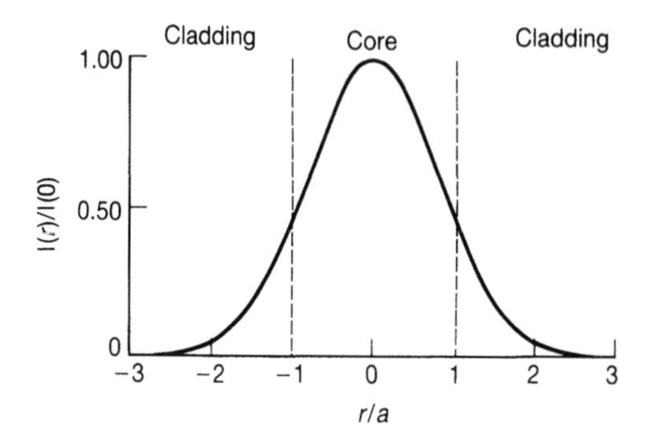

Figure 2.13 Variation of irradiance across a section of a single-mode optical fiber when $V = 1.6$.

much more sensitive to bending losses. Because of this, single-mode fibers are normally operated with a V value of about 2 (values close to 2.405 should also be avoided to minimize the possibility that more than one mode might propagate).

Another mode formalism that may be encountered results when the approximation is made that $n_1 - n_2 \ll 1$ (usually valid). The resulting modes are then referred to as *Linearly Polarized* (LP) modes. The lowest-order LP mode is the LP_{01} mode.

2.1.8 Reflection of light from metal surfaces

In this section we examine the reflection of light from a metal surface. The Fresnel equations developed in Section 2.1.6 may still be used, provided we allow for the possibility that the refractive index of the metal can be complex. This is reasonable, since metals are highly absorbing in the visible region and we know that the presence of absorption is associated with the imaginary part of the refractive index. (It may be noted that an alternative but equivalent approach is to introduce a finite conductivity into Maxwell's equations; see Problem 2.13.) For simplicity, we consider reflection at normal incidence from an air–metal interface where the refractive index of the metal, n, is written $n_r + jn_i$. From Eqs (2.60) and (2.58) we have

$$R = \frac{(1-n)(1-n^*)}{(1+n)(1+n^*)} \tag{2.76}$$

Thus

$$R = 1 - \frac{2n_r}{1 + 2n_r + n_r^2 + n_i^2} \tag{2.77}$$

We may note that if the material is highly absorbing (i.e. $n_i \gg n_r$) then the reflectivity will be high, indeed if the refractive index is wholly imaginary then $R = 1$.

To determine the real and complex parts of the refractive index we use the results of Section 2.1.5, in particular Eqs (2.46) and (2.47). These give the real and imaginary refractive indices assuming that atomic electrons behave like damped harmonic oscillators when subject to an oscillating electric field. In a metal we may assume that the conduction electrons, since they are not bound, have no natural oscillation frequency, that is, $\omega_0 = 0$. Equation (2.47) then becomes

$$n_{\mathrm{i}} = \frac{Ne^2}{4\varepsilon_0 m\omega}\left(\frac{\gamma}{\omega^2+\gamma^2}\right) \tag{2.78}$$

We see that at frequencies where $\omega \gg \gamma$, n_{i} will be relatively small and hence the material fairly transparent. Similarly, we may neglect the terms in ω_0 and γ in Eq. (2.46) and obtain

$$n_{\mathrm{r}}^2 = 1 - \frac{\omega_{\mathrm{p}}^2}{\omega^2} \tag{2.79}$$

where

$$\omega_{\mathrm{p}}^2 = \frac{Ne^2}{\varepsilon_0 m} \tag{2.80}$$

ω_{p} is called the *plasma frequency*. When $\omega < \omega_{\mathrm{p}}$ the refractive index is real, while when $\omega > \omega_{\mathrm{p}}$ the refractive index is wholly imaginary. Thus at the frequency ω_{p} we expect the reflection coefficient suddenly to become unity. Undoubtedly this treatment of the refractive index is very approximate (and indeed omits any contributions from the bound electrons). Nevertheless, a rapid change in reflectivity with frequency is observed in the alkali metals in the ultraviolet region (see Example 2.2).

Example 2.2 Plasma frequency for sodium

The electron density in sodium is $2.5 \times 10^{28}\ \mathrm{m}^3$, and using this value in Eq. (2.80) gives

$$\omega_{\mathrm{p}}^2 = \frac{2.5 \times 10^{28}\ (1.6 \times 10^{-19})^2}{8.85 \times 10^{-12}\ 9.1 \times 10^{-31}}$$

or

$$\omega_{\mathrm{p}} = 8.9 \times 10^{15}\ \mathrm{Hz}$$

This corresponds to a wavelength of 211 nm. In fact, in sodium the sudden change in reflectivity with wavelength occurs at 210 nm.

Generally, the reflectivity of most metals remains high through the visible and into the infrared. Beyond a wavelength of 10 μm or so the reflectivity is in quite good agreement with that calculated on the basis of Maxwell's equations, that is (see Problem 2.13),

$$r = 1 - \left(\frac{8\omega\varepsilon_0}{\sigma\mu_r}\right)^{1/2} \tag{2.81}$$

where σ is the conductivity.

Example 2.3 Reflectivity of copper

We may use Eq. (2.81) to calculate the reflectivity of copper at a wavelength of 10 μm (i.e. $\omega = 1.88 \times 10^{14}\ \text{s}^{-1}$). Copper has a DC conductivity of $5.8 \times 10^7\ \text{S m}^{-1}$, and if we assume this value is still valid at 10 μm and also that $\mu_r = 1$, then we have

$$R = 1 - \left(\frac{8 \times 1.88 \times 10^{14} \times 8.85 \times 10^{-12}}{5.8 \times 10^7}\right)^{1/2}$$

or

$$R = 0.985$$

Figure 2.14 shows the variation in reflectivity for several common metals as a function of wavelength.

2.1.9 Optically anisotropic media

So far in this chapter we have assumed that the dielectric media we are dealing with are isotropic, so that the polarization produced by an applied electric field is independent of the field direction. We must now consider optically anisotropic media where this is not the case. The anisotropy may be present because of the particular crystalline structure of the material or it may be induced by the presence of external forces. We deal first with the consequences of natural anisotropy.

In general, when a field is applied along three arbitrary directions x, y, z, then the polarization induced in a medium along these directions may be described by a susceptibility tensor χ_{ij} where

$$\begin{aligned}
\mathscr{P}_x &= \varepsilon_0(\chi_{11}\mathscr{E}_x + \chi_{12}\mathscr{E}_y + \chi_{13}\mathscr{E}_z) \\
\mathscr{P}_y &= \varepsilon_0(\chi_{21}\mathscr{E}_x + \chi_{22}\mathscr{E}_y + \chi_{23}\mathscr{E}_z) \\
\mathscr{P}_z &= \varepsilon_0(\chi_{31}\mathscr{E}_x + \chi_{32}\mathscr{E}_y + \chi_{33}\mathscr{E}_z)
\end{aligned}$$

Fortunately it is always possible to choose particular x, y and z axes so that all the 'off-diagonal' elements of the tensor vanish, and therefore

$$\left.\begin{aligned}
\mathscr{P}_x &= \varepsilon_0\chi_{11}\mathscr{E}_x \\
\mathscr{P}_y &= \varepsilon_0\chi_{22}\mathscr{E}_y \\
\mathscr{P}_z &= \varepsilon_0\chi_{33}\mathscr{E}_z
\end{aligned}\right\} \tag{2.82}$$

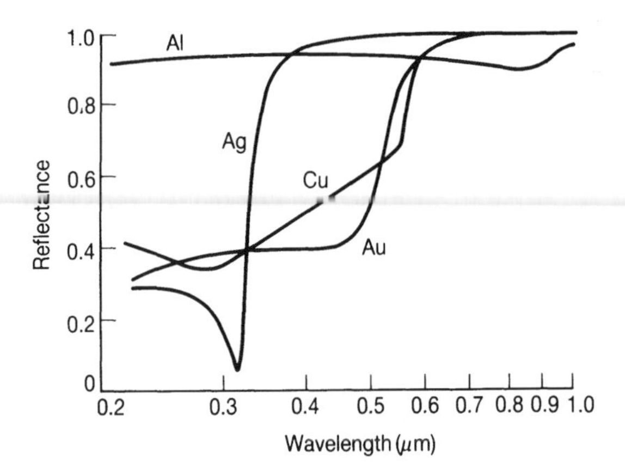

Figure 2.14 Reflectance as a function of wavelength for a number of common metals

These particular x, y and z axes are known as the *principal axes* of the crystal. For our purposes it is more convenient to describe the anisotropy in terms of an electric permittivity tensor ε_{ij} where $\varepsilon_{ij} = \varepsilon_0(1 + \varkappa_{ij})$ so that (again using the principal axes of the crystal)

$$\left.\begin{aligned} \mathscr{D}_x &= \varepsilon_{11}\mathscr{E}_x \\ \mathscr{D}_y &= \varepsilon_{22}\mathscr{E}_y \\ \mathscr{D}_z &= \varepsilon_{33}\mathscr{E}_z \end{aligned}\right\} \tag{2.83}$$

These equations imply that the phase velocity $(=(\mu_0\varepsilon)^{1/2})$ of a plane polarized wave travelling through the crystal will depend on the direction of the electric field. We define three principal refractive indices n_x, n_y and n_z, where $n_x = (\varepsilon_{11}/\varepsilon_0)^{1/2}$ etc, which are the refractive indices experienced by the wave when the field directions are along the principal axes. It is important to realize that the suffices x, y and z refer to the direction of the electric field vector of the wave and *not* to the direction of propagation.

Anisotropic crystals may be divided into two main classes, *uniaxial* and *biaxial*. Uniaxial crystals have only two different electrical permittivities and, by convention, the x and y axes are chosen such that $\varepsilon_{11} = \varepsilon_{22}$ (so that $n_x = n_y$). It is also conventional (for reasons which will become apparent later) to refer to n_x as n_o (the *ordinary* refractive index) and n_z as n_e (the *extraordinary* refractive index). The z axis is known as the *optic* axis, and any radiation propagating along this axis experiences a refractive index (n_o) that is independent of the direction of polarization. This is the only direction in the crystal for which this is true. For all other directions the refractive index depends on the direction of polarization, a property known as *birefringence*. For radiation propagating along the z or y axes, for example, the refractive index will be either n_o or n_e depending on the polarization direction. The magnitude of the birefringence is determined by the difference between n_e and n_o and is said to be positive or negative depending on whether n_e is greater or smaller than n_o. Calcite, for example, is a negative uniaxial crystal while quartz is a positive uniaxial crystal.

Biaxial crystals are crystals of lower symmetry than uniaxial crystals. They have two optic axes which do not coincide with any of the principal axes. Generally, most anisotropic crystals of interest to us are uniaxial, and we need not pursue the biaxial case any further.

The solution of Maxwell's equations in anisotropic media is rather too lengthy to be attempted here and we restrict ourselves to a consideration of the outcomes (for a full discussion see, for example, Ref. 7). It turns out that, for propagation along any direction other than the optic axis, only two directions of polarization are allowed. These are orthogonal, and the two components experience different refractive indices and hence travel at different speeds.

The polarization directions and the corresponding refractive indices may be determined from a construct known as the *index ellipsoid*. For a uniaxial crystal the equation for this is

$$\frac{x^2}{n_0^2}+\frac{y^2}{n_0^2}+\frac{z^2}{n_e^2}=1 \tag{2.84}$$

To illustrate the use of the index ellipsoid consider light propagating along a direction given by the vector **r** which is at an angle θ to the optic axis where the projection of **r** on the xy plane is along the y axis[11] (Figure 2.15). The plane normal to **r** which passes through the origin intersects the index ellipsoid along the perimeter of an ellipse. The two directions of polarization allowed by Maxwell's equations then lie along the principal axes (OP and OQ) of this ellipse. In addition, the corresponding refractive indices are just the lengths of these principal axes. One of these (n_O) is independent of θ (and corresponds to an *ordinary* ray) while the other, $n_e(\theta)$, depends on the ray angle θ (and corresponds to an *extraordinary* ray). An expression for $n_e(\theta)$ is readily derived (see Problem 2.14) thus

$$\frac{1}{n_e^2(\theta)}=\frac{\cos^2\theta}{n_0^2}+\frac{\sin^2\theta}{n_e^2} \tag{2.85}$$

The amount of birefringence exhibited will depend on the value of θ, and it will be maximum when **r** is in the xy plane (i.e. $\theta = 90^\circ$), when the two refractive indices will be n_O and n_e.

When an electric field is applied to a dielectric medium the effect may be to introduce birefringence into naturally isotropic materials (e.g. GaAs), or to induce new principal axes in naturally anisotropic materials (such as $LiNbO_3$). As far as we are concerned, the latter situation is more interesting and here the effect is to deform the index ellipsoid. The shape is changed as are the directions of the principal axes. The new ellipsoid may be described in general by the equation

$$\left(\frac{1}{n^2}\right)_1 x^2+\left(\frac{1}{n^2}\right)_2 y^2+\left(\frac{1}{n^2}\right)_3 z^2+2\left(\frac{1}{n^2}\right)_4 yz+2\left(\frac{1}{n^2}\right)_5 xz+2\left(\frac{1}{n^2}\right)_6 xy=1 \tag{2.86}$$

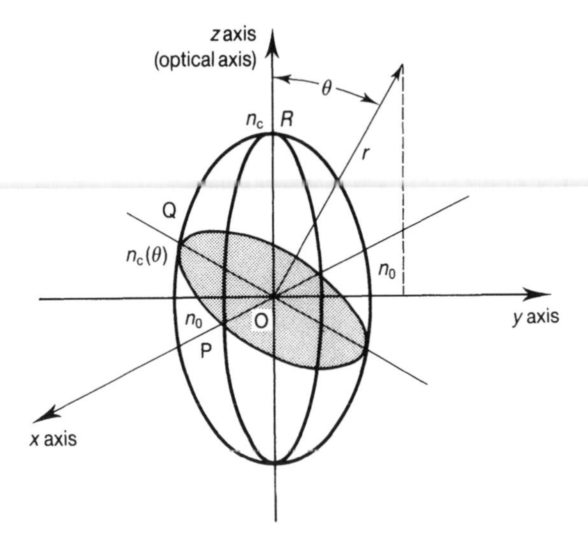

Figure 2.15 Refractive index ellipsoid for a uniaxial crystal. Waves with polarizations parallel to the z axis experience the refractive index n_e, while those polarized parallel to the x and y axes experience the refractive index n_0. For propagation along the general direction **r**, there are two allowed directions of polarization: parallel to OP (with index n_0) and parallel to OQ (with index $n_c(\theta)$).

The first three coefficients in Eq. (2.86) are given by equations similar to the following:

$$\left(\frac{1}{n^2}\right)_1 = \frac{1}{n_x^2} + \Delta\left(\frac{1}{n^2}\right)_1 \quad \text{etc.}$$

and the next three are given by

$$\left(\frac{1}{n^2}\right)_4 = \Delta\left(\frac{1}{n^2}\right)_4 \quad \text{etc.}$$

where (writing Δ_j for $\Delta(1/n^2)_j$)

$$\begin{vmatrix} \Delta_1 \\ \Delta_2 \\ \Delta_3 \\ \Delta_4 \\ \Delta_5 \\ \Delta_6 \end{vmatrix} = \begin{vmatrix} r_{11} & r_{12} & r_{13} \\ r_{21} & r_{22} & r_{23} \\ r_{31} & r_{32} & r_{33} \\ r_{41} & r_{42} & r_{43} \\ r_{51} & r_{52} & r_{53} \\ r_{61} & r_{62} & r_{63} \end{vmatrix} \begin{vmatrix} \mathscr{E}_x \\ \mathscr{E}_y \\ \mathscr{E}_z \end{vmatrix} \tag{2.87}$$

The quantity r_{ij} is the *electro-optic tensor* and has, in general, 18 elements.

In practice, the number of nonzero elements is considerably reduced by the symmetry of the material under consideration. In this respect an important class of materials for some of the later discussions in the book are those with the symmetry group $\bar{4}2m$[12], of which potassium dihydrogen phosphate (KH_2PO_4) is an example.[13] Here the only nonvanishing elements are r_{41}, r_{52} (which is equal to r_{41}) and r_{63}. The equation for the index ellipsoid then becomes

$$\frac{x^2}{n_0^2}+\frac{y^2}{n_0^2}+\frac{z^2}{n_e^2}+2r_{41}\mathscr{E}_x yz+2r_{41}\mathscr{E}_y xz+2r_{63}\mathscr{E}_z xy=1 \tag{2.88}$$

To simplify the analysis we assume that the field is applied along the original optic axis (i.e. z axis). $\mathscr{E}_x$ and $\mathscr{E}_y$ are then zero and Eq. (2.88) becomes

$$\frac{x^2}{n_0^2}+\frac{y^2}{n_0^2}+\frac{z^2}{n_e^2}+2r_{63}\mathscr{E}_z xy=1 \tag{2.89}$$

We now require to find new axes x', y' and z' such that Eq. (2.89) can be written

$$\frac{x'^2}{n_{x'}^2}+\frac{y'^2}{n_{y'}^2}+\frac{z'^2}{n_{z'}^2}=1 \tag{2.90}$$

It is evident that, because the field is along the z axis, $z'=z$ and $n_{z'}=n_z$. As far as the x' and y' axes are concerned, they turn out to be at 45° to the x and y axes, so that

$$\left.\begin{aligned} x&=x'\cos 45^\circ+y'\sin 45^\circ\\ y&=-y'\sin 45^\circ+y'\cos 45^\circ\end{aligned}\right\} \tag{2.91}$$

It is then easily verified that the index ellipsoid equation can be written

$$x'^2\left(\frac{1}{n_0^2}-r_{63}\mathscr{E}_z\right)+y'^2\left(\frac{1}{n_0^2}+r_{63}\mathscr{E}_z\right)+\frac{z'^2}{n_{z'}^2}=1 \tag{2.92}$$

Evidently the effect of applying the field is to rotate the xy plane ellipse and to change the lengths of the major and minor axes. Thus the length of the major axis, $2n_{x'}$, is given by

$$\frac{1}{n_{x'}^2}=\frac{1}{n_0^2}-r_{63}\mathscr{E}_z$$

Assuming that any field-induced changes are small, we have

$$n_{x'}=n_0+\frac{n_0^3}{2}r_{63}\mathscr{E}_z \tag{2.93}$$

and also

$$n_{y'}=n_0-\frac{n_0^3}{2}r_{63}\mathscr{E}_z \tag{2.94}$$

Figure 2.16 illustrates the refractive index changes induced by an external field. Equations similar to (2.93) and (2.94) can be derived for other classes of crystal, and in general the changes in refractive index can be written as $\pm\frac{1}{2}n_0^3 r\mathscr{E}$. The actual

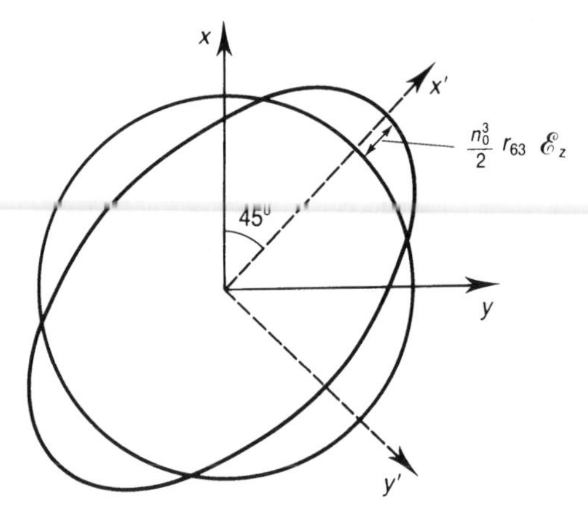

Figure 2.16 The effect of applying a field along the z axis of a uniaxial material is to create new axes (x', y') at 45° to the old (x, y) and also to induce a field-dependent anisotropy in the refractive indices.

matrix element r that must be used varies according to the crystal class. Typical values for the linear electro-optic coefficients are of the order of 10^{-11} m V^{-1} (values for a few common electro-optic materials are given in Table 8.1) so that the changes in the refractive indices are relatively very small (see Example 2.4).

Example 2.4 Refractive index change due to the Pockels effect
For KDP the value of r_{63} is 10.6×10^{-12} m V^{-1} while $n_0 = 1.51$ (Table 8.1). With a voltage of 1000 V across a thickness of 1 mm the electric field within the material is $10^3/10^{-3}$ or 10^6 V/m. Thus using Eq. (2.93) the change in refractive index is given by

$$n_{x'} - n_0 = \frac{(1.51)^3}{2} 10.6 \times 10^{-12} \times 10^6$$
$$= 1.8 \times 10^{-5}$$

Many isotropic media, both solids and liquids, when placed in an electric field behave as uniaxial crystals with the optic axis parallel to the electric field. The difference between the refractive indices for light polarized parallel to and perpendicular to the induced optic axis is given by

$$\Delta n = K\lambda\mathscr{E}^2 \tag{2.95}$$

where K is a constant. The effect is known as the *Kerr* effect. It is interesting in that the electric field of a beam of light itself can cause refractive index changes (the *optical* Kerr effect).

2.2 Black-body radiation

A *black-body* is defined as a body which will absorb all radiation falling on it no matter what the wavelength. Such a concept may at first seem somewhat abstruse and of limited practical importance. However, many sources of radiation with which we are familiar approximate reasonably closely to black bodies, the prime example being the sun (others would include quartz-halogen lamps and arc lamps). From this it will be apparent that black bodies not only absorb radiation, they also emit it.

It is in fact quite easy to construct an object which behaves like a black body. Consider, for example, a hollow sphere with a hole in it. If the hole diameter is very much smaller than the diameter of the sphere then any light entering the hole from the outside will suffer a large number of internal reflections before finding its way out again (Figure 2.17). Even if only a small amount of energy is lost at each reflection, practically no light will re-emerge from the hole. Thus the hole acts as an approximation to a black body, and the smaller the hole, the better the approximation will be.

The character of the radiation emitted by such a black body will be determined by the radiation within the container. Since a 'perfect' black-body would involve an infinitesimally small hole, we are led to investigate the radiation inside a totally sealed container. The interior walls of such a container will be continually emitting and absorbing radiation, but when the container is in thermal equilibrium the emission and absorption rates must be equal. Let us suppose that when the walls of the container are in thermal equilibrium and that the resulting flux of energy throughout the container can be described in terms of a *spectral energy density* $\rho(\nu, T)$. The amount of electromagnetic energy per unit volume which has a frequency between

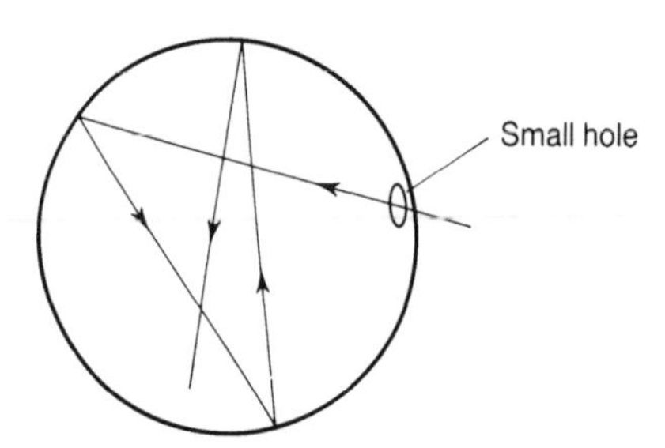

Figure 2.17 Light entering a hollow sphere through a small hole will undergo many reflections before it can re-emerge.

ν and $\nu + \mathrm{d}\nu$ is then given by $\rho(\nu, T)\,\mathrm{d}\nu$. The total amount of radiation per unit volume is given by $\rho(T)$ where

$$\rho(T) = \int_0^\infty \rho(\nu, T)\,\mathrm{d}\nu \tag{2.96}$$

A simply thermodynamic argument shows that $\rho(\nu, T)$ (and hence $\rho(T)$) is a 'universal' function in that it is independent of the nature of the walls of the container or its shape, or indeed of anything else except the temperature of the walls and the value of ν. Thus we consider two such containers in thermal contact to be connected optically by means of a small hole with a window that only allows radiation within a narrow frequency band centred on ν' to pass. If the radiation densities at frequency ν' are different then radiation will flow from one container to another, thereby changing the total energy content of the containers. However, when two bodies at the same temperature are placed in thermal contact no energy exchange can take place (this is sometimes referred to as the 'zeroth law of thermodynamics'). We must conclude therefore that the radiation densities in the two containers must be identical for all values of ν'.

The function $\rho(\nu, T)$ can be investigated experimentally by measuring the characteristics of the radiation emerging from a small hole in the side of such a container (which is, of course, just the radiation emitted from a black body as defined at the beginning of this section). It may be shown (see Problem 2.15) that the total amount of energy emitted per unit time per unit area by a black body is related to $\rho(\nu, T)$ by

$$I(\nu, T) = \rho(\nu, T)c/4 \tag{2.97}$$

The functional dependence of $\rho(\nu, T)$ on ν and T was first derived by Max Planck, who, in order to obtain the correct form, had to introduce the concept of light quanta. The standard derivation is well known and we need not reproduce it here (see, for example, Ref. 9). The result is that

$$\rho(\nu, T) = \left(\frac{8\pi h\nu^3}{c^3}\right)\frac{1}{\exp(h\nu/kT) - 1} \tag{2.98}$$

Equation (2.98) is in excellent agreement with experimental observations. A graph $\rho(\nu, T)$ as a function of ν is shown in Figure 2.18 for the case where the enclosure is at a temperature of 3000 K.

For some purposes it is useful to change the argument of the radiation density function. Thus if $\rho(\lambda, T)\,\mathrm{d}\lambda$ is the density of black-body radiation within the wavelength range λ to $\lambda + \mathrm{d}\lambda$ and $\rho(\omega, t)\,\mathrm{d}\omega$ is the density of black-body radiation within the angular frequency range ω to $\omega + \mathrm{d}\omega$, then it is easy to show that

$$\rho(\lambda, T) = \left(\frac{8\pi hc}{\lambda^5}\right)\frac{1}{\exp(hc/\lambda kT) - 1} \tag{2.99}$$

and

$$\rho(\omega, T) = \left(\frac{\hbar\omega^3}{\pi^2 c^3}\right)\frac{1}{\exp(\hbar\omega/kT) - 1} \tag{2.100}$$

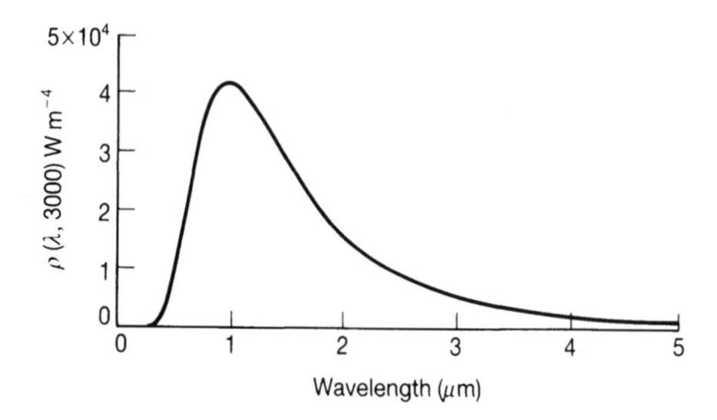

Figure 2.18 Radiation spectral density within a black body at 3000 K.

In future when referring to black-body radiation we drop the explicit dependence on T, but note must be taken as to whether the argument is ν, λ or ω.

Before leaving this section we note two results that may be derived from the above formulae. First, the total irradiance, I_{BB}, of a black body can be obtained by integrating its spectral irradiance over all frequencies (see Problem 2.16a), the result being

$$I_{BB} = \frac{2\pi^5 k^4}{15c^2h^3} T^4$$

or

$$I_{BB} = \sigma T^4 \tag{2.101}$$

where σ ($= 5.67 \times 10^{-8}$ W m^{-2} K^{-4}) is the *Stefan–Boltzmann constant*. Equation (2.101) is known as the *Stefan–Boltzmann law*.

The second result, which follows from Eq. (2.99), is that the wavelength (λ_{max}) at which peak emission takes place is given by (see Problem 2.16b)

$$\lambda_{max} T = \frac{hc}{4.965k}$$

or

$$\lambda_{max} T = 2.987 \times 10^{-3} \text{ mK} \tag{2.102}$$

This result is known as the *Wien displacement law*. Typical calculations involving the Stefan–Boltzmann and Wein displacement laws are given in Example 2.5.

Most incandescent sources are some way from being black bodies, and the extent to which they deviate from ideality is described by the *emissivity* function, $\varepsilon(\lambda, T)$. Thus $\varepsilon(\lambda, T)$ is the ratio between the amount of energy emitted by the unit area of the source in question and by the unit area of a black body. When $\varepsilon(\lambda, T)$ is more or less independent of either wavelength and temperature, the source is called a *grey* body.

Example 2.5 Radiation from a filament lamp
Consider a filament lamp where the filament at a temperature T has a radius of r and a length l. The amount of power radiated, W, is given by

$$W = 2\pi r l \sigma T^4$$

where we have assumed an emissivity of unity. Thus

$$T = \left(\frac{W}{2\pi r l \sigma}\right)^{\frac{1}{4}}$$

Taking a 100 W bulb with $r = 12\ \mu$m and $l = 0.3$ m gives

$$T = \left(\frac{100}{2\pi \times 12 \times 10^{-6}\ 0.3 \times 5.67 \times 10^{-8}}\right)^{\frac{1}{4}}$$

$$= 2972 \text{ K}$$

In practice the emissivity will be somewhat less than unity and also not all of the 100 W will be radiated. However, the corrections required will, to some extent, tend to cancel each other out.

Using Eq. (2.102) a black body with a temperature of 2972 K will have its peak emission at a wavelength given by

$$\lambda_{max} = 2.987 \times 10^{-3}/2972 \text{ m}$$
$$= 1\ \mu\text{m}$$

2.3 The Einstein coefficients

In 1917 Einstein showed that the Planck formulae, obtained in the previous section from essentially thermodynamic arguments, could also be obtained by considering absorption and emission rates between energy levels provided that *two* different kinds of emission process were present. To understand his argument we consider a collection of atoms, where each atom has just two energy levels E_1 and E_2 ($E_2 > E_1$) (Figure 2.19). We assume that there are then N_1 atoms per unit volume with energy E_1 and N_2 atoms with energy E_2 (note that we do not assume initially that this is necessarily an equilibrium distribution). The radiation that is absorbed or emitted by the atoms will have a frequency given by

$$h\nu_{12}(=\hbar\omega_{12}) = E_2 - E_1$$

If there are no external influences (i.e. radiation) present then the only radiative process that can take place is that of *spontaneous emission*. In this, atoms at energy E_2 spontaneously emit radiation and decay back to the lower energy state. The decay process is very similar to that of radioactive decay in nuclei in that it is impossible to say when a particular atom will decay. All we can say is that it has a particular probability of decay per unit time. That is, the number of transitions

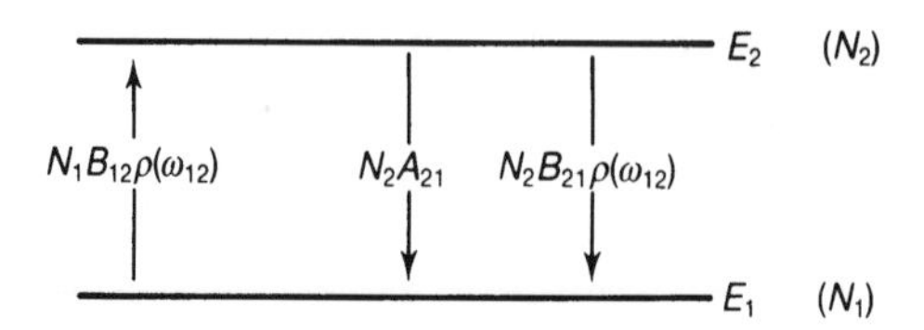

Figure 2.19 A simple energy level system consisting of just two levels at E_2 and E_1 with populations N_2 and N_1. The transition rates induced when radiation of density $\rho(\omega_{12})$ is present ($\hbar\omega_{12} = E_2 - E_1$) are also indicated.

taking place in the small time interval dt will be proportional to both N_2 and dt, therefore we can write that

$$dN_2 = -A_{21}N_2\, dt$$

or

$$\frac{dN_2}{dt} = -A_{21}N_2 \tag{2.103}$$

where A_{21} is the *Einstein A coefficient*. Equation (2.103) may be solved to give the variation of N_2 with time provided we know the values of N_2 at $t = 0$ and $t = \infty$. At long times the system must tend to the thermal equilibrium situation where Boltzmann statistics gives that

$$\frac{N_2}{N_1} = \frac{g_2}{g_1}\exp\left(-\frac{E_2 - E_1}{kT}\right) \tag{2.104}$$

Here g_1 and g_2 are the degeneracies of the lower and upper states, respectively. Provided $E_2 - E_1 \gg kT$, the equilibrium value for N_2 will be very small and Eq. (2.103) may then be solved to give

$$N_2(t) = N_2(0)\exp(-A_{21}t) \tag{2.105}$$

We see that the upper state population decays exponentially with time. An alternative way of describing the decay is in terms of an upper state *lifetime*, τ_{21}, where

$$N_2(t) = N_2(0)\exp(-t/\tau_{21}) \tag{2.106}$$

Comparing Eqs (2.105) and (2.106) we see that

$$A_{21} = \frac{1}{\tau_{21}} \tag{2.107}$$

The upper state lifetime is a quantity that can be deduced experimentally by artificially creating an excess of atoms in the excited state and then monitoring the resulting decay of spontaneous emission with time. A word of warning is needed here. In any atomic or molecular system there are often other (nonradiative) decay processes present (i.e. other mechanisms by which decay from levels E_2 to E_1 can take place). If each of these can be described in terms of lifetimes τ_a, τ_b etc., then the effective overall lifetime (τ_t) is given by (Problem 2.17).

$$\frac{1}{\tau_t}=\frac{1}{\tau_{21}}+\frac{1}{\tau_a}+\frac{1}{\tau_b}+\cdots \tag{2.108}$$

It is this *overall* lifetime which determines how the spontaneous emission will vary with time and not τ_{21}.

We turn now to the process of absorption. We would expect the absorption rate to depend both on the number of atoms in the lower energy state and on the density of radiation at the frequency ν_{12}. Accordingly, we write the number of absorptions taking place per unit time per unit volume as[14]

$$N_1B_{12}\rho(\omega_{12}) \tag{2.109}$$

where B_{12} is the the coefficient of proportionality involved (an *Einstein B coefficient*).

Both the emission and absorption processes described above were well known long before Einstein's work on radiation. The novel concept that he introduced was that of *stimulated* emission. In this, atoms are induced to make a transition from the higher to the lower energy level by the very presence of radiation with the frequency which will be emitted in the emission process (i.e. $\nu_{21}=(E_2-E_1)/h$). The stimulated emission rate depends not only on the number of atoms in the upper state but also on the density of radiation present at the required frequency ($\rho(\omega_{21})$). Thus the transition rate due to stimulated emission is written

$$N_2B_{21}\rho(\omega_{21}) \tag{2.110}$$

Einstein then examined what would happen if the above collection of atoms were to be placed within a closed container. Once equilibrium has been established, we know from our previous discussion that the radiation within the cavity will be black-body radiation. Within the collection of atoms equilibrium will be reached when the upward and downward transition rates are equal, that is, when

$$N_1B_{12}\rho(\omega_{21})=N_2A_{21}+N_2B_{21}\rho(\omega_{21})$$

or

$$\rho(\omega_{21})=\frac{A_{21}}{(N_1/N_2)B_{12}-B_{21}} \tag{2.111}$$

From Eq. (2.104) we have that

$$\frac{N_1}{N_2}=\frac{g_1}{g_2}\exp\left(\frac{E_2-E_1}{kT}\right)=\frac{g_1}{g_2}\exp\left(\frac{\hbar\omega_{21}}{kT}\right)$$

and using this relation in Eq. (2.111) gives

$$\rho(\omega_{21})=\frac{A_{21}}{(g_1/g_2)\exp(\hbar\omega_{21}/kT)\,B_{12}-B_{21}} \tag{2.112}$$

Equation (2.112) can be generalized for transitions between any two arbitrary energy levels by replacing ω_{21} by ω. If we compare the resulting equation with the Planck law expression (2.115) we see that they agree provided that

$$B_{21} = \frac{g_1}{g_2} B_{12} \tag{2.113}$$

and

$$\frac{A_{21}}{B_{21}} = \frac{\hbar\omega^3}{\pi^2 c^3} \tag{2.114}$$

Equations (2.113) and (2.114) determine the ratios of the Einstein coefficients but not their absolute values. This is as far as the thermodynamic argument allows us to go. The absolute values can only come from a suitable dynamic model of the transition process itself. In Chapter 3 we give a quantum-mechanical (or rather a 'semi-classical') analysis that enables a numerical calculation of the Einstein coefficients to be carried out from first principles.

Although the above relations between the Einstein coefficients were obtained by assuming thermal equilibrium there is nothing to stop us using the coefficients in nonequilibrium situations. It is interesting at this stage, for example, to investigate the initial dynamic behaviour within a group of atoms before equilibrium is reached. This result will be useful later when we come to compare the predictions of quantum theory with those of the present 'classical' treatment.

Suppose that at $t = 0$ all the atoms are in their lowest energy state and that they are then bathed in radiation of frequency ω_{21}. From Eqs (2.103), (2.109) and (2.110) we have

$$\frac{\mathrm{d}N_2}{\mathrm{d}t} = -N_2 A_{21} + N_1 \rho(\omega_{21}) B_{12} - N_2 \rho(\omega_{21}) B_{12} \tag{2.115}$$

To simplify the notation we assume the levels are nondegenerate so that we may put $B_{12} = B_{21} = B$. We also write A_{21} as A and $\rho(\omega_{21})$ as ρ. Equation (2.115) now becomes

$$\frac{\mathrm{d}N_2}{\mathrm{d}t} = -N_2 A + \rho B(N - 2N_2) \tag{2.116}$$

This equation is readily solved (Problem 2.18) to give

$$N_2(t) = \frac{NB\rho}{A + 2B\rho} (1 - \exp(-At - 2B\rho t)) \tag{2.117}$$

An illustration of the time behaviour of N_2 is shown in Figure 2.20.

For small values of t, expansion of the exponential in Eq. (2.117) gives

$$N_2(t) = NB\rho t \ (t \ll (A + B\rho)^{-1}) \tag{2.118}$$

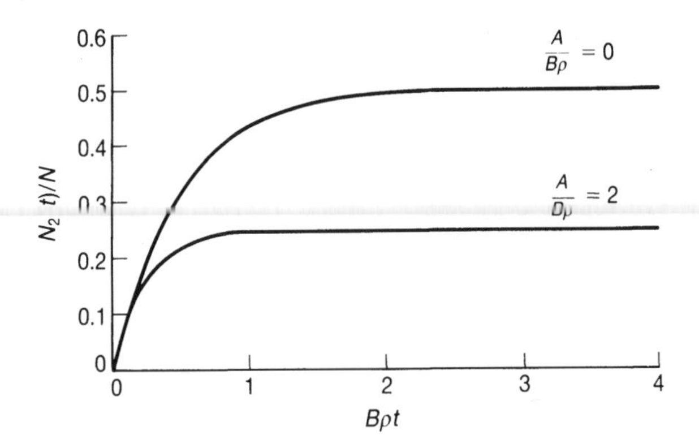

Figure 2.20 Plot of the relative numbers of atoms in the upper level of a simple two-level energy level system in the presence of black-body radiation of density ρ as given by Eq. (2.132). Two curves are shown for differing values of the ratio $A/B\rho$.

Thus we predict an initial linear increase with time in the excited state population. After long times the expression for $N_2(t)$ saturates at the value

$$N_2(t) = \frac{NB\rho}{A + 2B\rho} \quad (t \gg (A + B\rho)^{-1}) \tag{2.119}$$

Although we have introduced the idea of spontaneous and stimulated emission we have so far not mentioned any possible differences between the two types. These are considerable, although that is not apparent in the simple analysis presented so far. In fact, spontaneous emission results in radiation that is *incoherent*, whereas stimulated emission results in *coherent* radiation. We examine the concept of coherence in the next section, but make no attempt to show that coherent radiation results from stimulated emission. That connection requires a quantum-mechanical treatment of radiation that is beyond the scope of this book. At this stage suffice it to say that incoherent radiation consists of many 'beams' of light whose electric field oscillations are randomly related in-phase. In contrast, the oscillations of all the constituent light rays in a coherent beam are in-phase. Thus when stimulated emission takes place the emitted radiation is in-phase with that of the stimulating beam and it also takes place along the same direction as the stimulating beam.

If we imagine a 'single' beam of light of the correct frequency to be directed along a line of atoms which are all in their excited states then we can readily appreciate that the magnitude of the beam will grow as the atoms are stimulated to emit in turn (Figure 2.21). The original beam is amplified as a consequence of its passage through the atoms. As mentioned in the introduction, this amplification process lies at the heart of laser action.

Lasers then depend on stimulated emission and it is reasonable to ask why, since Einstein had postulated the existence of stimulated radiation in 1917, no-one

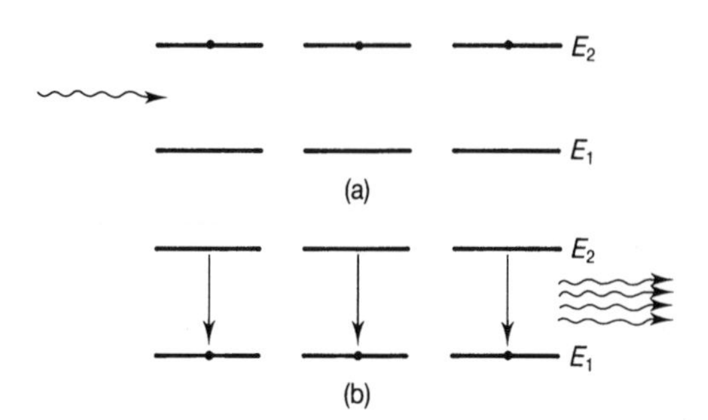

Figure 2.21 Illustrating the process of amplification through stimulated emission. In (a) a single photon approaches a line of atoms in the excited state E_2. Successive stimulated emission processes leave the atoms in a lower energy state (E_1) and the number of photons in the beam increases (b).

seriously followed up the idea until much later. This neglect can at least be partially understood if we calculate first, how common stimulated radiation is in the light sources we encounter and second, what the requirements are for us to observe a light-amplification process.

A number of common light sources approximate reasonably well to black bodies (or at least grey bodies). For example, the sun behaves like a black body with an apparent surface temperature of nearly 6000 K, while a domestic light bulb has a black-body temperature of nearly 3000 K (see Example 2.5). The ratio, R, of spontaneous emission events to stimulated emission events for an atom bathed in black-body radiation is easily seen (from Eqs (2.98) and (2.105)) to be given by

$$R = \frac{A_{21}}{\rho(\omega)B_{21}}$$

Substituting for $\rho(\omega)$ from Eq. (2.100) and from Eq. (2.114) for A_{21}/B_{21} then gives

$$R = \exp(\hbar\omega/kT) - 1 \tag{2.120}$$

Using values of T likely to be encountered in practice in black-body sources and values of ω corresponding to radiation in the visible region, Eq. (2.120) gives R values which are considerably larger than unity. On the other hand, for radiation at a much lower frequency (i.e. longer wavelength) the ratio can be much less than unity (see Example 2.6).

Example 2.6 Ratio of spontaneous and stimulated emission rates

We consider the light from a black-body source at 3000 K (this is the approximate filament temperature of quartz-halogen lamps) and take a wavelength of 0.5 μm (i.e. $\omega = 3.8 \times 10^{15}$ Hz). Substituting these values into Eq. (2.120) then gives:

$$R = \exp\left(\frac{1.06 \times 10^{-34} \times 3.8 \times 10^{15}}{1.38 \times 10^{-23} \times 3000}\right) - 1 = e^{9.7} - 1 \approx 1.7 \times 10^4$$

Evidently in this case stimulated emission is a fairly rare event compared with spontaneous emission.

If, however, we consider a much lower frequency, say 10^{10} Hz (i.e. microwaves) with an effective source temperature of 300 K, then we have

$$R = \exp\left(\frac{1.06 \times 10^{-34} \times 2\pi \times 10^{10}}{1.38 \times 10^{-23} \times 300}\right) - 1 = e^{0.0016} - 1 = 0.016$$

Here the radiation arises mostly from stimulated transitions.

Thus visible radiation from 'natural' sources arises almost wholly from spontaneous transitions while at much longer wavelengths, for example for microwaves or radio waves, the radiation arises mainly from stimulated emission. This is one of the reasons amplification using stimulated emission was first investigated using microwaves. The first MASER (i.e. *M*icrowave *A*mplification by the *S*timulated *E*mission of *R*adiation) involved transitions within the ammonia molecule resulting in radiation at a frequency of 24 GHz. In fact, lasers were initially referred to as 'optical masers' before the term 'laser' became widespread.

In introducing the idea of light amplification (i.e. lasing) we considered a light beam traversing a line of atoms which were all in their excited state. This, of course, is not the natural equilibrium state of affairs. Normally we would expect a Boltzmann distribution, that is (assuming the levels have the same degeneracy),

$$N_2/N_1 = \exp\left(-\frac{E_2 - E_1}{kT}\right)$$

Since $E_2 > E_1$ we have that $N_2 < N_1$. Now the rate of stimulated emission is proportional to the factor N_2B_{21} while the rate of stimulated absorption is proportional to N_1B_{12}. Since $B_{12} = B_{21}$ ($g_1 = g_2$) in this instance, we can see that the rates of stimulated absorption will exceed the rate of stimulated emission, and in fact a beam of light traversing the system will suffer a net loss instead of being amplified. According to this argument, we need to have $N_2 > N_1$ (a situation known as *population inversion*) before we can have amplification, and this is not a condition that can be met for any thermal equilibrium state of the system. This argument will be put on a rather more secure footing in Chapter 3, but the conclusion that we cannot have lasing action within a set of energy levels which are in thermal equilibrium remains valid. This conclusion provided a rather strong disincentive for further investigation. What was not realized was the relative ease with which the necessary population inversion can be achieved in systems involving more than two energy levels when external (nonequilibrium) stimuli are applied.

2.4 Coherence

The idea of coherence arose in Section 2.3 when stimulated emission was introduced but no very precise definition was attempted at the time, and we will now try to put it on a firmer mathematical footing. For convenience, we consider the two aspects of coherence, *temporal* coherence and *spatial* coherence separately.

2.4.1 Temporal coherence

If the field of a beam of light at a particular point in space were to exhibit a sinusoidal variation with time then we could predict the phase of the field at any future time. Such a beam is described as having perfect temporal coherence. If we split the beam into two then these will always produce sharp interference fringes when recombined, even if we introduce an arbitrarily long time delay in one of the beams. However, suppose now that the source of the field is an atom which after a fixed time (say, τ_c) undergoes a collision which results in a random change of phase. If we again split the beam into two and recombine them with a time delay which is greater than τ_c then any interference pattern produced will change randomly every τ_c seconds. Over a time long compared with τ_c the interference pattern will be completely 'washed out'. However, with a time delay less than τ_c any interference pattern will not be completely washed out when averaged over long times. Such a beam has a partial temporal coherence. As $\tau_c \Rightarrow 0$ the amount of coherence reduces. If $\tau_c = 0$, the beam would be completely incoherent and would be unable to form stable interference fringes under any circumstances.

We must now try to be more quantitative. We write the electric field at some particular point as $\mathscr{E}(t)$ and define the *autocorrelation function* $\Gamma_{11}(\tau)$ as

$$\Gamma_{11}(\tau) = \langle \mathscr{E}^*(t)\mathscr{E}(t+\tau)\rangle \tag{2.121}$$

Here the brackets $\langle \dots \rangle$ indicate an average over times much longer than τ and so we may write

$$\Gamma_{11}(\tau) = \lim_{T\to\infty} \frac{1}{2T}\int_{-T}^{T} \mathscr{E}^*(t)\mathscr{E}(t+\tau)\,\mathrm{d}t \tag{2.122}$$

It is convenient to introduce a normalized quantity $\gamma_{11}(\tau)$ called the *mutual coherence function* where

$$\gamma_{11}(\tau) = \frac{\Gamma_{11}(\tau)}{\Gamma_{11}(0)} \tag{2.123}$$

The actual degree of coherence is given by $|\gamma_{11}|$, and takes a value between 0 and 1. The latter value corresponds to perfect coherence, the former to complete incoherence.

To see how this works out we take the case of a perfect sine wave. Putting $\mathscr{E}(t) = \mathscr{E}_0 \exp(\mathrm{j}\omega t)$ we have

$$\mathscr{E}^*(t)\mathscr{E}(t+\tau) = \frac{\mathscr{E}_0^2}{2}\exp(\mathrm{j}\omega\tau)$$

so that

$$\Gamma_{11}(\tau) = \frac{\mathscr{E}_0^2}{2}\exp(\mathrm{j}\omega\tau) \tag{2.124}$$

It also follows that $\Gamma_{11}(0) = \mathscr{E}_0^2/2$ so that $\gamma_{11} = \exp(\mathrm{j}\omega\tau)$ and $|\gamma_{11}| = 1$ as expected.

We have seen above that for the situation where the phase of the beam changes randomly every τ_c seconds then we would not expect perfect coherence. How, though, do we calculate a value for the degree of coherence? In fact, this turns out to be not too difficult a problem since $\gamma_{11}(\tau)$ can be related to the spectral content of the light. Suppose that the irradiance due to frequencies between ω and $\omega + \mathrm{d}\omega$ can be written $I(\omega)\,\mathrm{d}\omega$. Then a normalized spectral irradiance, $g(\omega)$ can be defined by

$$g(\omega) = \frac{I(\omega)}{\int_{-\infty}^{\infty} I(\omega)\,\mathrm{d}\omega} \tag{2.125}$$

In the next chapter we will use $g(\omega)$ to refer to the frequency content of transitions between energy levels. Such transitions give rise to so-called *line* spectra (since their frequency widths are comparatively small), and hence $g(\omega)$ is also referred to as the *lineshape function*.

In Appendix 4 we show that $\gamma_{11}(\tau)$ and $g(\omega)$ are related by the Wiener–Khintchine theorem so that:

$$\gamma_{11}(\tau) = \int_{-\infty}^{\infty} g(\omega)\exp(\mathrm{j}\omega\tau)\,\mathrm{d}\omega \tag{2.126}$$

We will be considering the forms $g(\omega)$ can take in the next chapter, but for the purposes of illustration we consider here a rectangular frequency distribution centered on ω_0 where

$$g(\omega) = 1/\Delta\omega \qquad \omega_0 - \Delta\omega/2 < \omega < \omega_0 + \Delta\omega/2$$

Thus

$$\gamma_{11}(\tau) = \frac{1}{\Delta\omega}\int_{\omega_0 - \Delta\omega/2}^{\omega_0 + \Delta\omega/2} \exp(\mathrm{j}\omega\tau)\,\mathrm{d}\omega \tag{2.127}$$

$$= \frac{\exp(\mathrm{j}\omega\tau_0)}{\mathrm{j}\tau\,\Delta\omega}\left[\exp\left(\frac{\mathrm{j}\,\Delta\omega\tau}{2}\right) - \exp\left(-\frac{\mathrm{j}\,\Delta\omega\tau}{2}\right)\right]$$

$$= \frac{\sin(\Delta\omega\tau/2)}{\Delta\omega\tau/2}\exp(\mathrm{j}\omega_0\tau)$$

$$= \mathrm{sinc}(\Delta\omega\tau/2)\exp(\mathrm{j}\omega_0\tau) \tag{2.128}$$

and so

$$|\gamma_{11}(\tau)| = |\mathrm{sinc}(\Delta\omega\tau/2)| \tag{2.129}$$

The *sinc* function is plotted in Figure 2.22.

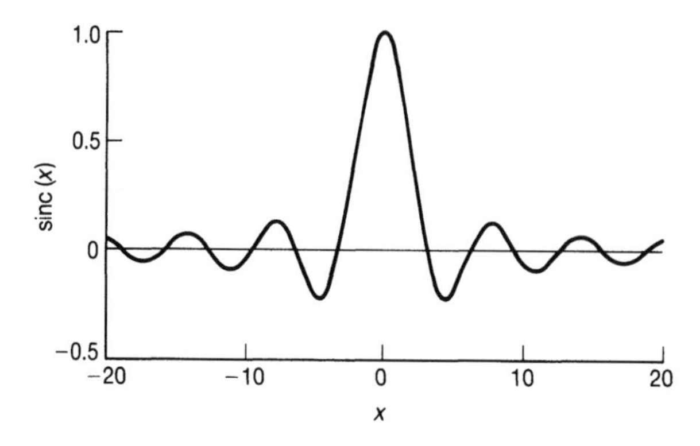

Figure 2.22 The sinc function.

It is convenient to define a *coherence time* τ_c as a value of τ at which the modulus of the coherence function falls to half its maximum value. Now sinc(x) is equal to 0.5 when $x = 1.895$, so that for the rectangular frequency spectrum

$$\tau_c = \frac{3.79}{\Delta\omega}$$

or

$$\tau_c = \frac{0.6}{\Delta f}$$

Generally (see also Problem 2.19), if Δf is the frequency halfwidth (i.e. frequency difference between points where the lineshape function falls to one half of its maximum value) then the coherence time is given, approximately, by

$$\tau_c \approx \frac{1}{\Delta f} \tag{2.130}$$

Another parameter often used is the *coherence length*, L_c. This is simply the distance travelled by the radiation during its coherence time, so that

$$L_c = c\tau_c \tag{2.131}$$

Example 2.7 Coherence lengths of conventional sources

The coherence lengths of most conventional sources is quite small. For example, in a low-pressure discharge of cadmium vapour the main emission is from a spectral line at 643.8 nm wavelength which has a half width of 0.045 nm. The equivalent frequency spread may be obtained from

$$\Delta f = -\frac{c}{\lambda^2}\,\Delta\lambda$$

whence for the cadmium lamp $\Delta f = 3.26 \times 10^{10}$ Hz. The coherence time is then (from Eq. (2.130)) $\approx 1/3.26 \times 10^{10}$ or 3×10^{-11} s, and the coherence length (from Eq. (2.131)) $\approx 3 \times 10^{-11}$ 3×10^{8} m or ≈ 10 mm.

For black-body sources, with their very wide spectral bandwidths, coherence lengths are extremely small. A black body at 2000°C (i.e. a typical heated filament lamp) has a frequency bandwidth of some 2×10^{14} Hz, leading to a coherence time of 5×10^{-15} s and a coherence length of only 1.5 μm.

So far, the coherence function has been presented as a theoretical parameter to be be evaluated from the linewidth. Fortunately, however, it is possible to measure it directly. This is most readily done using the Michelson interferometer (Figure 2.23). In this instrument a parallel beam of light falls onto a beam splitter (S). Both beams travel a certain distance and are then reflected back along their outgoing paths with plane mirrors (M_1 and M_2). Back at the beam splitter some of the light from each beam will travel along a direction at right-angles to the source beam and enter a telescope. At the focus of the telescope fringes can be seen which may be regarded as fringes as formed by the thin film of air between M_1 and M_2', the image of M_2 in S. The appearance of the fringes depends on the relative positioning of the mirrors. Usually M_1 and M_2' will be at a slight angle to each other, in which case straight-line fringes representing equal thickness contours will be formed.

In operation one of the mirrors is fixed and the other movable. As the mirror separation changes by an optical distance of $\lambda/2$ then one fringe will pass a given reference point in the plane of the fringe system. As the separation between M_1 and M_2' increases the fringes are observed to become less well defined, and at separations much in excess of the coherence length are not visible at all (Figure 2.24). The

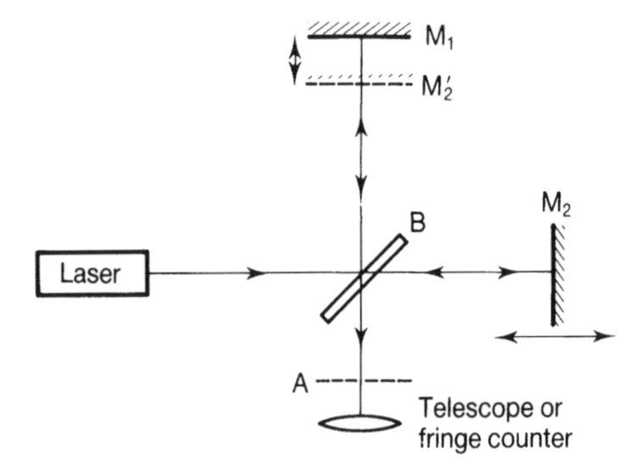

Figure 2.23 Schematic layout of the Michelson interferometer. The incoming beam is split into two by the beam splitter B, each then travels different distances (L_1 and L_2) before being recombined again at the beam splitter. One of the mirrors is usually movable so that the path difference between the two beams may be varied.

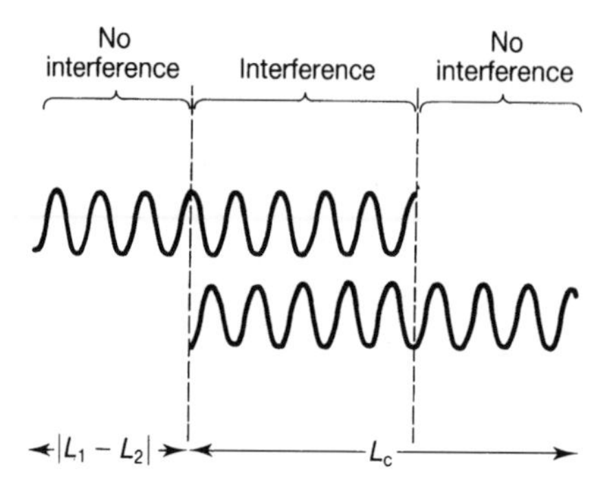

Figure 2.24 When two identical wavetrains of length L_c which have traversed different distances (L_1 and L_2) are recombined they can only interfere over a length $L_c - |L_1 - L_2|$.

sharpness of the fringes at any particular mirror separation may be defined by the *visibility* V, where

$$V = \frac{I_{max} - I_{min}}{I_{max} + I_{min}} \tag{2.132}$$

where I_{max} and I_{min} are the maximum and minimum beam irradiances. The fringe visibility can be shown to be directly related to the coherence function. Thus suppose the amplitudes of the two interfering beams can be written $f_1(t)$ and $f_2(t+\tau)$, where τ here is the time taken to traverse the air wedge between M_1 and M_2', that is, $\tau = 2\Delta x/c$, where Δx is the thickness of the air wedge. The total amplitude when the beams mix is just $f_1(t) + f_2(t+\tau)$ and hence the instantaneous irradiance can be written as

$$[f_1(t) + f_2(t+\tau)]\ [f_1(t) + f_2(t+\tau)]^*$$

or

$$f_1(t)f_1^*(t) + f_2(t+\tau)f_2^*(t+\tau) + f_2(t+\tau)f_1^*(t) + f_1(t)f_2^*(t+\tau)$$

However, any detector used to monitor the fringes will record an irradiance $I(\tau)$ which is the average of this quantity over the response time of the detector, that is,

$$I(\tau) = \langle f_1(t)f_1^*(t)\rangle + \langle f_2(t+\tau)f_2^*(t+\tau)\rangle + \langle f_2(t+\tau)f_1^*(t)\rangle + \langle f_1(t)f_2^*(t+\tau)\rangle$$

The first two terms are simply the irradiances of the separate beams, while the second two are related to the autocorrelation function defined in Eq. (2.121). Thus

$$I(\tau) = I_1 + I_2 + \Gamma_{11}(\tau) + \Gamma_{11}^*(\tau) \tag{2.133}$$

Now the function $\Gamma_{11}(\tau)$ will always involve the term $\exp(j\omega\tau)$ (see, for example, Eq. (2.124) and Problem 2.19), so that we can write

$$I(\tau) = I_1 + I_2 = |\Gamma_{11}(\tau)|(\exp(j\omega\tau) + \exp(-j\omega\tau))$$
$$= I_1 + I_2 + 2|\Gamma_{11}(\tau)|\cos(\omega\tau) \tag{2.134}$$

Thus we would expect a cosinusoidal variation in fringe irradiance with τ, the maximum and minimum irradiances being $I_1 + I_2 + 2|\Gamma_{11}(\tau)|$ and $I_1 + I_2 - 2|\Gamma_{11}(\tau)|$, respectively. The fringe visibility can then be written

$$V(\tau) = 2\frac{|\Gamma_{11}(\tau)|}{I_1 + I_2}$$

Now $|\Gamma_{11}(0)| = |\langle f_1(t) f_2^*(t)\rangle|$

$$= (I_1 I_2)^{1/2}$$

Thus $|\gamma_{11}(\tau)| = \dfrac{|\Gamma_{11}(\tau)|}{|\Gamma_{11}(0)|}$

$$|\gamma_{11}(\tau)| = \frac{I_1 + I_2}{2(I_1 I_2)^{1/2}} V(\tau) \tag{2.135}$$

If the two irradiances are equal then

$$|\gamma_{11}(\tau)| = V(\tau) \tag{2.136}$$

2.4.2 Spatial coherence

In the previous section we discussed the coherence of two beams which derived from the same initial beam but which suffered a differential time delay before recombination. Suppose now we have spatially extended source (or beam) and we place a screen with two pinholes in front of it (Figure 2.25). Interference fringes can be

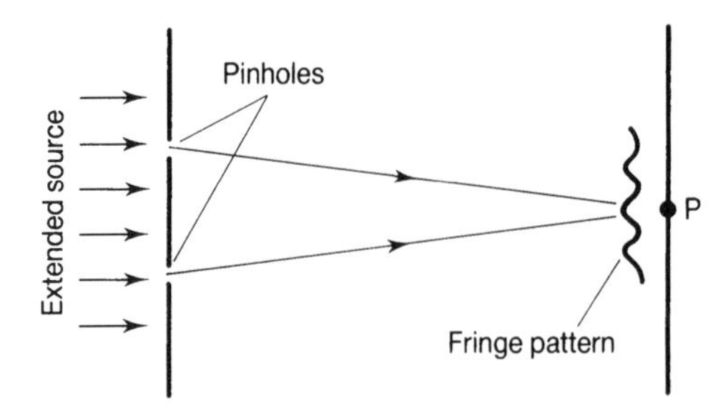

Figure 2.25 Layout that can be used to investigate the spatial coherence of an extended source. The visibility of the fringes at the point P (equidistant from the pinholes) depends on the spatial coherence.

observed on a further screen placed some distance beyond the first. This arrangement is, of course, similar to Young's slits experiment (see, for example, Ref. 10). If we look at the small group of fringes which are approximately equidistant from the two pinholes then their visibility gives information on the *spatial coherence* between two different points on the source (the two beams have exactly the same time delay with regard to their points of origin). We may readily extend the concept of the coherence function introduced in the previous section so that it covers both spatial and temporal coherence. Thus we define a *mutual correlation function* $\Gamma_{12}(\tau)$ where

$$\Gamma_{12}(\tau) = \langle \mathscr{E}_1^*(t)\mathscr{E}_2(t+\tau)\rangle \tag{2.137}$$

where the subscripts 1 and 2 refer to the two different spatial origins of the interfering beams.

Again a normalized coherence function $\gamma_{12}(\tau)$ can be defined by

$$\gamma_{12}(\tau) = \frac{\Gamma_{12}(\tau)}{\Gamma_{12}(0)} \tag{2.138}$$

The modulus of $\gamma_{12}(\tau)$ can be related to the visibility of the fringe pattern formed when the two beams interfere in exactly the same way as for $\gamma_{11}(\tau)$.

The functions $\Gamma_{12}(\tau)$ and $\gamma_{12}(\tau)$ are known as *first-order coherence functions.* More general coherence functions involving correlations between three or more source points can be defined. It is of interest to note that while it is not possible to differentiate between a laser source and a 'chaotic' source such as black-body radiation on the basis of the first-order coherence functions, it *is* possible to do so when higher-order functions are used. For a further discussion of this point the interested reader may consult Ref. 11.

Problems

2.1 Show that when **A** is any vector

$$\nabla\times\nabla\times\mathbf{A} = \nabla(\nabla.\mathbf{A}) - \nabla^2\mathbf{A}$$

2.2 Verify that the function

$$\psi = f\left(t \pm \frac{1}{c}\,\mathbf{n.r}\right)$$

satisfies the equation

$$\nabla^2\psi = \varepsilon_0\mu_0\,\frac{\partial^2\psi}{\partial t^2}$$

2.3 Consider a plane wave travelling in a vacuum along the z axis. Show that

$$\frac{\partial \mathscr{E}_z}{\partial z} = 0 \qquad (1)$$

(*Hint*: since we are dealing with a plane wave we must have

$$\frac{\partial \mathscr{E}_x}{\partial x} = \frac{\partial \mathscr{E}_y}{\partial y} = 0$$

Use this result together with Maxwell's first equation.)
Use Eq. (1) above in conjunction with Eq. (2.15a) to show that

$$\mathscr{E}_z = 0$$

Use similar arguments to show that

$$\mathscr{B}_z = 0$$

Starting from the assumption that the electric lies along the x axis, i.e. that

$$\boldsymbol{\mathscr{E}} = \mathscr{E}_0 \mathbf{a} \cos(\omega t - kz)$$

use Maxwell's third equation to show that

$$\boldsymbol{\mathscr{B}} = \mathscr{E}_0 \mathbf{b} \frac{k}{\omega} \cos(\omega t - kz)$$

2.4 Show that the amplitude profile of the field formed by superposing equal amplitude two waves which differ in angular frequency and wavenumber by the small amounts $\delta\omega$ and δk moves with the velocity $\delta\omega/\delta k$. In the limit when $\delta\omega$ and δk tend to zero, the velocity is then the *group velocity* v_g, where

$$v_g = \frac{\delta\omega}{\delta k}$$

2.5 A plane wave ($\boldsymbol{\mathscr{E}}_i = \boldsymbol{\mathscr{E}}_{i0} \exp(j(\omega_i t - \mathbf{k}_i.\mathbf{r}))$ is incident on the boundary between two different dielectric materials. The reflected and transmitted components may, in general, be described by

$$\boldsymbol{\mathscr{E}}_r = \boldsymbol{\mathscr{E}}_{r0} \exp(j(\omega_r t - \mathbf{k}_r.\mathbf{r}) \text{ and}$$
$$\boldsymbol{\mathscr{E}}_t = \boldsymbol{\mathscr{E}}_{t0} \exp(j(\omega_t t - \mathbf{k}_t.\mathbf{r})$$

Using the condition that the tangential components of $\boldsymbol{\mathscr{E}}$ must be the same either side of the interface show that

$$\omega_i = \omega_r = \omega_t \ (-\omega, \text{ say})$$

and

$$\mathbf{k}_i.\mathbf{r}_{int} = \mathbf{k}_r.\mathbf{r}_{int} = \mathbf{k}_t.\mathbf{r}_{int}$$

Thus show that all three $\mathbf{k}$ vectors must lie in the plane of incidence.
Bearing in mind that the incident beams travel in the same medium show that

$$k_{iz}^2 = k_{rz}^2$$

so that $k_{iz} = \pm\, k_{rz}$.

Since the reflected ray travels away from the interface the negative sign must be chosen here. This result can also be expressed by saying that the angle of reflection is equal to the angle of incidence.

Using the fact that in the second medium the transmitted wave will travel with a phase velocity of c/n_2, show that Snell's law can be obtained.

2.6 Light is incident on an interface between two media (refractive indices n_1 and n_2). Show that when the incident electric field (amplitude $\mathscr{E}_{0i}^{\perp}$) is perpendicular to the plane of incidence the reflected ($\mathscr{E}_{0r}^{\perp}$) and transmitted field ($\mathscr{E}_{0t}^{\perp}$) amplitudes are given by

$$\mathscr{E}_{0r}^{\perp} = \frac{n_1 \cos(\theta_i) - n_2 \cos(\theta_t)}{n_1 \cos(\theta_i) + n_2 \cos(\theta_t)} \mathscr{E}_{0i}^{\perp}$$

and

$$\mathscr{E}_{0t}^{\perp} = \frac{2n_1 \cos(\theta_i)}{n_1 \cos(\theta_i) + n_2 \cos(\theta_t)} \mathscr{E}_{0i}^{\perp}$$

For the situation where the electric field is in the plane of incidence show that:

$$\mathscr{E}_{0r}^{\parallel} = \frac{n_1 \cos(\theta_t) - n_2 \cos(\theta_i)}{n_1 \cos(\theta_t) + n_2 \cos(\theta_i)} \mathscr{E}_{0i}^{\parallel}$$

and

$$\mathscr{E}_{0r}^{\parallel} = \frac{2n_1 \cos(\theta)}{n_1 \cos(\theta_t) + n_2 \cos(\theta_i)} \mathscr{E}_{0i}^{\parallel}$$

2.7 Starting from Eq. (2.58), show that the Brewster angle, θ_B, is given by

$$\tan(\theta_B) = \frac{n_2}{n_1}$$

Demonstrate that at the Brewster angle the reflected and transmitted beams are at right-angles.

2.8 A glass plate which is inclined at the Brewster angle to a beam of radiation acts as a partial polarizer, since radiation whose electric field vector is parallel to the plane of incidence is wholly transmitted, whereas radiation whose electric field vector is perpendicular to the plane of incidence is partially reflected. A stack of such plates can thus act as quite an efficient polarizer. The efficiency of such a device is given by the extinction ratio, that is, the ratio between the emerging irradiances for a linearly polarized light beam which is polarized (a) along the transmitting direction and (b) along the blocking direction.

Derive an expression for the extinction ratio for a pile of N plates at the Brewster angle. Calculate a value for the extinction ratio for

(a) Twenty glass plates at visible wavelengths where the refractive index is ≈ 1.5, and
(b) Two germanium plates at a wavelength of 10 μm (i.e. the CO_2 laser wavelength) where the refractive index is ≈ 4.

It is evident that when the refractive index is high then just a few plates can make quite an efficient polarizer. This is the case with Ge at the CO_2 laser wavelength.

2.9 Use the Fresnell equations to show that at the critical angle the transmittance through an interface between an optically more dense and an optically less dense medium becomes zero.

2.10 Using Eq. (2.66) verify that the condition for only one mode to be able to propagate in a symmetric planar dielectric waveguide is that $V < \pi/2$.

2.11 (a) An optical fiber is made from core and cladding materials whose refractive indices are 1.47 and 1.46, respectively. If the core diameter is 200 μm estimate how many modes can propagate down the fiber when light of (vacuum) wavelength 1.3 μm is used.
(b) A single-mode fiber is to be constructed from the same core and cladding materials for use at 1.3 μm. Calculate the core diameter needed assuming $V = 2$. What is the cut-off wavelength for this fiber?
(c) Light of wavelength 1.4 μm is used in the fiber in part (b). Calculate the mode field diameter, and estimate what proportion of the energy within the fiber is travelling in the cladding.

2.12 At a wavelength of 0.5 μm, germanium has a refractive index of $3.47-1.4j$, calculate

(a) The normal reflectance from a plane surface of germanium
(b) The depth within the germanium at which the irradiance is a factor 0.01 less than at the surface.

2.13 In a magnetic conducting medium the total current density, **j**, in Eq. (2.4) has contributions from sources other than just the pure conduction current density ($\mathbf{j}_c$). It is possible to show that Maxwell's fourth equation can be rewritten:

$$\nabla \times \mathscr{H} = \mathbf{j}_c + \frac{\partial \mathscr{D}}{\partial t}$$

Using this equation, as well as the first three of Maxwell's equations (Eqs (2.1) to (2.3)), show that within a medium with conductivity σ

$$\nabla^2 \mathscr{E} = \mu_r \mu_0 \sigma \frac{\partial \mathscr{E}}{\partial t} + \mu_r \mu_0 \varepsilon_r \varepsilon_0 \frac{\partial^2 \mathscr{E}}{\partial t^2}$$

and that a similar equation holds for $\mathscr{H}$.

Assuming a linearly polarized wave travelling along the z axis of the form:

$$\mathscr{E}_x = \mathscr{E}_0 \exp \mathrm{j}\omega\left(t - \frac{n}{c} z\right)$$

show that in good conductors, where $\sigma/(\omega \varepsilon_r \varepsilon_0) \gg 1$ then

$$n = \left(\frac{\mu_r \sigma}{2\varepsilon_0 \omega}\right)^{1/2} (1 - \mathrm{j})$$

Using this result, derive the following expression for the reflectivity of a good conductor:

$$R = 1 - \left(\frac{8\omega\varepsilon_0}{\sigma \mu_r}\right)^{1/2}$$

2.14 A beam of light travels at an angle θ to the optic axis. Show that if the polarization is in the plane containing the optic axis and the direction of propagation (i.e. the ray is an extraordinary ray) then the refractive index experienced by the beam is given by $n_e(\theta)$, where

$$\frac{1}{n_e^2(\theta)} = \frac{\cos^2\theta}{n_0^2} + \frac{\sin^2\theta}{n_e^2}$$

2.15 Show that the amount of energy emitted per unit time per unit area by a black body is related to the density of radiation within a sealed enclosure $\rho(\nu, T)$ by

$$I(\nu, T) = \rho(\nu, T)c/4$$

(*Hint*: Consider the radiation emerging through a vanishingly small hole in the side of a 'sealed' cavity.)

2.16 (a) By integrating the spectral irradiance of a black body over all frequencies, show that the total irradiance, I_{BB}, of a black body is given by

$$I_{BB} = \sigma T^4$$

where

$$\sigma = \frac{2\pi^5 k^4}{15c^2h^3} = 5.67 \times 10^{-8} \text{ W m}^{-2}\text{K}^{-4}$$

(b) Show that the wavelength of peak emission from a black body, λ_{max}, is given by the Wein displacement law:

$$\lambda_{max} T = \frac{hc}{4.965k} = 2.987 \times 10^{-3} \text{ m K}$$

2.17 Normally for a particular energy level there are a number of different transitions that an electron in the level can make to leave that level. If these transitions can be described in terms of lifetimes τ_1, τ_2, etc. show that the effective overall lifetime, τ_t, is given by

$$\frac{1}{\tau_t} = \frac{1}{\tau_1} + \frac{1}{\tau_2} + \frac{1}{\tau_3} + \cdots$$

2.18 A system consisting of two energy levels is irradiated with radiation. Show that the population density of the upper level, N_2, will increase with time according to

$$N_2(t) = \frac{NB\rho}{A + 2B\rho}(1 - \exp(-At - 2B\rho t))$$

when ρ is (constant) the radiation density at the frequency corresponding to the transition between the two levels and where it is assumed that $N_2(0) = 0$.

2.19 Evaluate the mutual coherence function for:

(a) A sinusoidal waveform of constant amplitude but whose phase changes randomly every τ_0 seconds.

(b) Radiation which has a lineshape function that falls off exponentially either side of a frequency ω_0.

2.20 Show that a source emitting over a narrow range of wavelengths ($\Delta\lambda$) with a mean emission wavelength λ has a coherence length, l_c, given by

$$l_c \cong \lambda^2/\Delta\lambda$$

An LED emitting at 650 nm has a typical emission width of some 50 nm. Calculate its coherence length.

Notes

1. Strictly speaking, χ as defined by Eq. (2.20) should be a tensor quantity, since there is no *a priori* reason why a field applied along one direction should not result in the medium becoming polarized along another. Such a medium would be deemed anisotropic. For simplicity, we assume here a scalar susceptibility and an isotropic medium. Anisotropic media are dealt with in Section 2.1.9.
2. In deriving Eq. (2.23) we have tacitly assumed that the atoms/molecules are so far apart that they do not interact when they are polarized.
3. The term *intensity* is still sometimes used (incorrectly) for this quantity.
4. From a strictly physical point of view the normalization integral should have its lower limit at $x = -\omega_0$ (i.e. $\omega = 0$). Since ω_0 is so large compared with $\Delta\omega_L$, however, the extension to $-\infty$ makes no difference in practice and greatly simplifies the mathematics.
5. Other definitions of linewidth can be made and care must be exercised when comparing expressions in the literature that the same definitions are used.
6. For convenience we drop the suffix when describing θ_1 from now on.
7. In Appendix 3, Figure A3.5 shows the field distributions for the first four modes in a waveguide. There is an obvious tendency for the higher-order modes to have more of their field energy in the cladding. This is a direct consequence of their closer proximity to cut-off.
8. A waveguide is called single mode even when both TE_0 and TM_0 modes can propagate.
9. Technically the decay depends functionally on a particular type of Bessel function which is not a simple exponential.
10. Strictly speaking, two modes are allowed which have their fields polarized at right-angles.
11. The circular symmetry of the figure about the optic axis guarantees that we can do this without loss of generality.
12. This is a standard notation for crystal symmetry which will be discussed in any basic book on crystallography (for example Ref. 8).
13. The material is often known simply as KDP.
14. The coefficient B_{12} is defined here using the angular frequency black-body radiation density function $\rho(\omega)$. In some texts B_{12} is defined using $\rho(\nu)$ and will, accordingly, be a factor 2π larger.

References

[1] P. Lorrain, D. D. Corson and F. Lorrain, *Electromagnetic Fields and Waves*, 3rd edn, W. H. Freeman, New York, 1988, Chapter 27.

[2] P. Lorrain, D. D. Corson and F. Lorrain, *Electromagnetic Fields and Waves*, 3rd edn, W. H. Freeman, New York, 1988, Section 28.6.

[3] R. Loudon, *The Quantum Theory of Light*, 2nd edn, Clarendon Press, Oxford, 1983, Chapter 3.

[4] P. Lorrain, D. D. Corson and F. Lorrain, *Electromagnetic Fields and Waves*, 3rd edn, W. H. Freeman, New York, 1988, Chapter 27.

[5] A. Yariv, *Optical Electronics*, 4th edn, Saunders College Publishing, 1991, Chapter 3.

[6] Luc B. Jeunhomme, *Single Mode Fibre Optics*, 2nd edn, Marcel Dekker, New York, 1990, pp. 17–20.

[7] P. P. Banerjee and T-C. Poon, *Principles of Applied Optics*, Addison-Wesley, Reading, MA, 1991, Section 6.4.

[8] D. McKie and C. McKie, *Essentials of Crystallography*, Blackwell Scientific Publications, 1986.

[9] O. Svelto, *Principles of Lasers*, 3rd edn, Plenum Press, New York, 1989, Section 2.2.

[10] R. S. Longhurst, *Geometrical and Physical Optics*, 3rd edn, Longman, Harlow, 1973, Section 7.2.

[11] R. Loudon, *The Quantum Theory of Light*, 2nd edn, Clarendon Press, Oxford, 1983, Section 3.7.

3

Interaction of light with matter

3.1 Transition probabilities

In Chapter 2 we dealt with the emission and absorption of radiation by atoms using thermodynamic arguments which involved the spectral distribution of black-body radiation. This approach introduced the three Einstein coefficients, A_{21}, B_{21} and B_{12}, and enabled the ratios between them to be determined. Their absolute values, however, depend on the particular transition being considered. In this chapter we approach the problem from a quantum-mechanical viewpoint with the aim of calculating the Einstein coefficients from first principles. The student who has an (understandable!) aversion to quantum mechanics may, as in Chapter 1, skip some of the initial sections without unduly jeopardizing an understanding of the laser physics in later chapters. It is suggested that Sections 3.1.1–3.1.4 inclusive could be omitted along with the latter part of Section 3.2.1 dealing with the quantum-mechanical expression for the natural linewidth (the student will have to accept the result of Eq. (3.67)) and Section 3.2.2.

A full quantum-mechanical treatment of the interaction of light with matter would be beyond the scope of this book and we must content ourselves with a 'semi-classical' approach. Light is regarded as consisting of oscillating electric and magnetic fields. These fields interact with (or *perturb*) the atomic system of interest so that there is a finite probability that the system will change from one quantum state to another. We consider a system consisting of a collection of N (noninteracting) atoms which each have two energy levels separated by an energy $\hbar\omega_{12}$. The atoms are assumed to be all in their lower energy states at $t = 0$, and the system is then bathed in radiation of angular frequency ω_{12} and with a radiation density $\rho(\omega_{21})$. Our aim will be to derive an equation analogous to Eq. (2.118), that is,

$$N_2(t) = NB_{12}\rho(\omega_{21})t$$

where $N_2(t)$ is the number of atoms in the upper energy state after a time t.

In quantum-mechanical terms the ratio of $N_2(t)/N$ is just the probability, $P_{12}(t)$, that an atom, initially in state 1, will make a transition to state 2, so that we seek an equation of the form:

$$P_{12}(t) = B_{12}\rho(\omega)t \tag{3.1}$$

We hope, therefore, to obtain an initial transition probability that is linear with time (remember that Eq. (2.118) is only valid at small times). The constant of linearity should then enable an expression to be obtained for the parameter B_{12}.

3.1.1 Time-dependent perturbation theory

To simplify the analysis we assume that the atom has only two states with corresponding energies E_1 and E_2. The states have wavefunctions $\phi_1(\mathbf{r},t)$ and $\phi_2(\mathbf{r},t)$ which are functions of both space and time coordinates. We assume that the basic Hamiltonian of the system, $\hat{H}_0$, is time independent so that we can write (Appendix 1, Eq. (A1.16))

$$\phi_1(\mathbf{r},t) = \phi_1^0(\mathbf{r})\exp(-\mathrm{j}E_1t/\hbar)$$

and (3.2)

$$\phi_2(\mathbf{r},t) = \phi_2^0(\mathbf{r})\exp(-\mathrm{j}E_2t/\hbar)$$

Thus we have

$$\hat{H}_0\phi_1(\mathbf{r},t) = E_1\phi_1(\mathbf{r},t) \text{ and}$$
$$\hat{H}_0\phi_2(\mathbf{r},t) = E_1\phi_2(\mathbf{r},t) \tag{3.3}$$

We now consider the effect of introducing an external interaction whose Hamiltonian is $\Delta\hat{H}$. The total Hamiltonian for the system now becomes $\hat{H} = \hat{H}_0 + \Delta\hat{H}$. The new state wavefunction, $\psi(\mathbf{r},t)$, can always be expressed as a linear combination of ϕ_1 and ϕ_2 but the expansion coefficients may now be time dependent, so that we write

$$\psi(\mathbf{r},t) = a_1(t)\phi_1(\mathbf{r}) + a_2(t)\phi_2(\mathbf{r}) \tag{3.4}$$

If ψ is to remain normalized we must have

$$|a_1(t)|^2 + |a_2(t)|^2 = 1 \tag{3.5}$$

The time evolution of a state ψ is determined (see Appendix 1, Eq. (A1.14)) by the equation

$$\hat{H}\psi = \mathrm{j}\hbar \frac{\partial \psi}{\partial t}$$

Thus

$$(\hat{H}_0 + \Delta\hat{H})(a_1(t)\phi_1 + a_2(t)\phi_2) = \mathrm{j}\hbar \frac{\partial}{\partial t}(a_1(t)\phi_1 + a_2(t)\phi_2)$$

Using Eqs (3.2) and (3.3), this reduces to

$$\Delta\hat{H}(a_1(t)\phi_1 + a_2(t)\phi_2) = \mathrm{j}\hbar \left(\phi_1 \frac{\partial a_1(t)}{\partial t} + \phi_2 \frac{\partial a_2(t)}{\partial t}\right) \tag{3.6}$$

If this expression is multiplied from the left by ϕ_2^* and an integration carried out over all space we obtain

$$a_1(t)\int \phi_2^* \Delta\hat{H}\phi_1 \,\mathrm{d}\mathbf{r} + a_2(t)\int \phi_2^* \Delta\hat{H}\phi_2 \,\mathrm{d}\mathbf{r} = \mathrm{j}\hbar \frac{\partial a_2(t)}{\partial t} \tag{3.7}$$

Since the integration is only carried out over spatial coordinates we may take the time-varying parts of ϕ_1 and ϕ_2 outside the integrals. For example

$$\int \phi_2^* \Delta\hat{H}\phi_1 \,\mathrm{d}\mathbf{r} = \exp(\mathrm{j}(E_2 - E_1)t/\hbar) \int \phi_2^{0*}(r)\, \Delta\hat{H}\phi_1^0(r)\,\mathrm{d}\mathbf{r}$$

We write the quantity $\int\phi_2^{0*}(r)\,\Delta\hat{H}\phi_1^0(r)\,\mathrm{d}\mathbf{r}$ as H'_{21} and also put $(E_2 - E_1) = \omega_{12}\hbar$. Equation (3.7) can then be written

$$a_1(t)\exp(\mathrm{j}\omega_{12}t)H'_{21} + a_2(t)H'_{22} = \mathrm{j}\hbar \frac{\partial a_2(t)}{\partial t} \tag{3.8}$$

By multiplying Eq. (3.5) from the left by ϕ_1^* and then proceeding as above we obtain

$$a_1(t)H'_{11} + a_2(t)\exp(-\mathrm{j}\omega_{12}t)H'_{12} = \mathrm{j}\hbar \frac{\partial a_1(t)}{\partial t} \tag{3.9}$$

Unfortunately, it is not possible to obtain general solutions of Eqs (3.8) and (3.9), and to proceed further some approximations have to be made.

We assume that when the perturbation is first applied the atom is in state ϕ_1, so that

$$a_1(0) = 1, \;\; a_2(0) = 0$$

Now we recall that Eq. (2.118) is valid only at small times. Accordingly, if we are to calculate the Einstein coefficients using the results of this equation, we should restrict our analysis to times when $a_2(t)$ remains small (so that $a_1(t) \approx 1$). With this assumption Eqs (3.8) and (3.9) become

$$j\hbar \frac{\partial a_2(t)}{\partial t} = \exp(j\omega_{12}t)\, H'_{21} \tag{3.10}$$

and

$$j\hbar \frac{\partial a_1(t)}{\partial t} = H'_{11} \tag{3.11}$$

At present, we are only interested in the initial dependence of $P_{12}(t)$ (i.e. $|a_2(t)|^2$) on time, which can be obtained by solving Eq. (3.10). However, this requires a knowledge of the time dependence of H'_{21} which we deal with in the next section.

3.1.2 The effect of a harmonically varying perturbation

Whatever the detailed nature of the interaction between light and atomic energy states, because light basically consists of harmonically varying electric and magnetic fields it is reasonable to assume that we can also write the perturbation Hamiltonian as a harmonically varying quantity, that is, we can write

$$\Delta\hat{H} = \Delta\hat{H}^0 \sin \omega t$$

where ω is the frequency of the electromagnetic wave. We have therefore that

$$H'_{21} = H^{0\prime}_{2q} \sin(\omega t)$$

Since it is more convenient to deal with complex exponentials we put

$$H'_{21} = H^{0\prime}_{21} \left(\frac{\exp(j\omega t) - \exp(-j\omega t)}{2j} \right) \tag{3.12}$$

where $H^{0\prime}_{21} = \int \phi_2^{0*}(r)\, \Delta\hat{H}^0 \phi_1^0(r)\, \mathrm{d}\mathbf{r}$.

Integrating Eq. (3.10) with respect to time gives

$$j\hbar a_2(t) = \int_0^t H'_{21} \exp(j\omega_{21}t)\, \mathrm{d}t$$

Substituting for H'_{21} from Eq. (3.12) gives

$$j\hbar a_2(t) = \frac{H^{0\prime}_{21}}{2j} \int_0^t (\exp(j\omega t) - \exp(-j\omega t))\exp(j\omega_{21}t)\, \mathrm{d}t$$

The integration is readily carried out yielding

$$a_2(t) = \frac{H^{0\prime}_{21}}{2j\hbar} \left[\frac{1 - \exp j(\omega_{21} + \omega)t}{\omega_{21} + \omega} + \frac{1 - \exp j(\omega_{21} - \omega)t}{\omega_{21} - \omega} \right]$$

If we are dealing with radiation where ω is near to ω_{21} we may neglect the first term in comparison to the second (this is known as the *rotating wave approximation*). Use of the relationship $\exp(2j\theta) - 1 = 2j\exp(j\theta)\sin(\theta)$ then gives

$$a_2(t) = \frac{H_{21}^{0\prime}}{\hbar}\left[\frac{\sin(\omega_{21} - \omega)t/2)}{\omega_{21} - \omega}\right]\exp(j\theta) \tag{3.13}$$

The probability, $P_{12}(t)$, that a transition takes place within a time t is thus

$$P_{12}(t) = |a_2(t)|^2$$

or

$$P_{12}(t) = \frac{|H_{21}^{0\prime}|^2}{\hbar^2}\left(\frac{\sin^2((\omega_{21} - \omega)t/2)}{(\omega_{21} - \omega^2)^2}\right) \tag{3.14}$$

A plot of the function in brackets is shown in Figure 3.1. For a fixed value of t this has a maximum at $\omega = \omega_0$, as might be expected, but, perhaps surprisingly, there is still a finite probability of a transition taking place at frequencies away from ω_{21}. The range of frequencies concerned, however, is approximately equal to $4\pi/t$ and consequently becomes very narrow at large times.

Another unexpected feature is that the variation of $P_{12}(t)$ with time is *not* linear, even at very small times. Thus if we are 'off-resonance' (i.e. $\omega \neq \omega_2$) then the transition probability oscillates with a frequency of $(\omega_{21} - \omega)/2\pi$ between the values 0 and $|H^{0\prime}|^2/(\hbar^2(\omega_{21} - \omega)^2)$. At very short times the probability shows a quadratic time dependence (for small x, $\sin^2 x \approx x^2$).

On the other hand, if we take the 'resonance' situation (i.e. $\omega = \omega_{21}$) then we have

$$P_{12}(t) = \frac{|H_{21}^{0\prime}|^2}{4\hbar^2} t^2 \quad (\omega = \omega_{21}) \tag{3.15}$$

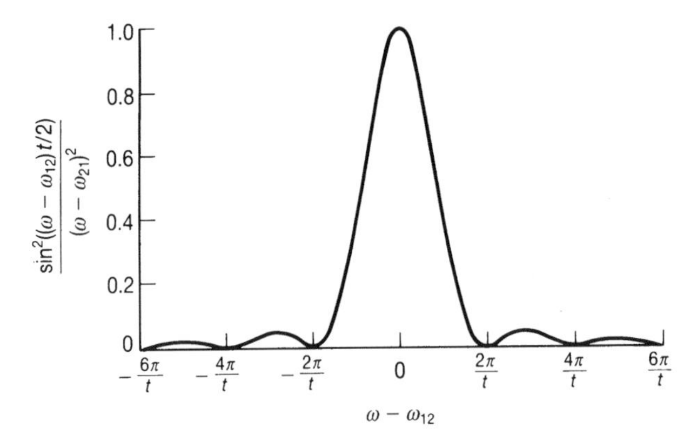

Figure 3.1 A plot of the function giving the probability that a transition has taken place between states separated in energy by $\hbar\omega_{12}$ in the presence of a perturbation oscillating at an angular frequency ω.

Again we have a quadratic time dependence (although this now applies for longer times than for the off-resonance situation). Both the on- and off-resonance transition probabilities are shown as a function of time in Figure 3.2.

Fortunately, it *is* possible to deduce a linear time dependence for the transition probability but to do so we must look more closely at the realities of the situation. We have assumed that the interacting radiation has a single frequency and that the energy levels are infinitely sharp. In practice, any radiation will have a finite frequency spread and the energy levels will never be infinitesimally sharp. Indeed, the aim of the present exercise is to derive explicit formulae for the Einstein coefficients which arose from a consideration of black-body (i.e. broadband) radiation. We therefore consider the effect of radiation which has a finite frequency range, from, say, $\omega_{21} - \Delta\omega$ to $\omega_{21} + \Delta\omega$, and which is characterized by an energy density $\rho(\omega)$.

In this case the total transition probability is obtained by summing (or rather integrating) the contributions from all the frequencies present. (Note: we are tacitly assuming here that the different frequencies are not related in phase, otherwise we would have to add the amplitude factors and then square the resultant.) To proceed further we need to consider the nature of the interaction between the atom and the radiation in more detail.

3.1.3 The electric dipole interaction

For simplicity, we take a single electron atom, and assume that the radiation is represented by an oscillating electric field which we write

$$\mathscr{E}(t) = \mathbf{d}\ \mathscr{E}_0 \sin(\omega t) \tag{3.16}$$

where $\mathbf{d}$ is a unit vector along the field direction. If the instantaneous position vector of the electron relative to the nucleus is $\mathbf{r}$ then we have an instantaneous dipole formed with a dipole moment $\mathbf{D}$ given by

$$\mathbf{D} = e\mathbf{r}$$

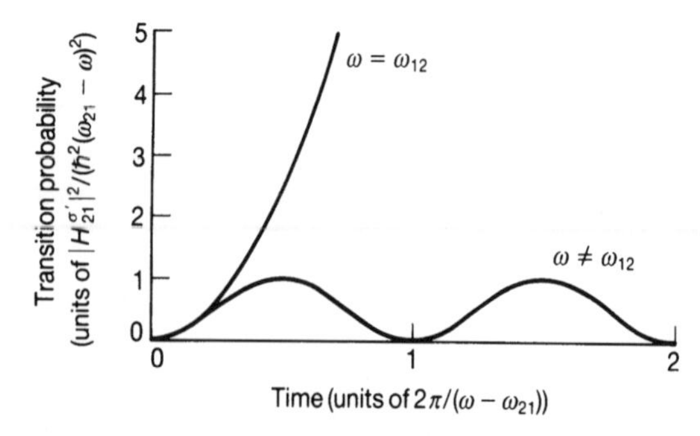

Figure 3.2 Time dependence of the transition probability between states for both the 'on-resonance' ($\omega = \omega_{12}$) and the 'off-resonance' ($\omega \neq \omega_{12}$) situations.

The energy E of the dipole in the field of Eq. (3.16) is then

$$E = \mathbf{D}.\mathscr{E}(t)$$
$$E = e\mathbf{r}.\mathbf{d}\ \mathscr{E}_0 \sin(\omega t)$$

The quantum-mechanical equivalent of this (classical) expression is essentially the same (see Appendix 1), and is called the *electric dipole operator* ($\hat{H}_{ed}$), thus

$$\hat{H}_{ed} = -e\mathscr{E}_0(\mathbf{d}.\mathbf{r}) \qquad (3.17)$$

In a multi-electron atom the vector $\mathbf{r}$ will represent the sum over all the electrons in the atom of their individual position vectors.

If a more exact (though still semi-classical) treatment of the interaction is carried out (see, for example, Ref. 1) it is found that the main interaction energy V_{int} be written

$$V_{int} = \frac{j\hbar e}{mc}(\mathbf{A}.\mathbf{p} + \mathbf{p}.\mathbf{A}) + \frac{e^2}{2mc^2}\mathbf{A}.\mathbf{A} \qquad (3.18)$$

where $\mathbf{A}$ is the so-called *vector potential.*[1] The term in $\mathbf{A}.\mathbf{A}$ here can usually be neglected compared to the others (in fact it gives rise to processes involving two photons). By assuming an interaction with a plane electromagnetic wave the operator equivalent of Eq. (3.18) may be obtained. This is conveniently expressed as an infinite series of terms which (fortunately!) rapidly decrease in importance. The first corresponds to the electric dipole term of Eq. (3.17). Next, after the electric dipole term comes the the *magnetic dipole* interaction which can be thought of as arising from the interaction between the magnetic field and the magnetic dipole moment of the atom. These higher-order interactions will be dealt with in more detail in Section 3.1.4.

Assuming that at present we need only concern ourselves with electric dipole interactions we may write

$$H_{21}^{0\prime} = -e\mathscr{E}_0 \int \phi_2^{0*}(r)\ (\mathbf{d}.\mathbf{r})\phi_1^0(r)\ \mathrm{d}^3r$$

Thus

$$|H_{21}^{0\prime}|^2 = e^2\mathscr{E}_0^2\mathscr{V}_{21}^2$$

where

$$\mathscr{V}_{21} = \int \phi_2^{0*}(\mathbf{r})\ (\mathbf{d}.\mathbf{r})\ \phi_1^0(\mathbf{r})\ \mathrm{d}^3r \qquad (3.19)$$

Now the average energy density associated with a plane electromagnetic wave in a vacuum is $\frac{1}{2}\varepsilon_0\mathscr{E}_0^2$ (see Eq. (2.35)), so that if, instead of a single frequency, we have a range of frequencies present from ω to $\omega + \delta\omega$ with an energy density $\rho(\omega)$, then the equivalent electric field can be written as

$$\left(\frac{2}{\varepsilon_0}\rho(\omega)\,\mathrm{d}\omega\right)^{1/2}$$

Thus we have

$$|H_{21}^{0\prime}|^2 = e^2\frac{2}{\varepsilon_0}\rho(\omega)\,\mathrm{d}\omega\;\mathscr{V}_{21}^2 \tag{3.20}$$

The total transition probability is then given by integrating the transition probability given by Eq. (3.14) over the whole range of frequencies involved. Making use of Eq. (3.19) we can then write

$$P_{12}(t) = \frac{2e^2}{\varepsilon_0\hbar^2}\int_{\omega_{12}-\Delta\omega}^{\omega_{12}+\Delta\omega}\left\{\frac{\sin^2((\omega_{12}-\omega)t/2)}{(\omega_{21}-\omega)^2}\right\}\mathscr{V}_{21}^2\rho(\omega)\,\mathrm{d}\omega \tag{3.21}$$

As time increases (i.e. t gets larger) that part of the integrand in Eq. (3.21) which is in curly brackets becomes increasingly sharply peaked about ω_{12} (see Figure 3.1). It is effectively zero outside a frequency range of $2\pi/t$ and we may assume therefore that both the factor $\mathscr{V}_{21}^2$ and the energy density $\rho(\omega)$ are approximately independent of ω over this range. Accordingly, we take them outside of the integral and also extend the limits of the integral out to infinity. Thus

$$P_{12}(t) = \frac{2e^2\mathscr{V}_{21}^2}{\varepsilon_0\hbar^2}\rho(\omega_{21})\left(\frac{t}{2}\right)^{1/2}\int_{-\infty}^{\infty}\frac{\sin^2((\omega-\omega_{21})t/2)}{((\omega-\omega_{21})t/2)^2}\,\mathrm{d}\omega$$

Making the substitution $x = (\omega_{21}-\omega)t/2$ and using the result that

$$\int_{-\infty}^{\infty}\frac{\sin^2 x}{x^2}\,\mathrm{d}x = \pi$$

(see, for example, Ref. 2) we have

$$P_{12}(t) = \frac{\pi e^2\mathscr{V}_{21}^2}{\varepsilon_0\hbar^2}\rho(\omega_{21})t \tag{3.22}$$

We have (finally!) succeeded in obtaining a transition probability that is linear in time.

Now the quantity $\mathscr{V}_{21}^2$ depends on the direction of **d** with respect to the particular atom being considered, and if we are to compare this result with Eq (3.1) we must bear in mind the angular homogeneity of black-body radiation. (The same conclusion will also hold when a collection of randomly oriented atoms, as, for example, in a gas, is irradiated with a plane beam of radiation.) In a situation where *all* angles between the vectors **d** and **r** in Eq. (3.19) are equally likely, the *average* value of $\mathscr{V}_{21}^2$ is given by

$$\langle\mathscr{V}_{21}^2\rangle = \langle\cos^2(\theta)\rangle\left|\int\phi_2^{0*}(\mathbf{r})r\phi_1^0(\mathbf{r})\,\mathrm{d}^3r\right|^2$$

Since $\langle \cos^2(\theta) \rangle = 1/3$ (see Problem 3.3).

$$\langle \mathscr{V}_{21}^2 \rangle = \tfrac{1}{3} | \mathscr{R}_{21} |^2 \tag{3.23}$$

where

$$\mathscr{R}_{21} = \int \phi_2^{0*}(\mathbf{r}) r \phi_1^0(\mathbf{r}) \, \mathrm{d}^3 r \tag{3.24}$$

and so

$$P_{12}(t) = \frac{\pi e^2}{3\varepsilon_0 \hbar^2} \rho(\omega_{21}) | \mathscr{R}_{21} |^2 t \tag{3.25}$$

Comparing Eqs (3.1) and (3.25) we obtain

$$B_{12} = \frac{\pi e^2}{3\varepsilon_0 \hbar^2} | \mathscr{R}_{21} |^2 \tag{3.26}$$

Thus we finally have an expression which enables us to calculate absolute values for the Einstein B coefficients. An equation for the A coefficients can be obtained by using the relation between A and B coefficients given by Eq. (2.114):

$$A_{21} = \frac{8\pi^2 e^2}{3\varepsilon_0 c^3 \hbar} \frac{g_1}{g_2} \nu^3 | \mathscr{R}_{21} |^2 \tag{3.27}$$

where we have used linear rather than angular frequency.

We should note that although Eq. (3.27) gives what appears to be a 'quantum mechanical' expression for A_{21}, the semi-classical theory used above provides no mechanism whereby spontaneous radiation can actually be generated. If no external perturbing field is present then the atoms remain in their stationary states and no radiation is emitted. A full description requires a theory that also considers the quantization of the light field.

It is found that when the radiation field by itself is quantized there is a fluctuation in the electric field which can be written

$$\Delta \mathscr{E} = (2)^{1/2} \mathscr{E}_a \, (n + 1/2)^{1/2} \tag{3.28}$$

where $\mathscr{E}_a$ is the 'average field' from one photon and n is the total number of photons present. There is apparently a fluctuating field even when *no* photons are present! This field is called the *vacuum field* and spontaneous emission can then be thought of as stimulated emission induced by the vacuum field. Fortunately, the resulting expression for A_{21} is identical to Eq. (3.27).

As we have seen in Section 2.3 (Eq. (2.107)), the A_{21} coefficient is related to the spontaneous (radiation) lifetime of the upper level τ_{21} by

$$\tau_{21} = 1/A_{21}$$

This relationship allows the experimental verification of Eq. (3.27), since the lifetime of a state can often be measured reasonably easily (though note the caveat following

Eq. (2.108)). An evaluation of a particular A_{21} coefficient is given in Example 3.1, but we may note here that we would expect typical values for $|\int \phi_2^{0*}(\mathbf{r}) r \phi_1^0(\mathbf{r})\, d^3r|^2$ to be the order of magnitude of the square of the Bohr radius, i.e. $\approx (0.5 \times 10^{-10})^2\ m^2$. Assuming that $g_1 = g_2$, Eq. (3.27) then gives $A_{21} \approx 7 \times 10^{-38} \nu^3\ s^{-1}$. For light in the middle of the visible range $\nu \approx 6 \times 10^{14}$ Hz so that $A_{21} \approx 10^7\ s^{-1}$. This would correspond to an excited state lifetime of about 100 ns.

Example 3.1 Lifetime of the 2P to 1S transition in hydrogen

We may readily determine the lifetime of any required transition from Eq. (3.27) provided we have a knowledge of the wavefunctions of the two states involved. In the case of the 2P → 1S transition in hydrogen we are concerned with transitions between the ($n = 2$; $l = 1$; $m_l = 1, 0, -1$; $m_s = \pm 1/2$) and ($n = 1$; $l = 0$; $m_l = 0$; $m_s = \pm 1/2$) states. The 2P state is then sixthfold degenerate and the 1S state doubly degenerate. Since, however, the dipole operator does not act on the electron spin, transitions can only take place between states with the same value of m_s. We suppose that the electric field acts along the z direction, so that we require the matrix element of z or $r\cos(\theta)$. As may be verified by the student, the matrix elements involving the $m_l = \pm 1$ initial states are both zero. Thus we have (in the notation of Appendix 2)

$$\phi_2^0(\mathbf{r}) = R_{2,1}(r)\, Y_{1,0}(\theta,\phi)$$

and

$$\phi_1^0(\mathbf{r}) = R_{1,0}(r)\, Y_{0,0}(\theta,\phi)$$

where

$$\begin{aligned} R_{2,1}(r) &= 2^{-3/2} 3^{-1/2} a_0^{-5/2} r \exp(-r/(2a_0)) \\ R_{1,0}(r) &= 2a_0^{-3/2} \exp(-r/a_0) \\ Y_{1,0}(\theta,\phi) &= (3/(4\pi))^{1/2} \cos(\theta) \\ Y_{0,0}(\theta,\phi) &= (1/(4\pi))^{1/2} \end{aligned}$$

and

$$a_0 = \frac{4\pi\varepsilon_0 \hbar^2}{e^2 m} \quad (= 0.53 \times 10^{-10}\ \mathrm{m})$$

We have therefore (bearing in mind that in spherical polar coordinates $d^3r = r^2 \sin(\theta)\, d\theta\, d\phi\, dr$)

$$\int \phi_2^{0*}(r) \mathbf{r} \phi_1^0(\mathbf{r})\, d^3\tau = (2^{5/2}\pi a_0^4)^{-1} \int_0^\infty r^4 \exp\left(-\frac{3r}{2a_0}\right) dr$$

$$\times \int_0^\pi \cos^2\theta \sin\theta\, d\theta \times \int_0^{2\pi} d\phi$$

The integrals are readily evaluated to give

$$\int_0^\infty r^4 \exp\left(-\frac{3r}{2a_0}\right) \, dr = 4 \times 3 \times 2 \times \left(\frac{2a_0}{3}\right)^5$$

$$\int_0^\pi \cos^2 \theta \sin \theta \, d\theta = 2/3$$

$$\int_0^{2\pi} d\phi = 2\pi$$

Thus $|\int \phi_2^{0*}(\mathbf{r})\mathbf{r}\phi_1^0(\mathbf{r}) \, d^3r|^2 = (2^{15/2}3^{-5}a_0)^2 = 0.555a_0^2$.

From Eq. (1.8) the frequency, ν, of radiation corresponding to a transition from an $n = 1$ state to an $n = 2$ state is given by

$$\nu = \frac{3me^4}{32\varepsilon_0^2 h^3} = 2.47 \times 10^{15} \text{ Hz}$$

Substituting for ν and $|\int \phi_2^{0*}(\mathbf{r})r\phi_1^0(\mathbf{r}) \, d^3r|^2$ in Eq. (3.27) we obtain (taking $g_1 = g_2 = 1$)

$$A_{21}(2\text{P} \rightarrow 1\text{S}) = 6.28 \times 10^8 \text{ s}^{-1}$$

The lifetime is the reciprocal of this quantity, that is, 1.59×10^{-9} s.

3.1.4 Selection rules and higher-order transitions

We have seen that the probability of spontaneous transitions due to the electric dipole interaction is directly proportional to the quantity $|\int \phi_2^{0*}(\mathbf{r})r\phi_1^0(\mathbf{r}) \, d^3r|^2$ which we have abbreviated to $|\mathscr{R}_{21}|^2$. There are a number of circumstances in which the matrix element $\mathscr{R}_{21}$ may be zero. If, for example, the functions $\phi_2^0(\mathbf{r})$ and $\phi_1^0(\mathbf{r})$ are both symmetric (that is, if they do not change sign when $\mathbf{r}$ is replaced by $-\mathbf{r}$), then the integral involved will have equal and opposite contributions from the points $\mathbf{r}$ and $-\mathbf{r}$, and in consequence will be zero. The same argument holds if the two functions are antisymmetric. Wavefunctions which are symmetric are said to have *even parity* while those which are antisymmetric are said to have *odd parity* (see Section 1.3.1). Thus electric dipole transitions can only take place between states of opposite parity.

The selection rules for the hydrogenic type wavefunctions (as listed in Appendix 2) can be fairly easily determined. As in Example 3.1, it is useful to consider applying the electric field along the x, y and z axes in turn. To enable a preferred direction to be established we assume a (vanishingly small) magnetic field is applied along the z axis. The rectangular components of the vector $\mathbf{r}$ are then

(a) $x = r \sin \theta \cos \phi$
(b) $y = r \sin \theta \sin \phi$
(c) $z = r \cos \theta$

The required matrix element $\mathscr{R}_{21}$ then breaks up into three integrals involving r, θ and ϕ, and it is the last which is of interest here. The ϕ dependence of a wavefunction is given simply by $\exp(-jm_l\phi)$, so that when a field is applied along the x direction the matrix element becomes proportional to

$$\int_0^{2\pi} \exp(jm_{l2}\phi) \cos\phi \exp(-jm_{l1}\phi)\, d\phi$$

or

$$\frac{1}{2}\int_0^{2\pi} \{\exp[j(m_{l2}-m_{l1}+1)\phi] + \exp[j(m_{l2}-m_{l1}-1)\phi]\}\, d\phi$$

This is equal to zero unless $m_{l2} = m_{l1} \pm 1$. An identical conclusion is reached when the field is along the y axis.

With the field along the z axis the above integral is simply

$$\int_0^{2\pi} \exp(jm_{l2}\phi)\exp(-jm_{l1}\phi)\, d\phi$$

which is zero unless $m_{l2} = m_{l1}$. Thus the selection rules for electric dipole radiation as far as m_l is concerned can be stated as:

1. Electric field parallel to magnetic field:

$$\Delta m_l = 0 \tag{3.29}$$

2. Electric field perpendicular to magnetic field:

$$\Delta m_l = \pm 1 \tag{3.30}$$

A similar analysis can be carried out for the θ integral (see Problem 3.5), and this results in the following selection rules for l:

$$\Delta l = \pm 1 \tag{3.31}$$

It is important to realize that electric dipole transitions are not the only type of transition that can take place. This is because we have not considered all the possible interactions between the light and the atom. For example, the *magnetic dipole interaction* arises from the interaction of the magnetic field of the radiation with the magnetic moment of the atom. This interaction, however, results in much smaller transition probabilities than are given by electric dipole interactions. A simple order of magnitude calculation can be carried out to verify this point.

First, we generalize the result obtained for electric dipole transitions (i.e. Eqs (3.25) and (3.24)) by saying that the transition probability is proportional to the quantity

$$\mathscr{M}_{12} = \left| \int \phi_2^{0*}(r)\, \Delta\hat{H} \phi_1^0(r)\, d^3\mathbf{r} \right|^2$$

where $\Delta\hat{H}$ is the operator representing the interaction between the atom and the electric field. In the case of electric dipole transitions $\Delta \hat{H}$ may be written as $-e\mathscr{E}_0\mathbf{r}.\mathbf{d}$ ($\mathbf{d}$ being a unit vector along the field direction), in which case

$$\mathscr{M}_{12}^{ed} = \left| -e\mathscr{E}_0 \int \phi_2^{0*}(\mathbf{r})\ \mathbf{r}.\mathbf{d}\ \phi_1^0(\mathbf{r})\ \mathrm{d}^3 r \right|^2$$

The integral will have the approximate value a_0 and so

$$\mathscr{M}_{12}^{ed} \approx (e\mathscr{E}_0 a_0)^2$$

This is just the square of the energy of an electric dipole of strength ea_0 in a field of $\mathscr{E}_0$. The corresponding quantity for a magnetic dipole interaction would then be the energy of the magnetic dipole moment of the atom in a field $\mathscr{B}_0$, where $\mathscr{B}_0 = \mathscr{E}_0/c$ (see Eq. (2.16)). The dipole moment of an atom is of the order μ_B (the Bohr magneton), and so

$$\mathscr{M}_{12}^{md} \approx (\mu_B \mathscr{E}_0/c)^2$$

Thus the ratio of the probabilities of magnetic dipole transitions to electric dipole transitions is expected to be given by

$$\frac{\mathscr{M}_{12}^{md}}{\mathscr{M}_{12}^{ed}} \approx \left(\frac{\mu_B}{ea_0 c}\right)^2 = 1.3 \times 10^{-5}$$

There exist yet further interactions which are associated with even smaller transition probabilities. For example, next in order of magnitude after the magnetic dipole interaction comes the *electric quadrupole* interaction. This arises from the interaction between the electron charge distribution and the gradient of the electric field. It assumes greater importance at very short wavelengths where there is a significant change in the electric field across the dimensions of the atom. For visible radiation electric quadrupole transition probabilities are some 10^8 times smaller than those for electric dipole transitions.

The selection rules governing magnetic dipole and electric quadrupole transitions are different from those governing electric dipole transitions, so transitions can take place that would not otherwise be allowed. However, the low probabilities associated with these transitions mean that the upper energy level lifetime is now the order of milliseconds or more (as compared to the 10^{-8} to 10^{-7} seconds for electric dipole transitions). Such relatively long-lived states are referred to as *metastable*.

3.1.5 Nonradiative transitions

It is important to remember that transitions are also possible between energy levels that do not necessarily involve the direct absorption or emission of photons. For example in nonradiative decay the energy difference $E_2 - E_1$ may be dissipated in the form of translational, rotational or vibrational energy given to the surrounding molecules, and there are many other possibilities. As discussed in Section 2.3, the overall lifetime for decay between an upper level and a lower level in a simple two-level system is characterized by a lifetime τ, given by

$$\frac{1}{\tau}=\frac{1}{\tau_r}+\frac{1}{\tau_{nr}} \tag{3.32}$$

where τ_r and τ_{nr} are the radiative and nonradiative lifetimes, respectively.

Thus in the presence of nonradiative decay processes the overall lifetime is *less* than it would be if the decay were purely radiative. This reduced lifetime will apply to all processes dependent on the variation of the excited state population with time. If, for example, the upper state lifetime is measured by monitoring the exponential decay of the spontaneous emission with time then the result will be the total lifetime τ rather than the radiative lifetime τ_r.

If we wish to measure τ_r, then we must have some means of allowing for the presence of nonradiative transitions. At any time during the radiation process the ratio of radiative transitions to the total number of transitions will be given by

$$\frac{N_2}{\tau_r}\bigg/\frac{N_2}{\tau} \text{ or } \frac{\tau}{\tau_r}$$

Since this remains true during the whole time the process takes place, the above ratio must represent the ratio of the initial number of atoms raised to the upper level to the total number of radiative transitions that take place. This ratio can often be measured and is known as the *quantum yield*, ϕ. Thus

$$\phi=\frac{\tau}{\tau_r} \tag{3.33}$$

3.1.6 Transitions in semiconductors

Semiconductor materials require a somewhat different treatment than do isolated atoms due to the rather different nature of their electronic wavefunctions. It is useful at this stage to consider the distribution of electrons among the energy levels in the conduction and valence bands. In thermal equilibrium the probability that an electron energy state at energy E is occupied is given by the Fermi function $F(E)$, where

$$F(E)=\frac{1}{1+\exp((E-E_F)/kT)} \tag{3.34}$$

Here E_F, the *Fermi level*, is a constant for a particular system at a particular temperature.

If the electron density of states function is $D(E)$ (Section 1.5.3) then the number of electrons per unit volume lying between E and $E+\mathrm{d}E$ is given by $N(E)\,\mathrm{d}E$, where

$$N(E)\,\mathrm{d}E=D(E)F(E)\,\mathrm{d}E$$

The total electron density in the system, N, will then be given by

$$N = \int_{-\infty}^{\infty} N(E)\,\mathrm{d}E$$
$$N = \int_{-\infty}^{\infty} D(E)F(E)\,\mathrm{d}E \tag{3.35}$$

Equation (3.35) then effectively defines E_F for the system.

The density of electrons in the conduction band of a semiconductor can then be written

$$n = \int_{E_c}^{E_T} D(E)F(E)\,\mathrm{d}E \tag{3.36}$$

where E_c and E_T are the energies of the bottom and top of the conduction band, respectively.

Now we have an expression for $D(E)$ (Eq. 1.59)) which is valid for states at the very bottom of the band and it is useful to examine the behaviour of the product $D(E)F(E)$ with energy using this. We may also approximate the Fermi function by assuming that E_F will lie roughly in the middle of the energy gap (justification for this will come later!). Since for typical semiconductors of interest here $E_g \cong 1$ eV and at room temperature $kT \cong 0.025$ eV, we see that $E - E_F \gg kT$, and so we have

$$F(E) \cong \exp\left(-\frac{E-E_F}{kT}\right) \tag{3.37}$$

Thus combining Eqs (1.59) and (3.37)

$$D(E)F(E) = \frac{4(2m_e^*)^{3/2}}{\pi^2\hbar^3}(E-E_c)^{1/2}\exp\left(-\frac{E-E_F}{kT}\right)$$
$$= \frac{4(2m_e^*)^{3/2}}{\pi^2\hbar^3}\exp\left(-\frac{E_c-E_F}{kT}\right)(E-E_c)^{1/2}\exp\left(-\frac{E-E_c}{kT}\right) \tag{3.38}$$

Figure 3.3 shows a plot of

$$(E-E_c)^{1/2}\exp\left(-\frac{E-E_c}{kT}\right)$$

as a function of $E - E_c$. We see that the electron density rises rapidly to peak at an energy of $kT/2$ above the bottom of the conduction band (see Problem 3.4). It then declines becoming relatively insignificant beyond an energy of $4kT$ above the band bottom. Most of the electrons are within $2kT$ (or 0.05 eV) of the bottom of the band and the halfwidth of the distribution is about $1.8kT$. The energy separation between the top and bottom of a band is typically much larger than this (e.g. 1 eV or more) and so we are reasonably justified in using the parabolic E-k relationship of Eq. (1.58). Similar arguments apply to the hole populations in the valence band, so that Figure 3.3 also represents the relative hole populations as a function of $E_v - E$.

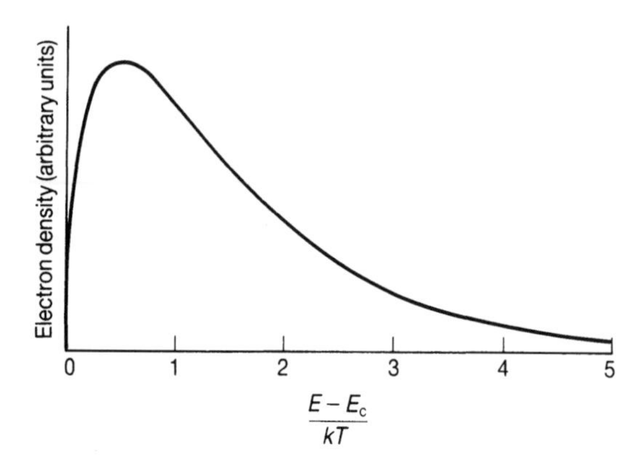

Figure 3.3 Relative electron population density within the conduction band of a semiconductor as a function of electron energy. The half width of the curve is about 1.8 eV.

We may then carry out the integration in Eq. (3.36) (which we further simplify by putting $E_T = \infty$) by substituting for $F(E)D(E)$ from Eq. (3.38) (see Problem 3.6). The result is

$$n = \frac{(8\pi m_e^* kT)^{3/2}}{h^3} \exp\left(-\frac{E_c - E_F}{kT}\right) \tag{3.39}$$

A similar calculation for holes yields

$$p = \frac{(8\pi m_h^* kT)^{3/2}}{h^3} \exp\left(-\frac{E_F - E_v}{kT}\right) \tag{3.40}$$

By multiplying these two expressions for n and p together we obtain

$$np = \frac{(8\pi kT)^3 (m_e^* m_h^*)^{3/2}}{h^6} \exp\left(-\frac{E_g}{kT}\right) \tag{3.41}$$

This expression is independent of E_F, and so the product of the electron and hole densities in a particular semiconductor at a particular temperature is constant even when the electron and hole populations are changed by doping.

In intrinsic material $n = p = n_i$, and by equating the right-hand sides of Eqs (3.38) and (3.39) we obtain

$$E_F = \frac{E_c - E_v}{2} + \frac{3kT}{4} \log_e(m_h^*/m_e^*) \tag{3.42}$$

From this latter equation we see that in intrinsic material the Fermi level is at an energy $3kT/4 \log_e(m_h^*/m_e^*)$ above the middle of the energy gap. Unless the effective masses are *very* different from each other this correction is usually relatively small (see Example 3.2).

Example 3.2 Fermi level position in intrinsic GaAs
We may use Eq. (3.42) to calculate the position of the Fermi level in intrinsic GaAs at room temperature (300k). The effective masses in GaAs are given by $m_e^* = 0.068 m_e$ and $m_h^* = 0.56\, m_e$. At room temperature kT is 0.025 eV, so that the Fermi level is at an energy of

$$\frac{3 \times 0.025}{4} \log_e \left(\frac{0.56}{0.068}\right) \text{ eV}$$

or 0.052 eV above the middle of the gap. Since the energy gap of GaAs is 1.43 eV, the Fermi level is not very far from the middle of the gap.

For extrinsic material the situation is rather more complicated. However, under normal circumstances it may be assumed that the density of majority carriers is equal to the doping density. As the majority carrier density increases, the Fermi level changes, so that the carrier populations of electrons and holes are still given by Eqs (3.39) and (3.40), respectively. The relative densities of carriers in the conduction and valence bands will, in thermal equilibrium, still be distributed in energy according to Figure 3.3.

We turn now to the types of transition that can take place between the conduction and valence bands. The simplest emission process is that of an electron falling from a state in the conduction band (energy E_2) into an empty state in the valence band (energy E_1) with the emission of a photon. Conservation of energy requires

$$E_2 - E_1 = h\nu$$

Another requirement (which can be thought of as a 'conservation of momentum') is that the k value be conserved. Thus the change in k value of the electron must be equal to the k value of the photon. Now a photon of wavelength λ will have a wavevector (or k value) of $2\pi/\lambda$. If we look at Figure 1.23 we see that electrons in the conduction and valence bands have a range of k values from $-\pi/a$ to π/a. Since we are normally dealing with near-optical radiation where $\lambda \cong 10^{-6}$ m, and since the interatomic spacing $\cong 0.5 \times 10^{-9}$ m, we have that

$$2\pi/\lambda \ll 2\pi/a$$

Thus, on the scale of Figure 1.23, the k values of the initial and final electron states can differ only by a very small amount. Another way of expressing this is to say that only vertical transitions are allowed between the bands in an E-k diagram.

A rather more rigorous proof (but perhaps giving less insight) involves examining the appropriate matrix elements for the transition, and is given in Appendix 5.

Thus transitions between states at the bottom of the conduction band and those at the top of the valence band present no problems in direct bandgap semiconductors such as GaAs, but obviously are forbidden in indirect band gap semiconductors such as Si and Ge (the terms 'direct' and 'indirect' bandgaps are defined at the end of

Section 1.5.2). Transitions can take place in indirect bandgap semiconductors, but then the process is rather more complicated. For example, the requirement for k conservation can be met if a *phonon* (a quantized lattice vibration) is involved. However, the involvement of another 'particle' makes the transition much less probable than in the direct case.

Let us now consider what happens when light falls on a direct bandgap semiconductor. Provided the frequency is such that $h\nu \geqslant E_g$ then excitation of an electron from the valence band to the conduction band can take place over a broad band of frequencies. The electron in the conduction band could then fall back down to the valence band and emit radiation. It is much more likely, however, to cascade down through states in the conduction band (losing energy by phonon emission) until it reaches the partially occupied states at the bottom of the band. A typical time scale for this relaxation process is $\approx 10^{-12}$ s. It can then make a radiative transition back to the valence band, the time scale for this process being of the order 10^{-9} s. Now, as we have seen, the equilibrium electron (and hole) distributions extend over an energy range of some 0.05 eV so that on recombination there is a possible energy spread in the emitted photon of some 0.1 eV. At a wavelength of 820 nm, for example, this energy spread corresponds to a wavelength spread of about 55 nm, and indeed this is approximately the linewidth seen in the emission from a GaAlAs LED as shown in Figure 3.4.

Now the peaks in the electron and hole distributions are both $kT/2$ away from the band edges, so we would expect the peak in the emission to correspond to an energy change of $E_g + kT$. In fact, in most LEDs emission occurs at the bandgap wavelength (corresponding to an energy change of E_g). The reason for this is that the energy level situation at the band edges is rather more complicated than we have assumed. In most semiconductors there are a large number of impurity states (some intentional, some not) at the band edges. These are sufficiently dense to make it appear that the bandgap is smaller than it actually is. This smearing out of the band edge is called *band tailing*. If the emission from semiconductors is examined at low

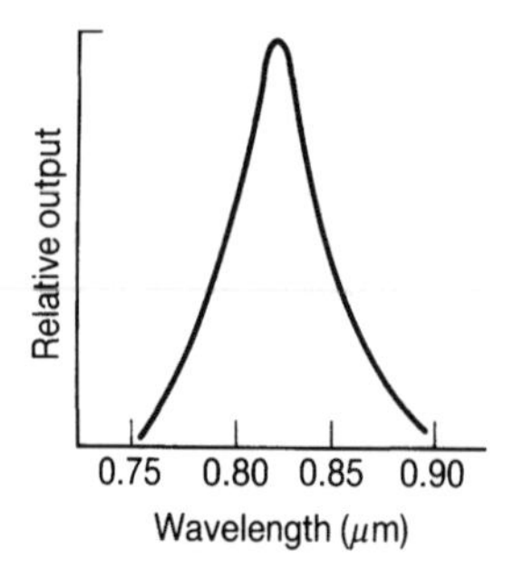

Figure 3.4 Emission spectrum from a near infrared emitting LED made from GaAlAs. The emission bandwidth extends over several tens of nm.

temperatures then the broad, featureless room-temperature curve develops a considerable fine structure arising from the energy levels at the band edge.

In the above discussion we have used arguments based on an assumption of thermal equilibrium to describe the emission from an LED. When an LED is emitting radiation, however, the electrons within the bands are *not* in thermal equilibrium. Light emission is observed because considerable excess concentrations of electrons and holes are injected into the p and n sides of the diode. However, because of the wide differences between the relaxation time within the bands and the recombination lifetime, the electrons and holes rapidly reach reach an 'equilibrium' within their own bands even though they are not in equilibrium with each other. Consequently, it is possible to describe their distributions in terms of two *quasi-Fermi levels* E_{Fc} and E_{Fv} where the probabilities of finding electrons and holes in states at energy E are given by the functions $F_c^e(E)$ and $F_v^h(E)$, respectively, where

$$F_c^e(E) = \frac{1}{1 + \exp((E - E_{Fc})/kT} \tag{3.43}$$

$$F_v^h(E) = \frac{1}{1 + \exp((E_{Fv} - E)/kT} \tag{3.44}$$

3.2 Line-broadening mechanisms

In all the discussion in Section 3.1 we have assumed that atomic energy levels are 'infinitely sharp' and that in consequence emission and absorption take place at a single frequency. (This is not quite true, since Eq. (3.14) gives rise to the possibility of transitions when $\omega \neq \omega_{12}$. However, as the measurement time increases, the probability function becomes more sharply peaked about ω_{12}.) Experimentally it is found that both absorption and emission 'lines' always have a finite spectral width. Further, the same transition can exhibit varying spectral widths depending on the atom's environment. The narrowest widths are exhibited when the atoms are far apart (i.e. in a dilute gas), at very low temperatures (i.e. when they are virtually stationary) and when a low density of radiation is present. The linewidth under these circumstances is known as the *natural* linewidth. However, if the gas pressure is increased the line broadens due to *pressure* broadening, if the temperature of the gas is raised we get *Doppler* broadening and if a high-powered beam of radiation is used to measure the absorption then *power* broadening may be observed. Whatever the broadening mechanism, we may describe the resulting line profile by the lineshape function $g(\omega)$. This function was introduced in Section 2.4.1, and represents the shape of the emission (or absorption) line normalized to unity. It can also be thought of as relating to the probability of emission. Thus $g(\omega)\,d\omega$ is the probability that radiation will be emitted within the frequency range ω to $\omega + d\omega$.

Line-broadening mechanisms can be divided into two main types termed *homogeneous* and *inhomogeneous* which give rise to somewhat different line profiles.

Homogeneous broadening results when each atom in the ensemble has the same individual lineshape function $g(\omega)$. The resultant lineshape of the radiation emitted by the whole ensemble will then just be that of an individual atom. However, in some situations the ensemble of atoms can be split into groups, with each group having different physical properties so that the atoms in different groups exhibit different lineshape functions (the usual situation is that each group has a different central frequency[2] ω_0). The lineshape function observed for the whole ensemble will then be given by averaging over all the groups and will thus be broader than that for each individual atom. This 'group dependent' broadening is called *inhomogeneous* broadening.

3.2.1 Radiative (or natural) broadening

As mentioned above, even an isolated atom with no external perturbations exhibits a finite linewidth in its spontaneous emission spectrum. Essentially this linewidth stems from the Heisenberg Uncertainty Principle, which, when applied to the two variables time and energy, yields

$$\Delta E \, \Delta t \approx \hbar$$

With an infinitely long measurement time ($\Delta t = \infty$) we have $\Delta E = 0$ and the energy of a state will be 'infinitely sharp'. However, spontaneous emission limits the observation time to approximately τ_2, the excited state lifetime. Thus we then have

$$\Delta E \tau_2 \approx \hbar \tag{3.45}$$

and the upper energy level will be broadened by approximately $\hbar/\tau_2$ or, using Eq. (2.107), $\hbar A_{21}$. In angular frequency terms this corresponds to a frequency spread $\Delta\omega (= \Delta E/\hbar)$ of just A_{21}.

To put this argument on a rather more precise footing, we consider absorption within a dilute gas of simple 'two-level' atoms as an infinitely narrow optical beam (in frequency terms) is scanned across the resonance condition. We may deduce the resulting absorption by first calculating an expression for the refractive index (via the dielectric constant) and then associating the complex part of the this with an absorption coefficient (see Section 2.1.5).

The dielectric constant is the ratio between the applied field and the polarization (or induced dipole moment per unit volume). To calculate the induced dipole moment we assume that the field is applied along the x direction. If an electron in a particular atom is in the state ψ, then the dipole moment of the atom, $d(t)$, along the x axis is given by

$$d(t) = -e \int \psi^*(\mathbf{r}, t) x \, \psi(\mathbf{r}, t) \, \mathrm{d}v \tag{3.46}$$

where the integration runs over all the coordinates of the Z electrons in the atom.

For our simple one-electron two-level system a general state of the system can be written

$$\psi(\mathbf{r}, t) = a_1(t)\phi_1(\mathbf{r})\exp(-\mathrm{j}E_1t/\hbar) + a_2(t)\phi_2(\mathbf{r})\exp(-\mathrm{j}E_2t/\hbar) \tag{3.47}$$

By substituting for ψ from Eq. (3.47) into Eq. (3.46) we obtain

$$d(t) = -e[a_1^*(t)a_2(t)X_{12}\exp(-\mathrm{j}\omega_{12}t) + a_2^*(t)a_1(t)X_{21}\exp(\mathrm{j}\omega_{12}t)] \tag{3.48}$$

where

$$X_{ij} = \int \phi_i^*(\mathbf{r})x\phi_j(\mathbf{r})\,\mathrm{d}\mathbf{r} \tag{3.49}$$

We have used the fact that $X_{11} = X_{22} = 0$ (this follows because the operator x has odd parity), and we have also put $X_{21} = X_{12}^*$ (which follows directly from Eq. (3.49)). We now need to use an equation dealing with the population dynamics. We start with Eq. (3.8). Since $X_{22} = 0$ we may put $H'_{22} = 0$ giving

$$\mathrm{j}\hbar\frac{\partial a_2(\mathrm{t})}{\partial t} = a_1(t)\exp(\mathrm{j}\omega_{12}t)H'_{21} \tag{3.50}$$

To allow for spontaneous emission we modify Eq. (3.50) by introducing an additional term on the right-hand side which is proportional to $a_2(t)$, so that

$$\mathrm{j}\hbar\frac{\partial a_2(t)}{\partial t} = a_1(t)\exp(\mathrm{j}\omega_{12}t)H'_{21} - \mathrm{j}\gamma_2\hbar a_2(t) \tag{3.51}$$

When no external perturbation is present (i.e. $H'_{21} = 0$) then this equation may be integrated to give

$$a_2(t) = a_2(0)\exp(-\gamma_2 t) \tag{3.52}$$

which implies an exponential decay for the upper energy level with a lifetime of $1/2\gamma_2$ (the numbers of atoms in the excited state being proportional to $a_2(t)a_2^*(t)$).

As before, we assume that the perturbation has a sinusoidal time dependence (i.e. as in Eq. (3.12)), and that the applied field is small so that $a_1(t) \approx 1$. Equation (3.51) can then be written in the general form

$$\frac{\partial a_2(t)}{\partial t} + \gamma_2 a_2(t) = f(t) \tag{3.53}$$

where

$$f(t) = -\exp(\mathrm{j}\omega_0 t)\,H_{21}^{0\prime}\left(\frac{\exp(\mathrm{j}\omega t) - \exp(-\mathrm{j}\omega t)}{2\hbar}\right) \tag{3.54}$$

By multiplying both sides of Eq. (3.53) by the factor $\exp(\gamma_2 t)$, we have

$$\frac{\partial}{\partial t}(a_2(t)\exp(\gamma_2 t)) = f(t)\exp(\gamma_2 t)$$

so that

$$a_2(t) = \exp(-\gamma_2 t)\int f(t)\exp(\gamma_2 t)\,\mathrm{d}t \tag{3.55}$$

As far as the integral is concerned

$$\int f(t)\exp(\gamma_2 t)\,\mathrm{d}t = \frac{\exp(\mathrm{j}\gamma_2 t)}{2\hbar}\left[\frac{\exp(\mathrm{j}(\omega_0-\omega)t)}{\mathrm{j}(\omega_0-\omega)+\gamma_2} - \frac{\exp(\mathrm{j}(\omega_0+\omega)t)}{\mathrm{j}(\omega_0+\omega)+\gamma_2}\right] + C$$

where C is a constant of integration. It may be determined from the requirement that $a_2(0) = 0$, and so

$$C = \frac{1}{2\hbar}\left[\frac{1}{\mathrm{j}(\omega_0+\omega)+\gamma_2} - \frac{1}{\mathrm{j}(\omega_0-\omega)+\gamma_2}\right]$$

Thus

$$a_2(t) = \frac{H_{21}^{0'}}{2\hbar\mathrm{j}}\left[\frac{\exp(\mathrm{j}(\omega_0-\omega)t) - \exp(-\gamma_2 t)}{\omega_0-\omega-\mathrm{j}\gamma_2} - \frac{\exp(\mathrm{j}(\omega_0+\omega)t) - \exp(-\gamma_2 t)}{\omega_0+\omega-\mathrm{j}\gamma_2}\right] \tag{3.56}$$

Before we substitute for $a_2(t)$ in Eq. (3.48) we make two changes to Eq. (3.56). First, since the dielectric constant relates to the situation after any initial transients have died away, we can neglect the exponentially decaying term. Second, we are explicitly applying the field along the x axis so that we may put

$$H_{21}^{0'} = e\mathscr{E}_0 X_{21}$$

Thus

$$a_2(t) = \frac{e\mathscr{E}_0 X_{21}}{2\hbar j}\left[\frac{\exp(\mathrm{j}(\omega_0-\omega)t)}{\omega_0-\omega-\mathrm{j}\gamma_2} - \frac{\exp(\mathrm{j}(\omega_0+\omega)t)}{\omega_0+\omega-\mathrm{j}\gamma_2}\right] \tag{3.57}$$

In addition, it is convenient to assume that we are applying a weak field so that few atoms will be excited from the ground state and we may put $a_1(t) = 1$. With these substitutions Eq. (3.51) becomes

$$d(t) = \frac{e^2\mathscr{E}_0|X_{21}|^2}{2\hbar\mathrm{j}}\left[\frac{\exp(\mathrm{j}\omega t)}{\omega_0+\omega-\mathrm{j}\gamma_2} - \frac{\exp(-\mathrm{j}\omega t)}{\omega_0-\omega-\mathrm{j}\gamma_2} - \frac{\exp(-\mathrm{j}\omega t)}{\omega_0+\omega+\mathrm{j}\gamma_2} + \frac{\exp(\mathrm{j}\omega t)}{\omega_0-\omega+\mathrm{j}\gamma_2}\right] \tag{3.58}$$

This expression gives us the dipole moment of a single atom. For a collection of N atoms which occupy a volume V the polarization is

$$P = N\,\mathrm{d}(t)/V \tag{3.59}$$

Since we are writing our fields as two component quantities (see Eq. (3.12)) we take a 'two-component' susceptibility $P(t)$ where

$$P(t) = \frac{\varepsilon_0 \mathscr{E}_0}{2\mathrm{j}}\left[\varkappa(\omega)\exp(\mathrm{j}\omega t) - \varkappa(-\omega)\exp(-\mathrm{j}\omega t)\right] \tag{3.60}$$

We are now in a position to derive an expression for $\varkappa(\omega)$. Using Eqs (3.60), (3.59) and (3.58), and also making allowance for the random orientation of the gas atoms (i.e. see Eq. (3.23)) we have finally

$$\varkappa(\omega) = \frac{Ne^2 |X_{21}|^2}{3\varepsilon_0 \hbar V}\left[\frac{1}{\omega_0 + \omega - \mathrm{j}\gamma_2} + \frac{1}{\omega_0 - \omega + \mathrm{j}\gamma_2}\right] \tag{3.61}$$

As before, we adopt the rotating wave approximation whereby we neglect the denominator containing the term $\omega_0 + \omega$, so that

$$\varkappa(\omega) = \frac{Ne^2 |X_{21}|^2}{3\varepsilon_0 \hbar V}\left[\frac{1}{\omega_0 - \omega + \mathrm{j}\gamma_2}\right]$$

$$\varkappa(\omega) = \frac{Ne^2 |X_{21}|^2}{3\varepsilon_0 \hbar V}\left[\frac{\omega_0 - \omega - \mathrm{j}\gamma_2}{(\omega_0 - \omega)^2 + \gamma_2^2}\right] \tag{3.62}$$

Our aim in this section is to obtain an expression for the (complex) refractive index. We know that the susceptibility and refractive index are related by (Eqs (2.26) and 2.31)):

$$n^2 = 1 + \varkappa \tag{3.63}$$

Putting $n = n_\mathrm{r} + \mathrm{j}n_\mathrm{i}$ (see Eq. (2.32)) and $\varkappa = \varkappa' + \mathrm{j}\varkappa''$ into Eq. (3.63) and equating real and imaginary parts yields

$$n_\mathrm{r}^2 - n_\mathrm{i}^2 = 1 + \varkappa' \tag{3.64}$$

and

$$2n_\mathrm{r}n_\mathrm{i} = \varkappa'' \tag{3.65}$$

From Eq. (2.39) the absorption coefficient, α, can then be written

$$\alpha = \frac{\omega\varkappa''}{cn_\mathrm{r}} \tag{3.66}$$

Since we are dealing with a dilute gas we may assume that the real part of the refractive index will be reasonably close to unity, so that, by taking the imaginary part of Eq. (3.63), we have

$$\alpha = \frac{Ne^2 |X_{21}|^2 \omega_{12}}{3\varepsilon_0 \hbar Vc} \left[\frac{\gamma_2}{(\omega_0 - \omega)^2 + \gamma_2^2} \right] \tag{3.67}$$

Here we have replaced ω in the numerator by ω_0, which is permissible since the function is only large when ω is close to ω_0. The variation of α with ω is thus Lorentzian (see Eq. (2.48) and Figure 2.5). The absorption will fall to one half of its peak value when $\omega = \omega_0 \pm \gamma_2$, thus the linewidth is given by $2\gamma_2$ or, simply, A_{21}. The calculation for the value of A_{21} for the 2P to 1S transition in hydrogen in Example 3.1 gave $6 \times 10^8\ \mathrm{s}^{-1}$. In frequency terms this corresponds to about 1.7×10^8 Hz. This is the *narrowest* linewidth the transition can have, and is called the *natural linewidth*.

3.2.2 Power broadening

In the derivation of Eq. (3.67) above we assumed a weak perturbing field, that is, when solving the coupled Eqs (3.53) and (3.9) we made the assumption that $a_1(t) \approx 1$. It is possible to solve these equations without making this assumption (at least to the next order of approximation), [Ref. 3]. The result of this is that the susceptibility can be written

$$\varkappa(\omega) = \frac{Ne^2 |X_{21}|^2}{3\varepsilon_0 \hbar V} \left[\frac{\omega_0 - \omega - \mathrm{j}\gamma_2}{(\omega_0 - \omega)^2 + \gamma_2^2 + \mathscr{F}^2} \right] \tag{3.68}$$

where

$$\mathscr{F}^2 = \frac{e^2 \mathscr{E}_0^2 |X_{21}|^2}{2\hbar^2} \tag{3.69}$$

The absorption coefficient can then be recalculated using this new value of $\varkappa(\omega)$. It is easy to see that the absorption linewidth is now increased to $2(\gamma_2^2 + \mathscr{F}^2)^{1/2}$. Since $\mathscr{F}^2$ is proportional to $\mathscr{E}_0^2$ (and hence to optical power), this result shows that the linewidth will broaden with increasing power.

3.2.3 Collision (or pressure) broadening

So far, we have considered collections of atoms which do not interact (i.e. dilute gases). In real gases atoms collide with each other. An atom undergoing spontaneous emission will emit radiation that is practically monochromatic between collisions but during the time of the collision the colliding atom will interact with the emitting atom and alter the energy levels so that the radiation changes frequency (Figure 3.5). The interaction time during the collision is much shorter ($\approx 10^{-13}$ s, see Problem 3.8) than the time over which the emission takes place (i.e. 10^{-8} s or longer). Thus as far as a 'long-term' view is concerned the emission consists of a wave of a single frequency whose phase changes randomly whenever there is a

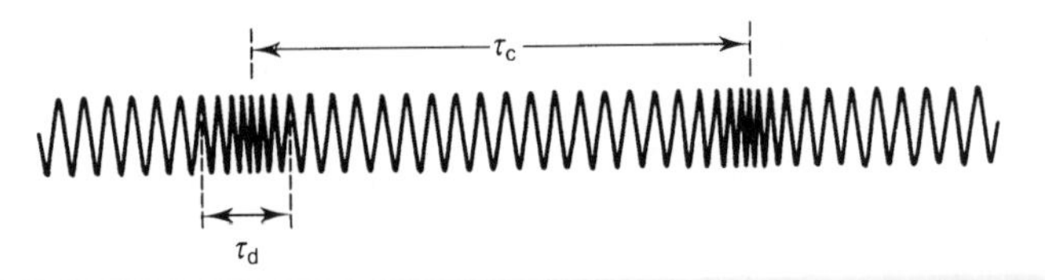

Figure 3.5 Schematic representation of the effect of a collision on the emission frequency of an atom. During the time τ_d the colliding atom is close enough to the emitting atom to affect significantly its emission frequency (in this case the frequency is shown as increasing). In reality the time between collisions (τ_c) is considerably larger than the time (τ_d) over which the collision takes place.

collision. We first assume that the time between collisions (τ_c) is constant. Because the wave changes its phase randomly after every τ_c seconds it has a frequency spectrum equivalent to that of a wave where

$$\mathscr{E}(t) = \mathscr{E}_0 \sin(\omega_0 t + \phi) \quad 0 \leqslant t < \tau_c$$
$$\mathscr{E}(t) = 0 \quad t < 0,\ t \geqslant \tau_c$$

Using Fourier analysis this can be shown to be equivalent to a continuous distribution (in frequency) of sinusoidal waves which have a frequency-dependent amplitude $\mathscr{E}_0(\omega)$ given by

$$\mathscr{E}_0(\omega) = \frac{1}{2\pi} \int_0^{\tau_c} \mathscr{E}_0 \exp(-\mathrm{j}\omega_0 t + \mathrm{j}\phi)\exp(\mathrm{j}\omega t)\,\mathrm{d}t$$

$$\mathscr{E}_0(\omega) = \frac{\mathscr{E}_0}{2\pi} \exp(\mathrm{j}\phi)\, \frac{\exp[(\mathrm{j}\omega - \omega_0)\tau_c] - 1}{\mathrm{j}(\omega - \omega_0)}$$

The power spectrum, $I(\omega)$, of the distribution is proportional to $|\mathscr{E}_0|^2$ so that

$$I(\omega) \propto \frac{\sin^2[(\omega - \omega_0)\tau/2]}{(\omega - \omega_0)^2}$$

Now this result applies for one particular value of τ, in fact a range of values will be encountered. If $P(\tau)\,\mathrm{d}\tau$ is the probability that an atom undergoes a transition during the time interval from τ to $\tau + \mathrm{d}\tau$, then the resulting frequency power distribution becomes

$$I(\omega) \propto \int_0^{\infty} \frac{\sin^2[(\omega - \omega_0)\tau/2]}{(\omega - \omega_0)^2} P(\tau)\,\mathrm{d}\tau \tag{3.70}$$

It is possible to show that (see Problem 3.9)

$$P(\tau) = \frac{\exp(-\tau/\tau_0)}{\tau_0}\,\mathrm{d}\tau$$

where τ_0 is the mean time between collisions.

The integral on the right-hand side of Eq. (3.70) is then readily carried out, the result being:

$$I(\omega) \propto \frac{\tau_0/2}{1+(\omega-\omega_0)^2\tau_0^2}$$

It is convenient to introduce a normalized power spectrum, $g(\omega)$, where

$$\int_{-\infty}^{\infty} g(\omega)\,\mathrm{d}\omega = 1$$

Thus

$$g(\omega) = \frac{\tau_0}{\pi}\frac{1}{1+(\omega-\omega_0)^2\tau_0^2} \tag{3.71}$$

As in the case of natural broadening, we obtain a Lorentzian lineshape, the half-width in this case being $2/\tau_0$.

To calculate a value for the linewidth requires an expression for τ_0. Unfortunately, this is by no means a trivial exercise and some rather drastic simplifications are in order. First, we assume that the gas is an 'ideal' monatomic gas whose atoms are hard spheres of radius r. Next, we focus our attention on one particular gas atom and suppose it is at rest and that all the other atoms are moving towards it with the average *relative* velocity of $v_{\mathrm{ave}}^{\mathrm{rel}}$. The moving atoms are taken to be distributed randomly within a column of unit cross-sectional area and with a number density (N) which is equal to that in the gas (Figure 3.6). In one second a column of atoms of length $v_{\mathrm{ave}}^{\mathrm{rel}}$ will have passed the stationary atom. If the centre of any of these atoms passes within a distance $2r$ of the stationary atom then a collision will take place. The probability of a collision is then simply $\pi(2r)^2$ or $4\pi r^2$ (remember, the atoms are contained within a unit cross-sectional area), and thus the total number of collisions per second is

$$4\pi r^2 v_{\mathrm{ave}}^{\mathrm{rel}} N$$

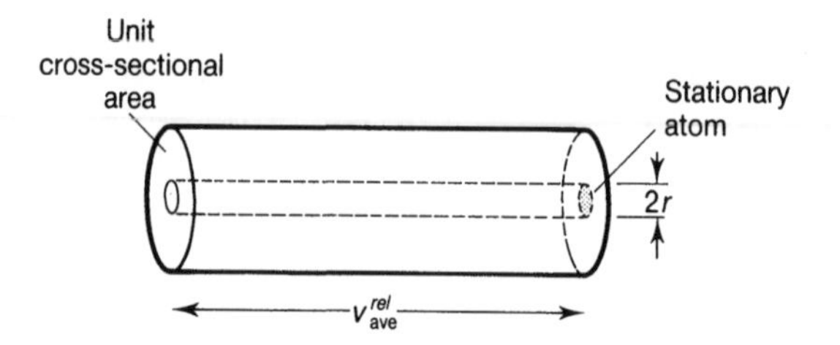

Figure 3.6 Diagram enabling an estimate to be made of the mean time between collisions among atoms in a gas.

The mean time between collisions is the reciprocal of this, so that

$$\tau_0 = \frac{1}{4\pi r^2 v_{\text{ave}}^{\text{rel}} N} \tag{3.72}$$

If we suppose that there are m moles of gas within a volume V, then the number density is given by

$$N = mN_A / V$$

where N_A is Avogadro's number. The universal gas law gives that

$$PV = mRT$$

where R is the universal gas constant ($= N_A k$, where k is Boltzmann's constant) so that

$$N = \frac{P}{kT} \tag{3.73}$$

and

$$\tau_0 = \frac{kT}{4\pi t^2 v_{\text{ave}}^{\text{rel}} P} \tag{3.74}$$

The halfwidth ($\Delta\nu_c$) of a collision-broadened line (i.e. $2/\tau_0$) should then be given by

$$\Delta\nu_c = \frac{8\pi r^2 v_{\text{ave}}^{\text{rel}} P}{kT} \tag{3.75}$$

Thus the linewidth is predicted to be directly proportional to P (hence the alternative name for collision broadening of *pressure broadening*). As a very rough guide it is found that the linewidth is given by

$$\Delta\nu_c / P \approx 10 \text{ MHz/torr}$$

An estimation of the magnitude of collision broadening in Ne using Eq. (3.75) is given in Example 3.3. For gases at pressures less than about 1 torr, line broadening due to collision broadening is often much less than the natural linewidth resulting from electric dipole transitions.

Example 3.3 Collision broadening in neon

To use Eq. (3.75) we need values for both the effective atomic radius and for the average relative velocity. The latter may be estimated by equating the average thermal energy of an atom ($3kT/2$) to the average kinetic energy ($Mv_{\text{av}}^2/2$ where M is the mass of the atom). Thus

$$v_{\text{av}} = \left(\frac{3kT}{M}\right)^{1/2}$$

The mass of the neon atom can be obtained using Avogadro's number together with the atomic mass (20). Thus $M = 20/6.022 \times 10^{26}$ kg. At a temperature of 400 K (a typical figure for a He–Ne laser), therefore, we have

$$\begin{aligned} v_{av} &= (3 \times 1.38 \times 10^{-23} \times 400/3.22 \times 10^{-26})^{1/2} \\ &= 717 \text{ m/s} \end{aligned}$$

This is *not*, of course, the value we require which can be shown to be $\sqrt{2}$ times the above value,[3] i.e. $\cong 1000$ m/s.

The 'hard sphere' radius can be taken to be about 1.2×10^{-10} m, so that in a neon gas at 400 K and at a pressure of 1 torr (i.e. 1/760 of an atmosphere or $10^5/760$ N/m^2) we have from Eq. (3.75) that

$$\Delta\nu_c = \frac{8\pi(1.7 \times 10^{-10})^2 \times 10^3 \times (10^5/760)}{1.38 \times 10^{-23} 400} \text{ Hz}$$

$$\Delta\nu_c = 17 \text{ MHz}$$

It should be noted that in the He–Ne laser the 633 nm transition is in fact broadened by about 70 MHz/torr due to collisions between He and Ne atoms.

A form of collision broadening can be observed in solids. Here electrostatic interactions (called *crystal field interactions*) between neighbouring ions can remove the degeneracies of the ionic energy levels, so that the position of the energy levels of each ion depends on the exact positions of its neighbours. During the presence of a phonon the neighbouring atoms will be in a state of oscillation. This will cause changes in the position of the energy levels and hence also in the emission frequency. Thus the 'collision' between an atom and a phonon in a solid and the collision between atoms in a gas have similar consequences, that is, a broadened spectral line with a Lorentzian profile.

In Section 2.4.1 we saw that the coherence function $\gamma_{11}(\tau)$ could be obtained by taking what is essentially the Fourier transform of the lineshape function. In the case of a Lorentzian lineshape we then have

$$\gamma_{11}(\tau) = \frac{\tau_0}{\pi} \int_{-\infty}^{\infty} \frac{\exp(j\omega\tau)}{1 + (\omega - \omega_0)^2 \tau_0^2} \, d\omega \tag{3.76}$$

Unfortunately, the evaluation of this integral involves a knowledge of integration over the complex plane (see, for example Ref. 4). The result is

$$\gamma_{11}(\tau) = \exp(j\omega\tau)\exp(-|\tau|/\tau_0) \tag{3.77}$$

so that

$$|\gamma_{11}(\tau_c)| = \exp(-|\tau_c|/\tau_0) \tag{3.78}$$

3.2.4 Doppler broadening

All the previous line-broadening mechanisms have applied identically to each atom in the group and therefore constitute homogeneous broadening mechanisms. We have seen that they all give rise to Lorentzian profiles and we now discuss a mechanism that (at any one time) affects all the atoms *differently*. Within a gas atoms are moving about with random thermal speeds and in random directions. We assume that each atom undergoing spontaneous emission emits the same frequency (ω_0) in its own frame of reference. If we observe the light emitted along the z axis from a particular atom whose component of velocity along the z axis is v_z, then the observed angular frequency (ω), according to classical wave theory, will be given by

$$\omega = \omega_0\left(1 - \frac{v_z}{c}\right)$$

so that

$$v_z = \frac{c(\omega - \omega_0)}{\omega_0} \tag{3.79}$$

To obtain the resulting distribution in emitted frequencies from a gas of such atoms we need to know the distribution of v_z within the gas. This is in fact given by the Maxwell distribution [Ref. 5]. Thus the probability $P(v_z)\,\mathrm{d}v_z$ that an atom has a z component of velocity between v_z and $v_z + \mathrm{d}v_z$ is given by

$$P(v_z)\,\mathrm{d}v_z = \left(\frac{M}{2\pi kT}\right)^{1/2} \exp\left(-\frac{Mv_z^2}{2kT}\right)\mathrm{d}v_z \tag{3.80}$$

The angular frequency probability distribution, $g(\omega)\,\mathrm{d}\omega$, is now given by substituting for v_z from Eq. (3.79) into Eq. (3.80). Thus

$$g(\omega) = \frac{c}{\omega_0}\left(\frac{M}{2\pi kT}\right)^{1/2} \exp\left(-\frac{Mc^2}{2kT}\frac{(\omega - \omega_0)^2}{\omega_0^2}\right)$$

or

$$g(\omega) = \frac{1}{\sqrt{\pi}w} \exp\left(-\frac{(\omega - \omega_0)^2}{w^2}\right) \tag{3.81}$$

where

$$w = \left(\frac{2kT\omega_0^2}{Mc^2}\right)^{1/2}$$

This distribution is called Gaussian, and it has its maximum value when $\omega = \omega_0$. The curve falls to half of its maximum value when

$$\exp\left(-\frac{Mc^2}{2kT}\frac{(\omega-\omega_0)^2}{\omega_0^2}\right)=\frac{1}{2}$$

and so the halfwidth, $\Delta\omega$, is given by

$$\Delta\omega = 2\omega_0\left(\frac{2kT}{Mc^2}\log_e 2\right)^{1/2} \tag{3.82}$$

Figure 3.7 shows both the Lorentzian and Gaussian distributions. They are both normalized to unit area under the curve and both have the same linewidth. It will be noticed that the Gaussian distribution is more sharply peaked than the Lorentzian.

A typical calculation of the linewidths in gases is given in Example 3.4. As this calculation shows, Doppler broadening in gases can be of much greater importance than either pressure or natural broadening. However, this is not always the case since at high pressures (e.g. atmospheric pressure) pressure broadening can become predominant (see Problem 3.10)

Example 3.4 Dopper-broadened linewidths in neon

We may calculate the Doppler-broadened linewidth for the 632.8 nm transition in neon (this, of course, is the transition responsible for the red output of a He–Ne laser). We assume a temperature of 400 K which is the effective temperature within the discharge of a He–Ne laser. The mass of the neon atom is 3.32×10^{-26} kg (see Example 3.3) so that Eq. (3.82) gives:

$$\Delta\nu = \frac{2c}{\lambda_{12}}\left(\frac{2kT}{Mc^2}\log_e 2\right)^{1/2}$$

$$= \frac{2\times3\times10^8}{632.8\times10^{-9}}\left(\frac{2\times1.38\times10^{-23}\times400\times0.693}{3.32\times10^{-26}\times9\times10^{16}}\right)^{1/2}$$

$$= 1.5\times10^9 \text{ Hz or } 1500 \text{ MHz}$$

Inhomogeneous broadening is observed in solids. In the previous section we saw that the electrostatic field due to the neighbouring ions can influence the position of the energy levels of the ion. Often this field varies slightly with the position of the ion in the lattice because of lattice imperfections or the presence of impurities, etc. This gives rise to a random distribution of transition frequencies for a particular transition and the result is a Gaussian-broadened line. Note that each ion only emits or absorbs at a 'single' frequency within the overall linewidth so that this is inhomogeneous broadening. A similar broadening mechanism is present in liquids.

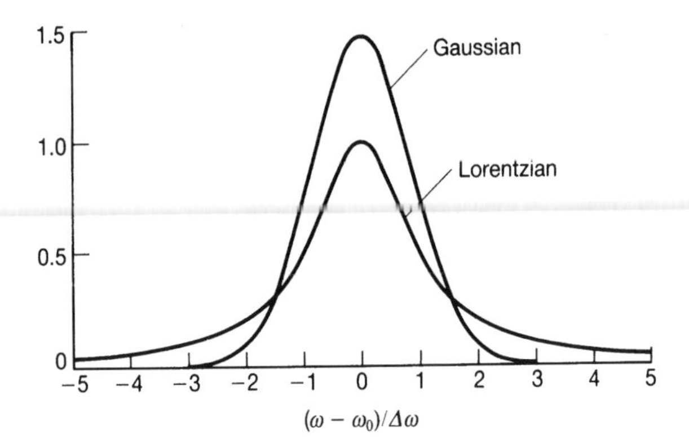

Figure 3.7 The Lorentzian and Gaussian distributions both scaled to the same halfwidth.

With a Gaussian line shape function the coherence function is given by

$$\gamma_{11}(\tau) = \frac{1}{\sqrt{\pi}w} \int_{-\infty}^{\infty} \exp\left(-\frac{(\omega-\omega_0)^2}{w^2} + \mathrm{j}\omega\tau\right) \mathrm{d}\omega \tag{3.83}$$

As in the case of the Lorentzian lineshape, a proper evaluation of this integral requires a knowledge of integration over the complex plane. However, it is possible to give a rather simplified derivation. If we make the substitution $x = (\omega - \omega_0)/w$ we obtain

$$\gamma_{11}(\tau) = \frac{\exp(\mathrm{j}\omega_0\tau)}{\sqrt{\pi}} \int_{-\infty}^{\infty} \exp\left(-x^2 + \mathrm{j}\omega x\tau\right) \mathrm{d}x$$

$$\gamma_{11}(\tau) = \frac{\exp(\mathrm{j}\omega_0\tau)}{\sqrt{\pi}} \int_{-\infty}^{\infty} \exp\left(-(x - \mathrm{j}\omega\tau/2)^2 - w^2\tau^2/4\right) \mathrm{d}x$$

Putting $q = x - \mathrm{j}w\tau/2$

$$\gamma_{11}(\tau) = \frac{\exp(\mathrm{j}\omega_0\tau)\,\exp(-w^2\tau^2/4)}{\sqrt{\pi}} \int_{-\infty}^{\infty} \exp(-q^2)\,\mathrm{d}q$$

As we have seen earlier, the integral has the value of $\sqrt{\pi}$ (provided we do not worry unduly about the complex nature of q!):

$$\gamma_{11}(\tau) = \exp(\mathrm{j}\omega_0\tau)\exp(-w^2\tau^2/4) \tag{3.84}$$

and so

$$|\gamma_{11}(\tau)| = \exp(-w^2\tau^2/4) \tag{3.85}$$

As may be readily verified (see Problem 3.11), both this result and the corresponding one for the Lorentzian profile (Eq. (3.78)) are in accord with the general assertion concerning linewidths and coherence times (Eq. 2.145)), that

$$\tau_c \approx 1/\Delta f$$

3.2.5 The Voigt profile

In the previous two sections we have discussed several different mechanisms for line broadening, which have given rise to either Lorentzian or Gaussian line profiles. Before proceeding it is useful to summarize the forms of the profiles for future reference.

In terms of the halfwidth $\Delta\omega_L$, the normalized profile function, $g_L(\omega)$, for the Lorentzian (homogeneous broadening) can be written

$$g_L(\omega) = \frac{2}{\pi\,\Delta\omega_L}\,\frac{1}{1 + [(\omega - \omega_0)/(\Delta\omega_L/2)]^2} \tag{3.86}$$

whence

$$g_L(\omega_0) = \frac{2}{\pi\,\Delta\omega_L} \tag{3.87}$$

The corresponding functions for the Gaussian profile (inhomogeneous broadening) are

$$g_G(\omega) = \frac{2}{\Delta\omega_G}\left(\frac{\log_e 2}{\pi}\right)^{1/2} \exp\left[-\left(\frac{\omega - \omega_0}{\Delta\omega_G/2}\right)^2 \log_e 2\right] \tag{3.88}$$

and

$$g_G(\omega_0) = \frac{2}{\Delta\omega_g}\left(\frac{\log_e 2}{\pi}\right)^{1/2} \tag{3.89}$$

In a particular atomic system one of the line-broadening mechanisms will often dominate and the profile will then be determined (at least over the central portion of the line) by this one alone. Sometimes, however, two or more of the broadening mechanisms may be present which give rise to comparable line broadenings. If there are two homogeneous mechanisms each causing Lorentzian linewidths of $\Delta\omega_1$ and $\Delta\omega_2$ then it can be shown that the resultant is a Lorentzian with a total linewidth of $\Delta\omega_1 + \Delta\omega_2$. On the other hand, if two inhomogeneous mechanisms are present which give Gaussian linewidths of $\Delta\omega_1$ and $\Delta\omega_2$ the resultant is a Gaussian line with a total width of $(\Delta\omega_1 + \Delta\omega_2)^{1/2}$ (see Problem 3.13). The argument can obviously be extended to more than two broadening mechanisms of each type.

The only problem that arises is when comparable homogeneous and inhomogeneous broadening mechanisms are both present. Suppose that we have a group of N atoms and that the inhomogeneous broadening profile is written $g_i(x)$, so that $Ng_i(\omega')\,d\omega'$ represents the number of atoms that have their centre resonance angular frequencies between ω' and $\omega' + d\omega'$. If, in addition, we have a homogeneous broadening mechanism present characterized by the function $g_h(x)$, then the number of atoms emitting (or absorbing) between ω and $\omega + d\omega$ originating from the atoms under consideration can be written

$$Ng_i(\omega')\, d\omega'\, g_h(\omega - \omega' + \omega_0)\, d\omega$$

Now emission can be obtained between ω and $\omega + d\omega$ for all possible values of ω' since (in theory) both functions $g_i(x)$ and $g_h(x)$ extend from $-\infty$ to $+\infty$, and therefore the total number of atoms emitting between ω and $\omega + d\omega$ can be written as $Ng_V(\omega)d\omega$ where

$$g_V(\omega) = \int_{-\infty}^{+\infty} g_i(\omega')g_h(\omega - \omega' + \omega_0)\, d\omega' \tag{3.90}$$

Thus the resultant lineshape profile is given by the function $g_V(\omega)$, which is known as the *Voigt* profile. Unfortunately, when the explicit functions for $g_i(\omega)$ and $g_h(\omega)$ (i.e. the Gaussian and Lorentzian functions) are substituted into the integral, it is not possible to obtain an analytical result (see Problem 3.14), although numerical values have been tabulated (see, for example, Ref. 6). In mathematical terms $g_V(\omega)$ is said to be the *convolution* of the functions $g_i(\omega)$ and $g_h(\omega)$.

The coherence function resulting from a Voigt line profile can be written

$$\gamma_{11}(\tau) = \exp(j\omega_0\tau)\, \exp(-w^2\tau^2/4 - |\tau|/\tau_0) \tag{3.91}$$

where the parameters w and τ_0 are those of the constituent Gaussian and Lorentzian profiles, respectively.

Far from line centre the Lorentzian line profile is proportional to $(\omega - \omega_0)^{-2}$, which results in a much less rapid fall off with frequency than does the Gaussian line profile, which is proportional to $\exp(-(\omega - \omega_0)^2)$. An interesting consequence of this is that all transitions, even if they are strongly inhomogeneous, exhibit homogeneous line profiles at frequencies sufficiently far removed from line centre.

For examples of lineshapes being fitted to Voigt profiles see Ref. 7 (the 10.6 μm transition in CO_2) and Ref. 8 (the ruby laser transition at low temperatures).

3.2.6 The Stokes shift

When an impurity ion is present in a solid host material we expect that its absorption and emission lines will be broadened (into 'bands') because of the interactions with the surrounding lattice. Another feature often observed is that the absorption and emission bands arising from the same electronic transition do not take place at the same wavelength. The emission band is always found to be at a longer wavelength than the absorption band. This wavelength shift is known as the *Stokes shift*.

We can understand this phenomenon if we refer back to Figure 1.15, which shows the ground state energy in a molecule as a function of atomic separation. A similar curve will apply for the energy states of an impurity ion in terms of the positions of the neighbouring atoms. For simplicity, we assume that these are all at a distance

R from the impurity ion, and so the variation of the ground state energy as a function of R will be as indicated in Figure 3.8, with a minimum of energy at R_0. Such a diagram is called a *configurational coordinate* diagram. When the impurity ion is in an excited state the electron interaction with the surrounding host ions will be different from when it is in the ground state. In particular, the energy minimum will occur at a different value of R (i.e. R_1).

The time taken for an electron transition to take place is much smaller than that for an oscillation of the neighbouring ions. Thus during the time of the transition the ions will hardly move at all and may be assumed to be stationary (this is called the *Franck–Condon principle*). On a configurational coordinate diagram electronic transitions will thus take place vertically.

We assume that when an electronic transition takes place R has the value R_0. Immediately after the transition R will still have the value R_0. Since this does not correspond to the equilibrium value of R for the excited state, the surrounding atoms will relax until R reaches its equilibrium value R_1. There is plenty of time for this relaxation process to take place since the lattice oscillation period ($\approx 10^{-12}$ s) is much shorter than the excited state lifetime ($\approx 10^{-9}$ s). When a transition to the ground state subsequently takes place we can see from Figure 3.8 that the emitted photon will have less energy than that of the absorbed photon.

In the above discussion we have assumed that when absorption and emission takes place the impurity atom is at the energy minimum for the state in question. In fact, the surrounding ions will be in a state of continuous thermal oscillation, and only their average positions will correspond to the minima. Thus instead of a single absorption (or emission) wavelength a band of wavelengths will be seen. Since the

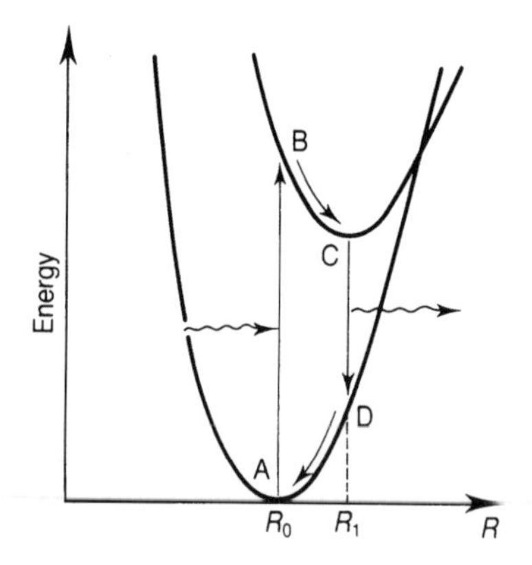

Figure 3.8 A configurational coordinate diagram illustrating the variation in the energy level positions of an impurity ion within a lattice as a function of the distance, R, away from the nearest neighbour atoms. The path ABCD shows the energy changes taking place during a photon absorption and subsequent emission process. It is readily seen that the energy of the emitted photon is less than that of the absorbed photon (the Stokes shift).

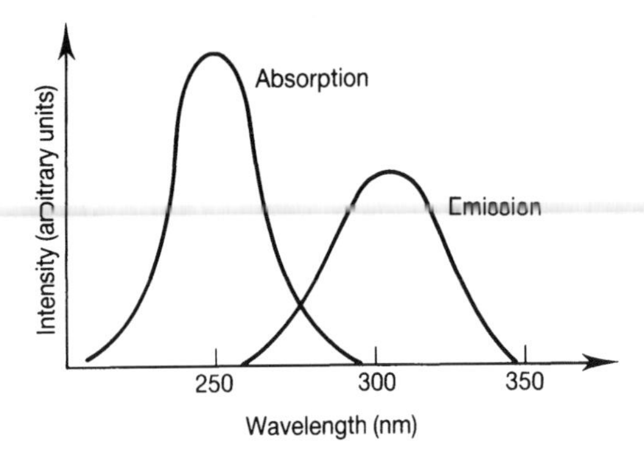

Figure 3.9 Absorption and emission spectra for thallium-activated potassium chloride (KCl : Tl) at room temperature. The emission peak occurs at a higher wavelength than that of the absorption curve. This is an example of the Stokes shift.

amplitude of oscillation will increase with increasing temperature, the absorption bandwidths are also expected to increase with increasing temperature. Figure 3.9 shows the absorption and emission spectrum for thallium in potassium chloride.

3.3 Optical gain

In our derivation of expressions for the Einstein coefficients in Section 3.1 we considered the interaction of broadband radiation with an infinitesimally narrow linewidth transition. In this section we consider the other extreme, that is the interaction of a very narrow linewidth beam (such as laser light!) with a broadened transition (by 'very narrow' we mean that the linewidth of the light beam is much narrower than that of the transition).

3.3.1 Absorption of a monochromatic beam

First, we examine the probability of a transition between the ground state and a group of densely packed excited states which have an average energy of $\hbar\omega_{12}$ and which are distributed, in terms of frequency, according to the line-broadening function[4] $g(\omega)$. The function $g(\omega)$ acts as an effective density of states function. Suppose our monochromatic light beam has a frequency of ω'. From Eq. (3.14) the probability of a transition from the ground state to the $g(\omega)\,\mathrm{d}\omega$ states which lie at energies of between $\hbar\omega$ and $\hbar\omega + \mathrm{d}(\hbar\omega)$ above the ground state can be written:

$$P_{1\omega}(\omega',t) = g(\omega)\,\frac{|H^{0\prime}_{\omega}|^2}{\hbar^2}\left[\frac{\sin^2((\omega-\omega')t/2)}{(\omega-\omega')^2}\right]\mathrm{d}\omega \tag{3.92}$$

The *total* transition probability to all possible final states is then obtained by integrating the right-hand side of Eq. (3.92) over all possible final states (i.e. all possible values of ω). Now we know that for reasonably large values of t, the function in square brackets is negligible unless ω is very close to ω'. It may thus be assumed that the functions $g(\omega)$ and $|H_{\omega}^{0\prime}|^2$ vary little over the range of ω values where the integrand is finite, and may be taken out of the integral and assumed to have the values appropriate to $\omega = \omega'$. We then have

$$P_{12}(\omega', t) = g(\omega') \frac{|H_{\omega}^{0\prime}|^2}{\hbar^2} \int_{-\infty}^{\infty} \frac{\sin^2((\omega - \omega')t/2)}{(\omega - \omega')^2} \, d\omega \tag{3.93}$$

The integral is exactly the same as that leading up to Eq. (3.22), so we may write

$$P_{12}(t) = g(\omega') \frac{|H_{\omega}^{0\prime}|^2}{\hbar^2} \frac{\pi}{2} t \tag{3.94}$$

To simplify the notation we now replace ω' by ω for the rest of this discussion. Assuming a sinusoidally oscillating field at angular frequency ω and a dipole interaction together with a random orientation of the field with the dipole moment gives

$$|H_{\omega}^{0\prime}|^2 = e^2 \mathscr{E}_0^2 \tfrac{1}{3} |\mathscr{R}_{12}|^2 \tag{3.95}$$

where

$$\mathscr{R}_{12} = \int \phi_1^{0*}(r) \mathbf{r} \phi_2^0(r) \, d^3 r \tag{3.96}$$

If we suppose that the interaction is with a plane beam of irradiance I_{ω}, then from Eq. (2.34) the associated electric field $\mathscr{E}_0$ has the value $(2I_{\omega}/c\varepsilon_0)^{1/2}$. The transition probability per unit time, $\mathscr{P}_{12}(\omega)$, can thus be written

$$\mathscr{P}_{12}(\omega) = g(\omega) |\mathscr{R}_{12}|^2 I_{\omega} \frac{\pi e^2}{3\hbar^2 c \varepsilon_0}$$

If we substitute for the coefficient B_{12} from Eq. (3.26), and also write the irradiance in terms of a radiation density[5] ρ_{ω} where $\rho_{\omega} = I_{\omega}/c$, we obtain

$$\mathscr{P}_{12}(\omega) = g(\omega) B_{12} \rho_{\omega} \tag{3.97}$$

The only difference between this and the 'normal' expression (i.e. Eq. 2.110) is the presence of a 'modulating' factor $g(\omega)$. For stimulated emission from level 2 to level 1 we can similarly write:

$$\mathscr{P}_{21}(\omega) = g(\omega) B_{21} \rho_{\omega} \tag{3.98}$$

We now examine the changes in irradiance that take place when such a beam travels through the medium. Initially, we make the assumption that the population densities of atoms in the lower and upper levels (i.e. N_1 and N_2) are maintained at constant values throughout the medium.

Consider a thin slab of the medium of thickness δz and cross-sectional area A. It contains $N_1 A \delta z$ atoms in the lower state and so in one second the number of upward transitions will be $N_1 A\delta z \mathscr{P}_{12}$. Similarly, the number of downward transitions per second will be $N_2 A\delta z \mathscr{P}_{21}$. Each of the former will reduce the number of photons in the beam by one while the latter will increase it by one. Thus in one second $A\delta z(\mathscr{P}_{12}N_1 - \mathscr{P}_{21}N_2)$ photons will have been lost from the beam. During this time $I_\omega A/\hbar\omega$ photons will have passed through the slab, and so we may write

$$\frac{\delta I_\omega}{I_\omega} = -\frac{A\delta z(\mathscr{P}_{12}N_1 - \mathscr{P}_{21}N_2)}{I_\omega A/\hbar\omega}$$

In the limit $\delta z \Rightarrow 0$,

$$\frac{\mathrm{d}I_\omega}{\mathrm{d}z} = -(\mathscr{P}_{12}N_1 - \mathscr{P}_{21}N_2)\hbar\omega \tag{3.99}$$

Substituting for $\mathscr{P}_{12}$ and $\mathscr{P}_{21}$ from Eqs (3.97) and (3.98),

$$\frac{\mathrm{d}I_\omega}{\mathrm{d}z} = -g(\omega)(B_{12}N_1 - B_{21}N_2)\hbar\omega\ \rho_\omega \tag{3.100}$$

or, expressing B_{12} in terms of B_{21} (Eq. (2.113)) and putting $\rho_\omega = I_\omega/c$,

$$\frac{\mathrm{d}I_\omega}{\mathrm{d}z} = -g(\omega)\left(\frac{g_2}{g_1}N_1 - N_2\right)B_{21}\hbar\omega I_\omega/c \tag{3.101}$$

If N_1 and N_2 are independent of z and I, then Eq. (3.101) has the solution

$$I_\omega(z) = I_\omega(0)\exp(-\alpha(\omega)z) \tag{3.102}$$

where

$$\alpha(\omega) = g(\omega)\left(\frac{g_2}{g_1}N_1 - N_2\right)B_{21}\hbar\omega/c \tag{3.103}$$

$\alpha(\omega)$ is, of course, the absorption coefficient (see Eq. (2.38)).

Another form of Eq. (3.103) can be obtained by expressing B_{21} in terms of A_{21}. Using the result of Eq. (2.114) we have

$$\alpha(\omega) = g(\omega)\left(\frac{g_2}{g_1}N_1 - N_2\right)A_{21}\frac{\pi^2c^2}{\omega^2}$$

It should be realized that we have tacitly assumed that the medium we are dealing with has a refractive index of unity. When this is not so we may generalize the last expression by replacing c by c/n, giving

$$\alpha(\omega) = g(\omega)\left(\frac{g_2}{g_1}N_1 - N_2\right)A_{21}\frac{\pi^2c^2}{\omega^2n^2} \tag{3.104a}$$

Alternatively, in terms of wavelength Eq. (3.104a) becomes

$$\alpha(\omega) = g(\omega)\left(\frac{g_2}{g_1}N_1 - N_2\right)A_{21}\frac{\lambda_0^2}{4n^2} \tag{3.104b}$$

Experimentally, $\alpha(\omega)$ is a relatively easy quantity to measure as a function of frequency (or wavelength) using, say, a spectrophotometer. Examples have already been shown in Figures 1.9 and 1.11. If the upper level is at an energy much greater than kT (i.e. 0.025 eV at 300k), then in thermal equilibrium its population will be negligible. Thus provided we know N_1, which may be approximated by N, the total number of atoms per unit volume, then we may determinc A_{21} for any particular transition by measuring $\alpha(\omega)$ across the whole of the line absorption profile and then determining the factor which will ensure the normalization of $g(\omega)$.

The absorption coefficient will only remain positive while $N_1(g_2/g_1) > N_2$. If the reverse is true then α will be negative and the beam irradiance will *grow* exponentially with distance. This result is at the very heart of laser action. When $N_2 > N_1(g_2/g_1)$ we have a condition known as *population inversion*. This, however, does not occur naturally within a system in thermal equilibrium. Thus from Eq. (2.104), since $\exp((E_2 - E_1)/kT)$ will always be greater than zero ($E_2 > E_1$), then $N_1 g_2 > N_2 g_1$. Under these conditions α will always be positive. We shall see in later chapters, however, that population inversion, and hence light amplification can be achieved (although not in a system in thermodynamic equilibrium).

Assuming population inversion *is* possible we may then write

$$I_\omega(z) = I_\omega(0)\exp(\gamma(\omega)z) \tag{3.105}$$

where $\gamma(\omega)$, the *small signal gain coefficient* and a positive quantity is given by

$$\gamma(\omega) = g(\omega)\left(N_2 - \frac{g_2}{g_1}N_1\right)A_{21}\frac{\pi^2c^2}{\omega^2n^2} \tag{3.106}$$

We should note that an exponential increase in $I(z)$ with z will only occur if the population inversion term is independent of $I(z)$ (and z). This is only valid provided $I(z)$ is sufficiently small, hence the term *small signal* gain coefficient. Situations where the population inversion is dependent on beam irradiance are dealt with in the next two sections.

It is sometimes convenient to give the strength of the absorption process in terms of a transition *cross-section*. Thus each atom is imagined as presenting to the incoming beam, of frequency ν, an actual cross-section area (the *stimulated absorption cross-section*) of $\sigma_{12}(\omega)$ (Figure 3.10). That part of the beam which overlaps with this area is absorbed and lost from the beam. If the beam area is A then the fractional change in the beam irradiance when passing through a thickness δz of material with a lower energy state density N_1 is given by

$$\frac{\delta I_\omega}{I_\omega} = -\frac{N_1 A\ \delta z \sigma_{12}(\omega)}{A}$$

In the limit of $\delta z \Rightarrow 0$

$$\frac{dI_\omega}{dz} = -I_\omega N_1 \sigma_{12}(\omega) \tag{3.107}$$

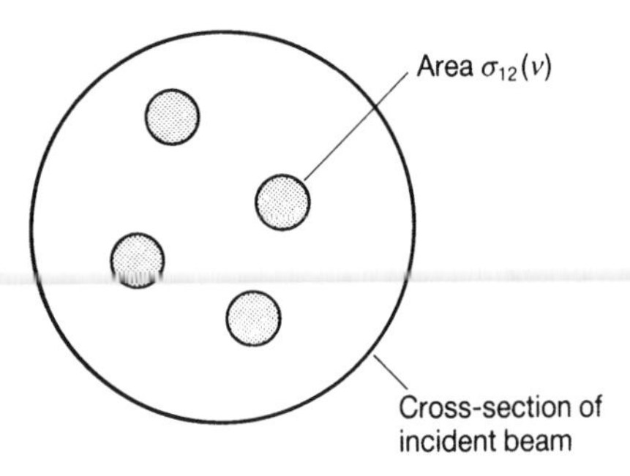

Figure 3.10 The concept of cross-section. As far as an atomic transition caused by radiation of frequency ν is concerned each atom in the path of the incident beam is regarded as having a physical cross-sectional area of $\sigma_{12}(\nu)$. Light which falls on this area is wholly absorbed in the transition process. It is assumed that the sample is sufficiently thin that there are no atoms lying behind others.

If the lower energy state refers to the ground state then the cross-section is related to the absorption coefficient by

$$\sigma_{12}(\omega) = \alpha(\omega)/N_1 \tag{3.108}$$

Using this in Eq. (3.104b), we obtain an explicit formula for the cross-section (we assume that all atoms are in the ground state, i.e. $N = N_1$ and $N_2 = 0$):

$$\sigma_{12}(\omega) = g(\omega)\,\frac{g_2}{g_1}\,A_{21}\,\frac{\lambda_0^2}{4n^2} \tag{3.109}$$

The maximum value of the cross-section will occur at the centre of the absorption line (i.e. using $g(0)$). If the line has a Lorentzian profile, then the value of $g(0)$ is

$$\frac{2}{\pi}\,\frac{1}{\Delta\omega}$$

where $\Delta\omega$ is the halfwidth (Eq. (3.87)). The smallest halfwidth (and hence the largest cross-sections) are obtained when the transition is subject to natural broadening only. The halfwidth then is just A_{21} (Section 3.2.1), so that the maximum cross-sections that can be expected are given by (taking $g_1 = g_2$)

$$\sigma_{12}^{\max} = \frac{\lambda_0^2}{2\pi}$$

For transitions in the visible (i.e. taking $\lambda_0 \approx 0.5$ μm) and assuming $n = 1$, we have therefore that

$$\sigma_{12}^{\max} \approx 4 \times 10^{-14}\ \mathrm{m}^2$$

As might be expected, because of the presence of line-broadening mechanisms over and above natural broadening, actual cross-sections are smaller than this value.

Typical values of the line centre cross-section for various laser transitions are given in Table 3.1 (see also Example 3.5). It is interesting to note that the radius of the outermost electron orbit in an atom is a few tenths of a nanometre, which would imply a physical 'cross-section' of $\approx 10^{-19}\ \mathrm{m}^2$, considerably smaller than the maximum estimated above.

Example 3.5 Cross-section of Nd:YAG laser transition

The laser transition in neodymium-doped YAG at a wavelength of 1.06 μm has an approximate Lorentzian lineshape with a linewidth, at room temperature, of 200 GHz. The upper state lifetime is 230 μs and the quantum yield for this transition is 0.42. The refractive index of YAG at the lasing wavelength is 1.82.

From Eq. (3.33), the value of the radiative lifetime, τ_r, is given by $230 \times 10^{-6}/0.42$ s, or 548×10^{-6} s. This gives a value for A_{21} of $1/548 \times 10^{-6}\ \mathrm{s}^{-1}$ or $1826\ \mathrm{s}^{-1}$. As explained above, the value of $g(0)$ for a Lorentzian is

$$\frac{2}{\pi}\frac{1}{\Delta\omega}$$

where $\Delta\omega$ is the (angular frequency) halfwidth. Thus using Eq. (3.109), we have that the cross-section at the peak absorption frequency is

$$\sigma = \frac{2}{\pi}\frac{1826}{2\pi \times 200 \times 10^9}\left(\frac{1.06 \times 10^{-6}}{2 \times 1.82}\right)^2$$

$$= 7.8 \times 10^{-23}\ \mathrm{m}^2$$

Another useful result is obtained by integrating both sides of Eq. (3.109) over all frequencies:

$$\int_{-\infty}^{\infty} \sigma_{12}(\omega)\,\mathrm{d}\omega = \frac{g_2}{g_1} A_{21} \int_{-\infty}^{\infty} g(\omega)\,\frac{\lambda_0^2}{4n^2}\,\mathrm{d}\omega$$

Table 3.1 Values of the cross-sections for transitions in various laser systems

System	Cross-section (m^2)
Gas lasers (visible-near-IR)	10^{-15}–10^{-17}
CO_2 laser (10.6 μm)	3×10^{-22}
Dye laser (rhodamine 6G)	1–2×10^{-20}
Nd:YAG	7×10^{-23}
$Cr:Al_2O_3$	2×10^{-24}

Assuming that the line is narrow enough so that both λ_0 and n can be regarded as constant at all frequencies where $g(\omega)$ gives a contribution to the integral, then

$$\int_{-\infty}^{\infty} \sigma_{12}(\omega)\,\mathrm{d}\omega = \frac{g_2}{g_1} A_{21} \frac{\lambda_0^2}{4n^2} \int_{\infty}^{\infty} g(\omega)\,\mathrm{d}\omega$$

Whatever the line profile,

$$\int_{-\infty}^{\infty} g(\omega)\,\mathrm{d}\omega = 1$$

so that

$$\int_{-\infty}^{\infty} \sigma_{12}(\omega)\,\mathrm{d}\omega = \frac{g_2}{g_1} A_{21} \frac{\lambda_0^2}{4n^2} \tag{3.110}$$

Thus a value of the A_{21} coefficient (or, equivalently, of the upper state lifetime) can be obtained directly from the integrated cross-section of the transition. This can be a useful technique in those situations where the lifetime is too small to be measured directly.

In the above discussion on cross-section the assumption was made that all the atoms were in the ground state. We can still use the concept of cross-section when this is not so. Thus if there are N_2 atoms in the excited state then stimulated emission will add to the beam intensity so that the effective absorption area can be written

$$N_1\sigma_{12}(\omega) - N_2\sigma_{21}(\omega)$$

where $\sigma_{21}(\nu)$ is the *stimulated emission cross-section.*

With degenerate energy levels we write

$$g_1\sigma_{12}(\omega) = g_2\sigma_{21}(\omega)$$

In future we will tend to use the stimulated emission cross-section rather than the stimulated absorption cross-section, and will refer to the former as σ_{SE}. In terms of σ_{SE} the absorption coefficient may then be written

$$\alpha(\omega) = \sigma_{SE}(\omega)\left(\frac{g_2}{g_1} N_1 - N_2\right) \tag{3.111}$$

If we have a situation of population inversion then the gain coefficient, $\gamma(\omega)$, is given by

$$\gamma(\omega) = \sigma_{SE}(\omega)\left(N_2 - \frac{g_2}{g_1} N_1\right) \tag{3.112}$$

3.3.2 Gain saturation in homogeneous media

Although we have not yet discussed mechanisms for obtaining population inversion, and hence optical gain, it is of interest to extend the analysis of the previous sections by removing the constraint of fixed population inversions. Thus we now assume that the upper level is populated by some unspecified process at a rate R_2, while the lower level is populated at a rate R_1 (Figure 3.11). The upper level can decay spontaneously to the lower level with a lifetime τ_{21}, while both levels decay nonradiatively to a lower level 0 with lifetimes τ_{20} and τ_{10}. For simplicity, we assume that the levels are nondegenerate. The rate equations governing the two populations N_1 and N_2 are then

$$\frac{dN_2}{dt} = R_2 - \frac{N_2}{\tau_{20}} - \frac{N_2}{\tau_{21}} - (N_2 - N_1)g(\omega)B_{21}\rho_\omega \tag{3.113}$$

$$\frac{dN_1}{dt} = R_1 - \frac{N_1}{\tau_{10}} + \frac{N_2}{\tau_{21}} + (N_2 - N_1)g(\omega)B_{21}\rho_\omega \tag{3.114}$$

Assuming a time-independent radiation density, the steady-state solution to these equations can be obtained by taking $dN_2/dt = dN_1/dt = 0$. It is convenient to introduce the total lifetime τ_2 of state 2 defined as

$$1/\tau_2 = 1/\tau_{20} + 1/\tau_{21}$$

and to write τ_{10} as simply τ_1. The beam irradiance,[6] I, is given by $I = \rho_\omega c$, so that we can write $g(\omega)B_{21}\rho_\omega$ as $G(\omega)I$, where $G(\omega) = g(\omega)B_{21}/c$. Thus in equilibrium

$$R_2 = -N_2(1/\tau_2 + G(\omega)I) + N_1G(\omega)I \tag{3.115}$$

$$R_1 = N_2(1/\tau_{21} + G(\omega)I) - N_1(1/\tau_1 + G(\omega)I) \tag{3.116}$$

These two equations may be solved for N_2 and N_1, with the result that

$$N_2 - N_1 = \frac{R_2\tau_2(1 - \tau_1/\tau_{21}) - R_1\tau_1}{1 + G(\omega)I(\tau_1 + \tau_2 - \tau_1\tau_2/\tau_{21})} \tag{3.117}$$

If the second term in the denominator of Eq. (3.117) is small compared to unity (i.e. for small values of I) a small signal population inversion density ΔN_0 can be introduced where

$$\Delta N_0 = R_2\tau_2(1 - \tau_1/\tau_{21}) - R_1\tau_1 \tag{3.118}$$

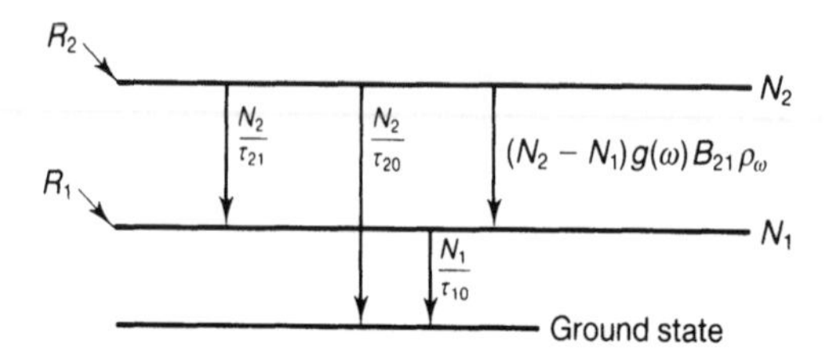

Figure 3.11 Population/depopulation rates of the two energy levels in a four-level system between which lasing occurs.

In terms of ΔN_0 Eq. (3.117) can then be written

$$N_2 - N_1 = \frac{\Delta N_0}{1 + \bar{g}(\omega) I/I_s} \tag{3.119}$$

where $\bar{g}(\omega)$ is given by $g(\omega)/g(\omega_0)$, that is, the lineshape function scaled so that

$$\bar{g}(\omega_0) = 1$$

and where

$$I_s = \frac{c}{g(\omega_0) B_{21}} (\tau_1 + \tau_2 - \tau_1 \tau_2/\tau_{21})^{-1} \tag{3.120}$$

I_s is known as the *saturation irradiance*. As we shall see in a moment, it represents the irradiance at which the gain starts to be significantly reduced from its small signal value. It is convenient to express I_s in terms of the stimulated emission cross-section. Using Eqs (3.111) and (3.103), B_{21} and $\sigma_{SE}(\omega)$ may be related by

$$g(\omega) B_{21} = \frac{c}{\hbar\omega} \sigma_{SE}(\omega)$$

Thus

$$g(\omega_0) B_{21} = \frac{c}{\hbar\omega} \sigma_{SE}(\omega_0)$$

and so

$$I_s = \frac{\hbar\omega}{\tau_1 + \tau_2 - \tau_1\tau_2/\tau_{21}} \frac{1}{\sigma_{SE}(\omega_0)} \tag{3.121}$$

As we shall see later, in many laser systems $\tau_2 \gg \tau_1$ and $\tau_{21} \approx \tau_2$, so that

$$I_s \cong \frac{\hbar\omega}{\tau_2} \frac{1}{\sigma_{SE}(\omega_0)} \tag{3.122}$$

We now know how the population inversion in a material depends on the local beam irradiance and consequently we can also obtain an expression for the variation of the gain coefficient with irradiance. The gain coefficient, $\gamma(\omega, I)$, may be defined by

$$\frac{\partial I}{\partial z} = \gamma(\omega, I) I \tag{3.123}$$

Equation (3.101) shows that $\partial I/\partial z$ is proportional to the population inversion multiplied by the irradiance, so that, using Eq. (3.119) we may write the gain coefficient in the form

$$\gamma(\omega, I) = \frac{\gamma_0(\omega)}{1 + (I/I_s)\bar{g}(\omega)} \tag{3.124}$$

where $\gamma_0(\omega)$ is the gain coefficient when $I \ll I_s$. Figure 3.12 shows the relative gain

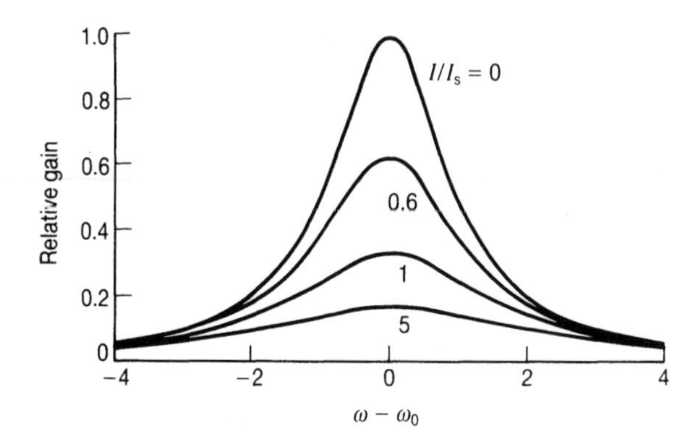

Figure 3.12 Gain saturation in a laser system subject to homogeneous broadening. As the irradiance increases, the gain over the entire curve is reduced.

as a function of frequency and beam irradiance as given by Eq. (3.124) for a Lorentzian profile and for several values of I/I_s. We note that as the beam irradiance is increased, the gain over the *whole* of the gain profile is reduced. This is a direct consequence of the fact that we are dealing with a homogeneous transition, so that every atom introduced into level 1 or level 2 can contribute to absorption and emission processes over the whole of the line profile. This is not so for an inhomogeneous line, and, as we shall see in the next section, the reduction in gain as beam irradiance is increased does not then follow Eq. (3.124). The overall reduction in gain with increasing beam irradiance is known as *gain saturation*.

Although the above equations were derived assuming a situation of population inversion and hence gain, they can readily be adapted to the case where there is no population inversion. They then predict that the absorption coefficient will behave in exactly the same way as does the gain coefficient in Eq. (3.124), that is, reduce as the beam irradiance increases thus giving rise to an increased transparency. The reduction in absorption coefficient with increasing beam irradiance in certain materials can be used to generate narrow laser pulses as described in Chapter 8 (see also Example 3.6).

Example 3.6 Saturation irradiance in dye solutions

A solution of crypotocyanine in methanol is often used in the *Q*-switching and mode-locking of ruby lasers. The absorption cross-section of the dye at the laser operating wavelength of 694.3 nm is $8.1 \times 10^{-20}\ \text{m}^2$, while the upper state lifetime is 2.2×10^{-11} s. The saturation irradiance can then be obtained from Eq. (3.122). We have that

$$I_s \cong \frac{\hbar\omega}{\tau_2}\frac{1}{\sigma(\omega_0)} = \frac{hc}{\tau_2\lambda}\frac{1}{\sigma_{21}(\omega_0)}$$

$$= \frac{6.6 \times 10^{-34}\ 3 \times 10^{8}}{2.2 \times 10^{-11}\ 6.94.3 \times 10^{-9}\ 8.1 \times 10^{-20}}\ \mathrm{W\ m^{-2}}$$

$$= 1.6 \times 10^{11}\ \mathrm{W\ m^{-2}}$$

Reverting to the situation where we have population inversion, we may now determine how the beam irradiance depends on distance. To do this we must solve the equation

$$\frac{\partial I}{\partial z} = \frac{\gamma_0(\omega)}{1 + (I/I_s)\bar{g}(\omega)} I \tag{3.125}$$

When $I \ll I_s$ we have the familiar solution

$$I(z) = I(0)\exp(\gamma_0(\omega)z)$$

so that while the beam irradiance is small it will grow exponentially. If, at the other extreme $I \gg I_s$, then

$$\frac{\partial I}{\partial z} = \gamma_0(\omega)I_s/\bar{g}(\omega)$$

or

$$I(z) = I(0) + \frac{\gamma_0(\omega)I_s}{\bar{g}(\omega)} z \tag{3.126}$$

Thus when the beam irradiance becomes large it will only increase linearly with distance.

While it is not possible to solve Eq. (3.125) explicitly for the intermediate situation (i.e. where the magnitudes of I and I_s are comparable), we may make some general observations. For simplicity, we assume $\omega = \omega_0$ and replace $\gamma_0(\omega)$ by γ_0. Rearranging the equation and then integrating gives

$$\int_{I=I_i}^{I_f}\left(\frac{1}{I} + \frac{1}{I_s}\right) \mathrm{d}I = \gamma_0 \int_{z=0}^{L} \mathrm{d}z$$

where I_i and I_f are the initial and final irradiances and L is the length of the gain medium. Thus

$$\log_e\left(\frac{I_f}{I_i}\right) + \frac{I_f - I_i}{I_s} = \gamma_0 L$$

We can write $\gamma_0 L$ as $\log_e G_0$ where G_0 is the small signal power gain of the medium, so that

$$\log_e\left(\frac{I_f}{I_i}\right) + \frac{I_f - I_i}{I_s} = \log_e G_0 \tag{3.127}$$

With a large signal the power gain (G) is simply I_f/I_i, so that Eq. (3.127) may be written

$$G = G_0 \exp\left(-(G-1)\frac{I_i}{I_s}\right) \tag{3.128}$$

To make further progress we must resort to numerical solutions. Figure 3.13 shows the variation in G as a function of I_i/I_s for a particular value of G_0. A point to note is that the gain starts to fall significantly below G_0 well before I_i becomes equal to I_s.

The power that is extracted (P_{ex}) from the gain medium is given by $I_f - I_i$. From Eq. (3.127)

$$I_f - I_i = I_s \log_e\left(\frac{G_0}{G}\right) \tag{3.129}$$

Maximum power will be extracted when G attains its minimum value, that is, unity, giving a maximum extracted power (P_{max}) of $I_s \log_e G_0$. Thus the ratio of extracted power to maximum possible extracted power can be written

$$\frac{P_{ex}}{P_{max}} = \frac{\log_e(G_0/G)}{\log_e G_0}$$

$$\frac{P_{ex}}{P_{max}} = 1 - \frac{\log_e G}{\log_e G_0} \tag{3.130}$$

We see that there is a (linear) trade-off between the logarithm of the gain (i.e. the gain in dB) and the efficiency of power extraction. For example, if we have a medium that exhibits a small signal power gain of, say, 20 dB (i.e. a linear gain of 100) then to extract half the stored power means having to accept an overall power gain of only 10 dB (or 10 in linear terms). This illustrates one of the main problems of the single-pass optical amplifier. High amounts of power can be extracted only if low overall gains can be tolerated.

3.3.3 Gain saturation in inhomogeneous media

For simplicity, we assume we are dealing with a gaseous medium where inhomogeneous broadening results from the velocity distribution of atoms (or molecules). When a monochromatic beam travels through such a medium it will interact differently with those atoms which have different velocity components along the beam direction. To begin with, therefore, we consider only those atoms in the upper and lower states which have velocities along the beam direction of between v and $v + dv$;

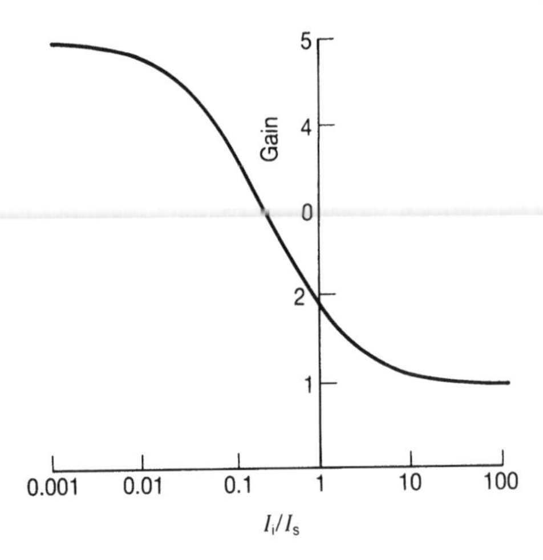

Figure 3.13 Variation in gain of a laser amplifier as a function of input irradiance to the amplifier. The small signal gain has been set at 5.

these are designated $N_2(v)$ and $N_1(v)$, respectively. As in the homogeneous case, we assume that the upper and lower levels are being populated at rates R_2 and R_1, respectively. As far as the groups $N_2(v)$ and $N_1(v)$ are concerned, the rates may be written $P(v)R_2$ and $P(v)R_1$, where $P(v)$ is the Maxwellian distribution of Eq. (3.80). We may now write the equivalent of Eqs (3.111) and (3.112) which applies to the groups $N_2(v)\,\mathrm{d}v$ and $N_1(v)\,\mathrm{d}v$. These are

$$\frac{\mathrm{d}N_2(v)}{\mathrm{d}t} = R_2P(v) - \frac{N_2(v)}{\tau_{20}} - \frac{N_2(v)}{\tau_{21}} - (N_2(v) - N_1(v))g(\omega, v)B_{21}\rho_\omega \tag{3.131}$$

and

$$\frac{\mathrm{d}N_1(v)}{\mathrm{d}t} = R_1P(v) - \frac{N_1(v)}{\tau_{10}} - \frac{N_2(v)}{\tau_{21}} + (N_2(v) - N_1(v))g(\omega, v)B_{21}\rho_\omega \tag{3.132}$$

The lineshape function $g(\omega, v)$ will now also depend on v. Thus as far as an atom travelling at v is concerned, the beam will appear to have an angular frequency of $\omega(1 - v/c)$, so that, assuming a Lorentzian profile for the transition, we may write

$$g_L(\omega, v) = \frac{1}{2\pi}\frac{\Delta\omega}{(\omega_0 - \omega(1 - v/c))^2 + (\Delta\omega/2)^2} \tag{3.133}$$

We now proceed in a very similar manner to that in the previous section. Once equilibrium has been established both $\mathrm{d}N_1(v)/\mathrm{d}t$ and $\mathrm{d}N_2(v)/\mathrm{d}t$ will be zero. The resulting equations enable $N_2(v)$ and $N_1(v)$ to be determined with the result that we can write

$$N_2(v) - N_1(v) = P(v)\frac{R_2\tau_2(1 - \tau_1/\tau_{21}) - R_1\tau_1}{1 + G(\omega, v)I(\tau_1 + \tau_2 - \tau_1\tau_2/\tau_{21})} \tag{3.134}$$

where $G(\omega, v) = g(\omega, v)\tau_{21}/c$. For small values of I

$$N_2(v) - N_1(v) = P(v)\ (R_2\tau_2(1 - \tau_1/\tau_{21}) - R_1\tau_1)$$

Since $P(v)$ is normalized, this expression is readily integrated over all velocity components to give the inversion density for low irradiance signals, so that

$$\int_{-\infty}^{\infty} (N_2(v) - N_1(v))\, dv = R_2\tau_2(1 - \tau_1/\tau_{21}) - R_1\tau_1$$

As before, it is convenient to put $R_2\tau_2(1 - \tau_1/\tau_{21}) - R_1\tau_1 = \Delta N_0$. We again define a function $\bar{g}_L(\omega, v)$, which now is the Lorentzian scaled to its maximum value, so that

$$\bar{g}_L(\omega, v) = \frac{(\Delta\omega/2)^2}{(\omega_0 - \omega(1 - v/c))^2 + (\Delta\omega/2)^2} \tag{3.135}$$

Equation (3.134) can then be written in terms of ΔN_0 and $\bar{g}_L(\omega, v)$, thus

$$N_2(v) - N_1(v) = P(v)\frac{\Delta N_0}{1 + (I_\omega/I_s)\bar{g}_L(\omega, v)} \tag{3.136}$$

A plot of the resulting population inversion for groups of atoms with different values of v is given in Figure 3.14. So long as $I_\omega \ll I_s$ then the population inversion has the same form as $P(v)$, that is, Gaussian. However, for larger values of I_ω a dip appears in the curve, which has its maximum depth when $v = c(1 - \omega_{12}/\omega)$, that is, for those atoms which 'see' the resonant frequency ω_{12}. The beam is said to 'burn a hole' in the velocity distribution. As I_ω increases, the hole widens and deepens.

Again, as in the case of homogeneous broadening, we may introduce an irradiance-dependent gain coefficient $\gamma(\omega, v, I)$, resulting from population inversion

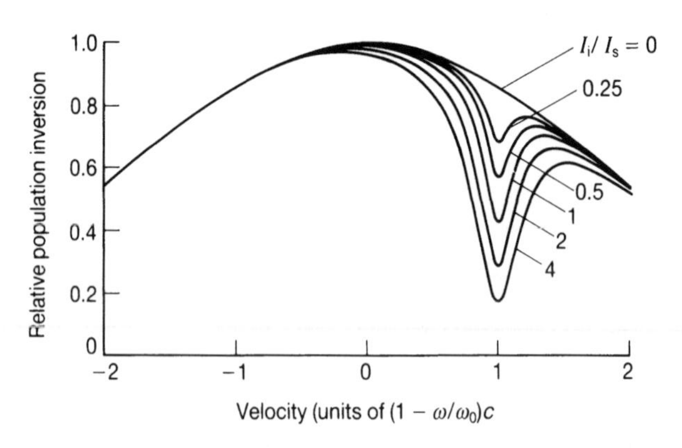

Figure 3.14 Effect of gain saturation on the curve of population inversion versus atomic velocity in an inhomogeneously broadened laser transition. A 'hole' is burnt in the curve which widens and deepens as the irradiance increases.

among the atoms travelling at speed v. Using the result of Eq. (3.106) and substituting from Eq. (3.136) for the population inversion, we then obtain

$$\gamma(\omega, v) = g_L(\omega, v)P(v)\,\frac{\Delta N_0}{1 + (I_\omega/I_s)\bar{g}_L(\omega, v)}\,B_{21}\,\frac{\hbar\omega}{c}$$

or

$$\gamma(\omega, v) = B_{21}\,\frac{\hbar\omega}{c}\,\Delta N_0 g(\omega_0)P(v)\,\frac{\Delta N_0}{1/\bar{g}_L(\omega, v) + I_\omega/I_s} \tag{3.137}$$

The contribution to $\gamma(\omega)$ from all the groups of atoms with different velocity components is given by integrating Eq. (1.137) over all velocities, thus

$$\gamma(\omega) = B_{21}\,\frac{\hbar\omega}{c}\,\Delta N_0 g(\omega_0)\,\Delta N_0 \int_{-\infty}^{\infty} \frac{P(v)}{1/\bar{g}_L(\omega, v) + I_\omega/I_s}\,\mathrm{d}v$$

The integral has no direct analytical solution. However, if we assume that the homogeneous broadening is much smaller than the inhomogeneous broadening, the integrand will only be appreciable over a relatively small range of velocities, and we may assume that the factor $P(v)$ can be replaced by $P(v_\omega)$, where $v_\omega = c(1 - \omega_0/\omega)$, and taken out of the integral. Under these circumstances,

$$\gamma(\omega, I) = B_{21}\,\frac{\hbar\omega}{c}\,\Delta N_0 g(\omega_0)\,\Delta N_0 P(v_\omega) \int_{-\infty}^{\infty} \frac{1}{1/\bar{g}_L(\omega, v) + I_\omega/I_s}\,\mathrm{d}v$$

$$\gamma(\omega, I) = B_{21}\,\frac{\hbar\omega}{c}\,\Delta N_0\; g(\omega_0)\,\Delta N_0 P(v_\omega)$$

$$\times \int_{-8}^{\infty} \frac{(\Delta\omega/2)^2}{(\omega_0 - \omega(1 - v/c))^2 + (\Delta\omega/2)^2(1 + I_\omega/I_s)}\,\mathrm{d}v \tag{3.138}$$

By putting $\omega_0 - \omega(1 - v/c) = x$ and $(\Delta\omega)^2(1 + I_\omega/I_s) = a^2$, the integral reduces to

$$\frac{\Delta\omega^2 c}{4\omega} \int_{-\infty}^{\infty} \frac{1}{x^2 + a^2}\,\mathrm{d}x = \frac{\Delta\omega^2 c}{4\omega}\,\frac{\pi}{a}$$

Thus

$$\gamma(\omega, I) = B_{21}\,\frac{\Delta\omega\hbar\pi}{2}\,\Delta N_0 g(\omega_0)\,\Delta N_0 P(v_\omega)\,\frac{1}{(1 + I_\omega/I_s)^{1/2}}$$

Therefore the small signal gain coefficient may be written

$$\gamma(\omega, I) = \frac{\gamma_0(\omega)}{(1 + I_\omega/I_s)^{1/2}} \tag{3.139}$$

We see that the gain coefficient for an inhomogeneously broadened line profile saturates more slowly than for a homogeneously broadened one (compare Eq. (3.124)). This behaviour arises because the hole that is burned in the velocity distribution for inhomogeneous broadening widens as well as deepens with increasing beam irradiance. Thus the number of atoms that can interact with the beam (and hence supply gain) increases as the irradiance increases. This is in contrast to the homogeneous situation, where the number of interacting atoms is always the same. It is this increasing supply of interacting atoms in the inhomogeneous case which causes the gain coefficient to saturate at a slower rate.

The widening of the hole burned in the velocity distribution with increasing beam irradiance can be seen directly from an inspection of Eq. (3.138). We see that the denominator within the integral has a Lorentzian frequency dependence with an intensity-dependent linewidth. Thus the hole linewidth, $\Delta\omega_{\mathrm{H}}$, can be written in terms of the homogeneously broadened linewidth, $\Delta\omega_{\mathrm{h}}$

$$\Delta\omega_{\mathrm{H}} = \Delta\omega_{\mathrm{h}}\left[1 + \frac{I_{\omega}}{I_{\mathrm{s}}}\right]^{1/2} \tag{3.140}$$

It should be pointed out that in most gas lasers both homogeneous and inhomogeneous broadening is present and neither Eq. (3.124) nor Eq. (3.140) accurately represents the gain saturation process.

3.3.4 Gain in semiconductor media

The discussion of optical gain in the previous sections has been in terms of transitions between two energy states (possibly degenerate) and, as such, typifies transitions in isolated atomic systems such as gases and impurity atoms in solids. In semiconductors transitions are between energy bands where the individual energy levels are so closely packed as to form a quasi-continuum. This situation requires a different approach.

We assume that we are dealing with a direct bandgap semiconductor, so that when an electron makes a transition from conduction to valence band its k value remains unchanged. At a particular value of k, the upper state energy E_2 can be written

$$E_2 = E_{\mathrm{c}} + \frac{\hbar^2 k^2}{2m_{\mathrm{e}}^*} \tag{3.141}$$

where m_{e}^* is the electron effective mass, while for the lower state

$$E_1 = Ev - \frac{\hbar^2 k^2}{2m_{\mathrm{h}}^*} \tag{3.142}$$

where m_{h}^* is the hole effective mass. Thus the photon energy ($\hbar\omega$) is related to k by

$$\hbar\omega = E_2 - E_1$$

$$= E_{\mathrm{g}} + \frac{\hbar^2 k^2}{2}\left(\frac{1}{m_{\mathrm{e}}^*} + \frac{1}{m_{\mathrm{h}}^*}\right) \tag{3.143}$$

It is convenient to introduce the electron- and hole-reduced mass m_r^* where

$$\frac{1}{m_r^*} = \frac{1}{m_e^*} + \frac{1}{m_h^*} \qquad (3.144)$$

so that

$$\hbar\omega = E_g + \frac{\hbar^2 k^2}{2m_r^*} \qquad (3.145)$$

By differentiating Eq. (3.145) we obtain

$$k\mathrm{d}k = \frac{m_r^*}{\hbar}\mathrm{d}\omega \qquad (3.146)$$

Now the number of electron states per unit volume in either the conduction or valence bands which have k values between k and $k + \mathrm{d}k$ is given by $(2k^2/\pi^2)\,\mathrm{d}k$ (see Eq. (1.54)). Thus the density of electron states $(\rho_{es}(\omega)\,\mathrm{d}\omega)$ in the conduction (and valence) band which can be 'connected' by a photon with frequency between ω and $\omega + \mathrm{d}\omega$ is given by

$$\rho_{es}(\omega)\,\mathrm{d}\omega = \frac{2km_r^*}{\pi^2\hbar}\mathrm{d}\omega \qquad (3.147)$$

Substituting for k from Eq. (3.145) into Eq. (3.147) we then have

$$\rho_{es}(\omega) = \frac{(2m_r^*)^{3/2}}{\hbar^2\pi^2}(\hbar\omega - E_g)^{1/2} \qquad (3.148)$$

Now the probability that a state at E_2 in the conduction band is occupied by an electron is given by the function $F_c^e(E_2)$ (Eq. (3.43)), while in the valence band the corresponding probability for a state at E_1 is $F_v^e(E_1)$, where $F_v^e(E_1) = 1 - F_v^h(E_1)$. Using Eq. (3.47) for $F_v^h(E_1)$, we have

$$F_c^e(E) = \frac{1}{1 + \exp((E - E_{Fc})/kT} \qquad (3.149)$$

$$F_v^e = \frac{1}{1 + \exp((E - E_{Fv})/kT} \qquad (3.150)$$

where E_{Fc} and E_{Fv} are the quasi-Fermi levels for the conduction and valence bands.

As far as emission is concerned, the transition rate at frequency ω will be proportional to both the number of electrons available in the conduction band and the number of empty states in the valence band. That is

$$R(2 \Rightarrow 1, \omega) \propto \rho_{es}(\omega)F_c(E_2)[(1 - F_v(E_1)]$$

Similarly, for the absorption rate

$$R(1 \Rightarrow 2, \omega) \propto \rho_{es}(\omega)F_v(E_1)[(1 - F_c(E_2)]$$

The net emission rate will then be proportional to the difference between these two, namely, to

$$\rho_{es}(\omega)(F_c(E_2) - \rho_s(\omega)F_v(E_1))$$

We may thus think of the quantity $\rho_{es}(\omega)(F_c(E_2) - F_v(E_1))$ as being the equivalent in semiconductors of the population inversion, $N_2 - (g_2/g_1)N_1$, in atomic systems.

The natural linewidth of each individual transition between conduction and valence bands is of the order of 10^{12} Hz, and this is usually negligible compared to the frequency range over which we have population inversion. Consequently when determining the gain at a particular frequency ω we may assume that the lineshape function, $g(\omega_0 - \omega)$, while still being normalized to unity, has negligible width. Using Eq. (3.106) for the gain at frequency ω and replacing $g(\omega_0 - \omega)$ by unity, we have

$$\begin{aligned}\gamma(\omega) &= \rho_{es}(\omega)[F_c(E_2) - F_v(E_1)]\, A_{21}\, \frac{\pi^2 c^2}{\omega^2 n^2} \\ &= \frac{A_{21}c^2}{2\omega^2 n^2}\left(\frac{2m^*}{\hbar}\right)^{3/2}\left(\omega - \frac{E_g}{\hbar}\right)^{1/2}[F_c^e(E_2) - F_v^e(E_1)]\end{aligned} \tag{3.151}$$

The condition for $\gamma(\omega)$ to be positive is then

$$F_c^e(E_2) > F_v^e(E_1) \tag{3.152}$$

that is,

$$\frac{1}{1 + \exp((E_2 - E_{Fc})/kT} > \frac{1}{1 + \exp((E_1 - E_{Fv})/kT)}$$

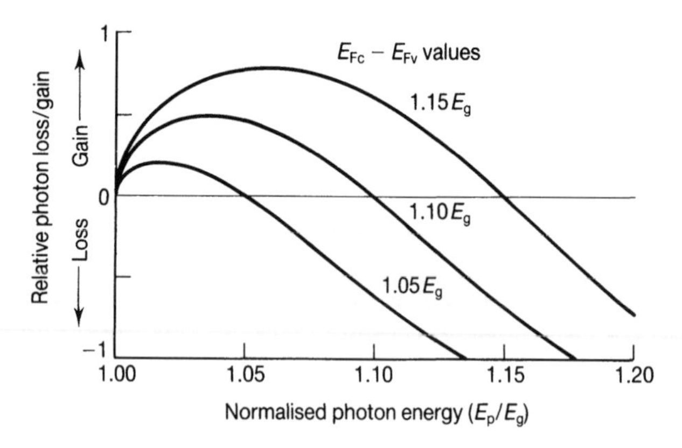

Figure 3.15 Curves of relative gain as a function of photon energy in a semiconductor. The three curves illustrated correspond to differing values of the quantity $E_{Fc} - E_{Fv}$. For simplicity, it has been assumed that the quasi-Fermi levels are symmetrically disposed about the centre of the energy gap.

or

$$E_2 - E_1 < E_{Fc} - E_{Fv}$$

or

$$\hbar\omega < E_{Fc} - E_{Fv} \tag{3.153}$$

Thus only those photons whose energy is less than the quasi-Fermi level separation will undergo gain. Since we must also have $\hbar\omega > E_g$ it is evident that at least one of the quasi-Fermi levels must lie within its respective band. Figure 3.15 illustrates the dependence of the gain coefficient on photon energy as predicted by Eq. (3.151).

Problems

3.1 A constant perturbation (i.e. one that is independent of time) acts on a simple two-level atomic system, where the atom is initially in the ground state. Show that, provided the probability of the atoms being excited to the upper state remains small, this probability can be written as

$$P_{12} = \frac{4\,|H_{21}^{0\prime}|^2}{\hbar^2\omega_{12}^2}\sin^2\left(\frac{\omega_{12}t}{2}\right)$$

3.2 For an electric dipole interaction between an external field and an electron in an atom, show that the matrix elements are such that

$$H_{22}' = H_{11}' = 0$$

and

$$H_{21}' = (H_{12}')^*$$

Hence show that in the presence of such an interaction the expansion coefficients ($a_1(t)$ and $a_2(t)$) for a simple two-level atom must obey the equations

$$a_1(t)\;H\exp(\mathrm{j}\omega_{12}) = \mathrm{j}\,\frac{\partial a_2(t)}{\partial t}$$

$$a_2(t)H^*\exp(-\mathrm{j}\omega_{12}) = \mathrm{j}\,\frac{\partial a_1(t)}{\partial t}$$

where $H = H_{21}'/\hbar$.
Show that if H is constant with time, then $a_2(=a_2(t))$ must satisfy

$$\frac{\partial^2 a_2}{\partial t^2} - \mathrm{j}\omega\,\frac{\partial a_2}{\partial t} + \frac{a_2}{\hbar^2}|H|^2 = 0$$

Prove that if $a_1(0) = 1$, then

$$|a_2(t)|^2 = \frac{4\,|H|^2}{\omega_{12}^2 + 4\,|H|^2}\sin^2\left[\frac{t}{2}\,(\omega_{12}^2 + 4\,|H|^2)^{1/2}\right]$$

3.3 If θ represents the angle between two vectors in three-dimensional space, show that the average value of $\cos^2(\theta)$, when θ takes all possible values, is 1/3.

3.4 Show that the maximum in the electron density in the conduction band of an ideal semiconductor occurs at an energy of $kT/2$ above the bottom of the conduction band while the average electron energy is $3kT/2$. Show also that the halfwidth of the distribution is approximately 1.8 kT.

3.5 Using the hydrogenic wavefunctions of Appendix 2, show that the θ part of the integral of the matrix element for electric dipole transitions gives rise to the selection rule

$$\Delta l = \pm 1$$

3.6 Carry out the integration in Eq. (3.36) to determine the number of electrons in the conduction band of a semiconductor to show that

$$n = \frac{(8\pi m_e^* kT)^{3/2}}{h^3} \exp\left(-\frac{E_c - E_F}{kT}\right)$$

Use the expressions for $D(E)$ and $F(E)$ as given by Eqs (1.59) and (3.34), and assume that the top of the energy band (E_T) is effectively at infinity. *Hint*: To evaluate the integral you may find it useful to make the substitution $x = (E - E_c)/kT$, and to use the result that

$$\int_0^\infty x^{1/2} \exp(-x)\, dx = \sqrt{\pi}/2$$

3.7 The lowest energy state of the Nd^{3+} ion is $^4I_{9/2}$. When the ion is incorporated into the YAG lattice the crystal field splits this term up into five levels, and the energies of the four states above the ground state are: 0.0166 eV, 0.0244 eV, 0.0386 eV and 0.1051 eV. Calculate the fraction of the total number of ions that is actually in the ground state at room temperature (ignore atoms in states other than $^4I_{9/2}$).

3.8 Estimate the time it takes for the collision process between atoms in a gas to take place (assume that atoms significantly affect one another if their nuclei come within twice an atomic diameter of each other).

3.9 Show that the probability, $P(\tau)$, that an atom undergoes a transition during the time interval from τ to $\tau + d\tau$ is given by

$$P(\tau) = \frac{\exp(-\tau/\tau_0)}{\tau_0} d\tau$$

where τ_0 is the lifetime of the transition.

3.10 Estimate the contributions to the linewidth of the 10.6 μm transition in the CO_2 molecule from Doppler broadening and pressure broadening at a temperature of 400 K and a gas pressure of 1 atmosphere. Is the line essentially homogeneous or inhomogeneous? Estimate the pressure at which the two contributions will be equal. Take the hard sphere radius for the CO_2 molecule to be 2×10^{-10} m.

3.11 Verify that for both the Lorentzian and Gaussian line profiles the values for the coherence time, τ_c, and linewidth, $\Delta\omega$, are in agreement with the general relationship (Eq. (2.130))

$$\tau_c \approx 1/\Delta f$$

3.12 Verify that the Lorentzian and Gaussian functions given in Eqs (3.86) and (3.88) are normalized to unit area.

3.13 Show that if two line-broadening mechanisms are present that would separately give rise to Lorentzian profiles with linewidths $\Delta\omega_1$ and $\Delta\omega_2$, then the resultant is a Lorentzian with a linewidth of $\Delta\omega_1 + \Delta\omega_2$. Show that if two line-broadening mechanisms are present that give rise to Gaussian profiles with linewidths $\Delta\omega_1$ and $\Delta\omega_2$ then the resultant will be a Gaussian but now with a linewidth of $(\Delta\omega_1 + \Delta\omega_2)^{1/2}$.

3.14 Show that when both homogeneous and inhomogeneous broadening mechanisms are present, then the resulting line profile (i.e. the Voigt profile) can be written in the form

$$g_V(x) = \frac{2}{\pi^{3/2}\Delta\omega_L}\int_{-\infty}^{\infty} \exp(-y^2)\,\frac{r^2}{r^2 + (x-y)^2}\,dy$$

where

$$x = \frac{(\log_e 2)^{1/2}}{\Delta\omega_G}(\omega - \omega_0)$$

and

$$r = \frac{\Delta\omega_L}{\Delta\omega_G}(\log_2 2)^{1/2}$$

3.15 Using the data of Figure 1.11, estimate the value for the A_{21} coefficient for the $^4A_2$ to $^4F_2$ transition in ruby. The multiplicities of the $^4A_2$ and $^4F_2$ states are 4 and 12, respectively. Take the refractive index of ruby at this wavelength to be 1.8.

3.16 The R_1 transition in ruby (i.e. from the lower of the crystal field split ^2E levels to the ground state) has a Lorentzian linewidth of 330 GHz halfwidth and a peak cross-section of 2.5×10^{-24} m^2. The observed lifetime of the transition is 3 ms. Calculate the fluorescent yield (take the refractive index to be 1.76).

3.17 Dye solutions are used in Q-switching and mode locking (Chapter 8). Estimate the saturation irradiance for such a solution which has a peak absorption cross-section of 8×10^{-20} m^2 at 0.7 μm and an upper state lifetime of 2×10^{-11} s. Repeat the calculation for the Nd:YAG laser transition where the cross-section at 1.06 μm is 6.5×10^{-23} m^2 and the upper state lifetime 2.3×10^{-4} s.

Notes

1. The vector and scalar potentials (**A** and ϕ) are alternative ways of writing the electric and magnetic fields which are given by

$$\mathscr{E} = -\nabla(\phi) - \frac{\partial \mathbf{A}}{\partial t}$$

and

$$\mathscr{B} = \nabla \times \mathbf{A}$$

However, these equations do not uniquely specify **A**. In the *Coulomb gauge* **A** is subject to the additional constraint $\nabla.\mathbf{A} = 0$.

2. In the interests of notational simplicity from now on we designate the central frequency of a transition by ω_0, rather than ω_{12}.
3. Essentially the reduced mass, M_r, of the two colliding molecules should have been used in the equation for v_{av}, instead of the mass of one of them (M), for equal masses $M_r = M/2$. This then leads to a factor $\sqrt{2}$ in the expression for v_{av}.
4. For consistency throughout this section angular frequency ω is used as the argument of functions such as $g(\omega)$, $\alpha(\omega)$, $\sigma_{12}(\omega)$. It would, of course, be equally valid to use linear frequency ν.
5. The radiation is here considered to have a 'single' frequency rather than a range of frequencies, hence the use of ρ_ω rather than $\rho(\omega)$.
6. To be consistent with previous usage we should use I_ω here rather than I. However, this would make the notation unduly cumbersome.

References

[1] S. Gasiorowicz, *Quantum Physics*, John Wiley, New York, 1974, Chapter 13.

[2] H. B. Dwight, *Tables of Integrals*, Macmillan, New York, 1961, p. 221.

[3] R. Loudon, *The Quantum Theory of Light*, 2nd edn, Clarendon Press, Oxford, 1983, Section 2.9.

[4] M. R. Spiegel, *Theory and Problems of Advanced Mathematics for Engineers and Scientists* (Schaum's outline series), McGraw-Hill, New York, 1980.

[5] W. H. Giedt, *Thermophysics*, Van Nostrand Reinhold, New York, 1971, Chapter 8.

[6] B. D. Fried and B. D. Conte, *The Plasma Dispersion Function*, Academic Press, New York, 1961. See also W. J. Surtrees, 'Calculation of combined Doppler and collision broadening', *J. Opt. Soc. Am.*, **55**, 893–4, 1965.

[7] E. T. Garry and D. A. Leonard, 'Measurement of 10.6 μm CO_2 laser probability and optical broadening cross sections', *Appl. Phys. Lett.*, **8**, 227, 1966.

[8] D. F. Nelson and M. D. Sturge, 'Relation between absorption and emission in the region of the *R* lines of ruby' *Phys. Rev.*, **137**, A1117–30, 1965.

[9] P. W. Miloni and J. H. Eberly, *Lasers*, John Wiley, New York, 1988, Sections 3.5 and 3.6.

4

Optics of Gaussian beams

4.1 Introduction

Light beams diverge as they propagate and, consequently, some focusing element such as a lens or concave mirror is required to compensate for this and to contain the light within the gain medium. In many lasers the light suffers no wall reflections and is entirely contained by the end mirrors. We will refer to this mirror structure (i.e. without the gain medium) as an open resonator or cavity. Some lasers, on the other hand, confine the light using wall reflections together with end mirrors. They might use, for example, the waveguiding action of the gain medium or a waveguide to contain the light. In this chapter we confine our discussion to the open-resonator laser cavity and this will give rise to a discussion of the generation and propagation of Gaussian beams.

4.2 Ray tracing analysis using matrices

In geometrical optics the propagation of a wave is shown by a ray which is simply an arrow normal to the wavefront showing the direction of energy flow at a given point. The ray itself gives no information on the wave amplitude, its state of polarization or the distribution of energy in the beam. The phase change along the path of the ray can be evaluated from the number of oscillations of the electromagnetic field along the optical path, and this is used to determine the conditions for fringe formation in interferometers. Thus any analysis using rays will only be a partial solution but is nevertheless a very useful one. The position and slope of a ray is the necessary information required to calculate its progress through an optical system. Although ray tracing is done by drawing, it will be shown that the procedure can be achieved analytically by matrix multiplication and the optical system characterized by a 2×2 matrix. In the analysis the rays will be considered to remain close to the optical axis and to have small divergent angles; this is referred to as the *paraxial approximation*.

First, we consider a length d of free space. Figure 4.1 shows a ray inclined at a small angle θ to the z direction so that its slope x' and its displacement x are small. The rays are therefore almost parallel to the z axis and their slope is given by

$$\frac{dx}{dz} = x'(z) = \tan\theta \cong \theta$$

In this case the ray is in the xz plane. However, if the ray is a skew ray then Figure 4.1 would represent its projection onto the xz plane with a similar diagram for the projection in the yz plane. The position and slope of the ray at $z = d$ can be written

$$x_2 = 1x_1 + dx_1' \tag{4.1}$$

$$x_2' = 0x_1 + 1x_1' \tag{4.2}$$

Since the rays are straight lines, the displacement of the ray in Eq. (4.1) has changed due to its slope and in Eq. (4.2) the slope is unchanged. These two equations may be written in matrix notation as

$$\begin{bmatrix} x_2 \\ x_2' \end{bmatrix} = \begin{bmatrix} 1 & d \\ 0 & 1 \end{bmatrix} \begin{bmatrix} x_1 \\ x_1' \end{bmatrix} \tag{4.3}$$

Extending the ideas to a convex lens, Figure 4.2 illustrates the ray diagram used

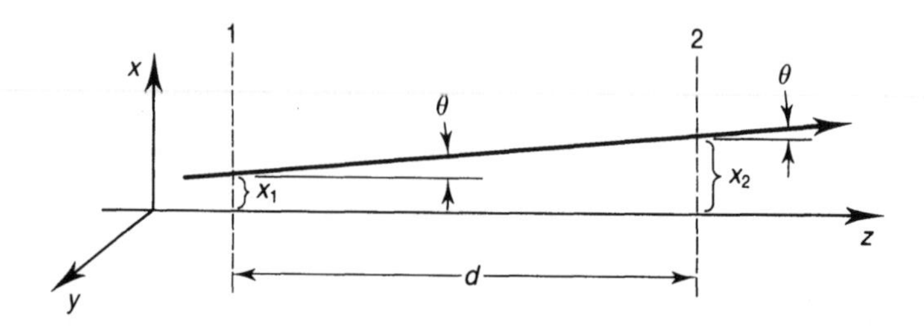

Figure 4.1 Ray in a length d of free space for evaluation of the matrix.

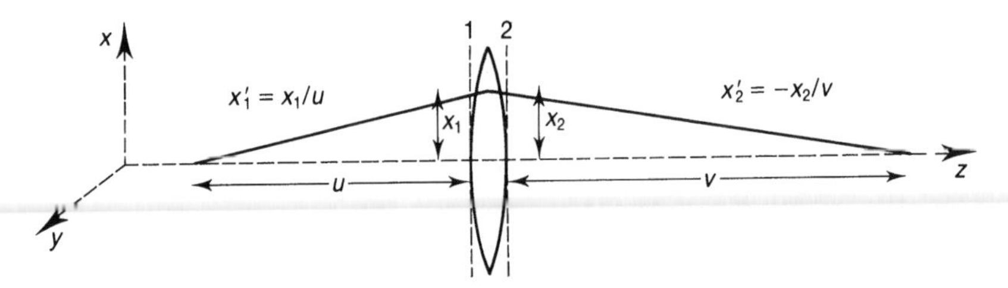

Figure 4.2 Ray transformation by a thin lens.

in geometrical optics to show the image formation with a convex lens. Since the lens is considered thin, the positions of the ray are equal on either side of the lens, so that

$$x_2 = 1.x_1 + 0x_1' \tag{4.4}$$

Using the lens equation

$$\frac{1}{u} + \frac{1}{v} = \frac{1}{f}$$

we can obtain the change in gradient of the ray. Multiplying across by x_1 gives

$$\frac{x_1}{u} + \frac{x_1}{v} = \frac{x_1}{f}$$

Alternatively, since $x_1 = x_2$ for a thin lens, this becomes

$$-\frac{x_2}{v} = \frac{x_1}{u} - \frac{x_1}{f}$$

The slopes of the rays on either side of the lens are given by

$$x_2' = -x_2/v$$

and

$$x_1' = +x_1/u$$

Substitution for the slopes gives

$$x_2' = x_1' + \left(-\frac{1}{f}\right)x_1 \tag{4.5}$$

Equations (4.4) and (4.5), respectively, give the output position and slope of the ray in terms of the input position and slope. In matrix form they are

$$\begin{bmatrix} x_2 \\ x_2' \end{bmatrix} = \begin{bmatrix} 1 & 0 \\ -1/f & 1 \end{bmatrix} \begin{bmatrix} x_1 \\ x_1' \end{bmatrix} \tag{4.6}$$

Consider now the combination of a length of free space followed by a lens. With these two optical elements cascaded together, the output of the length of free space becomes the input to the lens. The output of the complete system thus becomes the product of the matrices of the two optical elements:

$$\begin{bmatrix} x_2 \\ x_2' \end{bmatrix} = \underbrace{\begin{bmatrix} 1 & 0 \\ -1/f & 1 \end{bmatrix}}_{\text{Lens}} \underbrace{\begin{bmatrix} 1 & d \\ 0 & 1 \end{bmatrix} \begin{bmatrix} x_1 \\ x_1' \end{bmatrix}}_{\text{Output from the length of free space}}$$

Using the laws for matrix multiplication:

$$\begin{bmatrix} x_2 \\ x_2' \end{bmatrix} = \begin{bmatrix} 1 & d \\ -1/f & 1 - d/f \end{bmatrix} \begin{bmatrix} x_1 \\ x_1' \end{bmatrix} \tag{4.7}$$

The progress of any paraxial ray through the optical system can now be reduced to the product of the ray matrices of the constituent components of the system with two important points to be noted:

1. The order of the matrices is in the *reverse order* to the light path through the system.
2. Inspection of the matrices for the examples chosen shows that if the refractive index is the same at the output as at the input then the determinant of the matrix is unity.

In general, the matrix equation is written in the form

$$\begin{bmatrix} x_{\text{out}} \\ x_{\text{out}}' \end{bmatrix} = \begin{bmatrix} A & B \\ C & D \end{bmatrix} \begin{bmatrix} x_{\text{in}} \\ x_{\text{in}}' \end{bmatrix} \tag{4.8}$$

where the ray matrix representing the optical system is called the system *ABCD* matrix. Another matrix can be determined for the paraxial rays in the plane determined by the y and z axes. It is possible that this matrix will have different parameters, meaning that the optical system is astigmatic. This matrix formalism finds wide application in ray tracing in very complex optical systems (for example, camera lenses) and computer ray tracing software packages based on this matrix formalism are now available for the optical designer.

4.3 Resonator stability condition

We may use the ray matrix to trace the rays in an open resonator used for a laser cavity. Figures 4.3 and 4.4 show two possible cases for the progress of an arbitrary input paraxial ray as a result of multiple reflections in a cavity with two concave mirrors. In Figure 4.3 it is seen that the cavity may simply capture the ray, so that after a number of reflections the ray finds itself back at its starting position and so continues to circulate around the cavity. In this case the cavity is said to be *stable* and the ray is said to be *re-entrant*. On the other hand, as shown in Figure 4.4,

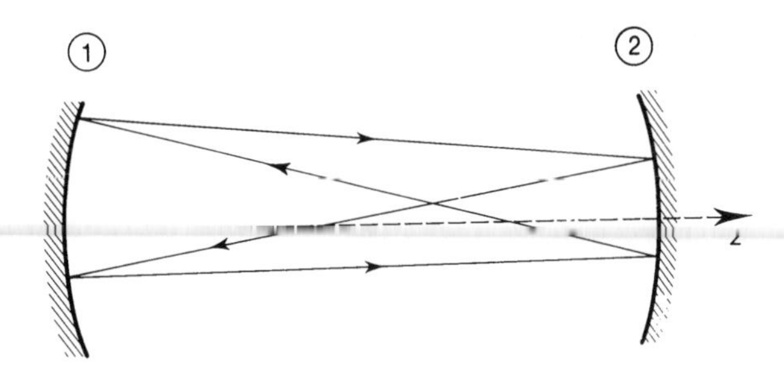

Figure 4.3 Ray diagram for a typical stable cavity showing the ray to be re-entrant after four transits of the cavity.

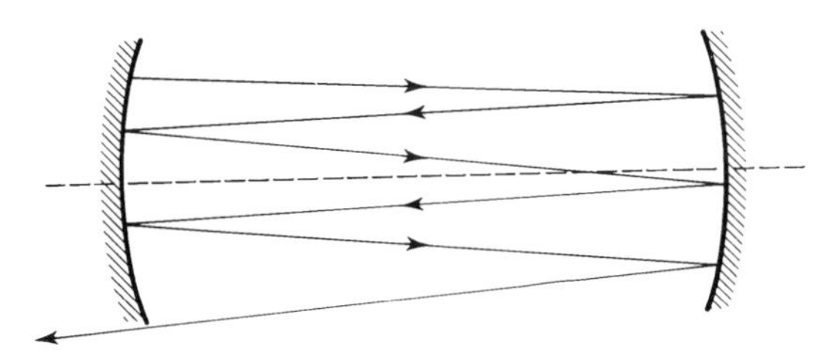

Figure 4.4 Ray diagram showing an unstable cavity.

a ray may find itself leaving the cavity after a few roundtrips, and this gives rise to an *unstable* cavity. Low-gain lasers such as gas lasers require stable cavities while unstable cavities can be used for lasers with high gain where a few roundtrips are sufficient to saturate the gain. It is obviously important to find the conditions for cavity stability and we now determine this using the *ABCD* matrices.

The concave mirror has the same focusing action as a convex lens save that it redirects the light. Thus each mirror can be simulated by a convex lens and the transit of light in the cavity treated as a sequence of lenses called a *lens waveguide*. This lens waveguide stretches to infinity and is similar to a graded index optical fiber. Such a lens waveguide is shown in Figure 4.5 in which one unit cell represents one

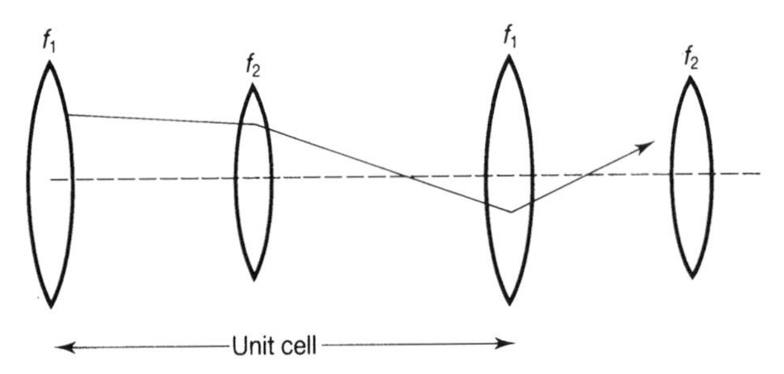

Figure 4.5 Equivalent lens waveguide of optical resonator showing the unit cell where the ray starts at mirror 1.

roundtrip of the cavity. The ray shown in Figure 4.5 starts at one mirror and traverses a length of free space plus a lens of focal length $f_2 = b_2/2$. It then goes another length of free space plus a lens of focal length $f_1 = b_1/2$, where b_1 and b_2 are the radii of curvature of the concave mirrors. The *ABCD* matrix for this unit cell is obtained from the product of two matrices (like those in Eq. (4.7)) in the reverse order of light transmission:

$$\begin{bmatrix} A & B \\ C & D \end{bmatrix} = \begin{bmatrix} 1 & d \\ -1/f_1 & (1-d/f_1) \end{bmatrix} \begin{bmatrix} 1 & d \\ -1/f_2 & (1-d/f_2) \end{bmatrix}$$

$$= \begin{bmatrix} (1-d/f_2) & d + d(1-d/f_2) \\ \left[-1/f_1 - \frac{1}{f_2}(1-d/f_1)\right] & (1-d/f_1)(1-d/f_2) - d/f_1 \end{bmatrix} \qquad (4.9)$$

Figure 4.6 shows how the coordinates of the ray can change at each unit cell on successive transits of the cavity. Using the matrix (4.9) a difference equation can be derived giving the coordinates at successive cells. Thus for cells s and $s+1$ the matrix equation 4.8 gives

$$x_{s+1} = Ax_s + Bx_s' \quad \text{or} \quad x_s' = \frac{1}{B}(x_{s+1} - Ax_s)$$

Similarly, for cell $s+1$

$$x_{s+1}' = \frac{1}{B}(x_{s+2} - Ax_{s+1}) = Cx_s + Dx_s'$$

Substituting for x_s'

$$\frac{1}{B}(x_s - Ax_{s+1}) = Cx_s + \frac{D}{B}(x_{s+1} - Ax_s)$$

Multiplying by B and rearranging,

$$x_{s+2} - Ax_{s+1} = BCx_s + Dx_{s+1} - ADx_s$$

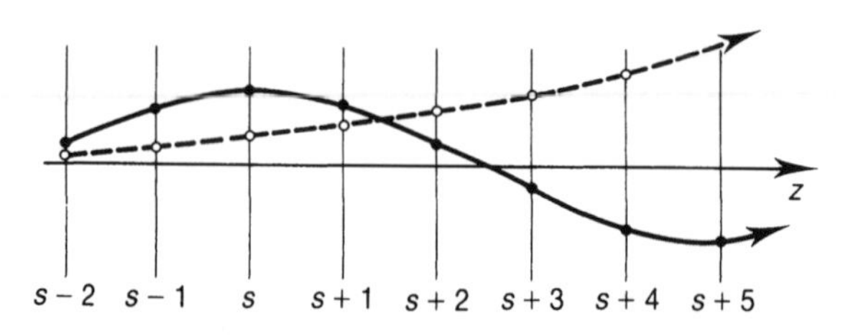

Figure 4.6 Diagram showing the position of rays at each unit cell. There are two possibilities: stable and unstable cavities.

Since the determinant of the matrix is unity (i.e. $AD - BC = 1$) then

$$x_{s+2} - (A + D)x_{s+1} + x_s = 0 \tag{4.10}$$

This difference equation relates the coordinates of the rays on their transit from one unit cell to the next. It gives no information on the rays inside the unit cells as the lines joining the points in Figure 4.6 are *not* rays.

There are two possible ray paths in the cavity:

1. A stable cavity in which the ray wanders to extreme positions and returns to the starting position. This path in Figure 4.6 is represented by the solid points and can be considered as sinusoidal.
2. The position of the ray gradually increases and never returns to its starting position. It escapes from the cavity when its radial location is larger than the radius of the mirror. The ray coordinate, x, can be considered as increasing exponentially.

A solution to Eq. (4.10) should be able to accommodate both these possibilities and in so doing define criteria for a stable cavity. We therefore assume a solution of the form

$$x_s = x_0 e^{j\theta s}$$

where x_0 = the initial coordinate of the ray (i.e. for $s = 0$)
θ = a geometric constant of the cavity which is real for a stable cavity.

Substituting the trial solution into Eq. (4.10) results in

$$x_0(e^{j\theta})^{s+2} + x_0(A + D)\ (e^{j\theta})^{s+1} + x_0(e^{j\theta})^s = 0$$

$$x_0 e^{j\theta s}(e^{j2\theta} - (A + D)e^{j\theta} + 1) = 0$$

For the most general case x_0 is not zero and also we note $e^{j\theta} \neq 0$. Hence we can put the bracketed part of the equation equal to zero:

$$(e^{j\theta})^2 - (A + D)e^{j\theta} + 1 = 0$$

This is seen to be a quadratic in $e^{j\theta}$ and has solutions

$$e^{j\theta} = \frac{(A + D) \pm \sqrt{(A + D)^2 - 4}}{2} = \frac{A + D}{2} \pm j\left(1 - \left(\frac{A + D}{2}\right)^2\right)^{1/2}$$

Comparing with the Euler relation,

$$e^{j\theta} = \cos\theta + j\sin\theta$$

we find that

$$\cos\theta = \frac{A + D}{2} \tag{4.11}$$

Thus the geometric factor θ introduced in the trial solution depends only on the *ABCD* matrix for the cavity. The complete solution for x_s must be real since it is an observable quantity and so, including the complex conjugate, the complete solution is

$$x_s = x_0 e^{j\alpha} e^{j\theta s} + x_0 e^{-j\alpha} e^{-j\theta s}$$

where α is an angular constant determining the starting position of the ray.

When θ is real then the solution will be oscillatory:

$$x_s = x_{max} \cos(s\theta + \alpha) \tag{4.12}$$

where x_{max} is the maximum displacement of the ray in the cavity. The condition for stability from Eq. (4.11) is

$$-1 \leqslant \frac{A + D}{2} \leqslant +1 \tag{4.13}$$

Adding 1 to each side and dividing by 2 gives

$$0 \leqslant \frac{A + D + 2}{4} \leqslant 1$$

Substituting from Eq. (4.9) for the values of A and D gives the stability condition for a two concave mirror cavity:

$$\frac{A + D + 2}{4} = \frac{1}{4}\left(1 - \frac{d}{f_2} - \frac{d}{f_1} + \left(1 - \frac{d}{f_2}\right)\left(1 - \frac{d}{f_1}\right) + 2\right)$$

$$= 1 - \frac{d}{2f_2} - \frac{d}{2f_1} + \frac{d^2}{4f_2 f_1}$$

$$= \left(1 - \frac{d}{2f_1}\right)\left(1 - \frac{d}{2f_2}\right) = \left(1 - \frac{d}{b_1}\right)\left(1 - \frac{d}{b_2}\right)$$

The cavity is stable when

$$0 \leqslant \left(1 - \frac{d}{b_1}\right)\left(1 - \frac{d}{b_2}\right) \leqslant 1$$

Introducing the cavity g parameters defined as

$$g_1 = 1 - \frac{d}{b_1} \text{ and } g_2 = 1 - \frac{d}{b_2} \tag{4.14}$$

the stability condition can be written in the form

$$0 \leqslant g_1 g_2 \leqslant 1 \tag{4.15}$$

The boundaries of the inequality in Eq. (4.15) are normally expressed on a stability diagram whose abscissa and ordinate are the cavity g parameters g_1 and g_2. One boundary of the inequality is $g_1 g_2 = 0$, which are the axes of the graph. The second boundaries are the hyperbolas $g_1 g_2 = \pm 1$.

Figure 4.7 illustrates this diagram and also indicates the g parameters which define some specific cavities. These are as follows:

1. The symmetrical confocal cavity in which the mirror foci coincide, i.e. $d = b_1 = b_2$ so that $g_1 = g_2 = 0$.
2. The plane parallel cavity, i.e. $b_1 = b_2 = \infty$ making $g_1 = g_2 = 1$.
3. Spherical cavity. Here the centre of curvature of the mirrors coincide in the centre of the cavity, i.e. $d = b_1 + b_2$, making

   $$g_1 = b_2/b_1 \text{ and } g_2 = b_1/b_2$$

   therefore

   $$g_1 g_2 = 1$$

   Again this is a marginally stable cavity and not often used in practice.
4. Quasi-hemispherical cavity. This consists of one plane mirror and one curved with its centre of curvature lying just beyond the plane mirror surface. The exact hemispherical cavity becomes unstable at $d = b$.

These cavities will be discussed in more detail in Section 4.8.

Example 4.1 Stability range of a cavity

Given two mirrors of radius of curvature 1 m and 1.5 m, find the values of mirror separations, d, which will make a stable cavity:

$$g_1 = 1 - \frac{d}{R_1} = 1 - \frac{d}{1.5} = 1 - \frac{2}{3}d$$

$$g_2 = 1 - \frac{d}{R_2} = 1 - \frac{d}{1} = 1 - d$$

The stability condition $0 \leqslant g_1 g_2 \leqslant 1$ becomes

$$0 \leqslant \left(1 - \frac{2}{3}d\right)(1 - d) \leqslant 1$$

There are two inequalities which we deal with separately:

$$0 \leqslant \left(1 - \frac{2}{3}d\right)(1 - d)$$

Hence $d \leqslant 1$ and $d \geqslant \frac{3}{2}$

$$1 - \frac{2}{3}d - d + \frac{2}{3}d^2 \leqslant 1$$

$$d\left(\frac{2}{3}d - \frac{5}{3}\right) \leqslant 0$$

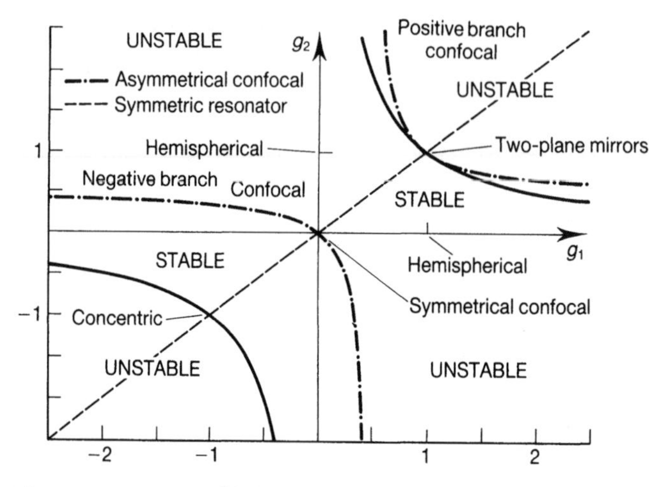

Figure 4.7 Cavity stability diagram.

Hence $d \geqslant 0$ and $d \geqslant \frac{5}{2}$

The stability range for d becomes

$$0 \leqslant d \leqslant 1$$

$$\frac{3}{2} \leqslant d \leqslant \frac{5}{2}$$

The cavity is unstable for values of d between 1.0 and 1.5 m, and for values of d greater than 2.5 m.

4.3.1 Re-entrant ray paths

The ray analysis has produced the cavity condition such that any paraxial ray of the cavity will return to its position after a number of roundtrips. The question arises of how many roundtrips actually occur? The analysis has been done for rays in an xz plane. However, any general skew ray will have projections onto both xz and yz planes. If the cavity has no astigmatism then the same condition applies simultaneously to both. The gyrations of the ray in a stable cavity have been shown in Eq. (4.12) to be sinusoidal and any general skew ray will be sinusoidal in both planes. This is described by the equations

$$x_s = x_{\max} \sin(s\theta + \alpha) \text{ and } y_s = y_{\max} \sin(s\theta)$$

The general case reduces to the addition of two sinusoidal oscillations at right-angles. This means that the position of any ray on, say, a mirror surface will fall on the surface of an ellipse. The eccentricity of this ellipse will be determined by the position and direction of the input ray. This now allows a physical interpretation of the parameter θ. From Eq. (4.11)

$$\theta = \cos^{-1}\left(\frac{A+D}{2}\right)$$

If the re-entrant ray performs N roundtrips before returning to its starting position then θ must have changed by a multiple of 2π. Thus

$$N\theta = 2K\pi \tag{4.16}$$

where K is another integer such that $0 < K < N/2$.

Since

$$\frac{A+D+2}{4} = g_1 g_2$$

so that

$$\frac{A+D}{2} = 2g_1 g_2 - 1$$

Eq. (4.11) then becomes

$$\cos\theta = 2g_1 g_2 - 1$$

Using the identity

$$\cos^2\theta/2 = \frac{1+\cos\theta}{2}$$

$$\cos^2\theta/2 = (1 + 2\,g_1 g_2 - 1)/2 = g_1 g_2$$

$$\theta = 2\cos^{-1}\sqrt{g_1 g_2} = 2\,K\pi/N \tag{4.17}$$

Figure 4.8 illustrates that θ represents the angular position of the ray on one of the mirrors after each roundtrip. After N roundtrips the angular rotation of the ray changes by 2π. Cases with $K = 1$ correspond to situations in which θ can change by 2π and the ray returns to its starting position. Higher values of K mean that θ gyrates through higher multiples of 2π before returning to its start, but note that $K < N/2$, otherwise the same ray path would be counted twice. Equation (4.17) shows therefore that a number of different cavity geometries (i.e. different g values) can have the same number of roundtrips for their re-entrant rays. From Eq. (4.17)

$$N = \frac{K\pi}{\cos^{-1}\sqrt{g_1 g_2}} \text{ with } 0 < K < N/2$$

This could be represented on the stability diagram (Figure 4.7) by the hyperbolas

$$g_1 g_2 = \left(\cos\left(\frac{K\pi}{N}\right)\right)^2$$

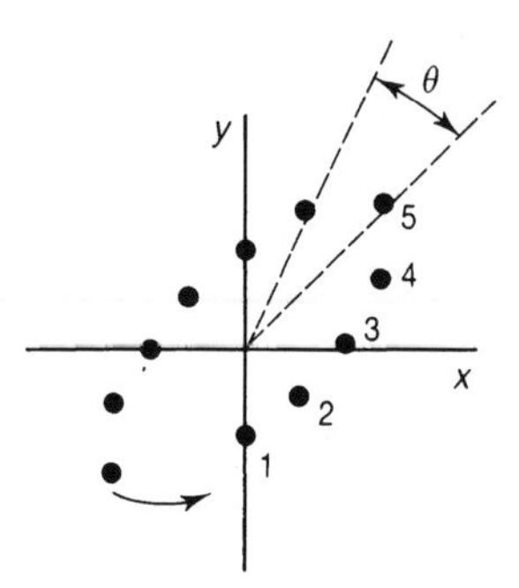

Figure 4.8 Diagram showing the points of impact of the ray with one of the mirrors in a two concave-mirror cavity. The locus of these points of impact is an ellipse.

in the stable regions of the diagram. Since K/N is *not* an integer the stable regions of the diagram are therefore a continuum of hyperbolas each representing a particular value for K/N and hence a particular re-entrant ray path.

Example 4.2 The confocal cavity
Find the number of roundtrips for the re-entrant rays in a symmetrical confocal cavity. The symmetrical confocal cavity has mirrors of equal radius of curvature spaced a distance apart equal to their radius of curvature. Hence

$$g_1 = g_2 = 0$$

and so

$$\cos^{-1}\sqrt{g_1 g_2} = \frac{\pi}{2}$$

Thus from Eq. (4.16) we have

$$\frac{K\pi}{N} = \frac{\pi}{2}$$

giving

$$N = 2 \text{ and } K = 1$$

The ray diagram for this cavity is shown in Figure 4.9 and shows the predicted two roundtrips of the cavity for the rays to be re-entrant.

Experimental demonstration of re-entrant rays can be obtained by using the narrow beam of light from a gas laser to simulate a ray. Injecting this beam into

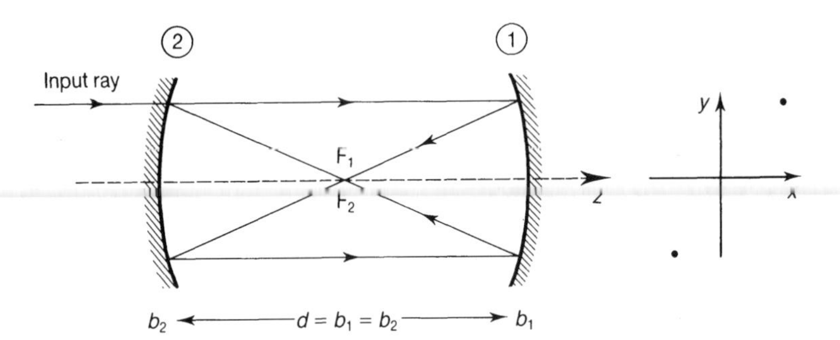

Figure 4.9 Ray diagram for confocal cavity. Rays parallel to the principal axis pass through the focus. If the rays were narrow beams of light then two spots would be seen exiting from mirror 1.

the cavity when g_1g_2 have been chosen to give, say, six roundtrips for the re-entrant rays (i.e. $g_1g_2 = 0.75$) will display six spots of light arranged around an ellipse on both mirrors, each spot being produced on each roundtrip. The effect is illustrated schematically in Figure 4.8 while photographs of these patterns for $N = 19$ are shown in Ref. 1. Cells with this kind of ray path are known as *Herriott cells* and can be used to contain light for amplification by stimulated emission [Ref. 2].

4.4 Diffraction

The ray tracing analysis makes no allowance for diffraction since the rays are assumed to travel in straight lines. It gives no information on any curvature of the wavefronts due to diffraction or any amplitude distribution of the light beam. A more complete analysis of the resonator requires taking diffraction into account. This will be referred to as the *mode analysis* and will give rise to large amplitude variations of the light irradiance in the transverse plane of the beam. The ray and mode analysis produce results which complement each other even though they are conceptually different.

Diffraction normally concerns the bending of light at edges or at circular or rectangular apertures. It is a wave phenomenon and inexplicable from the ray approach. The term 'divergence diffraction' refers to the spreading of a beam of light as it propagates as if it had come through an aperture. Huygen's principle serves as a start for the development of diffraction theory:

> Each point on a wavefront can be treated as a source of a spherical wavelet called a secondary (or Huygen's) wavelet. The envelope of these wavelets, at some time later, is constructed by finding the tangent to the Huygen's wavelets. The envelope is taken to be the new wavefront.

This principle will now be shown to predict the wavefront at a plane (x, y) located at a point z, from the wavefront at the plane (x_0, y_0) located at $z = 0$ as illustrated in Figure 4.10. The total field at point P will result from the mutual interference of the spherical wavelets emanating from all the different points Q. The field at P due to a spherical wavelet from Q is

$$\Delta\mathscr{E}(x, y, z) = \mathscr{E}(x_0, y_0, 0)\left(\frac{e^{-jkr}}{r}\right) dx_0\, dy_0$$

From the geometry of Figure 4.10 the distance QP is given by:

$$r = [(x - x_0)^2 + (y - y_0)^2 + z^2]^{1/2} \tag{4.18}$$

The total field at P is the summation of the fields from all the elemental points Q on the plane $(x_0 y_0)$ and in the limit this summation becomes the integral

$$\mathscr{E}(x, y, z) = C \iint \mathscr{E}(x_0 y_0, 0)\left(\frac{e^{-jkr}}{r}\right) dx_0\, dy_0 \tag{4.19}$$

C is a constant whose value can be obtained by using the fact that a plane wave should be unchanged under this transformation:

$$C \int_{-\infty}^{\infty} \int_{-\infty}^{+\infty} \mathscr{E}\left(\frac{e^{-jkr}}{r}\right) dx_0\, dy_0 = \mathscr{E}\, e^{-jkz}$$

The integration is carried out with the assumption that most of its contribution comes from the domain: $|x - x_0|, |y - y_0| \ll z$. Then, on evaluation,

$$C = j/\lambda$$

The integral, which is known as the Huygen–Fresnel diffraction integral, becomes

$$\mathscr{E}(x, y, z) = \frac{j}{\lambda} \iint_{\substack{\text{input}\\ \text{field}}} \mathscr{E}(x_0, y_0, 0)\left(\frac{e^{-jkr}}{r}\right) dx_0\, dy_0 \tag{4.20}$$

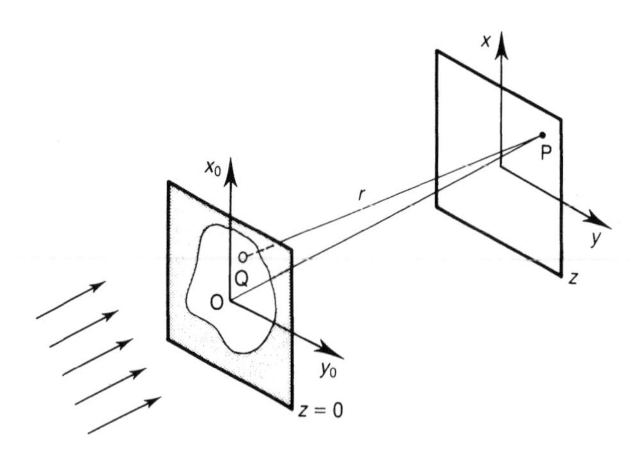

Figure 4.10 Propagation of a wavefront through an arbitrary aperture.

In most cases this integral must be solved numerically but for an analytic approach we are forced to make some appropriate approximations. Rewriting Eq. (4.18), which gives the radius of the Huygen wavefront, we obtain

$$kr = kz\left[1 + \left(\frac{x - x_0}{z}\right)^2 + \left(\frac{y - y_0}{z}\right)^2\right]^{1/2}$$

Expanding this expression using the binomial theorem gives

$$kr = kz + \frac{k}{2z}\,[(x - x_0)^2 + (y - y_0)^2] - \frac{k}{8z^3}\,[(x - x_0)^2 + (y - y_0)^2]^2 + \ldots$$

Two different approximations are used each describing a different optical arrangement:

1. *Fresnel diffraction*. The first two terms of the binomial expansion are retained. The third term becomes small when

$$z^3 \gg \frac{\pi}{4\lambda}\,[(x - x_0)^2 + (y - y_0)^2]^2 \tag{4.21}$$

We substitute for kr into Eq. (4.19) and put $r \cong z$ except when it is multiplied by the large number $k\ (=2\pi/\lambda)$. This latter approximation will be referred to as *the paraxial approximation*. It implies that the spherical wavelet is only being considered over a small angle in the observation plane. Then

$$\mathscr{E}(x, y, z) = \frac{\mathrm{j}}{\lambda z}\,\mathrm{e}^{-\mathrm{j}kz} \int \mathrm{d}x_0 \int \mathrm{d}y_0\ \underbrace{\mathscr{E}(x_0, y_0, 0)}_{\text{input field}}\ \mathrm{e}^{-\mathrm{j}(k/2z)[(x-x_0)^2+(y-y_0)^2]} \tag{4.22}$$

Equation (4.22) is known as *the Fresnel diffraction integral in the paraxial approximation*.

2. *Fraunhofer diffraction*. This is the limit of Fresnel diffraction at large distances between the input plane, where $\mathscr{E}(x_0, y_0)$ is specified, and the observation plane. In this case a further approximation is made to the Fresnel integral Eq. (4.22). The argument of the exponential is approximated as

$$\frac{k}{2z}\,[(x - x_0)^2 + (y - y_0)^2] \cong \frac{k}{2z}\,[(x^2 + y^2) - 2xx_0 - 2yy_0]$$

Where the term $(k/2z)(x_0^2 + y_0^2)$ is taken, for large values of z, to be small compared with the other terms in the square brackets. If the input plane extends over a distance a then this approximation is written as $(ka^2)/2z \ll 1$:

or $\qquad \pi a^2/\lambda z \ll 1 \qquad$ (4.23)

The Fresnel diffraction integral (Eq. (4.21)) is now written with the Fraunhofer approximation:

$$\mathscr{E}(x,y,z) = \frac{\mathrm{j}}{\lambda z}\,\mathrm{e}^{-\mathrm{j}kz}\,\mathrm{e}^{-\mathrm{j}k(x^2+y^2)/2z}\int \mathrm{d}x_0 \int \mathrm{d}y_0\, \mathscr{E}(x_0,y_0,0)\,\mathrm{e}^{\mathrm{j}k(xx_0+yy_0)/z} \tag{4.24}$$

For readers familiar with Fourier analysis Eq. (4.24) is seen to be the Fourier transform in two dimensions of the input field distribution $\mathscr{E}(x_0, y_0, 0)$. Fraunhofer diffraction requires the source to be at infinity relative to the screen. In practice this is achieved by using two lenses; one to produce parallel light at the input and a second to image the parallel light on the screen. Examples of the evaluation of this integral for different input functions or aperture shapes are available in most optics texts [Refs 3, 4].

4.5 The paraxial wave equation

The diffraction integrals can be applied, using numerical techniques, to the determination of the mode or self-reproducing field pattern of an open resonator. In this the field distribution is evaluated on successive roundtrips of the cavity after assuming an initial field distribution. Modes are defined as the steady-state field distributions which result. An approximate analytic solution is possible, using very advanced mathematical techniques, for the symmetric confocal cavity [Ref. 5]. A more convenient approach which can give an analytic solution for any stable open resonator has been developed by obtaining the differential equation of the system and then looking for appropriate solutions. For laser beams the paraxial approximation used for the Fresnel diffraction integral will be used, but this time in differential form.

We start with the time-dependent scalar wave equation for the electric field in a medium of refractive index, n (Eq. 2.11):

$$\nabla^2\mathscr{E}(x,y,z,t) - \frac{n^2}{c^2}\frac{\partial^2}{\partial t^2}\mathscr{E}(x,y,z,t) = 0$$

The scalar wave equation does not account for polarization but is quite adequate for most applications. An oscillatory monochromatic field will have a solution of the form

$$\mathscr{E}(x,y,z,t) = \mathscr{E}(x,y,z)\,\mathrm{e}^{-\mathrm{j}\omega t}$$

With this substitution the wave equation can be written in the form known as the *Helmholtz equation*

$$\nabla^2\mathscr{E}(x,y,z) + k^2\mathscr{E}(x,y,z) = 0 \tag{4.25}$$

where $k = \omega/v$.

Introducing ∇_t^2, Laplacian operator, in the transverse plane in either Cartesian (x, y) or cylindrical (r, ϕ) coordinates we can write Eq. (4.25) as

$$\nabla_t^2 \mathscr{E} + \frac{\partial^2 \mathscr{E}}{\partial z^2} + k^2 \mathscr{E} = 0 \tag{4.26}$$

At this point it is necessary to summarize the important features of the beam for which a solution of the Helmholtz equation is being sought. The solution for our narrow beam cannot be a plane wave since this would have infinite transverse extent. It cannot be a spherical wave since this is not unidirectional and is not defined at the point from which it appears to radiate. In practice, our laser beam has small divergence and when focused with a convex lens converges to a minimum radius and then diverges again. Thus at some finite singular point the wavefronts will be plane. As the wave propagates it maintains its irradiance profile into the far field where the Fraunhofer diffraction integral, Eq. (4.24), implies that the amplitude distribution must transform into itself.

This combination of a plane wave and some transverse distribution which does not change appreciably as the wave propagates can be written in a trial form:

$$\mathscr{E}(x, y, z) = \mathscr{E}_0 \psi(x, y, z)\ \mathrm{e}^{-\mathrm{j}kz} \tag{4.27}$$

where $\mathscr{E}_0$ is the amplitude of the field at $x = y = z = 0$ and $\psi(x, y, z)$ determines how the wave differs from plane wave. It gives the transverse variation of the field having a weak dependence on z and some characteristics of a spherical wave.

Putting $\mathscr{E}_0 = 1$ for algebraic ease and evaluating the derivatives for substitution of Eq. (4.27) into Eq. (4.26)

$$\nabla_t^2 \mathscr{E} = [\nabla_t^2 \psi]\ \mathrm{e}^{-\mathrm{j}kz}$$

$$\frac{\partial \mathscr{E}}{\partial z} = \left(-\mathrm{j}k\psi + \frac{\partial \psi}{\partial z}\right)\ \mathrm{e}^{-\mathrm{j}kz}$$

$$\frac{\partial^2 \mathscr{E}}{\partial z^2} = \left[-k^2\psi - \mathrm{j}2k\frac{\partial \psi}{\partial z} + \frac{\partial^2 \psi}{\partial z^2}\right]\ \mathrm{e}^{-\mathrm{j}kz}$$

Substitution into Eq. (4.26) gives

$$\nabla_t^2 \psi - 2\mathrm{j}k\frac{\partial \psi}{\partial z} + \frac{\partial^2 \psi}{\partial z^2} = 0 \tag{4.28}$$

Since the laser beam has a low divergence we expect that ψ should not vary rapidly with z. The wave vector k is a large number for optical waves and the paraxial approximation in differential form can be expressed as

$$\left|\frac{\partial^2 \psi}{\partial z^2}\right| \ll \left|2\mathrm{k}\frac{\partial \psi}{\partial z}\right| \tag{4.29}$$

Hence Eq. (4.28) approximates to

$$\nabla_t^2\psi - j2k\frac{\partial\psi}{\partial z} = 0 \tag{4.30}$$

This equation is known as the *paraxial wave equation* and can be considered as the differential form of the Fresnel diffraction integral. A periodic solution will be forced by appropriate boundary conditions to be a mode of an open resonator.

4.6 Lowest-order TEM$_{00}$ mode solution

The transverse Laplace operator in Eq. (4.30) can be expressed in either cylindrical (r, ϕ, z) or Cartesian (x, y, z) coordinates, thus two sets of solutions exist. It is convenient, initially, to find a solution which fits the experimental observations of the 'simplest' beam which we will see has a Gaussian profile. For a reason that will become apparent later this will be referred to as the TEM$_{00}$ mode solution. Diffraction theory can show that a Gaussian amplitude distribution will maintain its profile as it propagates into the far field (Fraunhofer diffraction). This is a result of the fact that the Fourier transform of a Gaussian function is another Gaussian function. A more complete mathematical treatment [Ref. 6] shows that Hermite–Gaussian amplitude distributions propagate without change in the near field (i.e. Fresnel diffraction). This self-transforming property ensures that the Gaussian beam profile has the lowest divergence for a given diameter for any radially restricted beam.

Writing the transverse Laplacian in cylindrical coordinates, Eq. (4.30) becomes

$$\frac{1}{r}\frac{\partial}{\partial r}\left(r\frac{\partial\psi}{\partial r}\right) + \frac{1}{r^2}\left(\frac{\partial^2\psi}{\partial\phi^2}\right) - j2k\frac{\partial\psi}{\partial z} = 0 \tag{4.31}$$

The Gaussian beam has circular symmetry, so in Eq. (4.31) any variation with ϕ can be equated to zero. It must be realized that this gives only one particular solution of Eq. (4.30). Different solutions exist which do not have cylindrical symmetry and these will be dealt with later. Using the subscript $_0$ to remind us that we are seeking the simplest solution the differential equation (4.31) becomes

$$\frac{1}{r}\frac{\partial}{\partial r}\left(r\frac{\partial\psi_0}{\partial r}\right) - j2k\frac{\partial\psi_0}{\partial z} = 0 \tag{4.32}$$

Using the observations about the laser beam noted above a trial solution is proposed:

$$\psi_0 = \exp(-jP(z))\exp\left(\frac{-jkr^2}{2q(z)}\right) \tag{4.33}$$

The two factors $P(z)$ and $q(z)$ will be found from substitution into Eq. (4.32). $P(z)$ will determine how the longitudinal phase of the wave differs from a plane wave. The second term (with $q(z)$) will involve the Gaussian radial amplitude distribution which is real and the radial phase which is imaginary. Thus $q(z)$ will have real and imaginary parts.

Using the prime to denote differentiation with respect to z:

$$-\mathrm{j}2k\frac{\partial\psi_0}{\partial z} = -\mathrm{j}2k\left[-\mathrm{j}P'(z) - \frac{-\mathrm{j}r^2kq'(z)}{2q^2(z)}\right]\psi_0$$

$$-\mathrm{j}2k\frac{\partial\psi_0}{\partial z} = -2k\left[P'(z) - \frac{r^2kq'(z)}{2q^2(z)}\right]\psi_0 \tag{4.34}$$

$$\frac{\partial\psi_0}{\partial r} = -\mathrm{j}\frac{kr}{q(z)}\psi_0$$

Hence

$$r\frac{\partial\psi_0}{\partial r} = -\mathrm{j}\frac{kr^2}{q(z)}\psi_0$$

$$\frac{\partial}{\partial r}\left(r\frac{\partial\psi_0}{\partial r}\right) = -\mathrm{j}\frac{kr^2}{q(z)}\left[-\mathrm{j}\frac{kr}{q(z)}\right]\psi_0 - \mathrm{j}\frac{2kr}{q(z)}\psi_0 \tag{4.35}$$

$$\frac{1}{r}\frac{\partial}{\partial r}\left(r\frac{\partial\psi_0}{\partial r}\right) = -\left[\frac{k^2r^2}{q^2(z)} - \mathrm{j}\frac{2k}{q(z)}\right]\psi_0$$

Substituting Eqs (4.34) and (4.35) into (4.32) and rearranging gives

$$\left\{\left[\frac{k^2}{q^2(z)}(q'(z)-1)\right]r^2 - 2k\left[P'(z) + \frac{\mathrm{j}}{q(z)}\right]\right\}\psi_0 = 0 \tag{4.36}$$

Since this equation must hold true for any r and providing ψ_0 is not zero we can equate the coefficients in r on either side of the equation, then

$$q'(z) = 1 \text{ and } P'(z) = \frac{-\mathrm{j}}{q(z)} \tag{4.37}$$

Solving the simpler uncoupled equation which only involves $q(z)$ results in

$$q(z) = q_0 + z \tag{4.38a}$$

where q_0 is the constant of integration given by the value of q at $z = 0$.

Now in the trial solution (4.33), the term $\mathrm{e}^{-\mathrm{j}kr^2/2q(z)}$ determines the amplitude distribution through its real part and the radial phase (or wavefront curvature) with its imaginary part. When a laser beam is focused with a lens the diameter of the beam reduces to a minimum at the focal point, and the wavefront curvatures as a consequence must change sign on passing through this focus. In geometrical optics the rays come to a point but this neglects diffraction, and it is this very diffraction which our solution is including. Thus at this point where the beam diameter is minimum the wavefront must be plane. It is convenient to choose this point as the origin, i.e. $z = 0$. At $z = 0$ we force the solution to be that of a plane wavefront and make the radial term real. This is done by making q_0 imaginary.

Putting $q_0 = jz_R$, where z_R is a real parameter with the dimensions of length, we write

$$q(z) = z + jz_R \tag{4.38b}$$

We will obtain $1/q(z)$ which is required for the trial solution

$$\frac{1}{q(z)} = \frac{1}{z + jz_R} = \underbrace{\left(\frac{z}{z^2 + z_R^2}\right)}_{\text{real}} - j\underbrace{\left(\frac{z_0}{z^2 + z_R^2}\right)}_{\text{imaginary}} \tag{4.39}$$

Introducing two parameters which are functions of z, $R(z)$ and $w(z)$, Eq. (4.39) is rewritten in the form

$$\frac{1}{q(z)} = \frac{1}{R(z)} - j\frac{2}{kw^2(z)} \tag{4.40}$$

where

$$R(z) = \frac{z^2 + z_R^2}{z} = z\left[1 + \left(\frac{z_R}{z}\right)^2\right] \tag{4.41}$$

and

$$w^2(z) = \frac{2}{kz_R}(z_R^2 + z^2) = \frac{2z_R}{k}\left[1 + \left(\frac{z}{z_R}\right)^2\right] \tag{4.42}$$

Let us now write down our solution obtained so far by substituting Eq. (4.40) into Eq. (4.33) to produce

$$\psi_0 = \exp(-jP(z))\exp\left(-\frac{r^2}{w^2(z)}\right)\exp\left(-\frac{jkr^2}{2R(z)}\right) \tag{4.43}$$

Introducing $w(z)$ in the form of Eq. (4.42) makes the real term in Eq. (4.43) a Gaussian function, as was suggested by the trial solution. At any point z the field amplitude profile of the beam is Gaussian with width determined by $w(z)$. Hence $w(z)$ can be interpreted as the radial distance at which the field amplitude has dropped to e^{-1} of its value at the centre $r = 0$ and is known as the *spot size*. Equation (4.42) shows that at $z = 0$ the spot size is minimum. This point is referred to as the *waist* and here the spot size is given the symbol w_0. Hence from Eq. (4.42) the waist size is given by

$$w_0^2 = \frac{2z_R}{k}$$

where

$$z_R = \frac{\pi w_0^2}{\lambda} \tag{4.44}$$

Returning to Eq. (4.37) it is now possible to find an expression for $P(z)$. Using (4.38b) for $q(z)$

$$P'(z) = \frac{-\mathrm{j}}{q(z)} = \frac{-\mathrm{j}}{z + \mathrm{j}z_\mathrm{R}}$$

Multiplying by j and integrating produces

$$\mathrm{j}P(z) = \log_\mathrm{e}(z + \mathrm{j}z_\mathrm{R}) + C$$

where C is a constant of integration. If the longitudinal phase of the wave is assumed to be zero at $z = 0$ then we can put

$$P(z = 0) = 0 \text{ and so } C = -\log_\mathrm{e} \mathrm{j}z_\mathrm{R}$$

Thus

$$\mathrm{j}P(z) = \log_\mathrm{e}\left(1 - \mathrm{j}\frac{z}{z_\mathrm{R}}\right) \tag{4.45}$$

We now use the Argand diagram method for displaying complex numbers to prove a useful identity for simplifying this equation. Introducing two parametric variables a and b,

$$\exp(\mathrm{j} \arctan(b/a)) = \cos(\arctan(b/a)) + \mathrm{j} \sin(\arctan(b/a))$$

$$= \frac{a}{\sqrt{a^2 + b^2}} + \mathrm{j}\frac{b}{\sqrt{a^2 + b^2}}$$

or

$$a + \mathrm{j}b = (a^2 + b^2)^{1/2} \exp(\mathrm{j} \arctan(b/a))$$

Now put $a = 1$ and $b = z/z_\mathrm{R}$. We obtain

$$1 - \mathrm{j}\frac{z}{z_\mathrm{R}} = \left[1 + \left(\frac{z}{z_\mathrm{R}}\right)^2\right]^{1/2} \exp\left(-\mathrm{j} \arctan \frac{z}{z_\mathrm{R}}\right)$$

Using this identity in the determination of $P(z)$ above,

$$\mathrm{j}P(z) = \log_\mathrm{e}\left(1 - \mathrm{j}\frac{z}{z_\mathrm{R}}\right) = \log_\mathrm{e}\left[1 + \left(\frac{z}{z_\mathrm{R}}\right)\right]^{1/2} - \mathrm{j} \arctan\left(\frac{z}{z_\mathrm{R}}\right)$$

$$\mathrm{e}^{-\mathrm{j}P(z)} = \frac{1}{[1 + (z/z_\mathrm{R})^2]^{1/2}} \mathrm{e}^{\mathrm{j} \arctan(z/z_\mathrm{R})}$$

Using Eqs (4.42) and (4.44),

$$\mathrm{e}^{-\mathrm{j}P(z)} = \frac{w_0}{w(z)} \mathrm{e}^{\mathrm{j} \arctan(z/z_\mathrm{R})} \tag{4.46}$$

Equation (4.46) can now be substituted into Eq. (4.43) to give the solution for ψ_0 and this in turn can be substituted into Eq. (4.27) to produce the total solution for our paraxial wave.

In summary, let us remind ourselves of the boundary information which has been used to achieve this solution:

1. The beam is paraxial, i.e. a low divergent spherical wave.
2. The beam has circular symmetry.
3. The field amplitude radial distribution is Gaussian.
4. There is a point, namely $z = 0$, where the spot size is minimum and the wavefronts planar.

The complete solution becomes

$$\begin{aligned}\mathscr{E}(r, \phi, z) = \mathscr{E}_0 \frac{w_0}{w(z)} &\exp(-r^2/w^2(z)) \text{ amplitude} \\ &\times \exp[-\mathrm{j}(kz - \arctan z/z_\mathrm{R})] \text{ longitudinal phase} \\ &\times \exp\left(-\mathrm{j}\frac{kr^2}{2R(z)}\right) \text{ radial phase}\end{aligned} \tag{4.47}$$

4.7 Physical description of the TEM$_{00}$ mode solution

A complete physical description and explanation of the beam parameters can be obtained by discussing in turn each of the named terms in Eq. (4.47).

4.7.1 The amplitude term and beam divergence

We identify the real term in Eq. (4.47) as giving amplitude information. Since the solution has circular symmetry a transformation to Cartesian coordinates can be made using $r^2 = x^2 + y^2$. Thus

$$\frac{w_0}{w(z)} \exp[-(r/w(z))^2] = \frac{w_0}{w(z)} \mathrm{e}^{-(x/w(z))^2} \mathrm{e}^{-(y/w(z))^2}$$

This shows that the amplitude variation is the same Gaussian in both the x and y directions. As the wave propagates away from $z = 0$ the spot size increases. This variation of the spot size as a function of z (Eq. (4.42)) is plotted in Figure 4.11.

We may write Eq. (4.42), using Eq. (4.44), as

$$w(z) = w_0\left[1 + \left(\frac{z}{z_\mathrm{R}}\right)^2\right]^{1/2} \tag{4.48}$$

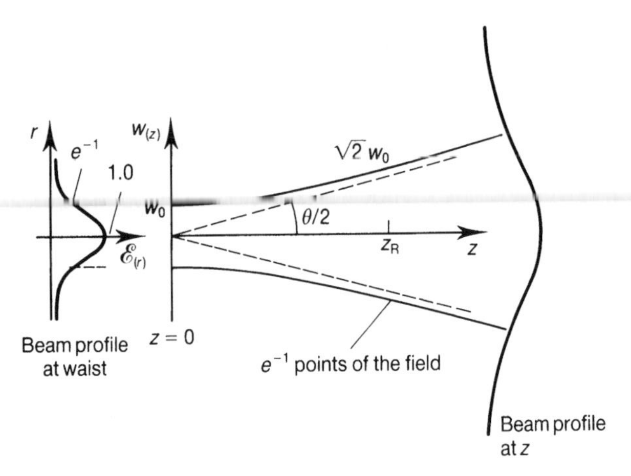

Figure 4.11 Variation of the spot size for the propagation of a Gaussian beam.

In the far field $z \gg z_R$ and Eq. (4.48) becomes

$$w(z) \cong w_0 \frac{z}{z_R} = \frac{\lambda z}{\pi w_0}$$

The spot size in the far field is seen to increase linearly with z or, more precisely, the spot size given by Eq. (4.42) tends asymptotically to the straight line shown dotted in Figure 4.11. The parameter z_R has the dimensions of length and from Eq. (4.42) at $z = z_R$ the spot size becomes

$$w(z_0) = \sqrt{2} w_0 \tag{4.49}$$

Thus at a distance of z_R from the waist the spot size has increased by $\sqrt{2}$ or the area occupied by the spot has doubled. The parameter z_R is known as the *Rayleigh range* or often $2z_R$ is referred to as the *confocal parameter*. From Eq. (4.48) the far-field divergence as defined by θ in Figure 4.11 is

$$\theta/2 \cong \frac{dw}{dz} = \frac{\lambda}{\pi w_0}$$

The full angular divergence, θ, becomes

$$\theta = \frac{2\lambda}{\pi w_0} = 0.636 \frac{\lambda}{w_0} \tag{4.50}$$

The full far-field angular divergence θ for a plane wave diffracted by an aperture is usually defined as the angular distance to the first minimum in the diffraction pattern. With this definition we can compare the divergence of circular and rectangular confined beams with the Gaussian beam as follows:

For a circular aperture of radius a	$\theta = 1.22\lambda/a$
For a slit of width $2a$	$\theta = 1.0\lambda/a$
Gaussian beam with waist w_0	$\theta = 0.636\lambda/w_0$

The Gaussian beam is seen to have the smallest divergence for the same radial size and is referred to as diffraction limited.

4.7.2 Power flux

As mentioned in Section 2.1.4, the power flux or irradiance is given by *the Poynting vector* **S** [Ref. 7]:

$$\mathbf{S} = \frac{1}{2}\,\mathscr{E} \times \mathscr{H}^*$$

where we use the complex conjugate form since the sinusoidal variation of the fields is written using the complex exponential notation. In free space this becomes

$$\mathbf{S} = \varepsilon_0 c \mathscr{E} \times \mathscr{E}^*$$

The average value of the magnitude of **S** over one cycle of oscillation of the field is given by

$$|\mathbf{S}_{\text{ave}}| = \frac{1}{2}\,\varepsilon_0 c \mathscr{E}\mathscr{E}^*$$

The total power, P, averaged over one cycle in a Gaussian beam can be obtained by integrating the irradiance, as given by the Poynting vector, over the total cross-section of the beam. This is written both in cylindrical and in Cartesian coordinates as

$$P = \frac{1}{2}\,\varepsilon_0 c \int_0^{2\pi}\int_0^{\infty} \mathscr{E}\mathscr{E}^* r\,\mathrm{d}r\,\mathrm{d}\phi \tag{4.51a}$$

$$= \frac{1}{2}\,\varepsilon_0 c \int_{-\infty}^{\infty}\int_{-\infty}^{\infty} \mathscr{E}\mathscr{E}^*\,\mathrm{d}x\,\mathrm{d}y \tag{4.51b}$$

Substituting Eq. (4.47) into Eq. (4.51a) where P is expressed in cylindrical coordinates gives

$$P = \frac{1}{2}\,\varepsilon_0 c \frac{\mathscr{E}_0^2 w_0^2}{w^2(z)} \int_0^{2\pi}\int_{-\infty}^{\infty} \exp\left(-\frac{2r^2}{w^2(z)}\right) r\,\mathrm{d}r\,\mathrm{d}\phi \tag{4.52}$$

which leads to

$$P = \frac{1}{2}\,\varepsilon_0 c \mathscr{E}_0^2 \left[\frac{\pi w_0^2}{2}\right] \tag{4.53}$$

For a particular Gaussian beam, Eq. (4.53) shows that the total power in the beam is constant as required to conserve energy in free space propagation. The term $w_0/w(z)$ in Eq. (4.47), which is included in the integral of Eq. (4.52), ensures that energy is conserved. As the beam diverges, i.e. the Gaussian broadens and its maximum must become smaller to conserve energy. This is illustrated in Figure 4.11, which shows the beam profile at two different distances, z, from the waist.

The spot size is defined in relation to the field amplitude but will normally be measured from the irradiance profile. The irradiance profile will be of the form $\exp(-2r^2/w^2(z))$. Thus the spot size may be defined as the radial distance where the irradiance has dropped to e^{-2} of its value at $r = 0$.

4.7.3 The radial phase term

The term

$$\exp\left(-j\frac{kr}{2R(z)}\right)$$

in Eq. (4.47) has an imaginary argument which determines the variation of the phase of the wave radially. We require our trial solution Eq. (4.27) for the paraxial wave to have spherical wavefronts and Figure 4.12 shows schematically the geometry for the propagation of an ideal spherical wave. The radius of curvature of its wavefront is

$$R = (r^2 + z^2)^{1/2}$$

Expanding using the binomial theorem,

$$R \cong z + \frac{r^2}{2z} + \frac{r^4}{8z^3} + \dots \tag{4.54a}$$

Using the *paraxial approximation* $z \gg r$,

$$R \cong z\left(1 + \frac{1}{2}\frac{r^2}{z^2}\right)$$

At large z we take $R \cong z$, then

$$R \cong z + \frac{r^2}{2R} \tag{4.54b}$$

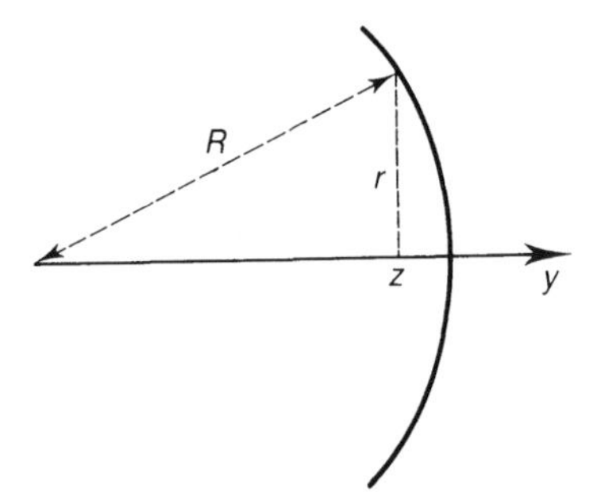

Figure 4.12 Diagram to show the propagation of an ideal spherical wave in the paraxial approximation.

A spherical wave as used in Eq. (4.18) has the form

$$\mathscr{E} \simeq \frac{1}{R} \, \mathrm{e}^{-\mathrm{j}kR}$$

and this can be written using the paraxial approximation given by Eq. ((4.54) as

$$\mathscr{E} \cong \frac{1}{z} \, \mathrm{e}^{-\mathrm{j}kz} \, \mathrm{e}^{-\mathrm{j}k \, \frac{r^2}{2R}}$$

We note that this form for the spherical wave ensures conservation of energy through the inverse square law while the Gaussian beam does this with the amplitude factor, $w_0/w(z)$.

Let us now compare the field terms of the Gaussian beam and the paraxial spherical wave. Using Eq. (4.48) for the ratio $w_0/w(z)$ for the Gaussian beam we have

Gaussian beam

$$\mathrm{e}^{-r^2/w^{-2}}(z) \sqrt{\frac{1}{(1+(z/z_\mathrm{R})^2}} \, \mathrm{e}^{-\mathrm{j}kz} \, \mathrm{e}^{-\mathrm{j}kr^2/2R(z)} \, \mathrm{e}^{-\mathrm{j} \arctan(z/z_\mathrm{R})}$$

Spherical wave

$$\frac{1}{z} \, \mathrm{e}^{-\mathrm{j}kz} \, \mathrm{e}^{-\mathrm{j}kr^2/2R}$$

The Gaussian beam will be a paraxial spherical wave if we interpret the parameter $R(z)$ as the radius of curvature of the wavefront. A spherical wave simply has $R = z$ while the algebra demands that, for the Gaussian beam, R is a function of z (Eq. (4.41)) such that the centre of curvature effectively changes its position – a novel concept. At $z = 0$ the spherical wave is not defined, but the algebra has allowed the Gaussian beam to exist, thereby forcing it to be a plane wave, i.e. $R_{(z=0)} = \infty$. Our Gaussian beam is therefore a low-divergence narrow spherical wave which *is* defined at $z = 0$ and so is often referred to as a *Gaussian-spherical wave*.

4.7.4 Longitudinal phase factor

From Eq. (4.47) we have that the longitudinal phase of the Gaussian beam is $[kz - \arctan(z/z_\mathrm{R})]$. It differs from that of a plane wave by the factor $\arctan(z/z_\mathrm{R})$. Seigman [Ref. 8] has called this term the *Guoy Shift* after Guoy, who first observed an extra phase shift of 180° for an optical beam as it passes through a focus. In the case of the Gaussian beam the extra phase shift comes on passing through the waist as is shown graphically in Figure 4.13. This phase shift occurs as a result of large transverse field gradients similar to those which occur in a guided wave. Figure 4.14 compares the concentric wavefronts of a spherical wave with the Gaussian beam wavefronts. In free space propagation this Guoy shift is seldom of any great

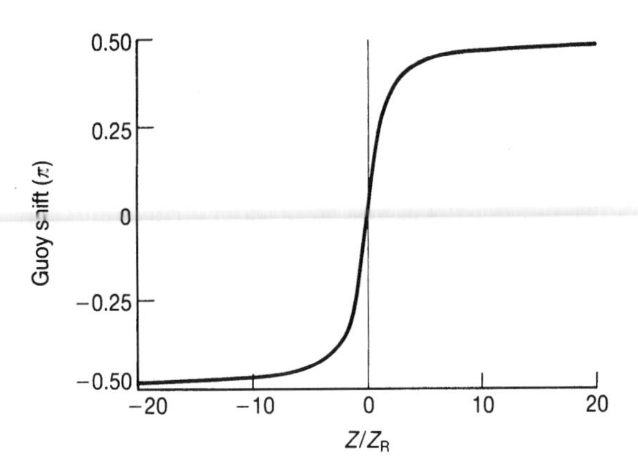

Figure 4.13 Plot of Guoy shift versus z/z_R.

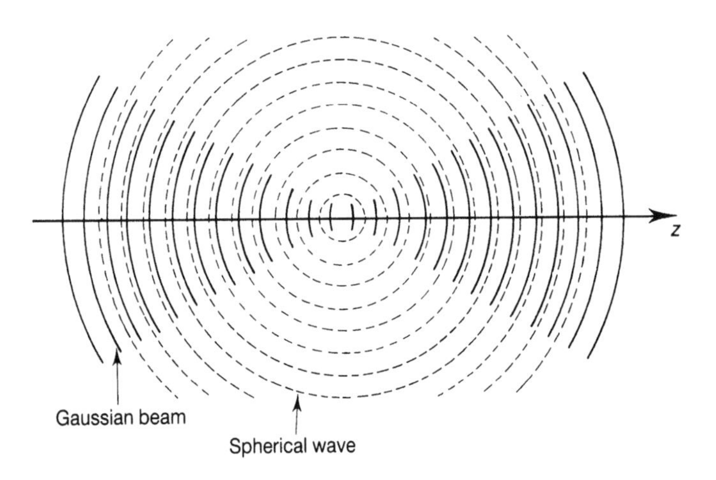

Figure 4.14 Comparison of the wavefronts of a spherical wave and a Gaussian beam.

importance. However, it will need to be included when we evaluate the resonant frequency of an open resonator in Chapter 5.

4.8 Gaussian beams as modes of curved-mirror open resonators

The paraxial wave equation solution discussed gives no information on how the Gaussian beam is generated. In fact it will be shown to be an eigenmode of a curved-mirror cavity. The fact that the wavefronts are very nearly spherical is more than fortuitous since the curved mirrors used for resonators are ground to be spherical with specific radii of curvature. Figure 4.15 shows a typical linear resonator in which a spherical Gaussian beam can become a cavity mode if its wavefront curvatures match those of the mirrors.

The spherical Gaussian interpretation of solution (4.47) and the identification of $R(z)$ as the wavefront radius of curvature was obtained by dropping terms higher than quadratic in the binomial expansion of a spherical wave. Thus the spherical Gaussian wave will be a good fit to the cavity if the third term in Eq. (4.54a) is much smaller than a wavelength, meaning that it departs from a spherical wavefront by less than a wavelength:

$$\frac{r^4}{8z^3} \ll \lambda$$

For a typical cavity we may take z to be the separation of the mirrors, d, and r to be the internal aperture, a. The condition then becomes

$$\frac{a^2}{\lambda d} \ll \frac{d^2}{a^2}$$

In terms of a parameter $N_F = a^2/\lambda d$, known as the *Fresnel number*, this condition for spherical wavefronts matching the mirror spherical surfaces can be written

$$N_F \ll \left(\frac{d}{a}\right)^2 \tag{4.55}$$

It can be shown that this is the same condition as Eq. (4.20), which gives the Fresnel diffraction integral in the paraxial approximation.

A second restriction arises from the requirement that the light is not diffracted out of the aperture of the cavity when propagating between the mirrors. This can be

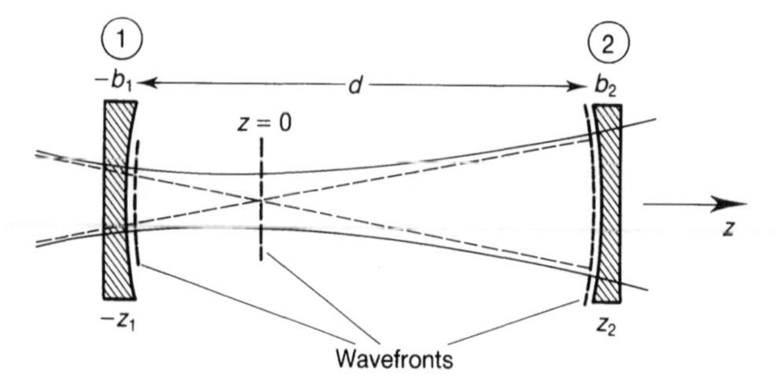

Figure 4.15 Schematic diagram of an open resonator showing the wavefront curvatures matching the mirror surfaces at the two concave mirrors. The waist occurs somewhere between them.

approximately quantified by requiring that the angular diffraction, as observed at each mirror, does not fall outside the cavity aperture. Thus

$$\frac{\lambda}{a} \ll \frac{a}{d}$$

or in terms of the Fresnel number

$$N_F \gg 1 \tag{4.56}$$

Comparison of this inequality with Eq. (4.23) implies that the situation in the cavity is given by the Fresnel diffraction integral.

The conditions for a Gaussian beam modes to be modes of the curved mirror cavity can be summarized as follows:

$$\underset{\text{Low diffraction loss}}{1 < N_F} < \underset{\text{Wavefront spherical}}{\left(\frac{d}{a}\right)^2} \tag{4.57}$$

The aperture of radius, a, will be determined in a number of ways. It can simply be determined by the physical size of the mirrors while in most laser cavities it will either be an iris placed in the cavity or the radius of the gain medium, e.g. a plasma tube. For light injected into a passive cavity, in addition to the above this aperture could be given by the radius of the input beam (see Section 5.6). Only one Gaussian beam will satisfy the wavefront-matching condition since once one of the parameters of the Gaussian beam is defined so then are all the others. There is only one Gaussian beam which is a mode of a given cavity.

From Figure 4.15 it is seen that if a Gaussian beam is a cavity mode then the waist must occur somewhere between the mirrors. Taking the waist location to be $z = 0$ and distances to its left to be negative, then, for consistency, mirror 1 must have a negative curvature. Using Eq. (4.41) for the Gaussian beam wavefront curvature three simultaneous equations can be generated:

$$z_2 - (-z_1) = d \tag{4.58}$$

Making the wavefront curvature equal to the mirror curvatures, b_1 and b_2, we obtain

$$\begin{aligned} R(z_2) &= z_2\left[1 + \left(\frac{z_R}{z_2}\right)^2\right] = b_2 \\ z_2 &+ \frac{z_R^2}{z_2} = b_2 \end{aligned} \tag{4.59}$$

$$R(z_1) = -z_1\left[1 + \left(\frac{z_R}{-z_1}\right)^2\right] = -b_1$$

$$z_1 + \frac{z_R^2}{z_1} = b_1 \tag{4.60}$$

After some algebra we can solve Esq (4.58), (4.59) and (4.60) for z_1, z_2 and z_R in terms of the mirror spacing and curvatures:

$$z_1 = \frac{d(b_2 - d)}{(b_1 + b_2 - 2d)} \tag{4.61}$$

$$z_2 = \frac{d(b_1 - d)}{(b_1 + b_2 - 2d)} \tag{4.62}$$

$$z_R = \sqrt{\frac{d(b_1 - d)\,(b_2 - d)\,(b_1 + b_2 - d)}{(b_1 + b_2 - 2d)^2}} \tag{4.63}$$

Equation (4.63) gives the size of the waist since

$$z_R = \pi w_0^2/\lambda$$

and, along with Eq. (4.48), allows the spot size anywhere inside or outside the cavity to be calculated. The location of the waist in the cavity can be obtained using the ratio

$$\frac{z_1}{z_2} = \frac{b_2 - d}{b_1 - d} \tag{4.64}$$

The importance of these equations to the design of a laser will be appreciated in Chapter 7, when the geometries of different gain media are discussed. At this stage we can see that the Gaussian beam occupies a hyperboloid of revolution known as the *mode volume*, whose geometry is determined by the cavity. It is the task of the laser design engineer to optimize the overlap of the gain medium and the Gaussian beam mode volume. Equations (4.61)–(4.63) can also be written in terms of the cavity g parameters defined in Eq. (4.14) as follows:

$$w_0 = \left(\frac{\lambda d}{\pi}\right)^{1/2}\left\{\frac{g_1 g_2(1 - g_1 g_2)}{(g_1 + g_2 - 2g_1 g_2)^2}\right\}^{1/4} \tag{4.65}$$

$$w_1 = \left(\frac{\lambda d}{\pi}\right)^{1/2}\left\{\frac{g_2}{g_1(1 - g_1 g_2)}\right\}^{1/4} \tag{4.66}$$

$$w_2 = \left(\frac{\lambda d}{\pi}\right)^{1/2}\left\{\frac{g_1}{g_2(1 - g_1 g_2)}\right\}^{1/4} \tag{4.67}$$

$$\frac{z_1}{z_2} = \frac{g_2(1 - g_1)}{g_1(1 - g_2)} \tag{4.68}$$

$$\frac{w_1}{w_2} = \sqrt{\frac{g_2}{g_1}} \tag{4.69}$$

Equation (4.65) shows that w_0 will only be real when

$$0 \leqslant g_1 g_2 \leqslant 1$$

which is exactly the same condition obtained for re-entrant rays (Eq. (4.15)) using the ray tracing approach. In our Gaussian beam analysis the unstable case occurs when w_0 becomes zero, and consequently the spot sizes on the mirrors becomes infinite – clearly impossible. The same limiting result (Eq. (4.15)) is obtained by both approaches, giving consistency between the ray and Gaussian beam analyses.

Example 4.3 Calculation of spot size in an open resonator

Using the two mirrors of Example 4.1 which had radii of curvature 1.0 m and 1.5 m, calculate the spot sizes on the mirrors, the waist size and its location if the mirror spacing is 0.5 m and the wavelength is 633 nm:

$$g_1 = 1 - \frac{0.5}{1} = 0.5$$

$$g_2 = 1 - \frac{0.5}{1.5} = 0.67$$

$$g_1 g_2 = 0.335$$

$$\left(\frac{\lambda d}{\pi}\right)^{1/2} = \left(\frac{6.33 \times 10^{-7} \times 0.5}{\pi}\right)^{1/2} = 3.17 \times 10^{-4}\ \text{m}$$

Using Eq. (4.65) for the waist size:

$$w_0 = 3.17 \times 10^{-4} \left(\frac{0.335\ (1 - 0.335)}{(0.5 + 0.67 - 2 \times 0.335)^2}\right)^{1/4} = 3.08 \times 10^{-4}\ \text{m}$$

Using Eq. (4.66) for the spot size on mirror 1:

$$w_1 = 3.17 \times 10^{-4} \left(\frac{0.667}{0.5\ (1 - 0.334)}\right)^{1/4} = 3.77 \times 10^{-4}\ \text{m}$$

Using Eq. (4.69) the spot size on mirror 2 becomes

$$w_2 = 3.26 \times 10^{-4}\ \text{m}$$

Using Eq. (4.68) for the location of the waist:

$$\frac{z_1}{z_2} = 2$$

Thus the waist occurs at 0.16 m from mirror 2 and 0.34 m from mirror 1, i.e. the waist is closer to the mirror with the larger radius of curvature.

4.8.1 Examples of stable resonators

It is now instructive to survey the properties of some commonly used laser resonators. The stability diagram (Figure 4.7) is used to illustrate cavities of interest using the cavity g parameter as the variable. A convenient way to illustrate cavities on this diagram is to imagine a hypothetical cavity in which the radii of curvature of the mirrors can be altered but their separation, d, is fixed, so allowing us to move on the $g_1 g_2$ diagram.

4.8.1.1 Symmetric resonators $b_1 = b_2$
Two mirrors of equal radius of curvature $b_1 = b_2 = b$. Hence

$$g_1 = g_2 = g = 1 - d/b$$

Equations (4.65)–(4.67) for the waist and spot sizes become

$$w_0^2 = \frac{d\lambda}{\pi} \sqrt{\frac{1+g}{4(1-g)}} \tag{4.70}$$

$$w_1^2 = w_2^2 = \frac{d\lambda}{\pi} \sqrt{\frac{1}{1-g^2}} \tag{4.71}$$

The spot sizes on the mirrors are equal and, from Eq. (4.68), the waist lies at the centre of the cavity. On the stability diagram, since $g_1 = g_2$, all symmetric resonators lie on a straight line through the origin at 45° to the axis as shown in Figure 4.7. At the concentric point ($g_1 = g_2 = -1$) the centres of curvature coincide. Increasing the mirror curvatures corresponds to moving along this line. At the origin the foci coincide and the cavity is referred to as confocal. This is a singular point and might appear to be a point of marginal stability. However, it is in practice a very stable configuration, since slight changes in d move the cavity along the straight line into a stable quadrant, provided, of course, the mirror curvatures are identical.

Moving further along the line into the positive quadrant takes the cavity to two plane parallel mirrors which again is a point of marginal stability. Figure 4.16 shows the variation of the spot size (Eq. (4.70) and waist, Eq. (4.71)) along this straight line as a function of the cavity g parameter. Notice that the ordinate here is in terms of the common scaling factor $\sqrt{\pi w^2/\lambda d}$ for the spot sizes which is the square root of the Fresnel number, N_F. When $g = -1$ (concentric) the waist approaches zero and the spot sizes on the mirrors become very large. When $g = +1$ (plane parallel) the waist and the mirror spot sizes both become very large. The mirror spot sizes are minimum for the confocal cavity ($g = 0$) and from Eq. (4.70) the waist size is then given by

$$w_0^2 = \frac{d\lambda}{2\pi}$$

giving

$$d = 2\pi w_0^2/\lambda = 2z_R$$

The length of the confocal cavity is twice the Rayleigh distance, which is the distance of tight confinement for the Gaussian beam. Thus the alternative name for $2z_R$ is the confocal parameter.

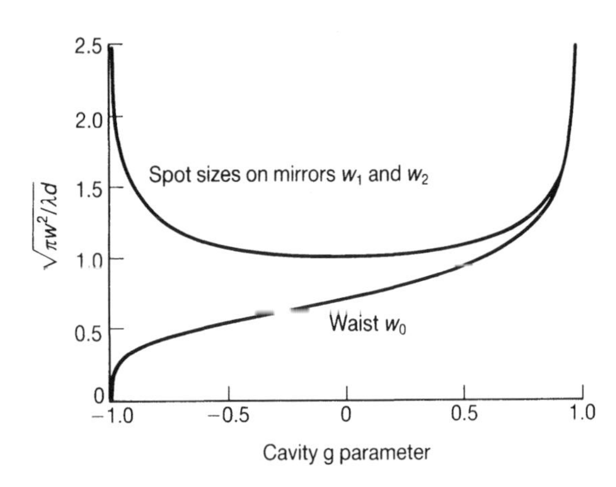

Figure 4.16 Plot of spot size and waist of symmetrical cavity as a function of cavity g. Ordinate units are $\sqrt{\pi w^2/\lambda d}$. The diagram shows that the minimum spot sizes on the mirrors occur for the confocal cavity ($g = 0$).

4.8.1.2 Quasi-hemispherical resonators $b_1 = \infty$; $b_2 > d$

In general Gaussian beam spot sizes vary quite slowly with the cavity parameters due to the fourth-root terms in Eqs (4.66) and (4.67). However, as the unstable condition is approached the spot sizes vary more rapidly and these regions are very sensitive to misalignment. The near-hemispherical cavity offers advantages of ease of alignment since it has one plane mirror and the ability to adjust the spot size by slight length changes, Δd, of the cavity. Figure 4.17 plots the spot size as a function of d/b for a quasi-hemispherical cavity. The spot size on the curved mirror increases rapidly as the exact hemispherical condition ($d = b$) is approached and, at the same time, the waist, which occurs on the plane mirror, approaches zero. Slight adjustment of the cavity length can match the mode volume of the Gaussian beam to the cavity aperture or gain medium. This cavity is widely used for gas and solid-state lasers.

4.8.1.3 Concave–convex resonators

In this type of cavity the mirror spot sizes can be made large by effectively placing the waist outside the cavity. For the convex mirror the g parameter is large (curvature is considered negative) and the g parameter for the concave mirror is small. These resonators require large radii of curvature mirrors and are sensitive to misalignment and so are rarely used in practice.

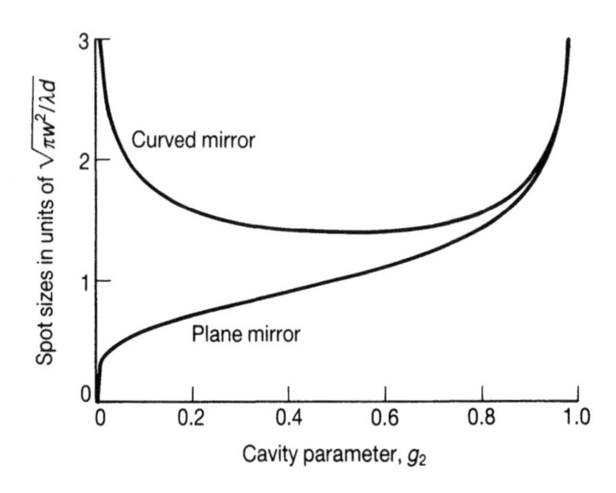

Figure 4.17 Plot of quasi hemispherical cavity spot size as a function of the cavity g parameters with $g_1 = 1$.

4.8.1.4 Asymmetric (unstable) confocal resonators

Two different radii mirrors with their foci coincident have a spacing given by

$$d = b_1/2 + b_2/2$$

which can be manipulated to give

$$g_1 + g_2 = 2g_1g_2 \tag{4.72}$$

This equation is plotted on the stability diagram as the dash-dotted line and shows two 'branches'. The positive branch has $g_1g_2 > +1$ and is always unstable save at the singular point $(1, 1)$, the plane parallel cavity. The negative branch is also always unstable except at the singular point $(0, 0)$, the symmetrical confocal cavity. Thus the stability of the symmetrical confocal cavity relies on the mirrors being identical, which can be achieved in manufacture by polishing simultaneously. These unstable branches are used to produce large spot sizes for high-gain lasers which require the light to make only a few transits of the cavity before saturating the gain medium.

4.9 Higher-order Gaussian beam modes

In Section 4.5 the TEM_{00} mode solution was obtained by assuming that the beam has circular symmetry. It was seen to be a mode of a curved-mirror open resonator when the wavefront curvatures match the mirror surfaces, requiring the wavefronts to be spherical over the extent of the beam. Other solutions of the paraxial wave equation exist with wavefront curvatures which match the mirror curvatures and with a sufficiently low diffraction loss to be a mode of the cavity. In some cavities the circular symmetry will be disturbed (for example, a Brewster angle window may be present or in a ring cavity as shown in Figure 5.20 the focusing conditions are different in the two planes of the ring). For these a solution of the paraxial wave equation in cartesian coordinates will be more appropriate than cylindrical coordinates. In this section we will discuss more general solutions of the paraxial wave equations and these will show radial phase reversals, causing nulls in the beam irradiance profiles.

4.9.1 Cartesian coordinate solution of the paraxial wave equation

Writing the transverse Laplacian in the paraxial wave Eq. (4.30) in cartesian coordinates gives

$$\frac{\partial^2 \psi}{\partial x^2} + \frac{\partial^2 \psi}{\partial y^2} - \mathrm{j}2k\frac{\partial \psi}{\partial z} = 0 \tag{4.73}$$

As before, the equation can be solved with the aid of a trial solution into which some experimential observations will be incorporated. The wavefront curvature must appear in the same form as for the TEM_{00} solution, i.e. it must still match the mirror curvature and take the spot size $w(z)$ as a common scaling parameter for all the solutions, indicating that the lowest-order Gaussian solution should only be a special case of the general solution. Experience with these cavities has shown that the higher-order modes have large amplitude variations in the transverse plane with rapid phase reversals at nodal points in the field amplitude.

Let us write the trial solution in the form

$$\psi(x, y, z) = Ag\left(\frac{x}{w(z)}\right)h\left(\frac{y}{w(z)}\right) \mathrm{e}^{-\mathrm{j}P(z)}\,\mathrm{e}^{-\mathrm{j}k(x^2+y^2)/2q(z)} \tag{4.74}$$

Here A is a constant which must be included in this solution since the field may be zero at $x = y = 0$; not the case for the lowest-order solution where A was taken as unity for algebraic convenience. The radial variation of the amplitude is included in the functions g and h to be determined. Making g and h functions of $(x/w(z))$ and $(y/w(z))$ ensures that the general solutions will scale according to the spot size $w(z)$. Forcing the solution to have the same wavefront curvature but a more complicated amplitude variation than the Gaussian means that the longitudinal phase constant $P(z)$ must be different from the lowest-order mode. Consequently along with g and h the other function to be determined is $P(z)$. On substitution of Eq. (4.74) into Eq. (4.73) two further differential equations for the functions g and h will result. It will be assumed that the complex beam parameter $q(z)$ will follow the same propagation law as for the TEM_{00} mode solution since it involves the common scaling parameter, $w(z)$, as well as the wavefront curvature.

From Eq. (4.37) we had

$$\frac{\mathrm{d}q(z)}{\mathrm{d}z} = 1$$

Introducing the new variables

$$\xi(x) = \frac{\sqrt{2}x}{w(z)} \text{ and } \eta(y) = \frac{\sqrt{2}y}{w(z)}$$

$$\frac{\partial g}{\partial x} = \frac{\mathrm{d}g}{\mathrm{d}\xi}\frac{\mathrm{d}\xi}{\partial x} = \frac{\sqrt{2}}{w(z)}\frac{\mathrm{d}g}{\mathrm{d}\xi}$$

$$\frac{\partial^2 g}{\partial x^2} = \frac{2}{w^2(z)}\frac{\mathrm{d}^2 g}{\mathrm{d}\xi^2}$$

$$\frac{\partial g}{\partial z} = \frac{\mathrm{d}g}{\mathrm{d}\xi}\frac{\mathrm{d}\xi}{\mathrm{d}z} = -\frac{\sqrt{2}x}{w^2(z)}\frac{\mathrm{d}w}{\mathrm{d}z}\frac{\mathrm{d}g}{\mathrm{d}\xi}$$

with analogous results for the partial derivatives of $h(\eta)$. Substitution into Eq. (4.73) and using Eqs (4.41) and (4.42) results in

$$\frac{1}{g(\xi)}\left\{\frac{\mathrm{d}^2 g}{\mathrm{d}\xi^2} - 2\xi\frac{\mathrm{d}g}{\mathrm{d}\xi}\right\} + \frac{1}{h(\eta)}\left\{\frac{\mathrm{d}^2 h}{\mathrm{d}\eta^2} - 2\eta\frac{\mathrm{d}h}{\mathrm{d}\eta}\right\} - \left\{\frac{2\,\mathrm{j}k}{q(z)} + 2k\frac{\mathrm{d}P}{\mathrm{d}z}\right\}w^2(z) = 0 \tag{4.75}$$

To obtain a solution of this equation it is necessary to recall that the beam is bound and retains its transverse profile as it propagates, since, to be a mode of a cavity, it must reproduce itself on each roundtrip. For this to be the case then the amplitude profiles determined by h and g must be independent of z as well as independent of each other. The differential equation Eq. (4.75) is then separated into three parts each containing independent variables. Each part can be equated to a constant:

$$\frac{1}{g(\xi)}\left\{\frac{\mathrm{d}^2 g}{\mathrm{d}\xi^2} - 2\xi\frac{\mathrm{d}g}{\mathrm{d}\xi}\right\} = -2m \tag{4.76}$$

$$\frac{1}{h(\eta)}\left\{\frac{\mathrm{d}^2 h}{\mathrm{d}\eta^2} - 2\eta\frac{\mathrm{d}h}{\mathrm{d}\eta}\right\} = -2n \tag{4.77}$$

$$-w^2(z)\left\{\frac{2\mathrm{j}k}{q(z)} + 2k\frac{\mathrm{d}P}{\mathrm{d}z}\right\} = 2(m+n) \tag{4.78}$$

The equation for g can then be written

$$\frac{\mathrm{d}^2 g}{\mathrm{d}\xi^2} - 2\xi\frac{\mathrm{d}g}{\mathrm{d}\xi} + 2mg = 0 \tag{4.79a}$$

Similarly,

$$\frac{\mathrm{d}^2 h}{\mathrm{d}\eta^2} - 2\xi\frac{\mathrm{d}h}{\mathrm{d}\eta} + 2mh = 0 \tag{4.79b}$$

These differential equations have a complete set of finite solutions if the constants m and n are integers. In quantum mechanics the Schrödinger wave equation when applied to the harmonic oscillator produces a similar equation to (4.79) [Ref. 10]. The solution to Eq. (4.79) is a complete set of orthogonal functions known as *Hermite functions*. The trial functions g and h become

$$g(\xi) = H_m\left(\sqrt{2}\,\frac{x}{w(z)}\right) \quad m = 0, 1, 2, 3 \ldots. \tag{4.80}$$

$$h(\eta) = H_n\left(\sqrt{2}\,\frac{y}{w(z)}\right) \quad n = 0, 1, 2, 3 \ldots. \tag{4.81}$$

where H_m and H_n are *Hermite polynomials* of order m and n, respectively, the properties of which will be discussed in the next section. Equation (4.78) can now be solved by rewriting it as

$$\frac{\mathrm{d}P}{\mathrm{d}z} = -\frac{\mathrm{j}}{q} - \frac{m+n}{kw^2(z)} = \frac{-\mathrm{j}z}{z^2 + z_R^2} - \frac{z_R(m+n+1)}{z^2 + z_R^2}$$

Integrating with the same boundary conditions as for Eq. (4.46) leads to

$$P(z) = -\mathrm{j}\,\log_e\sqrt{1 + \frac{z^2}{z_R^2}} - (m+n+1)\,\arctan\left(\frac{z}{z_R}\right)$$

or

$$\mathrm{e}^{-\mathrm{j}P(z)} = \frac{1}{\sqrt{1 + z^2/z_R^2}}\,\mathrm{e}^{\mathrm{j}(m+n+1)\,\arctan(z/z_R)}$$

Using Eq. (4.48) gives

$$\mathrm{e}^{-\mathrm{j}P(z)} = \frac{w_0}{w(z)}\,\mathrm{e}^{\mathrm{j}(m+n+1)\,\arctan(z/z_R)} \tag{4.82}$$

Collecting together the terms, writing $q(z)$ in its real and imaginary parts, using Eq. (4.40) and combining into Eq. (4.74), the complete solution finally becomes

$$\begin{aligned}\mathscr{E}_{m,n} = A\frac{w_0}{w(z)}\,H_m\left(\sqrt{2}\,\frac{x}{w(z)}\right)H_n\left(\sqrt{2}\,\frac{y}{w(z)}\right)\,\exp\left(-\frac{x^2+y^2}{w^2(z)}\right)\\ \times \exp[-\mathrm{j}(kz - (m+n+1)\arctan(z/z_R)]\\ \times \exp[-\mathrm{j}k(x^2+y^2)/2R(z)]\end{aligned} \tag{4.83}$$

The terms can be compared with Eq. (4.47), noting that in Eq. (4.83) the constant A is not necessarily the field at $x = y = 0$, since cases will arise where this is zero. It can be found from a normalization of the total power in the beam using the integral of Eq. (4.5l) in Section 4.6.2.

4.9.2 Physical properties of Hermite–Gaussian beams

The beams described by Eq. (4.83) have an amplitude distribution in the transverse plane which are the product of a Gaussian function and a Hermite polynomial and are referred to as *Hermite–Gaussian beams or modes*. A list of the first six Hermite polynomials is given below using ξ for their argument.

$$\left.\begin{aligned} H_0(\xi) &= 1 \\ H_1(\xi) &= 2\xi \\ H_2(\xi) &= 4\xi^2 - 2 \\ H_3(\xi) &= 8\xi^3 - 12\xi \\ H_4(\xi) &= 16\xi^4 - 48\xi^2 + 12 \\ H_5(\xi) &= 32\xi^5 - 160\xi^3 + 120\xi \end{aligned}\right\} \tag{4.84}$$

The following recursion relation can be used to calculate the higher-order polynomials:

$$H_{m+1}(\xi) = 2\xi H_m(\xi) - 2mH_{m-1}(\xi) \tag{4.85}$$

These functions and their properties are discussed in many mathematical texts [Ref. 9] and a few of their properties are summarized here without proof.

The Hermite–Gaussian function of order m is written

$$\phi_m(\xi) = C_m H_m(\xi)\, \mathrm{e}^{-\xi^2/2} \tag{4.86}$$

This can be normalized to unity

$$\int_{-\infty}^{\infty} \phi_m^*(\xi)\phi_m(\xi)\, \mathrm{d}\xi = 1$$

which can be shown [Ref. 6] to give

$$C_m = (2^m m! \sqrt{\pi})^{-1/2} \tag{4.87}$$

The functions are orthogonal, that is,

$$\int_{-\infty}^{\infty} \phi_n(\xi)\phi_m(\xi)\, \mathrm{d}\xi = 0 \tag{4.88}$$

for $m \neq n$. This property means that any arbitrary paraxial beam can be expanded as a summation of Hermite–Gaussian beams of the same complex beam parameter $q(z)$, the constant C_{mn} being determined by the physical constraints of the system. An example of this is the injection of a beam into a curved-mirror cavity. The input beam in general is not a mode of this cavity but can be considered as a summation of Hermite–Gaussian modes of this cavity. The number of modes included in the summation will depend on the diameter of the input beam compared with the TEM_{00} mode of the cavity together with its transverse irradiance profile and radial extent.

An important property of Hermite–Gaussian functions is that they are their own Fourier transforms and, in addition, the convolution of a Hermite–Gaussian with a Gaussian function yields a Hermite–Gaussian [Ref. 6]. This means that, in Fresnel diffraction, the Hermite–Gaussian amplitude distribution reproduces itself, since for this a convolution of the amplitude function with a Gaussian is necessary (Eq. (4.21)). Although other solutions for Eq. (4.75) can be produced [Ref. 8] the result is compatible with observations – lasers do oscillate with Hermite–Gaussian amplitude distributions. In fact, the form for the variables ξ and η was chosen to produce the differential equation Eq. (4.75) in the form which would produce Hermite–Gaussian functions as solutions.

The exponential term, $e^{-\xi^2/2}$, ensures that the function, $\phi_m(\xi)$ in Eq. (4.66) goes rapidly to zero as $\xi \Rightarrow \infty$. The order of the Hermite polynomial determines the number of quasi-sinusoidal oscillations of the function. The solid curves of Figure 4.18 give the amplitude profiles, in one dimension, for some low-order Hermite–Gaussian modes. As the amplitude passes through zero a rapid phase reversal occurs and thus the mode number determines the number of these reversals and the designation of

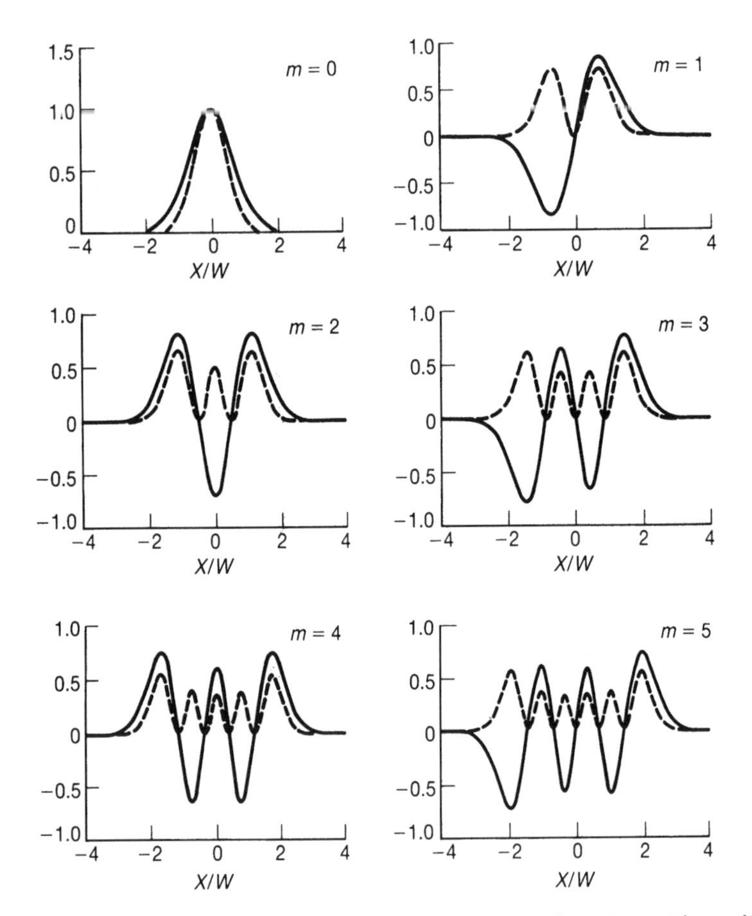

Figure 4.18 Plots of first six Hermite–Gaussian functions. The solid lines give the variation of the field amplitudes in the modes and the dashed lines the irradiance distributions. The curves are normalized for a constant power with $n = 0$.

the mode uses this number. Since the complete mode is the product of the functions describing the x and y variations of the field amplitude the mode must be designated by the two numbers m and n. These transverse modes are designated by $TEM_{m,n}$, the lowest-order mode discussed in Section 4.6 being the case $m = n = 0$.

The total power in the high-order mode is obtained as in Section 4.6.2 by integrating the Poynting vector over the amplitude distribution. The mode may then be normalized to unit power by using the identity of Eq. (4.87). The constant A is then given by

$$A = \sqrt{2\varepsilon_0 c}\left(\frac{2}{w_0^2 \pi 2^{m+n} m! n!}\right)^{1/2} \tag{4.89}$$

The irradiance distributions, given by the square of the functions is shown by the dashed curves in Figure 4.18. The total power in these plots is normalized to unity with $n = 0$.

The other factor which distinguishes the higher-order modes from the fundamental mode is the Guoy shift term $(m + n + 1)\arctan(z/z_R)$, which depends on the mode numbers. The fact that the radial distribution is larger than the TEM_{00} mode means that they require larger x and y components of the wave vector $\mathbf{k}$. This can be explained by consideration of the Fourier transform conjugate pairs x, y and k_x, k_y. To produce a given amplitude distribution requires a distribution of transverse wave vectors given by the Fourier transform of the amplitude distribution. Since the wavefront curvature is determined by the mirror curvatures so fixing the beam divergence and hence k_x and k_y, then higher values of k_z are required for the larger amplitude distributions associated with the higher-order modes.

4.9.3 Higher-order Laguerre–Gaussian modes

The transverse Laplacian in the paraxial wave equation Eq. (4.30) can be written in cylindrical coordinates instead of cartesian coordinates as in Eq. (4.73). This could be more appropriate in the following cases:

1. A high-gain laser in which the gain counteracts any astigmatic optical loss differences in the xz and yz planes
2. A low-gain laser with very well-aligned mirrors and no optical element in the cavity
3. Passive cavity (no gain) and no intra-cavity optics.

The trial solution in this case is

$$\psi(r, \phi, z) = Ag\left(\sqrt{2}\,\frac{r}{w(z)}\right)\, e^{-jl\phi}\, e^{-jP(z)}\, e^{-jkr^2/2q(z)} \tag{4.90}$$

The unknown radial function is g with the azimuth variation, assumed to be sinusoidal, determined by the term $e^{-jl\phi}$. Substitution into Eq. (4.30) and solving yields

$$\psi(r, \phi, z) = \left[\left(\sqrt{2}\,\frac{r}{w(z)}\right)^l L_p^l\left(2\,\frac{r^2}{w^2(z)}\right)\right] e^{-r^2/w^2(z)}\, e^{-jl\phi}\, e^{-jP(z)}\, e^{-jkr^2/2R(z)} \tag{4.91}$$

where l and p are the mode numbers determining the number of phase reversals, respectively, in the azimuthal and radial directions. The modes are designated as TEM_{pl} modes. In the solution $L_p^l(u)$ is a generalized *Laguerre polynomial* which obeys the differential equation

$$u\,\frac{d^2L_p^l}{du^2} + (l + 1 - u)\,\frac{dL_p^l}{du} + pL_p^l = 0$$

These polynomials are also documented in mathematical texts [Ref. 9]. Some low-order ones are

$$\left.\begin{aligned} L_0^l(u) &= 1 \\ L_1^l(u) &= l + 1 - u \\ L_2^l(u) &= \tfrac{1}{2}(l + 1)(l + 2) - (l + 2)u + \tfrac{1}{2}u^2 \end{aligned}\right\} \tag{4.92}$$

Two examples of these functions are shown by the solid lines in Figure 4.19 while the dotted lines indicate the irradiance distribution.

The amplitude term again is a Gaussian multiplied by the polynomial in the square brackets of Eq. (4.91) and the Guoy shift for these modes is given by

$$(2p + l + 1)\arctan(z/z_R) \tag{4.93}$$

4.9.4 Spot size for high-order transverse modes

The irradiance profiles of Figure 4.18 for the Hermite–Gaussian modes can be compared with photographs of some transverse modes produced using a helium–neon laser and shown in Figure 4.20. Each lobe of the modes is 180° out of phase with

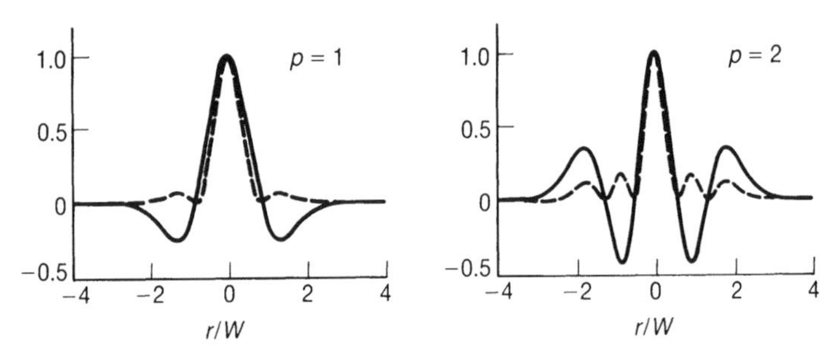

Figure 4.19 Plots for two Laguerre–Gaussian modes with the azimuthal number, $l = 0$. The solid lines are field and the dashed lines are irradiance profiles.

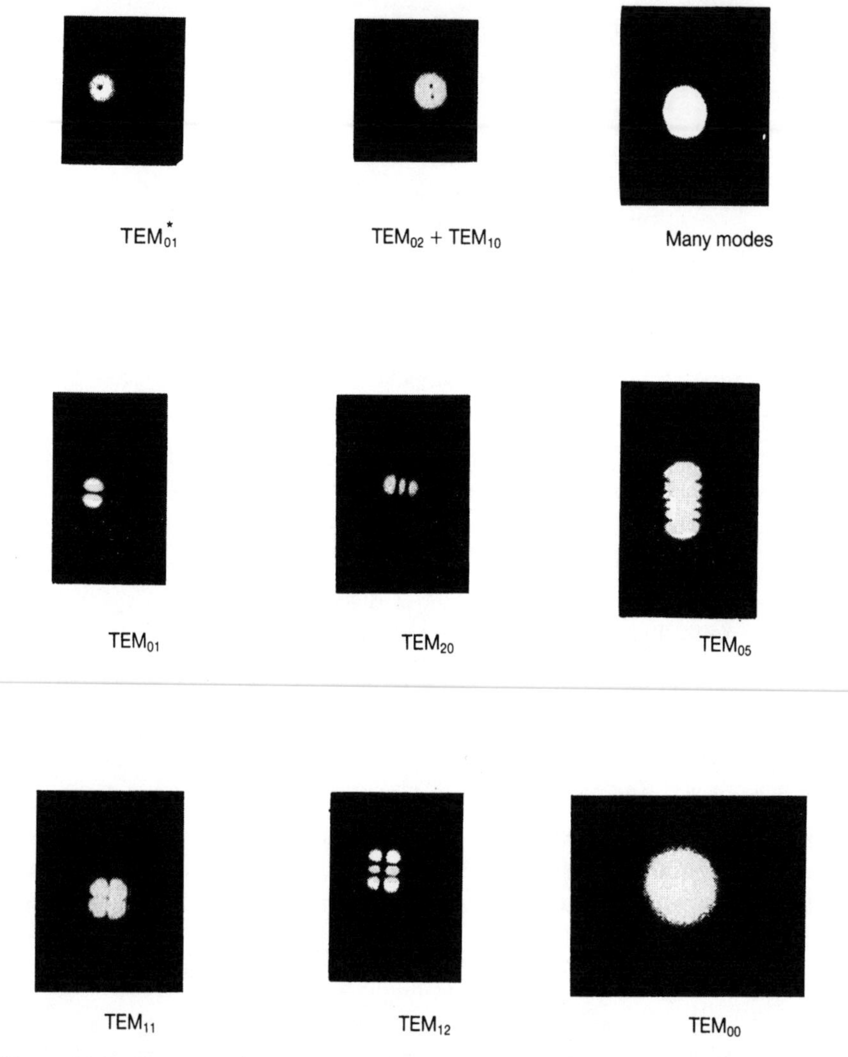

Figure 4.20 Photos of some Hermite–Gaussian modes taken from a helium–neon laser. All the modes are photographed to scale except the TEM_{00} mode which is magnified for clarity.

respect to its neighbour. All the modes except the TEM_{00} are photographed approximately to scale. In Figure 4.21 the transverse mode patterns for the circularly symmetrical modes of Figure 4.19 are shown. These modes are seen to be due to $(p+1)$ concentric circles together with $2l$ radial nodes. Since both the Hermite–Gaussian and Laguerre–Gaussian form complete sets then either can be used for an

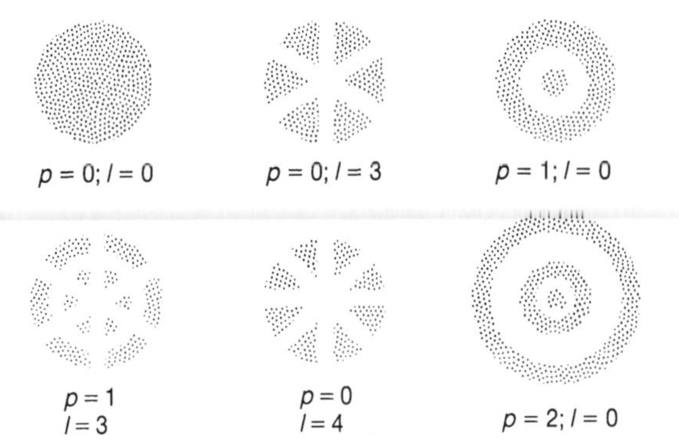

Figure 4.21 Sketches of some Laguerre–Gaussian mode patterns.

arbitrary optical beam, and so it must be possible to expand any one in terms of the the other.

The mode designated TEM^*_{01} in Figure 4.20 might appear at first to be in the wrong group. Reference to the amplitude factor in Eq. (4.91) shows that Laguerre–Gaussian modes with $l = 0$, i.e. zero azimuthal phase reversals, never have a null on axis as this mode has. It must be concluded that this mode is a combination of two Hermite–Gaussian modes TEM_{01} and TEM_{10} and is referred to as the *donut mode*. Other combinations in quadrature of Hermite–Gaussian modes to produce circularly symmetrical spots are possible.

It is clear from the the function plots of Figures 4.18 and 4.19 and the photographs of their physical appearance that the overall diameter of the field pattern increases as the mode numbers increase. When such a beam passes through a circular aperture of radius a some of the power at the edge of the beam will not be transmitted and the beam power will decrease. This is referred to as diffraction loss, and the aperture transmission can be readily calculated by integrating the mode irradiance profile over the extent of the apertures:

$$T_{m,n} = \frac{\displaystyle\int_0^{2\pi}\int_0^{a} \mathscr{E}_{m,n}\mathscr{E}^*_{m,n} r \, \mathrm{d}r \, \mathrm{d}\phi}{\displaystyle\int_0^{2\pi}\int_0^{\infty} \mathscr{E}_{m,n}\mathscr{E}^*_{m,n} r \, \mathrm{d}r \, \mathrm{d}\phi} \tag{4.94}$$

The integrals can be evaluated for both Hermite–Gaussian beams and for Laguerre–Gaussian beams using the appropriate amplitude functions. Thus for the (0, 0) and (0, 1) Hermite–Gaussian modes we obtain

$$T_{00} = 1 - \mathrm{e}^{-2(a/w(z))^2} \tag{4.95}$$

$$T_{01} = 1 - [1 + 2(a/w(z))^2]\, \mathrm{e}^{-2(a/w(z))^2} \tag{4.96}$$

Figure 4.22 shows these transmission functions plotted as a function of $a/w(z)$.The graph for T_{00} can be used to find the spot size of a given Gaussian beam

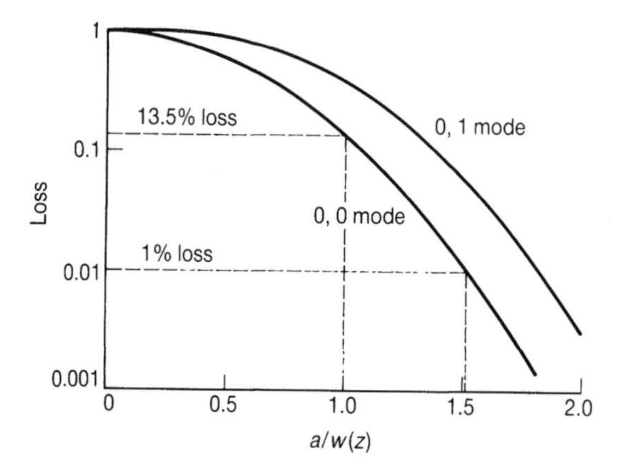

Figure 4.22 Plots of the aperture transmission function for TEM_{00} and TEM_{01} transverse modes.

by simply measuring the power transmission of an aperture of known radius. An alternative method is to adjust the diameter of an iris until 86.5 per cent (13.5 per cent loss) of the beam power is transmitted. As shown from the graph, the aperture radius now equals the spot size. A useful criterion used in the design of lasers is that 99 per cent of the beam is transmitted by an aperture of diameter equal to πw.

These high-order transverse modes can be excited in lasers having apertures which allow their diffraction loss to be less than the gain. Since the diameter of the mode increases with mode number so consequently will the diffraction loss. This is illustrated in Figure 4.22, in which the diffraction loss for the TEM_{00} and the TEM_{01} modes are compared. The condition of Eq. (4.56) expressing low diffraction loss in the cavity is now more precisely explained by saying that when $a \gg w(z)$ the aperture produces small truncation of the Gaussian beam. If oscillation in only the TEM_{00} mode is required then it is necessary to closely match the aperture size (normally the diameter of the gain medium) to the TEM_{00} mode spot size for oscillation only in this mode. If the diffraction loss is not high enough for higher-order modes many of them can oscillate simultaneously, producing a complex-looking spot (see Figure 4.20). The high-order transverse modes can be eliminated by insertion of an intra-cavity aperture.

It will be shown in Chapter 5 that the resonant frequency of the transverse modes depends on the mode numbers because of the dependence of the Guoy shift on mode number. If a paraxial beam of light is injected into a passive cavity then it can be considered as the summation of a number of Hermite–Gaussian (or Laguerre–Gaussian) modes of that passive cavity. The family of modes will be determined by their common scaling factor, the TEM_{00} mode waist size. We will show in Chapter 5 that in general they will all have different cavity resonant frequencies and they can be made to resonate for different mirror separations d. The transmitted light will take on the physical appearance of each mode as it resonates.

The concept of spot size as defined for the Gaussian beam cannot apply to the multimode beam and a number of ways to determine the 'diameter' of a multimode beam have been suggested [Ref. 1]. The most consistent of these is to define it from a measurement in which a knife edge is scanned across the beam. The multimode beam spot size, W, is defined as the distance between the locations of 16 per cent and 84 per cent transmission past the knife edge. It is left as an exercise to show that this definition is consistent with that for the TEM_{00} mode.

The high-order modes are a product of some function (Hermite polynomial) and a Gaussian function and thus they have an amplitude distribution larger than the fundamental Gaussian by a constant factor. It is a property of Hermite–Gaussian (and Laguerre–Gaussian) polynomials [Refs 8, 11] that the radial distance to the outer maximum is approximately $\sqrt{n+1}\, w(z)$, where n is the mode number. Hence to a good approximation the spot size, W, of the multi mode beam can be written

$$W(z) = Mw(z) \tag{4.97}$$

where M is a constant. In the far field the spot size of the fundamental Gaussian beam is given by Eq. (4.48)

$$w(z) = \frac{\lambda}{\pi w_0} z$$

Substituting for the multimode spot using Eq. (4.97)

$$W = \frac{M^2 \lambda}{\pi W_0} z \tag{4.98}$$

The far-field angular divergence as measured by the diameter of the spot divided by z is therefore

$$\Theta = \frac{2M^2 \lambda}{\pi W_0} \tag{4.99}$$

Comparison with the divergence of a Gaussian beam (Eq. (4.50)) shows the multimode beam to have an angular divergence M^2 times that for an ideal Gaussian if it had a waist equal to the multimode waist, W_0. For this reason, M^2 is called the *times diffraction limited number*.

Laser beams described as TEM_{00} in general are never perfectly Gaussian and the M^2 factor has been introduced [Ref. 2] to define the spot size of a multimode or imperfect Gaussian beam relative to the ideal Gaussian. An ideal Gaussian beam has the smallest angular divergence for any beam profile and has $M^2 = 1.0$. The M^2 number is used along with the equations for Gaussian beams to calculate the spot size at any distance from the waist for a multimode or non-exact Gaussian beam. The simple rule is to replace the wavelength, λ, in the Gaussian beam equations by $M^2\lambda$.

The M^2 factor can be measured by focusing the multimode beam to a waist using a convex lens and measuring the diameter of this waist by scanning a knife edge across it as described above. The angular divergence into the far field is obtained again from the spot size at another location and M^2 calculated using Eq. (4.99). If the beam is astigmatic then it must be scanned in two directions and two values produced for M^2.

4.10 Propagation of Gaussian beams in optical systems

An important requirement with laser beams is the ability to calculate the Gaussian beam parameters within an optical system. Two particular examples would be a focusing system for the delivery of a laser beam to a target or a beam in a multi-mirror cavity. A transformation law based on geometric optics has been developed for Gaussian beams. In this section this law will be derived for the optical elements which we need, namely lenses and free space. In general terms it uses the laws of geometric optics but allows for the divergence diffraction of the Gaussian beam.

4.10.1 Transformation of a Gaussian beam

The propagation of a Gaussian beam is described by the complex beam parameter, $q(z)$, which is defined in Eq. (4.38) as

$$q(z) = z + \mathrm{j}z_{\mathrm{R}}$$

in Eqs (4.39) and (4.40) it was written in the form

$$\frac{1}{q(z)} = \frac{1}{z + \mathrm{j}z_{\mathrm{R}}} = \frac{z}{z^2 + z_{\mathrm{R}}^2} - \mathrm{j}\,\frac{z_0}{z^2 + z_{\mathrm{R}}^2}$$

$$= \frac{1}{R(z)} - \mathrm{j}\,\frac{\lambda}{\pi w^2(z)}$$

For the simplest optical element, a length d of free space q transforms to $q' = q + d$ and the new wavefront curvature and spot size evaluated accordingly from the real and imaginary parts of q'.

Figure 4.23 shows how a convex lens transforms a Gaussian beam into another Gaussian beam having different beam parameters. To evaluate the effect of the lens on the wavefront we consider a plane wave incident on the lens as shown in Figure 4.24. The parallel rays are refracted and pass through the focus of the lens. The effect of this is that the wavefront therefore assumes a curvature, f. The field of the wave at point B is now ahead of A by an amount Δd. If the new wavefront is spherical and paraxial then

$$\Delta d = (r^2 + f^2)^{1/2} - f \cong f[1 + \tfrac{1}{2}(r/f)^2] - f$$

and

$$\Delta d \cong \frac{r^2}{2f} \qquad (4.100)$$

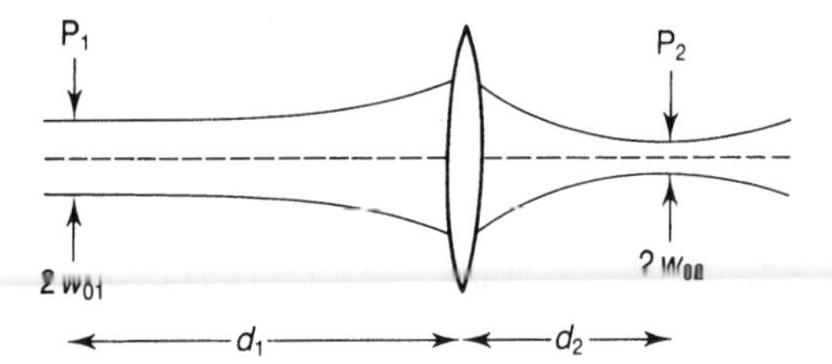

Figure 4.23 Focusing of a Gaussian beam by a convex lens.

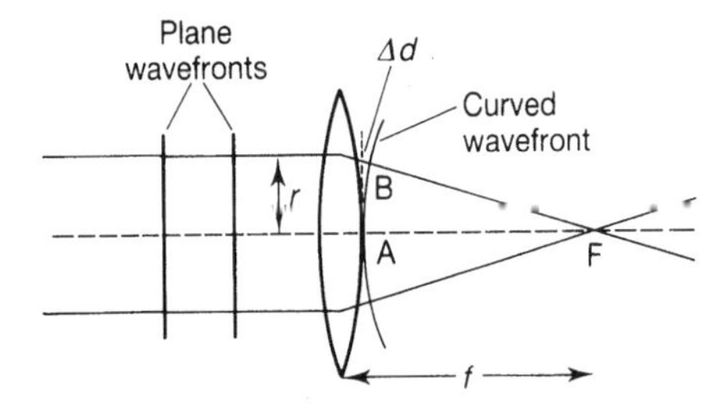

Figure 4.24 Effect of a convex lens on a plane wavefront.

If the input light to the lens is a Gaussian beam its wavefront curvature will be changed by the same amount and the radial phase factor is altered from

$$e^{-jkr^2/2R} \quad \text{to} \quad e^{j[-kr^2/2R + kr^2/2f]}$$

This means that the wavefront curvature is transformed from R to R' such that

$$-kr^2/2R + kr^2/2f = -kr^2/2R'$$

or

$$1/R' = 1/R - 1/f \tag{4.101}$$

Since the lens is thin, the spot size will be equal on either side of the lens. Referring to Eq. (4.40), this transformation (Eq. 4.101) of the wavefront curvature for a Gaussian beam can be written as a transformation of the complex beam parameter:

$$1/q' = 1/q - 1/f \tag{4.102}$$

Equation (4.101) above is the lens imaging law for geometric optics since an object point source would have a curvature R and produce an image point of curvature $-R'$. The same law is applied to a Gaussian beam by replacing the wavefront curvature with the complex beam parameter. Rearrangement of Eq. (4.102) produces

$$q' = \frac{q}{-q/f + 1} \tag{4.103}$$

This transformation law for a Gaussian beam can be shown to obey a general law known as the *ABCD law for Gaussian beams*:

$$q' = \frac{Aq + B}{Cq + D} \tag{4.104}$$

where A, B, C and D are the elements of the matrix which describes the optical system as discussed in Section 4.1. A rigorous proof of this law is possible [Ref. 6] but it is more usual to prove it for two examples which we are going to use:

1. *Length of free space*: Using matrix (4.3) for this optical element, the transformation law gives

$$q' = (1 \times q + d)/(0 \times q + 1) = q + d$$

2. *Thin lens*: Again using the matrix for the lens equation (4.6) and applying the law produces

$$q' = \frac{1 \times q + 0}{-(1/\mathrm{f})q + 1}$$

which leads to

$$1/q' = -1/f + 1/q$$

When optical components are cascaded then the complete matrix for the system may be used in the *ABCD* law. Care must be taken to use the *ABCD* matrix for the same start and end as the transformation.

4.10.2 Application of the *ABCD* law to optical cavities

The application of the *ABCD* law to open-cavity resonators provides a more general and rigorous approach to the determination of the Gaussian beam parameters of the resonator. Since a mode is a self-reproducing field pattern then the complex beam parameter of the Gaussian beam which is a mode of the cavity must also reproduce itself after a complete roundtrip. If the wave starts at some plane located at z this mode condition can be written

$$q(z + \text{roundtrip distance}) = q(z)$$

Using the $ABCD$ matrix for the unit cell of the cavity with the same starting location the mode condition becomes:

$$q(z) = \frac{A\, q(z) + B}{C\, q(z) + D}$$

Rearrangement of this equation results in

$$Cq^2 + Dq = Aq + B$$

Writing as a quadratic in $1/q$ gives

$$B(1/q)^2 + (1/q)(A - D) - C = 0$$

$$\frac{1}{q(z)} = -\frac{A - D}{2B} \pm \frac{1}{B}\left[\left(\frac{A - D}{2}\right)^2 + BC\right]^{1/2}$$

Using the fact given in Section 4.1 that the determinant of the matrix $[ABCD]$ is 1,

$$AD - BC = 1$$

Separating into real and imaginary parts:

$$\frac{1}{q(z)} = -\left(\frac{A - D}{2B}\right) - \mathrm{j}\left\{\frac{[1 - ((A + D)/2)^2]^{1/2}}{B}\right\} \tag{4.105}$$

Using Eq. (4.40) and equating the real and imaginary results in the radius of curvature being given by

$$R(z) = -\left(\frac{2B}{A - D}\right) \tag{4.106}$$

and the spot size by

$$w^2(z) = \frac{\lambda}{\pi}\frac{B}{[1 - ((A + D)/2)^2]^{1/2}} \tag{4.107}$$

Since the spot size is a real observable quantity then

$$-1 \leqslant \frac{A + D}{2} \leqslant 1$$

which is exactly the cavity stability condition (4.13).

Equation (4.105) gives the complex beam parameter at a chosen point in the cavity and the $ABCD$ matrix must be evaluated with this as the starting plane of the unit cell.

The *ABCD* matrix for the unit cell starting at mirror 1 in Figure 4.3 as given by Eq. (4.9) is used to evaluate the wavefront curvature in Eq. (4.106):

$$A - D = 1 - d/f_2 - [(1 - d/f_1)(1 - d/f_2) - d/f_1]$$

$$A - D = \frac{2d}{f_1} - \frac{d^2}{f_1 f_2}$$

$$B = d + d(1 - d/f_2)$$

giving

$$R(z) = \frac{-2d - 2d + 2d^2/f_2}{2d/f_1 - 2d/f_2} = -2f_1 = -R_1$$

showing that the curvature of the wavefront on the mirror equals the curvature of the mirror. It is left as an exercise to show that the same is true for the other mirror by using the unit cell which starts at mirror 2.

Equation (4.107) will now be used to evaluate the spot size on mirror 1. In the development of Eq. (4.17) it was shown that

$$\frac{A + D}{2} = 2g_1 g_2 - 1$$

which is rearranged to give

$$1 - \left(\frac{A + D}{2}\right)^2 = 1 - 4g_1^2 g_2^2 - 1 + 4g_1 g_2 = 4g_1 g_2 \,(1 - g_1 g_2)$$

Also from Eq. (4.9)

$$B = d(2 - d/f_2) = d(2 - 2 + 2g_2) = 2g_2 d$$

Substitution into Eq. (4.107) gives

$$\left(\frac{\pi w_1^2}{\lambda}\right)^2 = \frac{4g_2^2 d^2}{4g_1 g_2 \,(1 - g_1 g_2)}$$

This is identical to Eq. (4.66) obtained with the premise that the wavefront curvatures match the mirror curvatures. This transformation law is particularly useful in more complicated cavities such as rings which have more than two mirrors where wavefront curvatures do not match the mirror surfaces when the incident light is not at normal incidence.

4.10.3 Application of the *ABCD* law to focusing and mode matching a Gaussian beam

The *ABCD* law can now be applied to the determination of the beam parameters for a Gaussian beam transformed by a lens. A particular practical application might be the requirement to determine the size of a waist and its exact location for the purposes of heat treatment by focusing a laser beam. In another application it may be necessary to transform a Gaussian beam from a given laser to the TEM_{00} mode of another open resonator such as a curved mirror Fabry–Perot interferometer. This latter application is known as mode matching and both applications can be solved with the same algebra. The variables and the geometry are shown in Figure 4.23 with the input beam on the left of the lens and the desired beam on the right. The transformation law is applied from the waist P_1 to the waist P_2.

The complex beam parameter at P_1 is $q_{01} = \mathrm{j}\pi w_{01}^2/\lambda = \mathrm{j}z_{R1}$

The complex beam parameter at P_2 is $q_{02} = \mathrm{j}\pi w_{02}^2/\lambda = \mathrm{j}z_{R2}$

Note that a new Gaussian beam has been produced by the lens and so it has a new coordinate system, i.e. $z = 0$ is always at the waist. Applying transformation (4.104)

$$q_{02} = \frac{Aq_{01} + B}{Cq_{01} + D} \tag{4.108}$$

The *ABCD* matrix to be used for this transformation is for a length of free space d_1 + lens + length of free space d_2:

$$\begin{bmatrix} A & B \\ C & D \end{bmatrix} = \begin{bmatrix} 1 - d_2/f & d_1 + d_2 - d_1 d_2/f \\ -1/f & 1 - d_1/f \end{bmatrix} \tag{4.109}$$

Rearrangement of Eq. (4.108) gives

$$Cq_{01}q_{02} + Dq_{02} = Aq_{01} + B$$

$$-Cz_{01}z_{R2} + \mathrm{j}Dz_{R2} = A\mathrm{j}z_{R1} + B$$

Equating the real and imaginary parts

$$B + Cz_{R1}z_{R2} = 0 \tag{4.110}$$

$$Az_{R1} - Dz_{R2} = 0 \tag{4.111}$$

Substituting the matrix elements from Eq. (4.109) into (4.110) gives

$$d_1 + d_2 - d_1/d_2/f - \frac{1}{f}\frac{\pi w_{01}^2}{\lambda}\cdot\frac{\pi w_{02}^2}{\lambda} = 0 \tag{4.112}$$

$$d_1 + d_2 - d_1 d_2/f - \frac{1}{f} f_0^2 = 0 \tag{4.113}$$

where

$$f_0 = \frac{\pi w_{01} w_{02}}{\lambda} \tag{4.114}$$

Substituting the matrix elements from Eq. (4.109) into Eq. (4.111) gives

$$\frac{\pi w_{01}^2}{\lambda}\left(1 - \frac{d_2}{f}\right) - \frac{\pi w_{02}^2}{\lambda}\left(1 - \frac{d_1}{f}\right) = 0 \tag{4.115}$$

Rearrangement of Eqs (4.113) and (4.114) produces:

$$d_2 = \frac{((f_0^2/f) - d_1)}{(1 - (d_1/f))} \tag{4.116}$$

$$1 - \frac{d_2}{f} = \left(1 - \frac{d_1}{\mathrm{f}}\right) \frac{w_{02}^2}{w_{01}^2} \tag{4.117}$$

Two equations are obtained with four unknowns, i.e. d_1, d_2, w_{01} and w_{02}, and solutions can be found for the two applications stated above – beam focusing and mode matching.

4.10.3.1 Beam focusing

In this case the unknowns required from Eqs (4.116) and (4.117) are the new waist w_{02} and its location d_2 in terms of the input waist and its location. After some algebra it can be shown that

$$w_{02}^2 = \frac{f^2 w_{01}^2}{z_{R1}^2 + (f - d_1)^2} \tag{4.118}$$

$$d_2 = f + f^2 \left\{ \frac{(d_1 - f)}{z_{R1}^2 + (f - d_1)^2} \right\} \tag{4.119}$$

The location of the waist P_2 is seen from Eq. (4.119) not to be located precisely at the focus of the lens although in many cases it is close enough to make little difference. The following regions summarize Eq. (4.119):

When $d_1 < f$ the waist occurs inside the focus, i.e. $d_2 < f$
When $d_1 = f$ the waist occurs at the focus, i.e. $d_2 = f$
When $d_1 > f$ the waist occurs outside the focus, i.e. $d_2 > f$
When $d_1 = \infty$ the waist occurs at the focus, i.e. $d_2 = f$

Writing Eq. (4.119) in the form

$$1 - \frac{d_2}{f} = \frac{1 - d_1/f}{(z_{R1}/f)^2 + (1 - d_1/f)^2} \tag{4.120}$$

we can plot $(1 - d_2/f)$ as a function of $(1 - d_1/f)$ to produce the dispersion-like function shown in Figure 4.25. Also shown on this graph by the dashed curve is the geometrical optics imaging relation between d_1 and d_2:

$$(1 - d_1/f)(1 - d_2/f) = 1 \tag{4.121}$$

This figure is a clear indication of the difference between geometrical and Gaussian optics. The waist location becomes different from the focus of the lens when the value of $|z_{R1}|$ becomes comparable to $|d_1 - f|$. It is left as an exercise to show that the turning points where the maximum departure of d_2 from the focal position occurs are given by

$$|d_1/f| = |1 - z_{R1}| \tag{4.122}$$

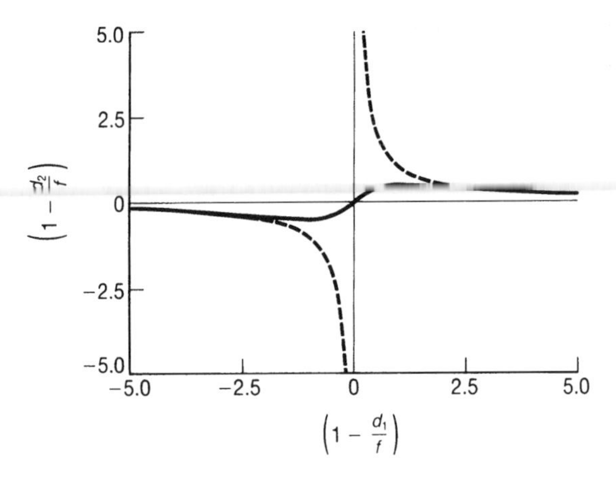

Figure 4.25 $(1 - d_1/f)$ plotted against $(1 - d_2/f)$ for geometric optics (dashed curves) and for Gaussian optics where $z_{R1} = f$.

The distance Δd_2 from the focus to the waist location at these positions is then given by

$$|\Delta d_2| = \frac{f^2}{2z_{R1}} \tag{4.123}$$

Using Eq. (4.48) the spot size on the lens, w_L, is obtained:

$$w_L^2 = w_{01}^2[1 + d_1^2/z_{R1}^2] = [z_{R1}^2 + d_1^2]w_{01}^2/z_{R1}^2$$

In many applications the input waist is far from the lens and so with $d_1 > f$ in Eq. (4.118) we obtain

$$w_{02}^2 \cong \frac{w_{01}^2 f^2}{z_{R1}^2 + d_1^2} = \frac{w_{01}^2 f^2}{w_L^2 z_{R1}^2/w_{01}^2} = \frac{w_{01}^4 f^2}{w_L^2 z_{R1}^2} = \frac{f^2\lambda^2}{\pi^2 w_L^2}$$

$$w_{02} \cong \frac{f\lambda}{\pi w_L} \tag{4.124}$$

Equation (4.124) is a more widely used approximation.

Small focused waist sizes are achieved by increasing the spot size on the lens. The limiting case is when the spot fills the lens with negligible spill-over and the criterion for this as obtained from Eq. (4.95) is $\pi w_L = D$, where D is the diameter of the lens. Then from Eq. (4.124)

$$w_{02} = \frac{f\lambda}{D} = F^{\#}\lambda \tag{4.125}$$

where $F^{\#}$ is the f-number of the lens, i.e. $F^{\#} = f/D$.

4.10.3.2 Mode matching a Gaussian beam to a cavity

Figure 4.26 shows how a lens can be used to transform a Gaussian beam from one cavity (e.g. a laser) to another (e.g. a Fabry–Perot interferometer). The requirement is to determine the location of the waists for a given lens with this time the waists known from the geometry of the cavities. Eliminating either d_1 or d_2 from Eqs (4.116) and (4.117) results in

$$d_1 = f \pm \frac{w_{01}}{w_{02}} [f^2 - f_0^2]^{1/2} \tag{4.126}$$

$$d_2 = f \pm \frac{w_{02}}{w_{01}} [f^2 - f_0^2]^{1/2} \tag{4.127}$$

The signs on the right-hand side of these equations must be both positive or both negative as summarized above.

Multiplying Eqs (4.126) and (4.127) together produces

$$(d_1 - f)(d_2 - f) = (f^2 - f_0^2)$$

Written in units of the lens focal length, this becomes

$$\left(\frac{d_1}{f} - 1\right)\left(\frac{d_2}{f} - 1\right) = 1 - \frac{f_0^2}{f^2} \tag{4.128}$$

Since the parameter f_0 is determined by particular combinations of waist sizes, w_{01} and w_{02} then the position of a lens of particular focal length for mode matching can be determined from Eq. (4.128). In some applications mode matching can be achieved using a variable focal length lens (zoom lens) and the two waist locations kept fixed. Mode matching will be discussed further in the next chapter when the Fabry–Perot interferometer is treated in detail.

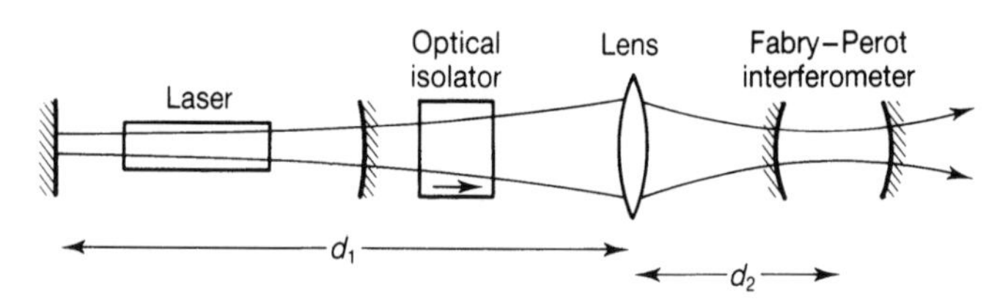

Figure 4.26 Experimental set-up for mode matching a Gaussian beam from a laser into the TEM_{00} mode of a passive cavity.

Example 4.4 Calculation of the waist size and its location after focusing a Gaussian beam by a convex lens

A laser operating in the TEM_{00} mode at 633 nm has its waist of size 0.308 mm located 0.5 m from a convex lens of focal length 0.1 m. Find the size and location of the focused waist.

Using Eq. (4.44) to find the Rayleigh range of the beam from the laser,

$$z_{R1} = \frac{\pi w_{01}^2}{\lambda} = \frac{\pi(3.08 \times 10^{-4})^2}{6.33 \times 10^{-7}} = 0.47 \text{ m}$$

Using Eq. (4.118) to calculate the waist size,

$$w_{02}^2 = \frac{f^2 w_{01}^2}{z_{R1}^2 + (f - d_1)^2} = \frac{(0.1)^2(3.08 \times 10^{-4})^2}{0.47^2 + (0.1 - 0.5)^2}\, m^2$$

$$w_{02} = 5 \times 10^{-5} \text{ m or } 50\ \mu\text{m}$$

Using Eq. (4.119) to find the location of the waist,

$$d_2 = f + f^2\left\{\frac{(d_1 - f)}{z_{R1}^2 + (f - d_1)^2}\right\} = 0.1 + \frac{(0.1)^2(0.4)}{0.47^2 + 0.4^2}$$

$$d_2 = 0.1 + 0.01 = 0.11 \text{ m}$$

The focused waist occurs 1 cm beyond the focus of the lens.

4.11 Elliptical Gaussian beams

In our Hermite–Gaussian beams it has been assumed that the waist size, w_0, is the same in both the xz and yz planes and the amplitude of the field, as given in Eq. (4.83), drops off according to

$$\mathscr{E}_{m,n} \propto \exp\left[-\frac{x^2 + y^2}{w^2(z)}\right] \tag{4.129}$$

There will be cases when the variation in the x and y directions is described by a different spot size and are given by

$$\mathscr{E}_{m,n} \propto \exp\left[-\frac{x^2}{w_x^2(z)} - \frac{y^2}{w_y^2(z)}\right] \tag{4.130}$$

Such beams appear elliptical in cross-section and are produced in cavities which have different focusing conditions in the two planes such as in ring cavities or when a circular Gaussian beam is focused by a cylindrical lens. Semiconductor lasers produce beams which are very good approximations to elliptical Gaussian. These are equally valid solutions to the paraxial wave equation and can be treated as two separate and independent Gaussian beams each with their own Gaussian beam parameters. This means that, if the elliptical beam is focused by a spherical convex lens, the waist in each plane will be located at different positions and the angular divergence in the two planes will be different. One consequence of this is that the beam will appear circular at one location. The *ABCD* transformation law can be applied separately to both planes of the beam and this can find application to the calculation of the spot sizes of an elliptical Gaussian beam produced in a ring cavity.

The complete solution to the paraxial wave equation (Eq. (4.73)) is similar to that given in Section 4.8.1. Without derivation we will quote it. The modifications to the field distribution can be seen by comparison with Eq. (4.83):

$$\mathscr{E}_{m,n} = A\sqrt{\frac{w_{0x}(z)w_{0y}(z)}{w_x(z)w_y(z)}}\, H_m\left[\sqrt{2}\,\frac{x}{w_x(z)}\right] H_n\left[\sqrt{2}\,\frac{y}{w_y(z)}\right] \exp\left[-\frac{x^2}{w_x^2(z)} - \frac{y^2}{w_y^2(z)}\right]$$
$$\times \exp[-jkx^2/2R_x(z) - jky^2/2R_y(z)]$$
$$\times \exp\left\{-j\left[kz - \left(m+\frac{1}{2}\right)\arctan\left(\frac{z - z_{0x}}{z_{Rx}}\right) - \left(n+\frac{1}{2}\right)\arctan\left(\frac{z - z_{0y}}{z_{Ry}}\right)\right]\right\} \tag{4.131}$$

Since the waists associated with the two planes occur in different positions we introduce the z coordinates of the waist locations. We define z_{0x} as the z coordinate of the waist in the xz plane and z_{0y} as the z coordinate of the waist in the yz plane. Hence with this notation the origin $z = 0$ can be taken anywhere. Hence the terms $(z - z_{0x})$ and $(z - z_{0y})$ give the distance from the two waists for calculation of the Guoy shift. We also note that the two planes contribute each one half-unit of Guoy shift to the longitudinal phase. Thus, since the Rayleigh ranges are different in the two planes, the axial phase shifts will not be degenerate for cases when the m and n mode numbers are reversed.

Problems

4.1 Determine the ray matrix for light incident at a plane dielectric interface where the refractive index changes from n_1 to n_2. From this determine the matrix for a slab of dielectric of thickness, d.

4.2 Show that the *ABCD* matrix for a ray incident on a concave spherical dielectric interface of curvature, b, where the refractive index changes from n_1 to n_2 is

$$\begin{bmatrix} 1 & 0 \\ \dfrac{1}{b}\left(\dfrac{n_2}{n_1} - 1\right) & \dfrac{n_1}{n_2} \end{bmatrix}$$

4.3 Using the matrix of Problem 4.2 and the matrix for a thin lens (Eq. (4.6)) show that the focal length of a lens having two closely spaced spherical surfaces of radii b_1 and b_2 is

$$\frac{1}{f} = \left(\frac{n_2}{n_1} - 1\right)\left(-\frac{1}{b_1} + \frac{1}{b_2}\right)$$

where n_2 is the refractive index of the lens material and n_1 is the refractive index of the material surrounding the lens.

4.4 A laser output mirror of thickness, d, has a concave inner surface with radius of curvature b_1 and a convex exit surface of curvature b_2. Determine its ray matrix.

4.5 Determine the ray matrix for a beam expander which consists of an input microscope objective of focal length, f_1, and a second convex lens of focal length, f_2, separated by the sum of their focal lengths.

4.6 A cavity has a convex lens placed between two plane mirrors The distance of the lens from one plane mirror is d_1 and from the other is d_2. Determine the ray matrix of the unit cell which starts at one of the plane mirrors.

4.7 Using the matrix derived in Problem 4.6, determine the range of location of the lens for stability of this cavity.

4.8 An equilateral triangular ring laser cavity has a concave mirror of radius of curvature, b, with two plane mirrors and sides of length, d. Determine the ray matrix for the unit cell which starts on the side of the ring midway between the two plane mirrors. Indicate the difference in the matrices for the tangential and sagittal planes of the ring.

4.9 Using the matrix for the ring cavity of Problem 4.8, determine the stability condition for this cavity.

4.10 Two concave mirrors of radii of curvature 7.5 mm and 8.0 mm form a small cavity. Determine the mirror spacings for which this cavity is stable. Illustrate this on the stability diagram.

4.11 A cavity has two concave mirrors of radius of curvature 1.7 m spaced by 49.8 cm. Show that this cavity is stable and determine the number of roundtrips made by the re-entrant rays.

4.12 A helium–neon laser operating on the TEM_{00} mode at a wavelength of 633 nm has a waist size of 0.18 mm. Evaluate the beam divergence, the Rayleigh range, the radius of curvature of the wavefront and the spot size 2 m from the waist. If the power of this beam is 1 mW calculate the peak irradiance at 2 m from the waist.

4.13 Calculate the spot sizes on the mirrors and the waist size in the cavities of Problems 4.10 and 4.11.

4.14 A carbon dioxide laser operating at 10.6 μm is 2 m in length and has an almost hemispherical cavity with a 6 m concave mirror. Calculate the waist size and the spot size on the concave mirror.

4.15 Determine the waist sizes and spot sizes on the curved mirror for the ring cavity in Problem 4.8. At what distance from the waist outside the cavity will the elliptical spot become circular?

4.16 An almost hemispherical cavity can alter the spot size on the curved mirror by small displacement, ε, of the plane mirror from the exact hemispherical condition. Derive an expression for the variation of this spot size in terms of the displacement, ε.

4.17 A small cavity has a plane mirror and a concave mirror of radius of curvature 10 cm separated by 7 cm. Between them is placed a 5 cm long crystalline rod of refractive index 1.8. Determine the spot sizes on the mirrors for this cavity.

4.18 If the crystalline rod in Problem 4.17 is cut at the Brewster angle at each end then the optical path length in the rod will be different for bundles of rays in the plane of incidence and for rays perpendicular to the plane of incidence. In the plane of incidence (xz plane) the optical path length is

$$t_{xz} = t\,\frac{(n^2+1)}{n^2}$$

while perpendicular to the plane of incidence (yz plane) it is

$$t_{yz} = t\,\frac{(n^2+1)}{n^4}$$

Using these equations calculate the spot dimensions on the plane mirror in both the xz and yz planes.

4.19 Show that a knife edge located at a distance equal to the spot size from the centre of a Gaussian beam will transmit 16 per cent or 84 per cent of the beam.

4.20 Using the data in Figure 4.18, sketch the Hermite–Gaussian mode patterns for the ($m = 5$, $n = 3$) and the ($m = 4$, $n = 2$) modes.

4.21 Determine the transmission of a circular aperture of radius, a, for the Laguerre–Gaussian mode $p = 2$, $l = 0$ in terms of the spot size to aperture radius ratio.

4.22 Determine the transmission of the circular aperture of radius, a, for the TEM_{00} Hermite–Gaussian mode in terms of the spot size to aperture radius ratio.

4.23 A particular beam is a mixture of TEM_{00} mode and the TEM_{10} modes. It has 2 W in the fundamental mode and 1 W in the higher order mode. Evaluate the ratio of their peak irradiances.

4.24 The beam from a carbon dioxide laser at 10.6 μm has an M^2 value of 1.15 and a waist size of 1.0 mm. Calculate the spot size at a distance of 2.5 m from the waist.

4.25 A beam from an argon ion laser operating at 514 nm is used to pump a dye by focusing with a 10 cm convex lens. If the divergence of the beam is 1 mradian and the lens is 1 m from the waist calculate the exact location of the focused waist and its size.

4.26 Show that the maximum departure of the waist position from the focus of a convex lens of a focused Gaussian beam is given by Eq. (4.122).

4.27 It is required to mode match a helium–neon laser beam operating at 633 nm and having a waist size of 0.15 mm to an almost hemispherical passive cavity. The cavity is 7 cm long and has a 10 cm radius of curvature concave mirror. If a 15 cm lens is used at a distance of 20 cm from the waist of the laser calculate the position of the lens in relation to the plane mirror of the passive cavity.

References

[1] D. Herriot, H. Kogelnik and R. Kompfer, 'Off-axis paths in spherical mirror interferometers', *Applied Optics*, **3**, 523, 1964.

[2] D. R. Hall and P. E. Jackson (Eds), *The Physics and Technology of Laser Resonators*, Adam Hilger, Bristol, 1989, Chapter 5.

[3] A. Ghatak and K. Thyagarajan, *Optical Electronics*, Cambridge University Press, Cambridge, 1989.

[4] R. Guenther, *Modern Optics*, John Wiley, Chichester, 1990.

[5] H. Koglenik and T. Li, 'Laser beams and resonators', *Applied Optics*, **5**, 1550–67, 1966.

[6] H. A. Haus, *Waves and Fields in Optoelectronics*, Prentice Hall, Englewood Cliffs, NJ, 1984.

[7] D. Corson and P. Lorain, *Introduction to Electromagnetic Fields and Waves*, Freeman, New York, 1962.

[8] A. Seigman, 'Lasers', University Science Books, 1986.

[9] M. Abramowitz and I. A. Stegun, *Handbook of Mathematical Functions*, Dover, New York, 799–802, 1965.

[10] P. W. Miloni and J. H. Eberly, *Lasers*, John Wiley, Chichester, 1988.

[11] W. B. Bridges, 'Divergence of higher order Gaussian modes', *Applied Optics*, **14**, 2346–7, 1975.

[12] A. Yariv, *Optical Electronics*, 4th edn, Saunders College Publishing, 1991.

5

Resonant optical cavities

5.1 Introduction

In Chapter 4 it was seen how a curved mirror cavity or open resonator could give rise to a self-reproducing field pattern which had particular amplitude profiles known as transverse modes. In this chapter we will show how the cavity can resonate at particular frequencies referred to as *longitudinal* or *axial modes*. One important implication of this is that we will show how a laser can resonate at precise frequencies or how a laser oscillating at one frequency can drive another cavity into resonance.

We begin the chapter by extending the classical view of the propagation of light in dielectrics from Maxwell's equations as given in Chapter 2 to the reflectance of thin films. These multilayer thin films coatings can have very high reflectance and are therefore a vital part of the cavity. An analysis of the Fabry–Perot interferometer using both the mode and ray theories of the cavities will be presented and shown to be consistent. The chapter will end by illustrating the discussion with experimental data.

5.2 The admittance and interface boundary conditions

In Section 2.1.6 the Fresnel equations were developed to give the reflectance and transmittance of electromagnetic waves at dielectric interfaces. Evaluation of the reflectance of a thin film, whose thickness is of the order of the coherence length of the light, must include interference effects between the waves reflected from the surfaces involved. Very often these thin films are made in a stack consisting of a number of layers, perhaps up to 50, with different refractive indices. Clearly, the calculation of the overall reflectance could become very daunting. Thus a matrix approach has been developed in which each layer is ascribed a matrix and the output electromagnetic field, for a given input field, from a number of layers is then determined by multiplication of the relevant matrices.

Let us begin by referring back to the vector equation (2.33) showing that $\mathscr{E}$ and $\mathscr{B}$ are mutually perpendicular. Writing the scalar relation we have

$$\mathscr{B} = \frac{\mathscr{E}}{v_p}$$

where v_p, the phase velocity, is given by

$$v_p = 1/(\varepsilon_0\mu_0\varepsilon_r\mu_r)^{1/2}$$

and

$$\mathscr{B} = \mu_r\mu_0\mathscr{H}$$

As in Chapter 2, we will use the boundary condition demanding continuity of $\mathscr{E}$ and $\mathscr{H}$ parallel to the interface [Ref. 1]. Writing Eq. (2.33) in terms of $\mathscr{H}$,

$$\mu_r\mu_0\mathscr{H} = \mathscr{E}(\varepsilon_0\mu_0)^{1/2}(\varepsilon_r)^{1/2}$$

For nonmagnetic dielectric materials $\mu_r = 1$, thus

$$\mathscr{H} = (\varepsilon_0/\mu_0)^{1/2}(\varepsilon_r)^{1/2}\mathscr{E}$$

Noting from Eq. (2.31) that the refractive index is given by

$$n = (\varepsilon_r)^{1/2}$$

the relationship between the magnitudes of $\mathscr{H}$ and $\mathscr{E}$ then becomes

$$\mathscr{H} = y\mathscr{E} \tag{5.1}$$

where we define the *admittance* of the medium as

$$y = \mathscr{Y}n \tag{5.2}$$

where $\mathscr{Y}$ is the admittance of free space defined as

$$\mathscr{Y} = (\varepsilon_0/\mu_0)^{1/2} \tag{5.3}$$

It is a significant simplification to write the boundary condition equations using admittance. We will do this first for light at normal incidence moving from a medium of admittance y_1 into a medium of admittance y_2 as shown in Figure 5.1. The continuity requirement for the electric and magnetic fields at the boundary may be written in terms of the field amplitudes, $\mathscr{E}_0$ and $\mathscr{H}_0$, as

$$\begin{aligned} \mathscr{E}_{0i} + \mathscr{E}_{0r} &= \mathscr{E}_{0t} \\ \mathscr{H}_{0i} - \mathscr{H}_{0r} &= \mathscr{H}_{0t} \end{aligned} \tag{5.4}$$

where the subscripts i, r, and t refer, respectively, to the incident wave, the reflected wave and the transmitted wave. Using the admittance to write the equation for $\mathscr{H}$ in terms of electric field gives

$$y_1\mathscr{E}_{0i} - y_1\mathscr{E}_{0r} = y_2\mathscr{E}_{0t} \tag{5.5}$$

Elimination of E_{0t} between Eqs (5.4) and (5.5) gives the field amplitude reflection coefficient:

$$r_{12} = \frac{\mathscr{E}_{0r}}{\mathscr{E}_{0i}} = \frac{y_1 - y_2}{y_1 + y_2} = \frac{n_1 - n_2}{n_1 + n_2} \tag{5.6}$$

Elimination of E_{0r} gives the field transmission coefficient:

$$t_{12} = \frac{\mathscr{E}_{0t}}{\mathscr{E}_{0i}} = \frac{2y_1}{y_1 + y_2} = \frac{2n_1}{n_1 + n_2} \tag{5.7}$$

We notice that when $n_2 > n_1$ the field experiences a phase change of π radians on reflection while there is never any phase change after transmission. The power reflectance coefficient is given by

$$R = r_{12}^2$$

When this argument is extended to angles of incidence other than 90° as discussed in Section 2.1.6 two states of electromagnetic wave polarization (s and p) must be considered separately. This can be simplified by introducing special symbols $\mathbb{E}$ and

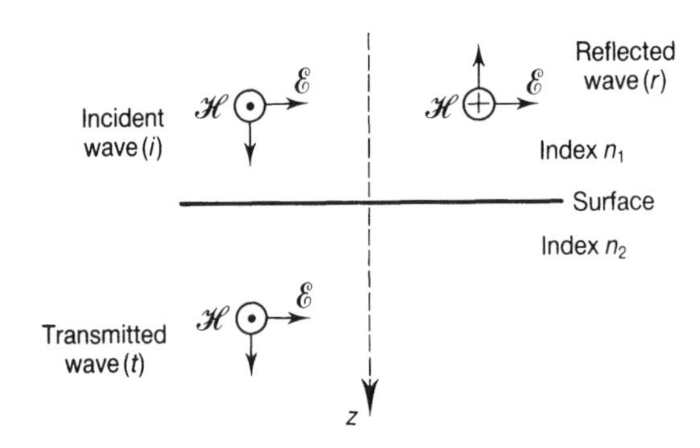

Figure 5.1 Normal incidence of an electromagnetic wave at a dielectric interface where the refractive index changes from n_1 to n_2 showing the field amplitudes and the sign conventions.

$\mathbb{H}$ for the components of the fields parallel to the boundary. The notation is shown in Figure 5.2.

For *p-polarized (TM)* waves these field components are written as

$$\mathbb{E}_i = \mathscr{E}_i \cos\theta_1 \tag{5.8a}$$

$$\mathbb{H}_i = \mathscr{H}_i = y_1 \mathscr{E}_i = \frac{y_1}{\cos\theta_1}\,\mathbb{E}_i \tag{5.8b}$$

$$\mathbb{E}_r = \mathscr{E}_r \cos\theta_1 \tag{5.9a}$$

$$\mathbb{H}_r = \frac{y_1}{\cos\theta_1}\,\mathbb{E}_r \tag{5.9b}$$

$$\mathbb{E}_t = \mathscr{E}_t \cos\theta_2 \tag{5.10a}$$

$$\mathbb{H}_t = \frac{y_2}{\cos\theta_2}\,\mathbb{E}_t \tag{5.10b}$$

The same boundary condition, i.e. continuity of electric and magnetic fields parallel to the surface, produces the equations

$$\mathbb{E}_i + \mathbb{E}_r = \mathbb{E}_t \tag{5.11}$$

$$\frac{y_1}{\cos\theta_1}\,\mathbb{E}_i - \frac{y_1}{\cos\theta_1}\,\mathbb{E}_r - \frac{y_1}{\cos\theta_1}\,\mathbb{E}_t \tag{5.12}$$

We note from Eqs (5.8a) and (5.9a) that the field reflection coefficient is given by

$$r_p = \mathbb{E}_r / \mathbb{E}_i$$

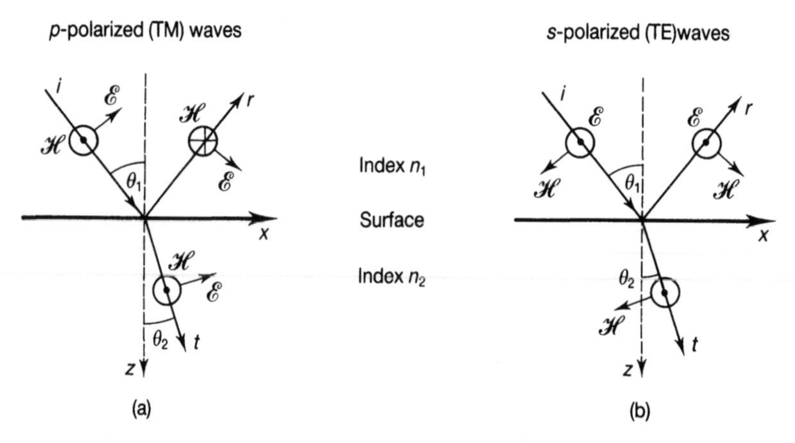

Figure 5.2 Plane electromagnetic wave at incidence angles other than normal. (a) *p*-polarized (TM) waves. (b) *s*-polarized (TE) waves.

and elimination between Eqs (5.11) and (5.12) gives for the power reflection coefficient

$$R_p = r_p^2 = \left(\frac{y_1}{\cos\theta_1} - \frac{y_2}{\cos\theta_2}\right)^2 \left(\frac{y_1}{\cos\theta_1} + \frac{y_2}{\cos\theta_2}\right)^{-2} \tag{5.13}$$

Substitution for the admittances from Eq. (5.2) and rearrangement will give the Fresnel equation (2.66).

For *s-polarized (TE)* waves the components of the field parallel to the boundary are

$$\mathbb{E}_i = \mathscr{E}_i \qquad \mathbb{H}_i = \mathscr{H}_i \cos\theta_1 = y_1 \cos\theta_1\, \mathbb{E}_i \tag{5.14}$$

$$\mathbb{E}_r = \mathscr{E}_r \qquad \mathbb{H}_r = \mathscr{H}_r \cos\theta_1 = y_1 \cos\theta_1\, \mathbb{E}_r \tag{5.15}$$

$$\mathbb{E}_t = \mathscr{E}_t \qquad \mathbb{H}_t = \mathscr{H}_t \cos\theta_2 = y_2 \cos\theta_2\, \mathbb{E}_t \tag{5.16}$$

A similar analysis shows that the power reflection coefficient for *s*-polarized waves is given by

$$R_s = r_s^2 = \left(\frac{y_1 \cos\theta_1 - y_2 \cos\theta_2}{y_1 \cos\theta_1 - y_2 \cos\theta_2}\right)^2 \tag{5.17}$$

Again, substitution from Eq. (5.2) will give the Fresnel equation (2.56).

If we define an optical admittance for oblique incidence as

$$\mathbb{H} = \eta \mathbb{E}$$

where for each polarization we have

$$\eta_p = y/\cos\theta \text{ and } \eta_s = y\cos\theta \tag{5.18}$$

then in both cases the power reflectivity is simply written

$$R = \left(\frac{\eta_1 - \eta_2}{\eta_1 + \eta_2}\right)^2 \tag{5.19}$$

In Eq. (5.19) the subscript indicates the medium in which the wave is travelling.

5.3 Reflectance of thin films

The reflectance and transmittance of thin films will now be considered and for this we will need to take into consideration interference effects which will occur in films whose thickness are of the order of optical wavelengths. When the thickness of the film is less than the coherence length of the light interference will affect the reflectance properties of the film. This will alter the admittance of the film and the reflectivity calculated using Eq. (5.19). Figure 5.3 shows the notation to be used for a thin film of index n_1 on a substrate of index n_{sub}. In the film there is one positive-

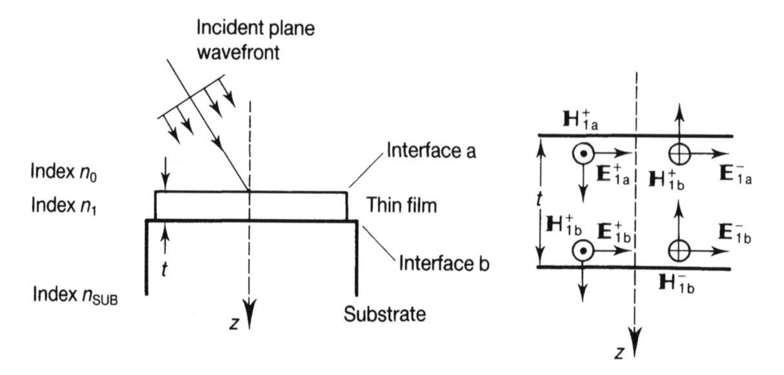

Figure 5.3 Plane wave incident on thin film of thickness t.

going and one negative-going wave and the resultant tangential components of the fields at the interface between the film and the substrate (interface b) are

$$\mathbb{E}_b = \mathbb{E}^+_{1b} + \mathbb{E}^-_{1b} \tag{5.20}$$

$$\mathbb{H}_b = \mathbb{H}^+_{1b} - \mathbb{H}^-_{1b} = \eta_1 \mathbb{E}^+_{1b} - \eta_1 \mathbb{E}^-_{1b} \tag{5.21}$$

where the superscripts + and − indicate the direction of propagation of the wave in the film. Equations for the four counter-propagating fields in the film are obtained by solving these two equations, resulting in

$$\mathbb{E}^+_{1b} = \tfrac{1}{2}(-\mathbb{H}_b/\eta_1 + \mathbb{E}_b) \tag{5.22}$$

$$\mathbb{E}^-_{1b} = \tfrac{1}{2}(-\mathbb{H}_b/\eta_1 + \mathbb{E}_b) \tag{5.23}$$

$$\mathbb{H}^-_{1b} = \eta_1 \mathbb{E}^-_{1b} = \tfrac{1}{2}(\mathbb{H}_b + \eta_1 \mathbb{E}_b) \tag{5.24}$$

$$\mathbb{H}^-_{1b} = -\eta_1 \mathbb{E}^-_{1b} = \tfrac{1}{2}(\mathbb{H}_b - \eta_1 \mathbb{E}_b) \tag{5.25}$$

The fields at the other interface, a, can be evaluated by allowing for the phase shift in the z direction from 0 (interface b) to $-t$. The positive-going wave is multiplied by $e^{j\phi/2}$ and the negative-going wave by $e^{-j\phi/2}$ where $\phi/2$ is the phase shift experienced by the wave in the film having a thickness, t, and index, n. This can be evaluated by referring to Figure 5.4. The path length difference, ΔL, between successive rays is given by

$$\Delta L = 2nt/\cos\theta - P$$

$$\Delta L = 2nt/\cos\theta - \frac{2nt\sin^2\theta}{\cos\theta}$$

$$\Delta L = 2nt\cos\theta$$

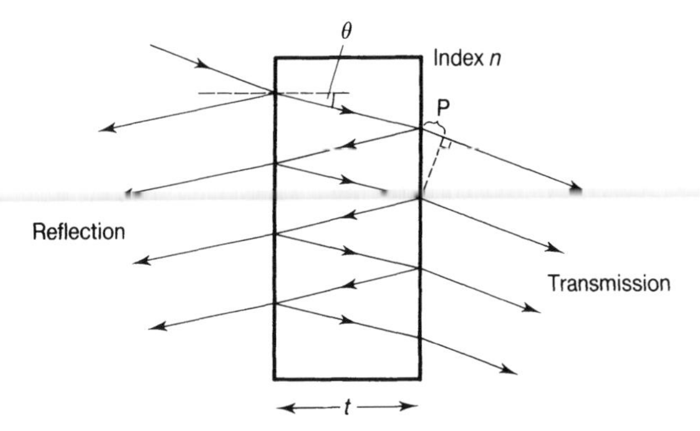

Figure 5.4 Diagram used to evaluate the phase difference between rays in a thin film. This diagram will also apply to the plane parallel Fabry–Perot interferometer. The diagram shows rays but remember, the wavefronts overlap.

Multiplying this path difference by $2\pi/\lambda$ to obtain the phase shift for a single pass of the film gives

$$\phi/2 = \frac{2\pi nt}{\lambda} \cos\theta \tag{5.26}$$

The values of the tangential components of the fields at the interface, a, are now obtained from Eqs (5.22)–(5.25):

$$\mathbb{E}_{1a}^{+} = \mathbb{E}_{1b}^{+}\, e^{j\phi/2} = \tfrac{1}{2}(\mathbb{H}_b/\eta_1 + \mathbb{E}_b)\, e^{j\phi/2}$$

$$\mathbb{E}_{1a}^{-} = \mathbb{E}_{1b}^{-}\, e^{j\phi/2} = \tfrac{1}{2}(-\mathbb{H}_b/\eta_1 + \mathbb{E}_b)\, e^{-j\phi/2}$$

$$\mathbb{H}_{1a}^{+} = \mathbb{H}_{1b}^{+}\, e^{j\phi/2} = \tfrac{1}{2}(\mathbb{H}_b + \eta_1\mathbb{E}_b)\, e^{j\phi/2}$$

$$\mathbb{H}_{1a}^{-} = \mathbb{H}_{1b}^{-}\, e^{-j\phi/2} = \tfrac{1}{2}(\mathbb{H}_b/\eta_1 + \mathbb{E}_b)\, e^{-j\phi/2}$$

After a little algebra the fields can be written

$$\mathbb{E}_a = \mathbb{E}_{1a}^{+} + \mathbb{E}_{1a}^{-} = \mathbb{E}_b \cos\phi/2 + \mathbb{H}_b\, \frac{j\sin\phi/2}{\eta_1}$$

$$\mathbb{H}_a = \mathbb{H}_{1a}^{+} + \mathbb{H}_{1a}^{-} = \mathbb{E}_b j\eta_1 \sin\phi/2 + \mathbb{H}_b \cos\phi/2$$

In matrix notation these equations for E_a and H_a become

$$\begin{bmatrix}\mathbb{E}_a\\ \mathbb{H}_a\end{bmatrix} = \begin{bmatrix}\cos\phi/2 & (j\sin\phi/2)/\eta_1\\ j\eta_1\sin\phi/2 & \cos\phi/2\end{bmatrix}\begin{bmatrix}\mathbb{E}_b\\ \mathbb{H}_b\end{bmatrix} \tag{5.27}$$

The propagation of the tangential components of the fields in the film can be reduced to multiplication by a 2×2 matrix. This matrix is known as the *characteristic matrix* of the thin film. The input admittance of the film is defined by

$$Y = \mathbb{H}_a / \mathbb{E}_a$$

and can be evaluated using Eq. (5. 27), which can be written

$$\mathbb{E}_a \begin{bmatrix} 1 \\ Y \end{bmatrix} = \begin{bmatrix} \cos \phi/2 & (\mathrm{j} \sin \phi/2)/\eta_1 \\ \mathrm{j}\eta_1 \sin \phi/2 & \cos \phi/2 \end{bmatrix} \begin{bmatrix} 1 \\ \eta_{sub} \end{bmatrix} \mathbb{E}_b \tag{5.28}$$

Evaluation of the matrix then gives

$$Y = \frac{\eta_{sub} \cos \phi/2 + \mathrm{j}\eta_1 \sin \phi/2}{\cos \phi/2 + \mathrm{j}(\eta_{sub}/\eta_1) \sin \phi/2} \tag{5.29}$$

The reflectance of the film on a substrate and in an incident medium of admittance η_0 is obtained by Eq. (5.19) as

$$R = \left(\frac{\eta_0 - Y}{\eta_0 + Y} \right)^2 \tag{5.30}$$

In summary, therefore, we can determine the power reflectivity of the film on a substrate using Eq. (5.30) when the admittance of the film is determined from Eq. (5.29). The wavelength dependence of the film reflectivity is contained in the single-pass phase shift $\phi/2$ given by Eq. (5.26).

The effect of another film (Figure 5.5) on the fields is given by multiplying the characteristic matrices to determine a new total admittance for the two films, thus:

$$\begin{bmatrix} \mathbb{E}_a \\ \mathbb{H}_a \end{bmatrix} = \begin{bmatrix} \cos \phi_1/2 & (\mathrm{j} \sin \phi_1/2)/\eta_1 \\ \mathrm{j}\eta_1 \sin \phi_1/2 & \cos \phi_1/2 \end{bmatrix} \begin{bmatrix} \cos \phi_2/2 & (\mathrm{j} \sin \phi_2/2)\eta_2 \\ \mathrm{j}\eta_2 \sin \phi_2/2 & \cos \phi_2/2 \end{bmatrix} \begin{bmatrix} \mathbb{E}_c \\ \mathbb{H}_c \end{bmatrix}$$

The total admittance and the reflection coefficient can then be determined as before.

The power of this matrix technique can be appreciated when the reflectance of a multilayer dielectric (MLD) stack is required. This is determined by multiplying the

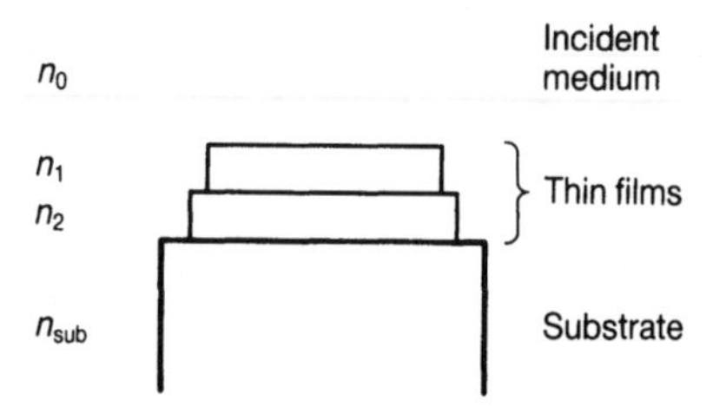

Figure 5.5 Notation for two films on a substrate.

characteristic matrix of each layer, in the correct order, to obtain the stack admittance and using Eq. (5.30) to find the stack reflectivity thus:

$$\mathbb{E}_a \begin{bmatrix} 1 \\ Y \end{bmatrix} = \left(\prod_{r=1}^{r=p} \begin{bmatrix} \cos \phi_r/2 & (\mathrm{j} \sin \phi_r/2)/\eta_r \\ \mathrm{j}\eta_r \sin \phi_r/2 & \cos \phi_r/2 \end{bmatrix} \right) \begin{bmatrix} 1 \\ \eta_{sub} \end{bmatrix} \mathbb{E}_b \tag{5.31}$$

where $\phi_r/2$ and η_r are the phase shift and admittance of layer r and p is the number of layers in the stack. The computation now begins to be tedious but well within the capabilities of modern desktop computers and, indeed, excellent software is available for MLD coating design.

Multilayer dielectric coatings play a vital part in laser optics. In any laser high-reflectance low-loss reflectors are required and in recent years it is the improvement in coating technology which has allowed both high-power lasers and lasers with extremely low gain to be developed. In addition, many optical systems have a large number of interfaces where Fresnel reflection losses can significantly decrease the transmitted beam power or produce multiple reflections. Such unwanted reflections can be reduced by using anti-reflection coatings on the optical surfaces. Other applications of MLD coatings include narrowband filters, polarizing beam splitters and circular polarizers.

5.3.1 Anti-reflection (AR) coatings

Figure 5.6 shows a quarter-wavelength thick film having a refractive index which is smaller than that of the substrate. Waves reflected from the top and bottom of the film will interfere destructively since there is a phase change of 180° at each surface and the path difference between successive rays is $\lambda/2$. At both interfaces the light is passing from a low to high index medium. Complete amplitude cancellation will occur if, neglecting absorption, the reflectance at each surface is identical. This condition requires

$$\frac{\eta_0 - \eta_1}{\eta_0 + \eta_1} = \frac{\eta_1 - \eta_{sub}}{\eta_1 + \eta_{sub}}$$

or, with rearrangement,

$$\eta_1 = (\eta_0 \eta_{sub})^{1/2} \tag{5.32}$$

where η_{sub} is the admittance of the substrate. For normal incidence this becomes

$$n_1 = (n_0 n_{sub})^{1/2} \tag{5.33}$$

A disadvantage of the single-layer AR coating is that it is difficult to find materials with refractive indices smaller than the substrate which exactly satisfy Eq. (5.33).

Far infra-red transmitting materials such as germanium have high refractive indices and so AR coatings are essential in most IR optical systems to avoid significant power loss. For broadband AR coatings multilayer combinations of high- and

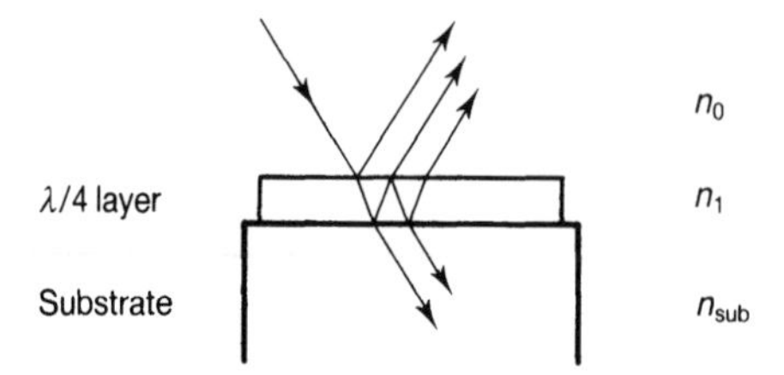

Figure 5.6 Low index quarter-wave-length thick layer on a substrate for AR coating.

low-index film materials are used, the best design often being obtained by trial and error using computer software.

5.3.2 Multilayer high-reflectance layers

For high reflectivity alternate layers each one quarter wavelength thick are used. Figure 5.7 shows how constructive interference with a high index quarter wave layer can increase reflectivity. The diagram illustrates that a maximum reflection coefficient will occur with an odd number of layers of high- and low-index material each a quarter wavelength thick with the high index layer on top and bottom. Using the matrix method, the reflectance at normal incidence of such a stack with $(2p+1)$ layers can be shown [Ref. 1] to be given by

$$R = \left(\frac{1 - (n_H/n_L)^{2p}(n_H^2/n_{sub})}{1 + (n_H/n_L)^{2p}(n_H^2/n_{sub})}\right)^2 \tag{5.34}$$

where n_H and n_L are the high and low refractive indices, respectively, and n_{sub} is the refractive index of the substrate. As the number of layers increases, the reflectivity approaches 100 per cent. The increase takes place more rapidly when the index ratio is larger.

Within any coating layer we are bound to get some scattering and absorption of the light. From conservation of energy

$$R + T + S + A = 1 \tag{5.35}$$

where S is the scattering loss and A is the absorption loss. It should be noted that we have neglected absorption in our theory for the coating reflectance. Absorption can be accommodated by using a complex refractive index (see Section 2.1.5) in the equations. The reader is directed to Ref. 2 for a complete discussion.

Figure 5.8 shows a typical result obtained from Eq. (5.34) for the reflectance as a function of wave number $(1/\lambda)$ for a high reflector for different number of layers [Ref. 3]. Notice how the value of the peak reflectance can be increased by increasing the number of layers. The centre wavelength is determined by the physical thickness of the quarter-wavelength thick layers.

These coatings are produced by successive thermal evaporation, under vacuum,

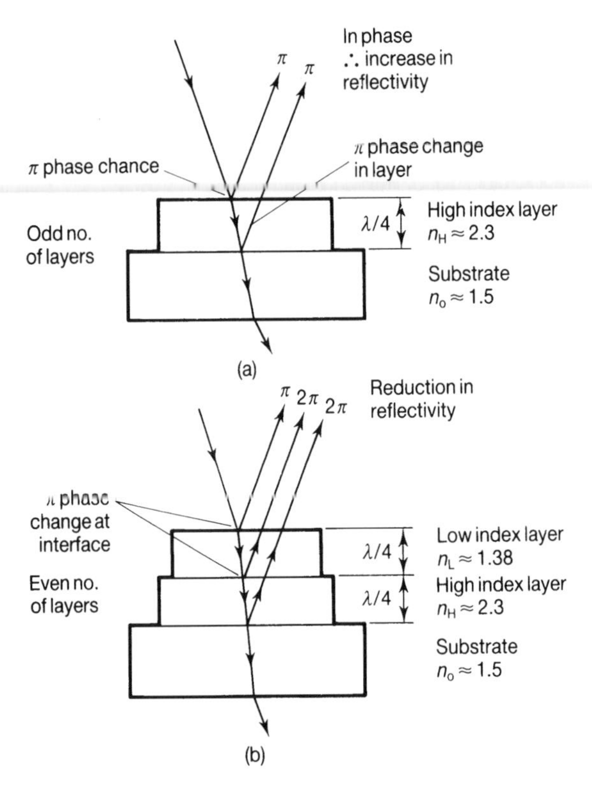

Figure 5.7 High reflecting multilayer quarter wave stack. (a) One *high layer* on the substrate increasing the reflectivity. (b) One *low layer* on top of the high layer so reducing the reflectivity.

of the high- and low-refractive index materials. The scattering loss and absorption of the coatings is very dependent on a number of factors, including the surface quality of the substrate, the parameters of the evaporation and the cleanliness of the whole process. Thermal evaporation techniques can produce reflectivities of up to 99.8 per cent and absorbance of less than 0.01 per cent. To improve on these figures ion beam sputtering techniques can be used to produce mirrors with total losses down to 20 parts per million. These high-reflectivity low-loss mirrors are vital components of ring laser gyroscopes [Ref. 4]. They also find application in high-power lasers where even small amounts of loss can cause catastrophic damage to the mirror, and in very low-gain lasers which require low-loss mirrors.

Other useful devices using coatings include:

1. The polarizing beam splitter. This has, at 45° angle of incidence, a high reflectivity for *s*-polarized light and a low reflectivity for *p*-polarized light.
2. The 50 per cent beam splitter. This has a 50 per cent reflectivity for both polarizations at 45° angle of incidence.

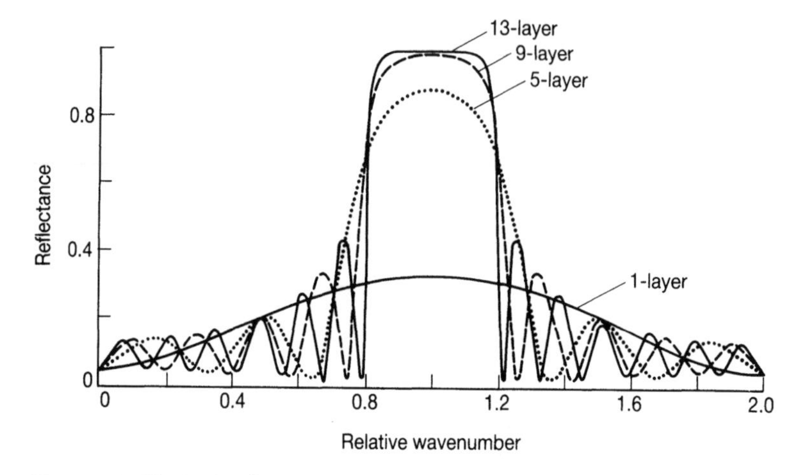

Figure 5.8 Typical reflectance versus wavelength for multilayer dielectric stack. The high refractive index material is CeO_2 ($n_H = 2.36$) and the low index material is MgF_2 ($n_L = 1.39$). The abscissa is in units of wavenumber $(1/\lambda)$ relative to the central wavenumber. Data reproduced by kind permission of Adam Hilger [Ref. 3].

5.4 The Fabry–Perot interferometer

The classical Fabry–Perot interferometer was developed jointly at the turn of the twentieth century by the two French scientists whose names it carries and has continued to evolve and find many new applications [Ref. 5]. Figure 5.9 shows a schematic diagram of the classical device. The mirrors are plane parallel and have an aperture, determined by the diameter of the mirrors, which is wide compared with their separation. In Section 4.7 we saw that the criterion for low diffraction was that the Fresnel number, N_F, should be larger than unity (Eq. (4.57)). The classical Fabry–Perot interferometer easily satisfies this criterion with Fresnel numbers of the order of 10^6 and we may accordingly use ray analysis. When N_F approaches unity, however, a mode analysis must be used and in this situation the device will then be referred to as an optical cavity and their frequencies which give a standing wave will be determined.

5.4.1 The plane parallel Fabry–Perot Interferometer (FPP)

The optical arrangements traditionally used for high-resolution spectroscopy using a Fabry–Perot interferometer with two plane mirrors are shown in Figure 5.9. In modern devices low-loss multilayer dielectric (MLD) coated mirrors are used. These allow reflectivities to be adjusted as required and, as we will see, give the instrument high resolution with good overall instrument transmission. The gap between the mirrors will often be filled with air but we can generalize by assuming this medium

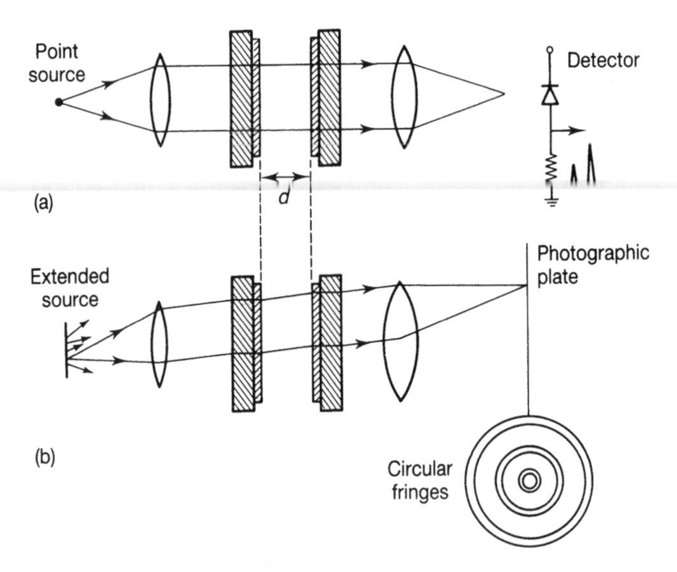

Figure 5.9 Diagram of classical Fabry–Perot interferometer showing two arrangements. (a) Input light parallel and device acting as a filter. (b) Extended source with rays incident at many angles producing circular fringes.

to have a refractive index of n, and a power absorption coefficient α. Light incident from outside onto the interferometer will pass through the input mirror and suffer multiple reflections at the mirrors, thereby making multiple transits in the gap. Consequently, multiple-beam interference will occur.

In the arrangement of Figure 5.9(b) light from an extended source will produce circular interference fringes. In Figure 5.9(a) light is restricted to normal incidence and so the central fringe only is detected. Whether this is bright or dark will depend on the exact mirror spacing.

If the latter is ramped then the fringe is scanned through the detector and the signal may be recorded electronically. Scanning can be achieved by moving the mirrors thermally, with a piezoelectric transducer or even by altering the refractive index of the material in the gap (by, for example, altering the pressure of the gas in the gap). This arrangement is known as the scanning mode and is used with a CW source. For a pulsed source the fringe pattern of Figure 5.9(b) can be recorded.

Figure 5.10 illustrates this situation for light at normal incidence. Light incident on the interferometer will be partially reflected and partially transmitted by mirror M1. The transmitted field travels to the second mirror where again it is partially reflected and partially transmitted. The many reflected components add to produce a total field travelling in each direction. These fields are the phasor sums of all the individual fields caused by the multiple reflections. Figure 5.10 also shows how the individual phasors of the electric fields may be added on a phasor diagram to produce the total field travelling in a particular direction. The angle between each

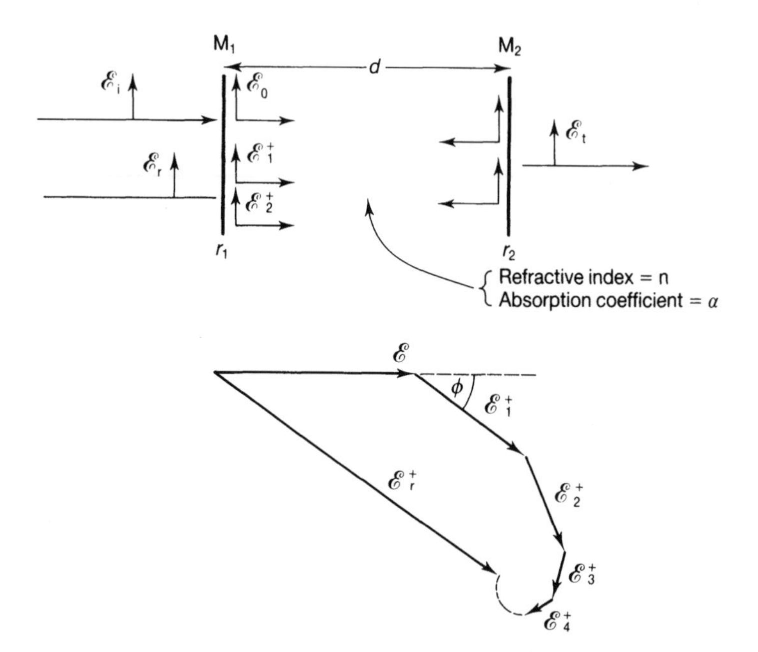

Figure 5.10 Detail of Figure 5.9(a) showing the travelling fields in the gap. Total phasor for the field going to the right is obtained by phasor addition of the individual fields with a phase angle ϕ between them.

phasor is ϕ, the *roundtrip phase shift*. As Figure 5.10 illustrates, the total field will be maximum when the roundtrip phase shift is a multiple of 2π, that is, when

$$\phi = q2\pi$$

where q is an integer which is referred to as the *longitudinal (or axial) mode number*.

The evaluation of ϕ is exactly as for the thin film except that in this case the mirror spacing is very much larger than the film thickness. We can replace t in Eq. (5.26) by d and for normal incidence $\theta = 0$ we have

$$\phi = 2dn\omega/c = 4\pi nd/\lambda_0 = \frac{4\pi\nu nd}{c} \tag{5.36}$$

where λ_0 is the vacuum wavelength of the light and c the velocity of light in vacuum.

Using the condition for the travelling fields to be maximum,

$$\phi = 4\pi nd/\lambda_0 = q2\pi$$

which gives

$$nd = q\lambda_0/2$$

Our condition for maximum travelling field in the gap is therefore the familiar resonant standing-wave condition that the gap spacing equals an integral number of half wavelengths.

The action of the interferometer on the light as shown in Figure 5.9(a) is obtained by evaluating its total transmission T_{FP}. The notation for the fields is shown in Figure 5.10. The total field travelling to the right at any point in the gap, $\mathscr{E}_{\mathrm{T}}^{+}$, can now be obtained by adding the phasors:

$$\mathscr{E}_{\mathrm{T}}^{+} = \mathscr{E}_0 + \mathscr{E}_1^{+} + \mathscr{E}_2^{+} + \mathscr{E}_3^{+} + \mathscr{E}_4^{+} + \ldots$$

Let us evaluate the total travelling field to the right just on the inside surface of the input mirror M_1. For this E_0 is the fraction of the incident field transmitted by M_1 and is given by

$$\mathscr{E}_0 = t_1 \mathscr{E}_{\mathrm{i}}$$

Each successive field in the summation differs in phase by ϕ and is smaller in amplitude due to absorption in the gap and the fact that the reflection coefficient at the mirrors is less than unity. We can write the above summation as

$$\mathscr{E}_{\mathrm{T}}^{+} = t_1 \mathscr{E}_{\mathrm{i}} [1 + r_1 r_2 \mathrm{e}^{-\mathrm{j}\phi}\, \mathrm{e}^{-(\alpha/2)2d} + (r_1 r_2)^2\, \mathrm{e}^{-2\mathrm{j}\phi}\, \mathrm{e}^{-(\alpha/2)4d} + \ldots]$$

Since the magnitude of the factor $\mathrm{e}^{-\mathrm{j}\phi}\,\mathrm{e}^{-\alpha d}$ is less than unity the summation is a convergent geometric progression, and we have

$$\mathscr{E}_{\mathrm{T}}^{+} = \frac{t_1 E_{\mathrm{i}}}{1 - r_1 r_2\, \mathrm{e}^{-\mathrm{j}\phi}\, \mathrm{e}^{-\alpha d}}$$

Allowing for the change of phase and absorption in going from M_1 to M_2 as well as the transmission through M_2, the total transmitted field is given by

$$\mathscr{E}_{\mathrm{t}} = \frac{t_1 \mathscr{E}_{\mathrm{i}} t_2\, \mathrm{e}^{-\alpha d/2}\, \mathrm{e}^{-\mathrm{j}\phi/2}}{1 - r_1 r_2 \mathrm{e}^{-\mathrm{j}\phi}\, \mathrm{e}^{-\alpha d}}$$

The total power transmitted by the interferometer, T_{FP}, is thus

$$T_{\mathrm{FP}} = \left(\frac{\mathscr{E}_{\mathrm{t}}}{\mathscr{E}_{\mathrm{i}}}\right)\left(\frac{\mathscr{E}_{\mathrm{t}}}{\mathscr{E}_{\mathrm{i}}}\right)^{*}$$

$$T_{\mathrm{FP}} = \frac{t_1^2 t_2^2\, \mathrm{e}^{-\alpha d}}{(1 - r_1 r_2\, \mathrm{e}^{-\mathrm{j}\phi}\, \mathrm{e}^{-\alpha d})(1 - r_1 r_2\, \mathrm{e}^{\mathrm{j}\phi}\, \mathrm{e}^{\mathrm{j}\phi}\, \mathrm{e}^{-\alpha d})}$$

$$T_{\mathrm{FP}} = \frac{t_1^2 t_2^2\, \mathrm{e}^{-\alpha d}}{1 - r_1 r_2\, \mathrm{e}^{-\mathrm{j}\phi}\, \mathrm{e}^{-\alpha d} - r_1 r_2\, \mathrm{e}^{\mathrm{j}\phi}\, \mathrm{e}^{-\alpha d} + r_1^2 r_2^2\, \mathrm{e}^{-2\alpha d}}$$

Using the identity,

$$\mathrm{e}^{\mathrm{j}\phi} + \mathrm{e}^{-\mathrm{j}\phi} = 2 \cos \phi = 1 - 2 \sin^2\phi/2$$

Changing to power reflectance and transmittance

$$\rho_1 = R_1^{1/2} \text{ and } \rho_2 = R_2^{1/2}$$

$$t_1^2 = T_1 \text{ and } t_2^2 = T_2$$

$$T_{FP} = \frac{T_1 T_2 \, e^{-\alpha d}}{1 - 2(R_1 R_2)^{1/2}(1 - 2 \sin^2\phi/2) \, e^{-\alpha d} + R_1 R_2 \, e^{-2\alpha d}}$$

$$T_{FP} = \frac{T_1 T_2 \, e^{-\alpha d}}{1 + 2(R_1 R_2)^{1/2} \, e^{-\alpha d} + R_1 R_2 \, e^{-2\alpha d} + 4(R_1 R_2)^{1/2} \, e^{-\alpha d} \sin^2\phi/2}$$

Factorizing gives

$$T_{FP} = \frac{T_1 T_2 \, e^{-\alpha d}}{[1 - (R_1 R_2)^{1/2} \, e^{-\alpha d}]^2 + 4(R_1 R_2)^{1/2} \, e^{-\alpha d} \sin^2\phi/2} \tag{5.37}$$

Since energy must be conserved

$$A_{FP} + R_{FP} + T_{FP} = 1$$

Assuming that absorption processes are small enough to be neglected we may put

$$\alpha = 0$$

Thus Eq. (5.37) simplifies to

$$T_{FP} = \frac{(1 - R_1)(1 - R_2)}{[1 - (R_1 R_2)^{1/2}]^2 + 4(R_1 R_2)^{1/2} \sin^2\phi/2} \tag{5.38}$$

The light reflected from the interferometer will be the phasor sum of the multiple reflected beams and, in the same way as for transmission, a total field reflection coefficient for the Fabry–Perot can be evaluated. When doing this a phase change of π on reflection must be included as discussed earlier. It can then be shown that

$$\frac{\mathscr{E}_r}{\mathscr{E}_i} = \frac{r_1 - r_2 \cos\phi - r_1^2 r_2 \cos\phi + r_1 r_2^2 + j(r_2 \sin\phi - r_1^2 r_2 \sin\phi)}{1 - 2\cos\phi + r_1^2 r_2^2} \tag{5.39}$$

As before, power reflectance is obtained by multiplying this field reflectance by its complex conjugate. Neglecting absorption we obtain

$$R_{FP} = \frac{R_1 - 2(R_1 R_2)^{1/2} \cos\phi + R_2}{1 - 2(R_1 R_2)^{1/2} \cos\phi + R_1 R_2} \tag{5.40}$$

Our discussion has assumed, for simplicity, that any coating or gap absorption may be neglected and hence it is reasonable to conclude that reflectance and transmittance must add to unity:

$$T_{FP} + R_{FP} = 1$$

It is left as an exercise for the student to show that Eqs (5.39) and (5.38) are consistent with this result.

The expression (5.39) for E_r is complex so it follows that the light experiences a phase change ψ on reflection at the input interface, which is a function of ϕ. By taking the ratio of the imaginary to real parts in Eq. (5.39) and after a little algebra we can be show that

$$\tan \psi = \frac{R_2^{1/2} \sin \phi - R_1 R_2^{1/2} \sin \phi}{R_1^{1/2} - R_2^{1/2}(1 + R_1) \cos \phi + R_1^{1/2} R_2} \tag{5.41}$$

Returning to interferometer transmission, we notice by inspection of Eq. (5.38) that a maximum occurs whenever the sine term becomes zero. This happens for $\phi/2 = q\pi$, which is just our standing-wave condition of Eq. (5.36). The maximum instrument transmission then becomes

$$T_{FP}(\text{Max}) = \frac{(1 - R_1)(1 - R_2)}{[1 - (R_1 R_2)^{1/2}]^2} \tag{5.42}$$

The minimum transmission will occur when the sine term in Eq. (5.38) is unity, giving

$$T_{FP}(\text{Min}) = \frac{(1 - R_1)(1 - R_2)}{[1 + (R_1 R_2)^{1/2}]^2} \tag{5.43}$$

Although our derivation has assumed normal incidence the equations still hold for angles other than normal, since the angle of incidence is contained in the roundtrip phase shift ϕ. We can now use these results to interpret the operation of the two arrangements of the interferometer shown in Figure 5.9.

When the light is incident as shown in Figure 5.9(b), circular fringes are formed by the cones of rays which are incident at angles, satisfying the condition

$$\phi = \frac{4\pi n \nu d \cos \theta}{c} = q2\pi \tag{5.44}$$

Light of one frequency incident at an angle θ which satisfies Eq. (5.44) for an integral value of q will produce bright circular fringes. The next ring will correspond to a different angle of incidence and so a change in q by one. In the arrangement of Figure 5.9(a) all the light is at normal incidence and is concentrated into the central fringe when Eq. (5.44) is satisfied.

The ratio of maximum to minimum transmission at normal incidence is called the *fringe contrast* and is given by

$$\frac{T_{FP}\text{Max}}{T_{FP}\text{Min}} = \frac{[1 + (R_1 R_2)^{1/2}]^2}{[1 - (R_1 R_2)^{1/2}]^2} \tag{5.45}$$

Example 5.1 Calculation of the fringe contrast

A Fabry–Perot interferometer has mirrors with MLD coating each having 99 per cent reflectivity and 1 per cent transmission:

$$\text{Fringe contrast } \frac{T_{\text{FP}}\text{Max}}{T_{\text{FP}}\text{Min}} = \frac{[1 + (0.99)]^2}{[1 - (0.99)]^2} = 390\,000$$

If the mirrors each had reflectivity of 90 per cent and transmission 10 per cent

$$\text{Fringe contrast } \frac{T_{\text{FP}}\text{Max}}{T_{\text{FP}}\text{Min}} = \frac{[1 - (0.90)]^2}{[1 + (0.90)]^2} = 360$$

The calculation illustrates the very high fringe contrast obtained using high mirror reflectivities.

When the mirrors are of equal reflectivity $R_1 = R_2$ the maximum transmission becomes unity. All the light incident on the interferometer is transmitted and none reflected, a point which will be reinforced later. This condition is known as matched and is analogous to a transmission line terminated by its characteristic impedance. Figure 5.11 shows the transmission, given by Eq. (5.38), plotted as a function of the single-pass phase change $\phi/2$. The smallest mirror reflectivity chosen here of 4 per cent is that for an uncoated flat of optical-quality glass. Increasing the mirror reflectivity increases the fringe sharpness and contrast while maintaining the transmission at unity.

The single-pass phase shift, $\phi/2$, can be varied in practice to produce the fringes by either changing the optical spacing, nd, or scanning the input optical frequency ν. A change in the single-pass shift of π corresponds to the separation between the resonances of Figure 5.11. We can interpret this either in terms of an interval $\lambda/4$ in the cavity length nd or, alternatively, a frequency interval $\Delta\nu$. From Eq. (5.44) the resonant frequency is given by

$$\frac{4\pi\nu_q nd}{c} = q2\pi$$

or

$$\nu_q = q\,\frac{c}{2nd} \tag{5.46}$$

Thus the frequency interval, $\Delta\nu$, between successive resonances is

$$\Delta\nu = \nu_{q+1} - \nu_q = (q+1)\,\frac{c}{2nd} - q\,\frac{c}{2nd}$$

$$\Delta\nu = \frac{c}{2nd} \tag{5.47}$$

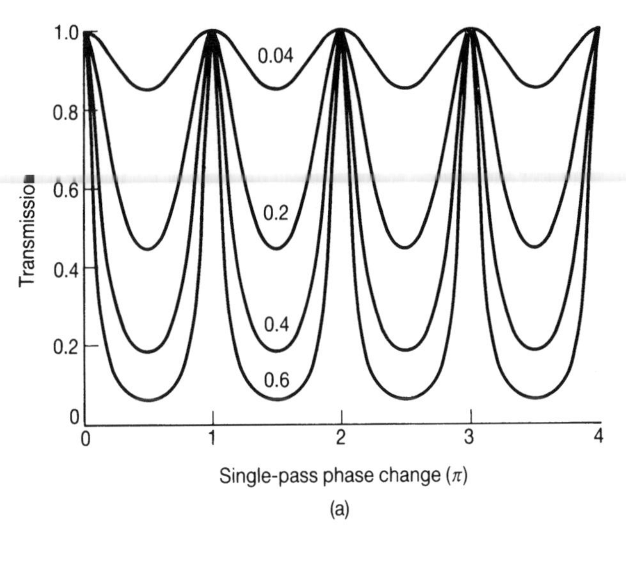

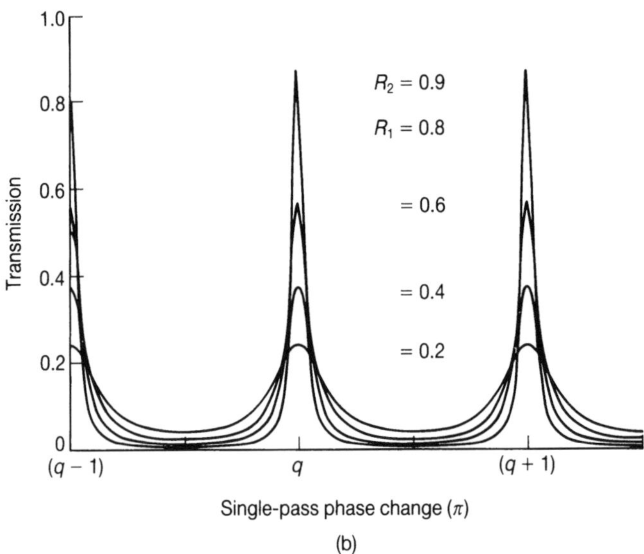

Figure 5.11 Plot of Fabry–Perot transmission function as a function of ϕ (a) for $R_1 = R_2 =$ 4 per cent, 20 per cent, 40 per cent and 60 per cent and zero intracavity loss, (b) for different mirror reflectivities.

This frequency interval $\Delta\nu$ is known as the *free spectral range* of the interferometer. It could be expressed in terms of wavelength but it will be seen that frequency intervals are dependent only on cavity length.

The accepted terminology to describe the width of the resonances is their frequency width at half-maximum $\Delta\nu_{1/2}$, and is referred to as the *resolution*. Defining ω_+ as the higher angular frequency at which the transmission has dropped to half its maximum value, we have

$$2\pi\ \Delta\nu_{1/2} = 2(\omega_+ - \omega_q) \tag{5.48}$$

From Eq. (5.42)

$$\frac{1}{2}\ T_{\mathrm{FP}}\mathrm{Max} = \frac{(1-R_1)(1-R_2)}{2[1-(R_1R_2)^{1/2}]^2}$$

Thus using eq 5.38 ω_+ is given by

$$\frac{(1-R_1)(1-R_2)}{2[1-(R_1R_2)^{1/2}]^2} = \frac{(1-R_1)(1-R_2)}{[1-(R_1R_2)^{1/2}]^2 + 4(R_1R_2)^{1/2}\sin(\omega_+ nd/c)}$$

This may be rearranged to give

$$\sin(\omega_+ nd/c) = \frac{1-(R_1R_2)^{1/2}}{2(R_1R_2)^{1/4}}$$

Substituting for ω_+ from Eq. (5.48)

$$\sin\frac{nd}{c}(\omega_q + \pi\ \Delta\nu_{1/2}) = \frac{(1-(R_1R_2)^{1/2})}{2(R_1R_2)^{1/4}}$$

Expanding using the sum angle identity

$$\sin\frac{nd}{c}\omega_q\cos\pi\frac{nd}{c}\Delta\nu_{1/2} + \cos\frac{nd}{c}\omega_q\sin\pi\frac{nd}{c}\Delta\nu_{1/2} = \frac{1-(R_1R_2)^{1/2}}{2(R_1R_2)^{1/4}}$$

For sharp resonances the angular variable $\pi(nd/c)\Delta\nu_{1/2}$ will be very small; also $(nd/c)\omega_q$ is a multiple of π since it is the condition for a cavity resonance. Thus writing in terms of $\Delta\nu_{1/2}$ we obtain

$$\Delta\nu_{1/2} = \frac{c(1-(R_1R_2)^{1/2})}{2nd\pi(R_1R_2)^{1/4}} \tag{5.49}$$

We may introduce a figure of merit known as the *reflectivity finesse*, $\mathscr{F}_{\mathrm{R}}$, defined as

$$\mathscr{F}_{\mathrm{R}} = \frac{\pi(R_1R_2)^{1/4}}{1-(R_1R_2)^{1/2}} \tag{5.50}$$

The resolution (Eq. (5.49)) can now be written

$$\Delta\nu_{1/2} = \Delta\nu/\mathscr{F}_{\mathrm{R}} \tag{5.51}$$

The reflectivity finesse[1] is a dimensionless quantity depending only on the mirror reflectivities and gives the width of the resonance as a fraction of the spacing between the resonances for an ideal monochromatic source. Being independent of the free spectral range it gives a measure of the resolving power of the instrument in terms of the mirror reflectivity only.

The result of driving the cavity through resonance can be observed in reflection as well as transmission. In the case of reflection the input light experiences a phase change as determined by Eq. (5.41). This phase shift is plotted in Figure 5.12, showing the rapid change in phase as the cavity passes through resonance. This rapid change in phase may be used to detect resonance and is often utilized to lock a laser frequency to that of an ultra-stable cavity (see Section 6.6.4.).

Example 5.2 Calculation of free spectral range, finesse and resolution
A plane parallel Fabry–Perot interferometer has a mirror spacing of 5 cm in air and the mirrors each have a reflectivity of 98 per cent. Calculate the free spectral range, resolution and finesse.

Free spectral range $\Delta\nu = c/2d = 3\times10^8/2\times0.05 = 3\ \text{GHz}$

Finesse $$\mathscr{F} = \frac{\pi R^{1/2}}{(1-R)} = \frac{\pi 0.99^{1/2}}{1-0.99} = 313$$

Resolution $\Delta\nu_{1/2} = \Delta\nu/\mathscr{F} = 3\times10^9/313 = 9.6\ \text{MHz}$

This interferometer can thus be used to analyse a spectrum over a range of 3 GHz with a resolution of 9.6 MHz.

5.4.2 The cavity lifetime

The reflectivity finesse is a measure of the losses in the interferometer or cavity attributable to the mirrors. In general, there will be other losses: for example, intra-cavity absorption (coefficient α) diffraction loss and scattering. These should be taken into account to determine the total finesse. It is left as an exercise (Problem 5.10) to show that when cavity absorption is included the finesse becomes

$$\mathscr{F}_\alpha = \frac{\pi[(R_1R_2)^{1/2}\,e^{-\alpha d}]^{1/2}}{1-(R_1R_2)^{1/2}\,e^{-2\alpha d}} \tag{5.52}$$

Including these lossy effects reduces the resolution and the total finesse can be written as

$$\frac{1}{\mathscr{F}} = \frac{1}{\mathscr{F}_R} + \frac{1}{\mathscr{F}_D} + \frac{1}{\mathscr{F}_S} + \frac{1}{\mathscr{F}_\alpha} \tag{5.53}$$

where the subscripts, respectively, refer to reflectivity, diffraction, scattering and absorption.

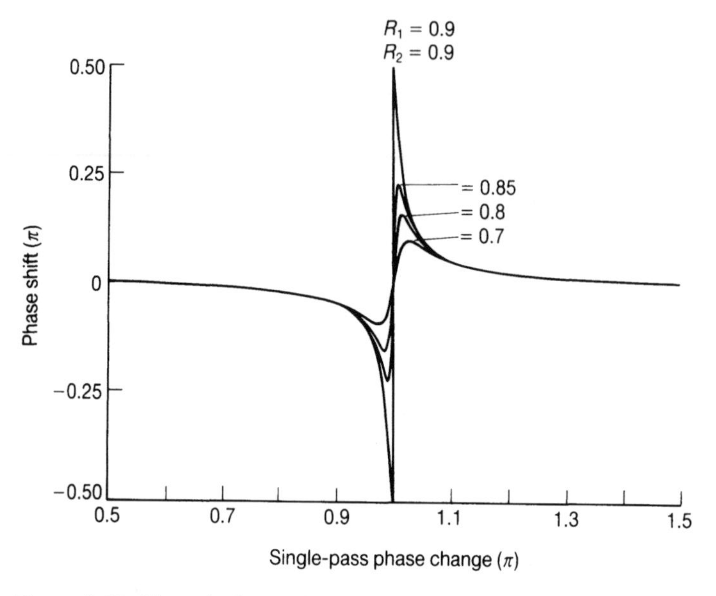

Figure 5.12 Plot of phase shift on reflection from a Fabry–Perot.

The losses can be expressed using another parameter, namely the *cavity lifetime*. Consider a cavity in which all the losses other than mirror transmission are included in α. The irradiance after a roundtrip is

$$I_0 R_1 R_2 \, e^{-2\alpha d}$$

where I_0 is some arbitrary starting irradiance at some arbitrary starting location in the cavity. The irradiance will decrease on each roundtrip of the cavity and it is convenient to regard the loss as occurring continuously as a function of time, even though some of it (in many cases, most of it) may occur at discrete positions (i.e. the mirrors). In addition, it is useful to express this loss in exponential terms since this allows the gain and loss to be compared directly when we come to consider a laser cavity. We may write the total decrease in the irradiance in one roundtrip of the cavity as

$$I_0 \exp(-\tau_{RT}/\tau_c)$$

In this form the cavity lifetime, τ_c, has been introduced as a time constant for the decay of the radiation within the cavity. The cavity roundtrip time is obtained directly from the wave velocity and the optical length of the cavity:

$$\tau_{RT} = 2nd/c \tag{5.54}$$

Equating both expressions for the roundtrip loss

$$I_0 R_1 R_2 \, e^{-2\alpha d} = I_0 \, e^{-\tau_{RT}/\tau_c}$$

Taking the natural logarithm of both sides and substituting from Eq. (5.54) for τ_{RT} gives

$$\frac{1}{\tau_c} = \frac{c}{2nd}(2\alpha d - \text{Log}_e(R_1 R_2)) \tag{5.55}$$

In many cases the term in α can be neglected and the cavity decay time expressed solely in terms of mirror reflectivity:

$$\frac{1}{\tau_c} = \frac{c}{2nd}(-\text{Log}_e(R_1 R_2)) \tag{5.56}$$

Rewriting Eq. (5.56)

$$\frac{1}{\tau_c} = \frac{c}{2nd}(-\text{Log}_e[1 - (1 - R_1 R_2)])$$

For cases when the mirror reflectivities approach unity we may make use of the relation

$$\text{Log}_e(1 + x) \cong x \text{ for } x \ll 1$$

to obtain

$$\tau_c \cong \frac{2nd}{c(1 - R_1 R_2)} \tag{5.57}$$

All these forms for the cavity lifetime will be used when we come to discuss the temporal growth and decay of light in a laser cavity in Chapters 6 and 8.

For high reflectivity the product $R_1 R_2$ will be close to unity so that we may write

$$(1 - R_1 R_2) = [1 - (R_1 R_2)^{1/2}][1 + (R_1 R_2)^{1/2}] \cong 2[1 - (R_1 R_2)^{1/2}]$$

Thus

$$[1 - (R_1 R_2)^{1/2}] \cong \tfrac{1}{2}(1 - R_1 R_2)$$

and the resolution given by Eq. (5.49) can be written as

$$\Delta\nu_{1/2} \cong \frac{c}{2nd}\frac{(1 - R_1 R_2)}{2\pi}$$

Comparing this expression with Eq. (5.57) we see that

$$\frac{1}{\tau_c} \cong 2\pi\, \Delta\nu_{1/2} \tag{5.57(a)}$$

The cavity lifetime and the resolution are seen to be different ways of expressing the cavity loss. However, we note that, in the derivation of the expression for the cavity lifetime, we have assumed the losses to be averaged over the length of the cavity while for the resolution the losses are taken at the mirrors. For this reason, the expressions for τ_c and $\Delta\nu_{1/2}$ are consistent only at low loss.

Example 5.3 Calculation of the cavity decay time
Evaluate the cavity decay time using the data in Example 5.2:

The roundtrip time $\tau_{RT} = 2d/c = 2 \times 0.05/3 \times 10^8 = 3.3 \times 10^{-10}$ s

The cavity decay time $\tau_c = \tau_{RT}/(1 - R_1R_2) = \dfrac{3.3 \times 10^{-10}}{(1 - 0.99^2)}$

$$= 1.66 \times 10^{-8} \text{ s}$$

The resolution $\Delta\nu_{1/2} = \dfrac{1}{2\pi\tau_c} = \dfrac{1}{2\pi \times 1.66 \times 10^{-8}}$

$$= 9.59 \text{ MHz}$$

Comparison with Example 5.2 shows the validity of the approximation.

At this stage it is appropriate to consider how it is that the cavity can have unity transmission even when the individual mirrors may have very low transmission. Figure 5.13 shows a schematic diagram of a cavity with two 99 per cent reflecting mirrors which has 1 W of optical power incident upon it. When the cavity resonates (i.e. Eq. (5.46) is satisfied) it will have 1 W of power leaving mirror 2, and hence there must be 100 W of power inside the cavity. The power inside the cavity is an enhanced version of the power incident on the cavity and it is possible to use this to detect very weak absorption or to enhance second harmonic generation (see Section 6.6.5). When light is first incident on the cavity it is not transmitted instantaneously by the cavity; it takes a time of the order of the cavity lifetime to build up. The cavity lifetime, τ_c, is the time constant for the build-up of the light as well as its decay. The resonance condition described by Eqs (5.38), (5.40) and (5.46) only apply when the steady state is established, that is, when the intra-cavity fields have been allowed to reach equilibrium.

Returning to Eq. (5.38) and including now the absorption of the mirror coatings but again taking $\alpha = 0$, the transmission at resonance for identical mirrors becomes

$$T_{FP}\text{Max} = \frac{T^2}{[1 - R]^2} = \frac{T^2}{[A + T]^2} = \left(1 + \frac{A}{T}\right)^{-2} \tag{5.58}$$

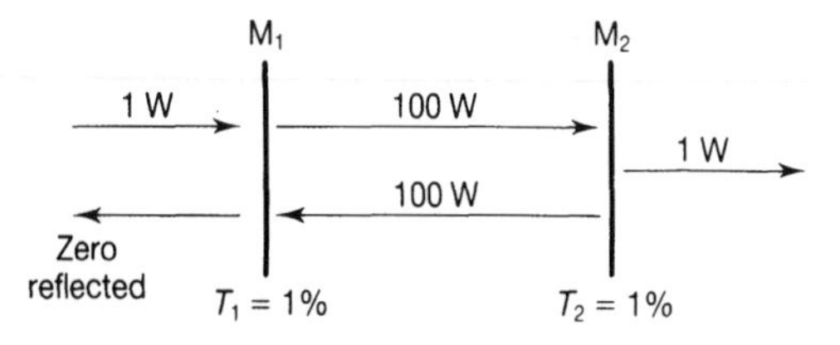

Figure 5.13 Diagram of a Fabry–Perot cavity in resonance showing input and output of 1 W with intra-cavity power of 100 W.

We now see that the maximum transmission will become less than unity when we have a coating absorption comparable to the mirror transmission. From Eq. (5.50) the cavity finesse, and so resolution, is improved by increasing mirror reflectivity. In practice, dielectric-coated mirrors have an inherent absorption loss, so by increasing the reflectivity by adding more layers in the stack we increase the ratio A/T. Thus improvements in finesse obtained by increasing the mirror reflectivity are paid for by incurring decreased cavity transmission. This is illustrated with the two graphs in Figure 5.14, which shows finesse as a function of reflectivity (Eq. (5.50)) and cavity transmission as a function of mirror transmission Eq. (5.58). In recent years the development of extremely low-loss dielectric mirrors for ring laser gyroscopes [Ref. 4] has allowed the development of Fabry–Perot interferometers with finesse values in excess of 10 000 which still exhibit tolerable transmission.

5.4.3 The Fabry–Perot interferometer with spherical mirrors (FPS)

Our mode analysis for optical cavities in Chapter 4 applies to the case of stable curved mirror cavities for which the Hermite–Gaussian (or Laguerre–Gaussian) have been shown to be eigenmodes. The plane wave approach (i.e. assuming no diffraction) to the Fabry–Perot interferometer assumed up to now in this chapter can be extended to incorporate the mode analysis. The roundtrip phase change used in Eq. (5.36) applies to a plane wave where diffraction divergence of the wave is negligible, which is the case for cavities having high Fresnel numbers. For Gaussian beams the longitudinal phase of the wave involves another term, i.e. the *Guoy shift*. The longitudinal phase change for a higher-order Gaussian beam travelling a distance z is given by Eqs (4.83) and (4.93) as

$$\text{TEM}_{m,n} \text{ Hermite–Gaussian mode } kz - (m+n+1)\tan^{-1}\left(\frac{z}{z_R}\right)$$

$$\text{TEM}_{l,p} \text{ Laguerre–Gaussian mode } kz - (2p+l+1)\tan^{-1}\left(\frac{z}{z_R}\right)$$

Consider the curved mirror cavity in Figure 4.15. It shows that the total single-pass phase shift is the sum of the phase shifts to each mirror from the waist and, for Hermite–Gaussian modes, this is given by

$$\phi(z_1)+\phi(z_2) = k(z_1+z_2) - \left[\tan^{-1}\left(\frac{z_1}{z_R}\right) + \tan^{-1}\left(\frac{z_2}{z_R}\right)\right](1+m+n)$$

$$= q\pi \text{ at resonance} \qquad (5.59)$$

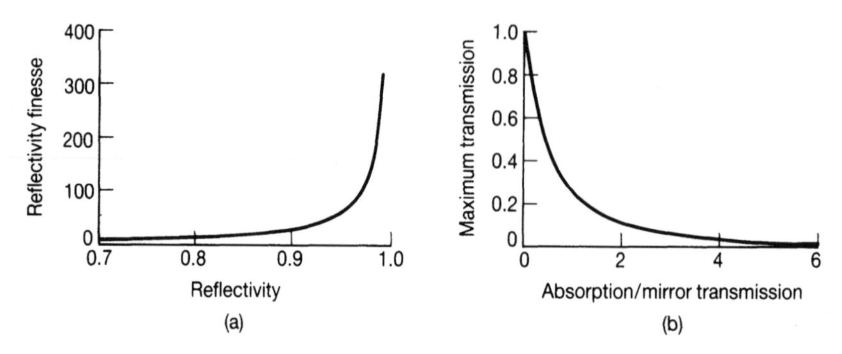

Figure 5.14 (a) Curve for reflectivity finesse as a function of mirror reflectivity. (b) Curve for cavity maximum transmission as a function coating absorption/transmission ratio.

Replacing k by the frequency in hertz using

$$k = 2\pi n\nu/c$$

we obtain

$$\nu_{mnq} = q\,\frac{c}{2nd} + \frac{c}{2nd}\,\frac{1}{\pi}\left\{\tan^{-1}\left(\frac{z_1}{z_R}\right) + \tan^{-1}\left(\frac{z_2}{z_R}\right)\right\}(1 + m + n)$$

Using the identity

$$\tan^{-1}a + \tan^{-1}b = \cos^{-1}\left(\frac{1 - ab}{\sqrt{1 + a^2 + b^2 + ab}}\right)$$

and Eqs (4.61), (4.62) and (4.63), after some lengthy algebra the resonant frequency in a curved mirror cavity of the qth longitudinal mode of the m, n transverse mode in terms of the cavity g parameters becomes

$$\nu_{mnq} = \frac{c}{2nd}\left(q + (1 + m + n)\,\frac{1}{\pi}\,\cos^{-1}(g_1 g_2)^{1/2}\right) \tag{5.60a}$$

The corresponding expression for Laguerre–Gaussian modes is

$$\nu_{plq} = \frac{c}{2nd}\left(q + (2p + l + 1)\,\frac{1}{\pi}\,\cos^{-1}(g_1 g_2)^{1/2}\right) \tag{5.60b}$$

We notice that the resonant frequency differs from Eq. (5.46) by the inverse cosine term which was obtained by including the Guoy shift in the cavity single-pass phase change, and is seen to be dependent on the transverse mode number. It shifts the resonances up in frequency from the plane parallel case. In Chapter 4 we saw how the mode analysis was used to calculate the transverse distribution of irradiance. Here we now use it to evaluate the resonant frequencies of the Fabry–Perot cavity when diffraction is significant, i.e. when the Fresnel number approaches unity. This mode approach will be important for laser cavities where normally the length is much greater than the diameter.

Example 5.4 Calculation of the transverse mode spacing

Calculate the frequency difference between successive transverse modes, i.e. the frequency difference for an increase in either of the transverse mode numbers in an air-filled 0.5 m cavity with mirrors of curvature 2 m and 1 m:

$$\nu_{m,n+1,q} - \nu_{m,n,q} = \frac{c}{2nd}\frac{1}{\pi}((1+m+n+1)-(1+n+m))\cos^{-1}\sqrt{g_1 g_2}$$

$$= \frac{c}{2nd}\frac{1}{\pi}\cos^{-1}\sqrt{g_1 g_2}$$

$$g_1 = 1 - \frac{d}{b_1} = 1 - \frac{0.5}{1} = 0.5$$

$$g_2 = 1 - \frac{d}{b_2} = 1 - \frac{0.5}{2} = 0.75$$

$$g_1 g_2 = 0.5 \times 0.75 = 0.375$$

$$\nu_{m,n+1,q} - \nu_{m,n,q} = \frac{3 \times 10^8}{2 \times 0.5 \times \pi}\cos^{-1}\sqrt{0.375} = 87\ \text{MHz}$$

Note that the axial mode spacing or free spectral range $c/2d$ is 300 MHz

Each transverse mode, designated by m and n or p and l, has a set (or comb) of longitudinal modes determined by q with their separation the same as for the plane parallel, i.e. $c/2nd$. Figure 5.15 illustrates this comb of transverse mode resonances which occur for a cavity when the product $g_1 g_2$ is close to unity, i.e. the near-plane parallel or concentric geometries as discussed in Figure 4.7. Decreasing the g parameter product from unity towards zero makes the transverse mode frequencies move apart, since the inverse cosine term then increases. When $g_1 g_2 = 0$ the transverse mode spacing becomes exactly half the longitudinal mode spacing. Thus in the confocal cavity, discussed in Section 4.2, alternate transverse mode resonances coincide with their next longitudinal resonance. For the confocal case Eq. (5.60a) becomes

$$\nu_{mnq} = \frac{c}{2nd}\left(q + \frac{m+n+1}{2}\right) \tag{5.61}$$

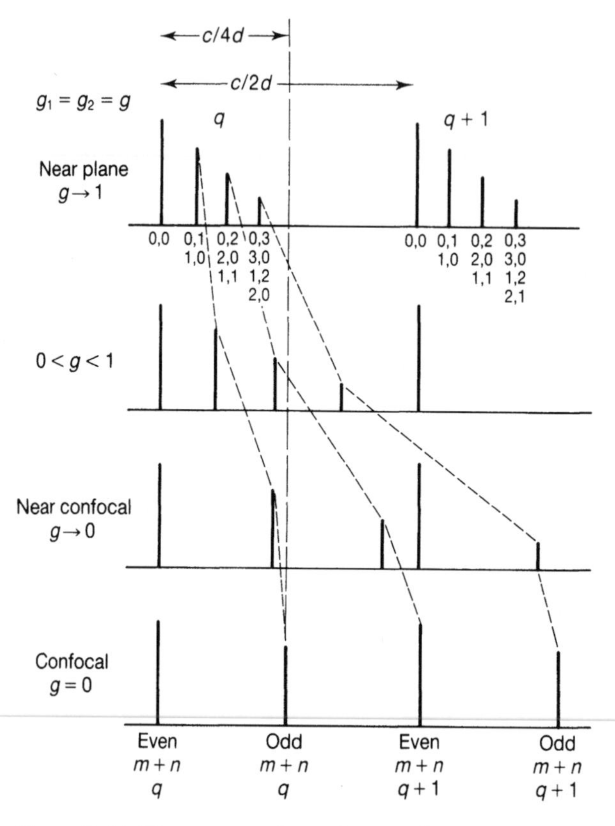

Figure 5.15 Diagram to show the comb of TEM_{mn} cavity resonant frequencies for (a) near plane parallel and (b) confocal cavities.

For completeness, let us see how this mode analysis gives a result consistent with the ray analysis of Section 4.2. Using Eq. (4.16) we write the inverse cosine term of Eq. (5.60a) in terms of the number of cavity roundtrips of the re-entrant rays:

$$\nu_{mnq} = \frac{c}{2nd}\left(q + (1 + n + m)\,\frac{K}{N}\right) \tag{5.62}$$

Equation (5.62) shows that the cavity resonant frequency will remain unchanged if $(m + n)$ is increased by N and q is decreased by K. This means that the cavity has every Nth transverse mode degenerate (the frequency shift of N transverse modes will correspond to one longitudinal mode) and is said to be N-fold degenerate. The confocal cavity discussed above is twofold degenerate, i.e. $N = 2$, consistent with Eq. (5.61). Since in an N-fold degenerate cavity the fields overlap after N roundtrips the cavity transmission will be the product of N terms of Eq. (5.58):

$$T_{\mathrm{FPS}}\mathrm{Max} = \frac{1}{N(1 + A/T)^2} \tag{5.63}$$

showing that the maximum transmission of the confocal cavity is half that of the plane parallel.

In addition, the reflectivity finesse is reduced due to the larger number of reflections and becomes

$$\mathscr{F}_{\text{FPS}} = \pi R^{N/2}/(1 - R^N) \tag{5.64}$$

For high reflectance this approximates to

$$\mathscr{F}_{\text{FPS}} = \pi/N(1 - R) \tag{5.65}$$

5.5 Application of the Fabry–Perot interferometer to optical frequency analysis

This chapter will be completed by a discussion of the application of the Fabry–Perot interferometer to the analysis of the spectra of light emitted by lasers. The reader may find it more appropriate to return to this section after Chapter 6, where laser oscillation is covered.

In general, most lasers produce light at discrete frequencies in Gaussian or higher-order transverse modes, and the Fabry–Perot interferometer is an important tool in determining the frequencies present. Figure 5.9 shows the two arrangements used with the plane parallel Fabry–Perot. Photographic recording of the fringe pattern (Figure 5.9(b)) is useful for pulsed lasers while with continuous-wave lasers the scanning method is more convenient since an electrical signal is more easily produced. The reflectivity finesse can only be achieved with the plane parallel Fabry–Perot when care is taken to reduce plate defect losses and to ensure parallel alignment of the plates. Some manufacturers ensure that the plates are held parallel using an electronic servosystem and because of these difficulties the spherical mirror Fabry–Perot, normally the confocal cavity, is preferred for laser mode analysis. The confocal cavity interferometer is particularly easy to manufacture since any relative angular misalignment of the mirrors results in a new optical axis for the interferometer and the only adjustment required is in their separation. The mode degeneracy of the confocal cavity ensures that the laser spectrum will be easily observed without the requirement of mode matching.

The input light from a laser may be a TEM_{00} mode but, in general, will not be a TEM_{00} mode of the analyser confocal cavity and it will therefore excite many transverse modes of the interferometer. From Eq. (5.60) it is seen that the resonances are separated by $c/4nd$. Another way of stating this is that the modes with $(m + n)$ even lie together separated by $c/2nd$ and between them lie the odd $(m + n)$ transverse modes. Thus although many high-order transverse modes are excited they divide themselves into pairs, i.e. odd and even.

Figure 5.16 is a system diagram for obtaining the spectrum of a CW laser. A linear voltage ramp is applied to the piezoelectric transducer which is attached to one of the mirrors and forces it to move in sympathy with the applied voltage. The movement of the mirror will scan the resonances through several free spectral ranges, so

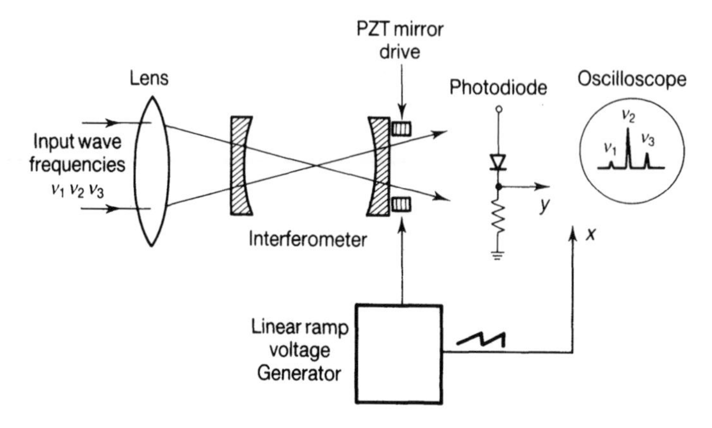

Figure 5.16 System diagram for confocal scanning Fabry–Perot interferometer.

moving about one wavelength. This will be small enough to ensure that the cavity is not significantly altered from a confocal configuration during the scan. The transmission of the interferometer is recorded using a photodetector with the appropriate wavelength response and amplifier with a frequency response high enough to faithfully record the true shape of the resonance.

Figure 5.17 displays the comb of resonances produced by a single-frequency (mode) CW helium–neon laser. When the input beam has good spatial coherence the interferometer, with care, can be aligned to excite only its symmetrical (even) transverse modes. In this situation the odd $(m + n)$ modes are extinguished and the effective free spectral range doubled, but, of course, the width of the resonances (resolution) stays the same.

Figure 5.18 shows how each alternative resonance can be extinguished by fine adjustment of the cavity axis relative to the input beam. In this figure further adjustment would be required to extinguish the small mode. Since these modes of the cavity can also be expressed by Laguerre–Gaussian functions this symmetrical

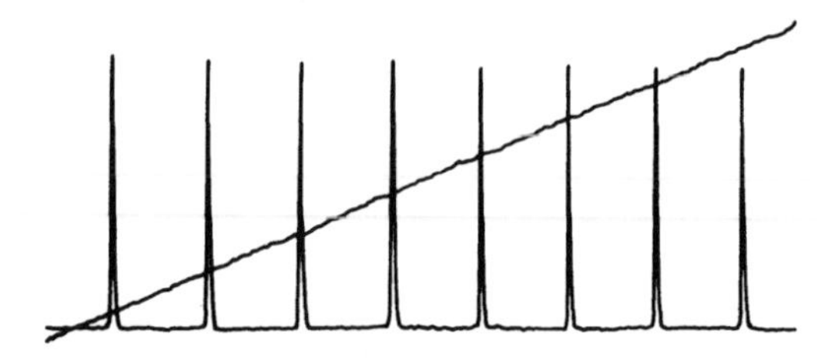

Figure 5.17 Comb of resonances for confocal Fabry–Perot driven by single-mode helium–neon laser. The finesse is obtained from this trace as the separation divided by the width of the peaks. The linear ramp voltage applied to the PZT is also shown.

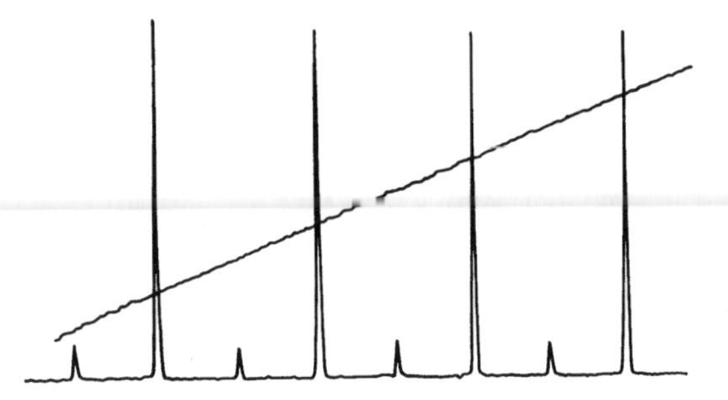

Figure 5.18 Effect of improved alignment showing the enhancement of the even transverse modes relative to the uneven modes. The linear ramp voltage is also shown.

alignment condition can be explained in the cylindrical coordinate system by saying that we are exciting modes with $l = 0$ as these have a maximum in the centre and no azimuthal phase reversals, i.e. they are concentric rings as shown in Figure 4.21.

In Figure 5.19 the effect of altering the cavity from the confocal separation by about 0.5 mm and removing the the confocal degeneracy is illustrated. Each resonance breaks into a number of resonances determined by the number of confocal cavity transverse modes which can be excited determined by the diameter of the input beam. In this case about 30 transverse modes are excited.

In any spectral analysis the spectral range to be analyzed must be less than the free spectral range of the interferometer, otherwise resonances with mode number q will become confused with $q - 1$ and $q + 1$. Since the mirror curvature in the confocal arrangement must equal the mirror separation, its major disadvantage is its fixed free spectral range. In addition, it is difficult to manufacture mirrors with

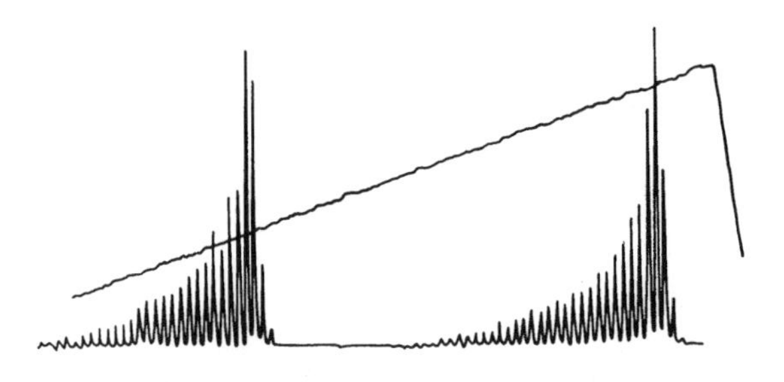

Figure 5.19 Effect of altering the cavity from confocal by 0.5 mm and removing the confocal degeneracy.

tight radii of curvatures. For large free spectral ranges (> 10 GHz) short cavities are required, and for these cases we use either plane parallel Fabry–Perot interferometers or large-radius curved mirrors and the input beam mode matched to the fundamental cavity mode, the conditions for which were discussed in Section 4.10.3.

5.6 Ring resonators

We conclude this chapter by considering the ring resonator, which is used in some of the laser systems to be discussed in Chapter 7. A particular application is the ring laser gyroscope which uses a triangular ring cavity[2] as shown in Figure 5.20, and this cavity has number of features which are different from the linear cavity.

If the cavity is used as a passive resonator and light is injected from outside then the reflected light from the ring, which also includes light having travelled round the ring, will not be returned back into the source. This is particularly useful if the source is a laser which is strongly effected by back-reflection.

The cavity shown in Figure 5.20 has two plane mirrors and one curved one. The cavity can be unfolded into a lens waveguide and the $[ABCD]$ matrix determined. It is left as an exercise to show (problem 4–7) using Eq. (4.13) that the resonator stability condition is

$$0 \leqslant 1 - \frac{L}{2b} \leqslant 1 \tag{5.66}$$

Because the light incident on the curved mirror is not at normal incidence there will

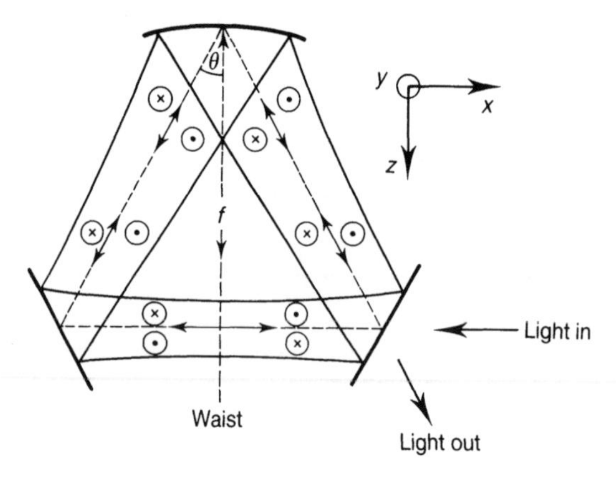

Figure 5.20 Triangular ring resonator showing the 180° phase shift which occurs for $(m, 0)$ modes due to lateral inversion on reflection.

be a different focusing condition in the two orthogonal planes. Referring to Figure 5.20, these two planes are (1) the plane of the ring (the *tangential plane*) and (2) perpendicular to the plane of the ring (the *sagittal plane*). It can be shown that the two focal lengths of the curved mirror for light at angle of incidence, θ are

$$\text{Tangential plane: } f_T = f\cos\theta \tag{5.67}$$

$$\text{Sagittal plane: } f_s = f/\cos\theta \tag{5.68}$$

The ring will therefore have different stability conditions and Gaussian beam parameters for the two planes. This condition means that the mode of this cavity is an elliptical Gaussian beam, the parameters of which are described in Section 4.11. In this symmetrical cavity the waist for both planes occurs in the same place – exactly between the two plane mirrors.

Another feature of the ring cavity compared with the linear cavity is the condition for resonance. In the ring there must be an integral number of whole wavelengths around the cavity. We can use this to determine the frequency at which the wave has a roundtrip phase change of an integral multiple of 2π. Using Eq. (4.131) the total phase shift for an elliptical Gaussian beam going once around the ring of perimeter, L, is, by symmetry, twice the phase shift from the waist to the curved mirror. Thus we obtain

$$kL - 2\left(m+\frac{1}{2}\right)\tan^{-1}\left(\frac{L}{2z_{RT}}\right) - 2\left(n+\frac{1}{2}\right)\tan^{-1}\left(\frac{L}{2z_{RS}}\right) = q2\pi$$

As with Eq. (5.60a), we obtain the frequency of the *mnq* transverse mode, this time in the form

$$\nu_{mnq} = \frac{c}{nL}\left\{q + \frac{m+\frac{1}{2}}{\pi}\tan^{-1}\left(\frac{L}{2z_{RT}}\right) + \frac{n+\frac{1}{2}}{\pi}\tan^{-1}\left(\frac{L}{2z_{RS}}\right)\right\} \tag{5.69}$$

This equation reveals that transverse modes with m and n reversed are not degenerate since the Rayleigh ranges in the two planes, z_{RS} and z_{RT}, are different. A further complication is that high-order modes in the tangential plane (i.e. higher value of m) suffer an additional phase change of π radians in one roundtrip for cavities with an even number of mirrors. This occurs due to lateral inversion of the mode after reflection as illustrated in Figure 5.20. The E-vectors for s-polarized light going around the ring is shown in the figure by the arrow symbol into or out of the page. Illustrated here is a transverse mode with one phase reversal in the tangential plane ($m = 1$) and after one roundtrip the phases in each lobe of the mode are seen to be π radians out of phase. Figure 5.21 shows the comb of resonant frequencies expected from a ring with an odd number of mirrors. Note that it shows the $(m, 0)$ modes shifted up from the centre of the free spectral range of the $(0, 0)$ modes. If light were p-polarized then an additional phase change of π occurs on each reflection and the p-polarized modes are shifted one half free spectral range from the s-polarized modes. This is due to lateral inversion of the electric field vector on reflection as illustrated in Figure 5.2.

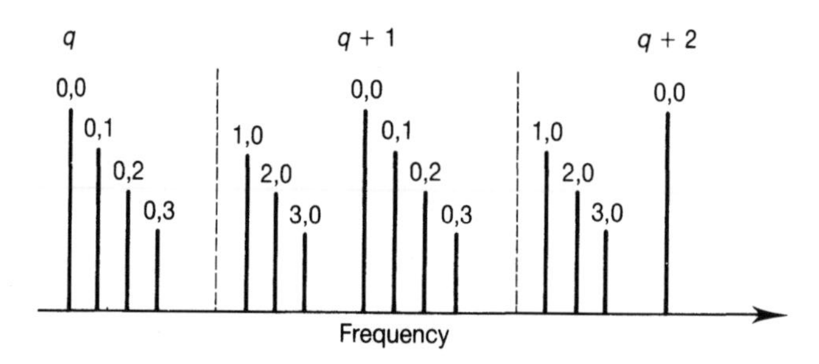

Figure 5.21 Comb of resonant frequencies for *s*-polarized light in a ring resonator with an odd number of mirrors.

Problems

5.1 Determine the Fresnel reflectance for infra-red radiation at normal incidence on silicon with refractive index 3.5. What should be the refractive index of the coating material for a quarter-wave thick anti-reflection coating?

5.2 Determine the Fresnel reflection coefficient of fused silica (refractive index 1.46) at normal incidence. Determine its reflectivity if a quarter-wave thick layer of magnesium fluoride with refractive index 1.38 is deposited on this substrate.

5.3 Calculate the reflectivities for *s*- and *p*-polarized light of a quarter-wave thick film of zinc sulphide ($n = 2.2$) on a germanium substrate ($n = 4.0$).

5.4 Evaluate the reflectivity of a multilayer coating of 11 layers on a substrate of fused silica ($n = 1.46$). The high layers are titanium dioxide ($n = 2.4$) and the low layers are silicon dioxide ($n = 1.46$). What would be the reflectivity if scandium oxide ($n = 1.8$) were used for the high layers?

5.5 Derive an expression for the reflectivity as a function of wavelength for a single layer of zinc sulphide ($n = 2.35$) on a fused silica substrate.

5.6 Calculate the free spectral range, finesse and resolution of a plane parallel Fabry–Perot interferometer with an air gap spacing of 100 μm and mirror reflectivities of 95 per cent.

5.7 Determine the free spectral range of a 5 cm long cavity containing a crystalline rod, length 4 cm, of refractive index 1.82.

5.8 Calculate the finesse of a 1 cm long cavity made using ultra-high reflectors having loss of 30 parts per million. What is the build-up time of light in this cavity?

5.9 It is desired to build a Fabry–Perot interferometer with a resolution of 1 MHz and a free spectral range of 1500 MHz. Calculate the reflectivity of the mirrors and the spacing required.

5.10 Derive expression (5.52) for the finesse of a plane parallel Fabry–Perot interferometer with intra-cavity absorption included.

5.11 An equilateral triangular ring cavity with sides 0.5 m has mirrors of reflectivity 0.985, 0.97 and 0.99. Calculate the free spectral range and resolution of this cavity.

5.12 Calculate the finesse of a rectangular ring cavity which has mirrors of reflectivity 0.98 and an absorption of 0.05 in each arm.

5.13 An almost hemispherical cavity has a mirror spacing of 25 cm and a concave mirror with radius of curvature 35 cm. Calculate the frequency spacing between neighbouring transverse modes.

5.14 A confocal cavity to be used for high-resolution spectroscopy can do so in the limit that the transverse mode spacing equals the resolution. Using this criterion, show that the tolerance on the mirror spacing is given by $\varepsilon \cong b(\pi/2\mathscr{F})$ where b is the radius of curvature of the mirrors and $\mathscr{F}$ is the finesse. Calculate this tolerance for a 10 GHz free spectral range cavity with a finesse of 500.

5.15 An equilateral triangular ring cavity has two plane mirrors and a concave mirror with radius of curvature of 1.25 m. The length of each side of the cavity is 0.3 m. Sketch the frequency comb for the transverse modes and determine the frequency separation between the modes. (See Problem 4.8).

Notes

1. In some texts the coefficient of finesse is defined:

$$F = 4(R_1R_2)^{1/2}/[(1-(R_1R_2)^{1/2}]^2$$

Hence

$$\mathscr{F}_R = \frac{2}{\pi}\sqrt{F}$$

2. Some manufacturers of ring laser gyroscopes prefer a square cavity.

References

[1] H. A. Macleod, *Thin Film Optical Filters*, Adam Hilger, Bristol, 1986, Chapter 5.
[2] *Ibid.*, page 37.
[3] H. M. Liddell, *Computer Aided Techniques for the Design of Optical Filters*, Adam Hilger, Bristol, 1988.
[4] J. Wilkinson, 'The ring laser gyroscope', *Progress in Quantum Electronics*, **11**, 1, 1987.
[5] J. M. Vaughan, *The Fabry–Perot Interferometer*, Adam Hilger, Bristol, 1989.

Further reading

D. J. Bradley and C. J. Mitchell, 'Characteristics of the defocused spherical Fabry–Perot interferometer as a quasi linear dispersion instrument for high resolution spectroscopy of pulsed lasers', *Phil. Trans. R. Soc. A*, **263**, 26, 1968.

M. Hercher, 'The spherical mirror Fabry–Perot interferometer', *Applied Optics*, **7**, 951, 1968.

D. Herriot, H. Kogelnik and R. Kompfer, 'Off-axis paths in spherical mirror interferometers', *Applied Optics*, **3**, 532, 1964.

I. A. Ramsay and J. J. Degnan, 'A ray analysis of optical resonators formed by two spherical mirrors', *Applied Optics*, **9**, 385, 1970.

6

Laser oscillation

6.1 Introduction

In Chapter 3 the concepts of gain and saturation of the gain were discussed while in Chapters 4 and 5 we covered in some detail the cavities which enhance the radiation fields so ensuring that stimulated emission rates are very much larger than spontaneous emission rates. In this chapter these ideas will be brought together and we will discuss the details of how energy in the form of population inversion is fed into the radiation fields. Two aspects of laser oscillation will be considered:

1. The initial build-up of the radiation at switch-on of the laser known as the threshold
2. The situation after the oscillation has built up to an equilibrium and saturated the gain.

The first of these determines whether or not the laser oscillates while the second determines the output parameters such as power, frequency and oscillation bandwidth. In this chapter the threshold, build-up of the radiation and the ultimate output parameters will be discussed.

6.2 The threshold condition

To use stimulated emission effectively for the amplification of radiation and to ensure that the amplified signal is significantly larger than the noise due to spontaneous emission, it is necessary that the radiation to be amplified has a high energy density. In Chapter 5 we saw how radiation can be contained in a cavity and how as modes of the cavity the radiation fields are enhanced. The laser can be considered as an active cavity in which the enhanced radiation density at the cavity resonant frequencies will make the stimulated emission rates much larger than the spontaneous emission rates and the cavity will significantly improve the spatial and temporal coherence of any output beam.

The action of a cavity can be considered in the same way as an electronic oscillator. Positive feedback can produce self-sustaining oscillations at a particular frequency determined by its resonant properties. Similarly, in a laser self-sustaining oscillations can occur at particular frequencies, the cavity modes, when the round-trip gain becomes larger than the optical losses. Below this threshold the atoms are radiating spontaneously while above it the energy in the excited atoms is emitted by stimulated emission.

Figure 6.1 is a schematic diagram of a linear laser. The gain medium can be a solid, liquid or gas and is excited to produce population inversion and so amplification. We need not worry at this stage exactly how this is done. In some systems the mirrors are attached to the end of the gain medium while in others they are remote. The latter case is shown in Figure 6.1 and for this, low-loss windows at the ends of the gain section are necessary. This can be achieved using anti-reflection coatings or having a window set at the Brewster angle. The optical loss in the cavity will be caused by factors which reduce the cavity radiation energy by absorption or optical elements which couple light out of the cavity. These include:

1. Fresnel reflective losses at surfaces
2. Bulk absorption of any intra-cavity material

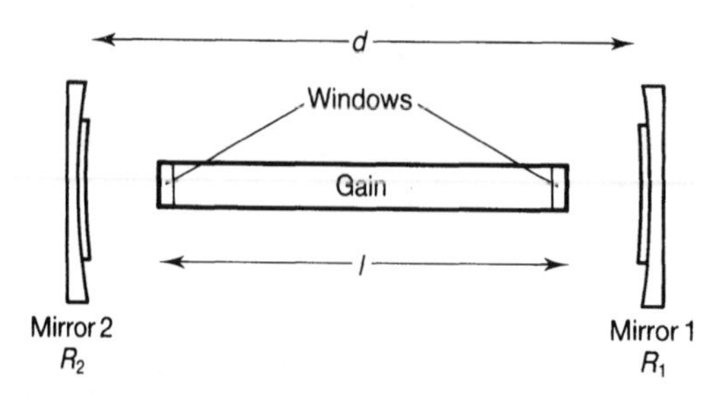

Figure 6.1 Schematic diagram of a laser with the mirrors remote from the gain medium.

3. Diffraction loss of the mode by intra-cavity apertures such as the gain medium itself
4. Transmittance and absorption of the mirror coatings.

The first three of these are lumped into a total roundtrip internal fractional loss, L, while we can consider mirror transmittance as an external loss since this is the method of coupling the intra-cavity radiation to the outside world.

If we start at any arbitrary point in the cavity with an irradiance I_0. Then, after one roundtrip and neglecting saturation of the gain, it becomes

$$I_0(1-L)R_1R_2\exp(2\gamma_0 l)$$

where R_1 and R_2 are the mirror reflectivities, l is the length of the gain medium and γ_0 is the unsaturated gain coefficient defined in Eq. (3.106). For self-sustained oscillation we require

$$(1-L)R_1R_2\exp(2\gamma_0(\omega)l) \geqslant 1$$

This condition for threshold can be written

$$2\gamma_{th}(\omega)l = \log_e(1/R_1R_2) + \log_e[1/(1-L)] \tag{6.1}$$

where $\gamma_{th}(\omega)$ is the smallest value of the gain coefficient required for self-sustaining oscillations and is known as the *threshold gain coefficient*. This will normally occur at the centre frequency, ω_0, of the gain curve where the gain is largest. Introducing the threshold population inversion density, N_{th}, and the stimulated emission cross-section $\sigma_{se}(\omega)$, this becomes

$$2\sigma_{SE}(\omega)N_{th}l = \log_e(1/R_1R_2) + \log_e[1/(1-L)] \tag{6.2}$$

$$N_{th} = \frac{\log_e(1/R_1R_2) + \log_e[1/(1-L)]}{2\sigma_{se}(\omega)l} \tag{6.3}$$

Rewriting Eq. (6.1) in terms of the threshold gain coefficient gives

$$\gamma_{th}(\omega) = \frac{\log_e(1/R_1R_2) + \log_e[1/(1-L)]}{2l} \tag{6.4}$$

The right-hand side of Eq. (6.4) represents the optical loss averaged over the length, l, of the gain medium and it is useful to define this as a total loss coefficient, α_T. By doing this we are conveniently taking an average of the loss over l even though the losses occur at specific positions in the laser. Having the total loss and the gain in the same units allows them to be represented on the same graph as shown in Figure 6.2. The variation of gain coefficient with frequency is determined by the lineshape function and is drawn here as a general bell-shaped curve while the total cavity loss is taken to be constant over the bandwidth of the gain and so is shown by a horizontal line. There are two frequencies where the threshold condition is satisfied and between them lies the frequency interval where gain exceeds loss. This frequency interval is known as the *gain bandwidth* and, in principle, oscillation is possible at any frequency within it. However, oscillation will preferentially take

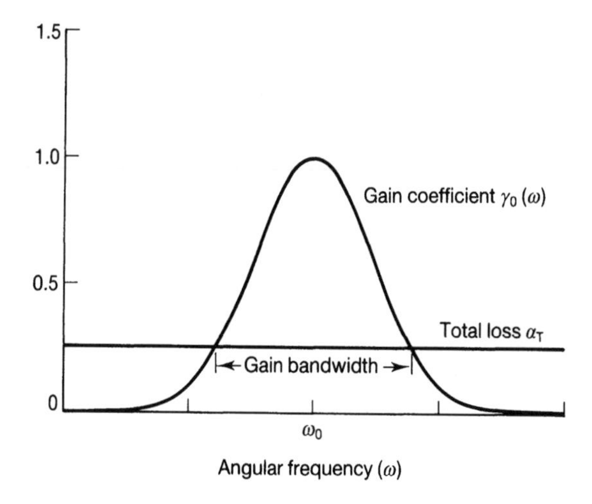

Figure 6.2 Gain coefficient and loss shown graphically.

place at frequencies for which the cavity fields are largest, so enhancing the stimulated emission. These frequencies are, of course, the cavity resonant frequencies.

Example 6.1 Estimation of the threshold gain coefficient
In a typical helium–neon laser operating at 633 nm the length of the gain medium is 1 m and the mirror reflectivities are 99.8 per cent and 98 per cent. The TEM_{00} mode suffers a single-pass loss of 1 per cent and each window has a loss of 0.2 per cent.

Since the internal losses are very small, they add and become

$$(1 - L) = 1 - 0.02 - 0.008 = 0.972$$

$$\log_e(1/(1 - L) = 0.028$$

$$\log_e(1/R_1 R_2) = \log_e(1/(0.998 \times 0.98) = 0.022$$

The threshold gain coefficient at the lasing frequency becomes

$$\gamma_{th}(\omega) = \frac{0.022 + 0.028}{2 \times 1} = 0.025\ \mathrm{m}^{-1}$$

The stimulated emission cross-section at line centre for this transition is $65 \times 10^{-22}\ \mathrm{m}^2$ (see Table 7.1), hence the threshold population inversion is

$$N_{th} = \frac{0.025}{64 \times 10^{-22}} = 3.86 \times 10^{18}\ \mathrm{m}^{-3}$$

In a typical helium–neon laser the atom density is approximately $3 \times 10^{22}\ \mathrm{m}^{-3}$ and in this laser only one atom in 10 000 is involved in the population inversion.

6.3 Laser power around threshold

Having established the population inversion required for the laser to reach threshold we now consider how the optical power changes at threshold. Below threshold there are only a few photons in each cavity mode while above there are many. It is this rapid and large change in the number of photons which we will now discuss.

Let us consider the four level homogeneous system as developed in Section 3.3.2 for the gain saturation. For simplicity, we assume that the relaxation of the lower state is so fast that its population is zero. The rate of stimulated emission at frequency ω from level 2 is

$$\frac{dN_2}{dt} = B_{21}\rho_\omega g(\omega)N_2V$$

where V is the volume of the cavity mode. Using Eq. (2.114) to write this in terms of the Einstein A coefficient we obtain

$$\frac{dN_2}{dt} = \frac{\pi^2c^3}{\omega^2} A_{21}g(\omega)\rho_\omega N_2V \tag{6.5}$$

The number of photons, N_p, in the cavity mode is given by

$$N_p = \rho_\omega V/\hbar\omega \tag{6.6}$$

and using the stimulated emission cross-section in the form

$$\sigma_{se}(\omega) = \frac{A_{21}\pi^2c^2}{\omega^2} g(\omega)$$

we can write Eq. (6.5) as

$$\frac{dN_2}{dt} = c\sigma_{SE}(\omega)N_pN_2 \tag{6.7}$$

The spontaneous relaxation rate from state 2 in the volume, V, is

$$A_{21}N_2V$$

Including the pumping rate R_2 per unit volume, and the decay rate of state 2 we can write for the rate of change of the total population of state 2

$$\frac{d}{dt}(N_2V) = -c\sigma_{SE}(\omega)N_pN_2 - A_{21}N_2V + R_2V \tag{6.8}$$

or

$$\frac{dN_2}{dt} = -\frac{c\sigma_{SE}(\omega)}{V} N_pN_2 - A_{21}N_2 + R_2 \tag{6.9}$$

This is one of a pair of coupled differential equations which describe the interplay between the population inversion (here we have neglected the lower state population and the inversion density is assumed to be N_2) and the number of photons. The number of photons in the cavity mode is determined by the stimulated emission and spontaneous emission rates. Since we have assumed no absorption then each photon in the cavity mode originates from either

1. A stimulated transition, given from Eq. (6.8) by the term

$$c\sigma_{SE}(\omega)N_pN_2$$

or

2. Excited atoms giving off photons by spontaneous emission isotropically and within the atomic linewidth but very few contributing to the cavity mode. From the definition of the line shape function in Section 3.2 we have for the rate of spontaneous emission between ω and $\omega + d\omega$

$$A_{21}N_2g(\omega)\,d\omega V$$

The number of oscillating modes in this frequency interval was shown for a cubical black body cavity from ref 9 – chapter 2 to be

$$n(\omega)V\,d\omega = \frac{\omega^2}{\pi^2c^3}V\,d\omega$$

We remind ourselves that the laser cavity is an open resonator with its length much larger than the transverse dimensions. Thus most of the off-axial modes will be too lossy to be above threshold but will be taking their share of the spontaneous photons.

The spontaneous emission rate, R_{sp}, into each mode at frequency ω is the spontaneous rate within the frequency interval $d\omega$ divided by the number of modes within the same interval. Thus

$$R_{sp} = \frac{\pi^2c^3}{\omega^2}g(\omega)A_{21}N_2 \tag{6.10}$$

In terms of the stimulated emission cross-section

$$R_{sp} = c\sigma_{SE}(\omega)N_2 \tag{6.11}$$

Comparison with Eq. (6.8) shows that when the number of photons, N_p, in the mode is one then the rate of stimulated emission is equal to the rate of spontaneous emission.

In the cavity any generated photons will be lost with a time constant equal to the cavity decay time and so the rate equation for the number of photons in a single mode becomes

$$\frac{dN_p}{dt} = c\sigma_{SE}(\omega)N_2N_p + c\sigma_{SE}(\omega)N_2 - N_p/\tau_c \tag{6.12}$$

This and Eq. (6.9) for the population inversion are our two coupled differential equations. They both involve the population inversion and the number of photons and can only be solved numerically unless certain constraints are imposed. The detailed temporal solutions will be reserved for Chapter 8. However, in our discussion here we are interested in the change in the number of cavity photons around threshold, so one constraint we can put on the system is that the time derivatives are zero. This applies to the two cases:

1. Below threshold where the energy pumped in is appearing as spontaneous emission, and
2. Above threshold where again the laser has reached a dynamic equilibrium between the population inversion and the photon density.

Putting $\mathrm{d}N_\mathrm{p}/\mathrm{d}t = 0$ in Eq. (6.12) we obtain for the steady state

$$N_2 = \frac{N_\mathrm{p}}{N_\mathrm{p}+1} \quad \frac{1}{c\sigma_{\mathrm{SE}}(\omega)\tau_\mathrm{c}} \tag{6.13}$$

Just below threshold there will be, at most, only one photon in the cavity mode since in this regime stimulated emission is less than spontaneous emission. Below threshold $n_\mathrm{p} \cong 1$ and

$$N_2 \leqslant 1/c\sigma_{\mathrm{SE}}(\omega)\tau_\mathrm{c} \tag{6.14}$$

When the laser is above threshold and oscillating under steady state then $n_\mathrm{p} \gg 1$ and

$$N_2 \cong 1/c\sigma_{\mathrm{SE}}(\omega)\tau_\mathrm{c} \tag{6.15}$$

Substituting for the cavity lifetime from Eq. (5.55) into Eq. (6.15) and putting, for simplicity, the gain length equal to the cavity length gives

$$N_2 \cong \frac{c\,[\log_\mathrm{e}(1/(1-L)) + \log_\mathrm{e}(1/R_1R_2)]}{2l\sigma_{\mathrm{SE}}(\omega)} \tag{6.16}$$

Comparison with Eq. (6.3) shows that this is equivalent to the threshold condition with $N_2 = N_\mathrm{th}$. Thus above threshold the population inversion is 'clamped' at its threshold value and, as a consequence, the spontaneous emission, which is proportional to N_2, will also be fixed. Pumping the laser harder above threshold channels all the energy of the excited atoms into the stimulated emission of the cavity mode.

Substitution of Eq. (6.13) back into the rate equation (6.9) for the inversion gives

$$\frac{\mathrm{d}N_2}{\mathrm{d}t} = \frac{N_\mathrm{p}^2}{V(N_\mathrm{p}+1)} - \frac{A_{21}VN_\mathrm{p}}{(N_\mathrm{p}+1)\sigma_{\mathrm{SE}}(\omega)c\tau_\mathrm{c}} + R_2 \tag{6.17}$$

Again at steady state (above or below threshold) we put $\mathrm{d}N_2/\mathrm{d}t = 0$ and obtain

$$\frac{c\sigma_{\mathrm{SE}}(\omega)}{VA_{21}}\,N_\mathrm{p}^2 + N_\mathrm{p}\left(\frac{R_2}{R_\mathrm{t}} - 1\right) + \frac{R_2}{R_\mathrm{t}} = 0 \tag{6.18}$$

where R_th, the threshold pumping rate, is given by

$$R_\mathrm{th} = \frac{A_{21}}{c\sigma_{\mathrm{SE}}(\omega)\tau_\mathrm{c}}$$

Equation (6.18) is a quadratic equation in N_p. Solving for N_p and taking the positive root (so that N_p remains positive) gives

$$N_p = \frac{VA_{21}}{2c\sigma_{SE}(\omega)}\left\{(r-1) + \left[(1-r)^2 + \frac{4c\sigma_{SE}(\omega)}{VA_{21}}r\right]^{1/2}\right\} \tag{6.19}$$

where r is the normalized pumping rate defined by

$$r = \frac{R_2}{R_{th}}$$

This equation can now be used to observe what happens to the photon density as the laser is pumped through threshold, that is, as we increase the pumping rate, r.

The rate of change with r of N_p in the vicinity of the threshold can be found by differentiating Eq. (6.19) with respect to r:

$$\frac{dN_p}{dr} = \frac{VA_{21}}{2c\sigma_{SE}(\omega)}\left\{1 + \frac{r-1+(2c\sigma_{SE}(\omega)/VA_{21})}{\sqrt{(r-1)^2+4r(2c\sigma_{SE}(\omega)/VA_{21})}}\right\} \tag{6.20}$$

Exactly at threshold $r = 1$ and Eq. (6.20) then gives

$$\frac{dN_p}{dr} = \frac{VA_{21}}{2c\sigma_{SE}(\omega)}\{1 + (c\sigma_{SE}(\omega)/VA_{21})^{1/2}\} \tag{6.21}$$

$$= \frac{N_m}{2}\left\{1 + \left(\frac{1}{N_m}\right)^{1/2}\right\}$$

where $N_m = VA_{21}/c\sigma_{se}(\omega)$ is the number of modes (see Problem 6.5) in the gain bandwidth of the transition.

Example 6.2 Calculation of the number of modes

A typical Nd : YAG laser has a stimulated emission cross-section of about 10^{-22} m^2, an A coefficient of 3×10^3 s^{-1} and an excited medium (crystal in form of a rod) of volume 10^{-3} m^3. Thus

$$N_m = \frac{VA_{21}}{c\sigma_{SE}(\omega)} \cong 10^{12} \tag{6.22}$$

As this example shows, N_m is a very large number. However, most of these modes have too much loss ever to reach threshold, especially if the cavity is an open resonator and not the box resonator commonly used with microwave devices. Thus the second term within the braces on the right-hand side of Eq. (6.21) will be small and the equation approximates to

$$\frac{dN_p}{dr} \cong \frac{VA_{21}}{2c\sigma_{SE}(\omega)} \tag{6.23}$$

The rate of rise therefore of the number of photons with respect to the pumping rate is very high due to the large number of modes lying within the gain bandwidth of laser cavities of normal laboratory dimensions.

Returning to Eq. (6.19) we have

$$N_p = \frac{VA_{21}}{2c\sigma_{SE}(\omega)} \left\{ (r-1) + \left[(1-r)^2 + \frac{4c\sigma_{SE}(\omega)}{VA_{21}} r \right]^{1/2} \right\} \tag{6.24}$$

which can be rearranged to give

$$N_p = \frac{VA_{21}}{2c\sigma_{SE}(\omega)} \left\{ (r-1) + (1-r)\left[1 + \frac{r}{(1-r)^2} \frac{4c\sigma_{SE}(\omega)}{VA_{21}} \right]^{1/2} \right\} \tag{6.25}$$

We can expand Eq. (6.25) using the binomial theorem provided r is not close to unity:

Below threshold: $r \ll 1$

$$N_p \cong \frac{r}{(1-r)} \tag{6.26}$$

Above threshold: $r \gg 1$

$$N_p \cong \frac{VA_{21}}{c\sigma_{SE}(\omega)} (r-1) \tag{6.27}$$

At threshold: $r \to 1$

$$N_p = \left(\frac{VA_{21}}{c\sigma_{SE}(\omega)} \right)^{1/2} \tag{6.28}$$

We are now in a position to plot the steady-state photon number as a function of pumping rate as given by Eq. (6.25). This is shown in Figure 6.3 for two values of the number of modes in the gain bandwidth. For a helium–neon laser this number is approximately 10^8; somewhat smaller than that calculated above for the Nd : YAG laser due to the lower gain bandwidth. For a semiconductor laser it is very much smaller ($\sim 10^5$), not due to its gain bandwidth being higher but because of the small gain volume of the laser. Figure 6.3 shows a very sharp change in the number of cavity photons indicated by the slope at threshold given by Eq. (6.21). The change can be as much as ten orders of magnitude for a Nd : YAG laser and three orders for the much 'softer' threshold of a semiconductor laser.

Let us now summarize the results of the above discussion:

1. At threshold an extremely rapid rise occurs in the power of the oscillating modes (Eq. (6.23)).

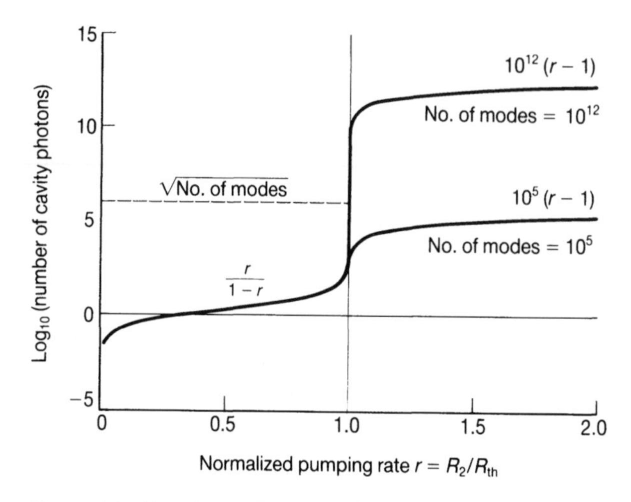

Figure 6.3 Number of cavity photons around threshold as a function of the pumping factor. Upper curve is for a laser mode volume with a high number of modes such as a Nd:YAG laser and the lower curve is for a small semiconductor laser.

2. Above threshold a linear rise in output power with pumping rate is expected (Eq. (6.27)).
3. Below threshold there averages less than one photon in each cavity mode.
4. The population inversion above threshold is clamped at its threshold value and this causes the spontaneous emission also to clamp at the threshold value. Above threshold any further energy pumped into the inversion is channelled into stimulated emission.

We can add three further points which will become apparent in Section 6.6 when the laser output frequency is discussed:

1. The output light changes from isotropic to a narrow beam with good spatial coherence.
2. The light changes from noisy incoherent radiation to coherent amplitude-stabilized oscillation.
3. Since the laser radiation occurs in only one or at least a few modes, a very pronounced spectral narrowing occurs from the broadband spontaneous emission to that of the narrow cavity resonances. Most of the modes have too much loss to be regenerative.

6.4 Oscillation and gain saturation

In the previous section it was shown that only those cavity modes whose frequencies fall within the gain bandwidth have a net positive gain and so will oscillate. It is the geometry of the laser (long and narrow) which restricts these. We now look more closely at other restrictions on the number of modes which oscillate, together with the precise frequency of oscillation and the bandwidth of this oscillation. In Figure 6.4 the passive cavity resonances or modes are shown as a comb of frequencies and it is seen how some of the axial modes fit into the gain bandwidth. Alteration of the cavity length, d, will shift this comb in one direction; for example, an increase in d of a half-wavelength will move the qth resonance to the position of the $q-1$ resonance since an extra half-wavelength is being accommodated in the standing wave. Clearly, the exact position of the modes in the gain bandwidth will depend strongly on the exact value of d. The gain profile is fixed and the comb of frequencies will move as d alters.

The effect of intra-cavity gain on the cavity resonances can be deduced from Eq. (5.37) which gives the transmission of a cavity having internal loss coefficient, α. By changing this coefficient from negative to positive we can observe the effect on the cavity resonances as the gain is increased. For simplicity, we can assume that $l = d$. The result of this calculation for mirror reflectivities $R_1 = R_2 = 0.40$[1] is shown in Figure 6.5. With a negative gain (i.e. intra-cavity absorption) the transmission is less than unity, as for the Fabry–Perot interferometer discussed in Chapter 5. This cannot be shown clearly on the scale of this graph (note the scale of Figure 5.11). As the gain is increased the transmission becomes much larger than unity and, just as important, the resonances become sharper. Eventually, when the gain balances the loss the transmission becomes infinite and the resonance width zero. The cavity is behaving like a classical driven oscillator and when the net loss is zero the oscillation amplitude is limited by the saturation of the gain. The bandwidth of the resonances will be seen to be finite due to the presence of spontaneous emission. This latter point will be explored further in the chapter, but in the first instance the effects

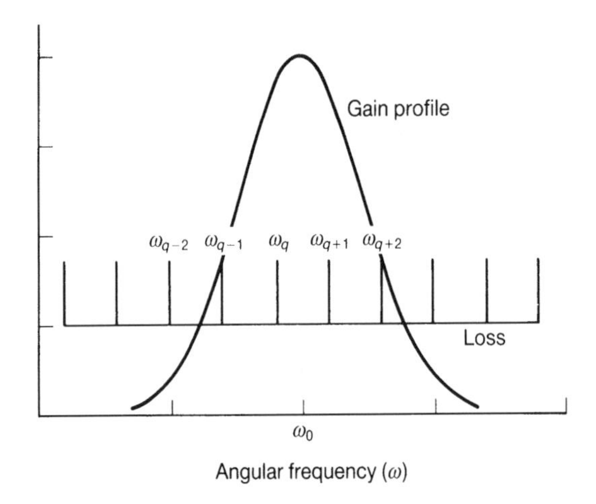

Figure 6.4 Gain curve with comb of passive cavity resonances shown.

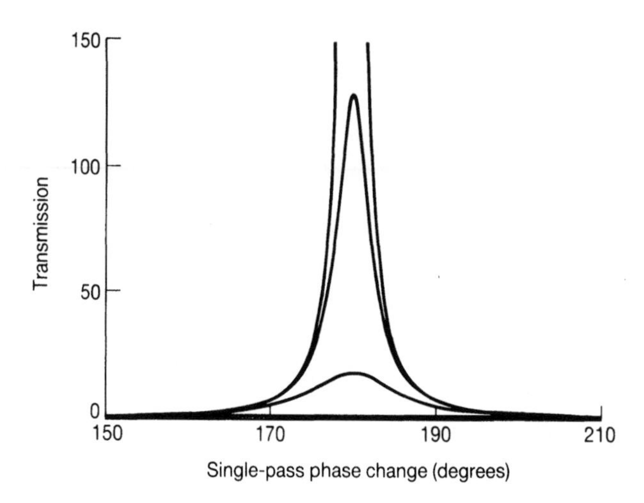

Figure 6.5 Curves showing regenerative power gain in cavity where loss changes to gain. Mirror reflectivities are each 40 per cent and the single-pass gain values are 1.0, 2.0, 2.3, 2.4. The threshold gain is 2.5.

of saturation of the gain for the two types of line broadening, i.e. homogeneous and inhomogeneous, will be discussed.

6.4.1 Homogeneous broadened transition

In Section 3.3.2 it was shown how the gain saturated as the irradiance of the light increases. Those modes above threshold, as shown in Figure 6.4, will experience a net gain and begin to grow from the spontaneous emission. However, as their irradiances become larger the population inversion and so the gain will be reduced at an increasing rate. If the transition is homogeneously broadened the modes with larger irradiance will reduce the inversion more rapidly (remember, the rate of stimulated emission is proportional to the irradiance). In a homogeneously broadened transition the atoms behave identically and if an atom is removed at one frequency it is not available to provide gain at another. The consequence of this is that those modes with larger irradiance (i.e. those closest to the centre of the gain curve) can dominate and eventually the smaller modes stop oscillating and single-mode oscillation results.

Figure 6.6 shows the saturation and the eventual gain curve after saturation when one mode is dominant. We see that the curve still retains its Lorentzian shape but has been pulled down so that, at the lasing mode frequency, the saturated gain equals the cavity loss. Thus the saturated gain becomes clamped at the value of the loss which is just the threshold gain value. It must be emphasized that the mode frequency need not necessarily be at the exact centre of the gain curve. However, the mode closest to the centre will experience the largest gain, and so will dominate.

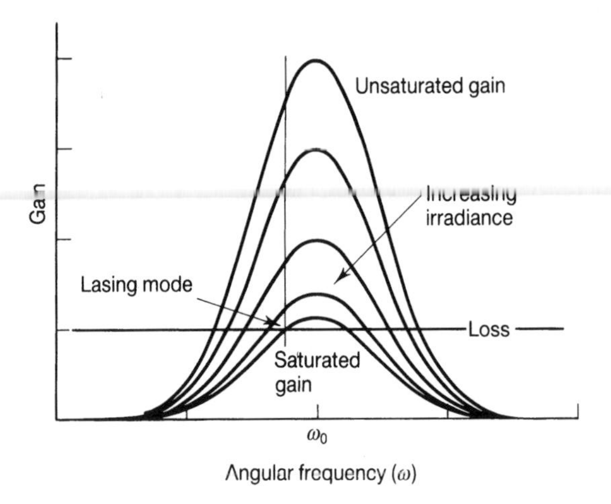

Figure 6.6 Saturation of the gain for a homogeneously broadened transition. The lasing mode frequency is not at line centre but the curves still maintain their Lorentzian form as they saturate.

The story is not complete without the inclusion of the spontaneous emission which still exists even though it is very small compared with the stimulated emission. A few spontaneously emitted photons add to the stimulated emission cavity field so ensuring that in a roundtrip the irradiance will remain unchanged at its steady-state value making the gain very slightly less than the loss. This gives the resonance a finite frequency width (see Section 6.6.2).

6.4.2 Inhomogeneous broadened transition and the Lamb dip

The saturation of the gain for inhomogeneously broadened transitions as occurs, for example, in a gas laser is quite different from the homogeneous case. As before, the oscillations begin from the spontaneous emission and build up most rapidly at the centre of the cavity resonances. We remember from Section 3.3.3 that the gain is only saturated within the vicinity of the mode frequency and that a 'hole' is burnt in the gain curve. This means that, in principle, each mode is taking its gain from a separate group of atoms and can saturate the gain independently of the other modes. Thus the laser is able to oscillate on more than one mode at a time. The build-up of the irradiance and burning of a hole in a Doppler-broadened gain curve is shown in Figure 6.7, and we note how the hole broadens as it deepens, since the width of the hole was shown to be given by (Eq. (3.140))

$$\Delta\omega_{\mathrm{H}} = \Delta\omega_{\mathrm{h}}\,(1 + I_\omega/I_{\mathrm{s}})^{1/2} \tag{6.29}$$

Again the gain saturates at a value slightly less than the loss but this time at the centre of each hole. Many gas lasers such as helium–neon and argon–ion do

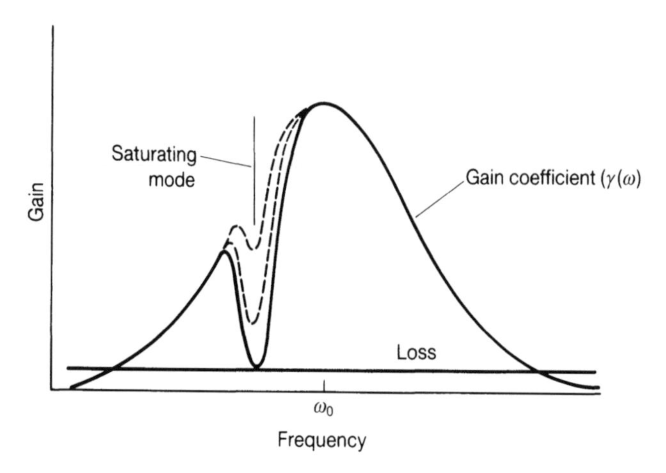

Figure 6.7 Hole burning and saturation for an inhomogeneously broadened transition. Note how the hole widens and deepens as saturation occurs.

oscillate multimode and are therefore considered as inhomogeneously broadened even though, in practice, the holes often overlap and the gain curve can become almost uniformly saturated.

The strongest evidence for hole burning in gas lasers comes from a dip in the optical power when the operating laser frequency is at the centre of the gain curve. This is known as the *Lamb dip* and was first predicted by Lamb in his classic paper in 1964 [Ref. 1]. In a standing wave, or linear laser, a single optical frequency will burn two holes in the gain curve. This is due to the fact that the wave travelling through the medium in one direction does not interact with the same atomic velocity group as the wave of the same frequency going in the opposite direction. A standing wave is considered as two oppositely directed travelling waves having the same frequency. Going in one direction, it will interact with and take its amplification from velocity group $+v_z$. After reflection, which reverses its direction, it must interact with velocity group $-v_z$. Since the gain curve can be considered to be a function of velocity as well as frequency we therefore have two holes burnt at two frequencies, symmetrically placed about the centre. One of these holes is at the oscillation frequency and so we will refer to the other one as the *image hole*. This is shown in Figure 6.8. If the cavity length, d, is changed so that the oscillation frequency moves closer to the line centre then these holes are brought closer to each other. Eventually at line centre they overlap and in this position the two travelling waves are interacting with the same group of atoms, i.e. those with velocity centred at zero. As a result, it has less roundtrip gain since only one group of atoms is providing this and consequently the power output drops.

In practice, the Lamb dip can be easily observed in a helium–neon laser which is short enough to allow only one mode in the gain bandwidth (i.e. where the mode

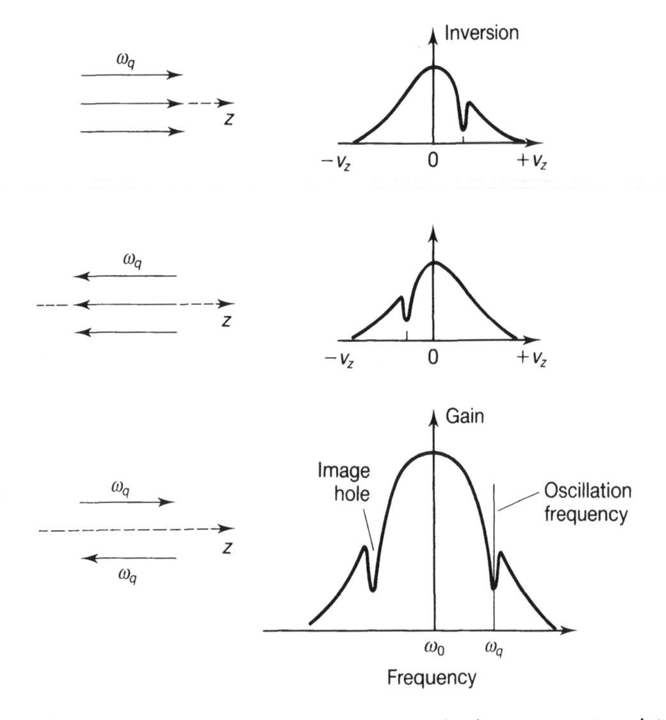

Figure 6.8 Holes produced by a single frequency (mode) in a standing-wave laser.

spacing is almost equal to the gain bandwidth), and allowing this mode to scan through the gain curve by altering d. This occurs quite naturally as the laser heats up and expands after switching on the discharge. In fact, with this arrangement a sequence of modes will move through the gain curve with the mode number, q, increasing by one for every half-wavelength change in length of the cavity. A typical experimental result is shown in Figure 6.9.

Increasing the gas pressure in the laser discharge tube leads to a broadening out of the Lamb dip and eventually it vanishes altogether. This is due to the pressure broadening described in Section 3.2.3 which increases the homogeneous linewidth and so, from Eq. (6.29), the hole width. With reference to Eq. (6.29) it is seen that the same effect is caused by increasing the irradiance.

6.4.3 Spatial hole burning

A single mode in a linear laser is a standing wave and is therefore equivalent to two travelling waves of the same frequency going in opposite directions. The total electric field at any point z in the standing wave pattern will be the sum of the

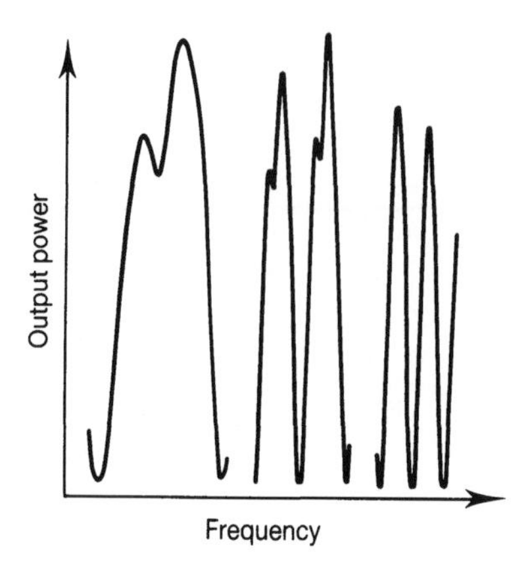

Figure 6.9 Typical result of monitoring the output power (power tuning curve) of a short helium–neon laser as the cavity length changes during warm-up illustrating the Lamb dip. Note the asymmetry in the curve due the gas mix being natural neon which has 90 per cent Ne^{20} and 10 per cent Ne^{22} whose gain curves are separated by 850 MHz. The gas pressures for the three traces are (from left to right) 0.65 torr, 0.9 torr and 5 torr. Note how the Lamb dip has disappeared at high pressure.

individual fields. Using the notation of Chapter 5 and overlooking the change in the field amplitudes due to the gain we have

$$\mathscr{E}_T(z) = \mathscr{E}^+(z)\, e^{j(\omega t - kz)} + \mathscr{E}^-(z)\, e^{j(\omega t + kz)} \tag{6.30}$$

The variation of the irradiance in the z direction is obtained by multiplying $\mathscr{E}_T(z)$ by its complex conjugate, thus:

$$I_T(z) \propto [\mathscr{E}^+(z)\, e^{j(\omega t - kz)} + \mathscr{E}^-(z)\, e^{j(\omega t + kz)}] \times [\mathscr{E}^+(z)^*\, e^{-j(\omega t - kz)} + \mathscr{E}^-(z)^*\, e^{-j(\omega t + kz)}]$$

$$I_T(z) \propto \mathscr{E}^+(z)\mathscr{E}^+(z)^* + \mathscr{E}^-(z)\mathscr{E}^-(z)^* + \mathscr{E}^+(z)^*\mathscr{E}^-(z)\, e^{2kzj} + \mathscr{E}^+(z)\mathscr{E}^-(z)^*\, e^{-2kzj} \tag{6.31}$$

In terms of irradiance we have

$$I_T(z) = I^+(z) + I^-(z) + 2(I^+(z)I^-(z))^{1/2} \cos 2kz \tag{6.32}$$

The saturation of the population inversion as a function of z can now be obtained by substituting for the saturating irradiance from Eq. (6.32) into Eq. (3.124), giving:

$$\frac{N(z)}{N_0} = \frac{1}{1 + \bar{g}(\omega)[I^+(z) + I^-(z) + 2(I^+(z)I^-(z))^{1/2} \cos 2kz]/I_s} \tag{6.33}$$

In Figure 6.10 we have used this expression to plot the variation of the population inversion, and so the gain, as a function of z taking the frequency to be at line

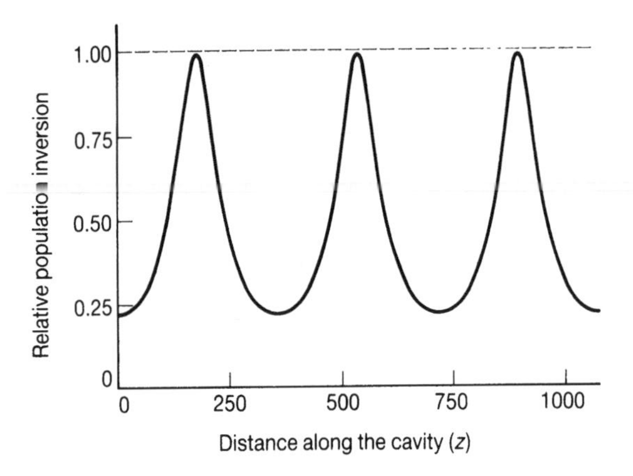

Figure 6.10 Saturation grating burnt in the population inversion by the single-frequency standing wave.

centre, i.e. $\bar{g}(\omega) = 1$, and assuming equal irradiances in both directions. It shows a strong volume grating-like structure formed in the inversion along the length of the cavity, the holes of which correspond to the antinodes in the standing wave field. This is referred to as *spatial hole burning*.

This spatial hole burning has a very fundamental effect on homogeneously broadened lasers. Since the whole medium is not uniformly saturated then maximum power is not being extracted from the population inversion. Just as hole burning in the gain curve discussed in the previous section allowed multimode oscillation, so spatial hole burning permits a homogeneously broadened laser

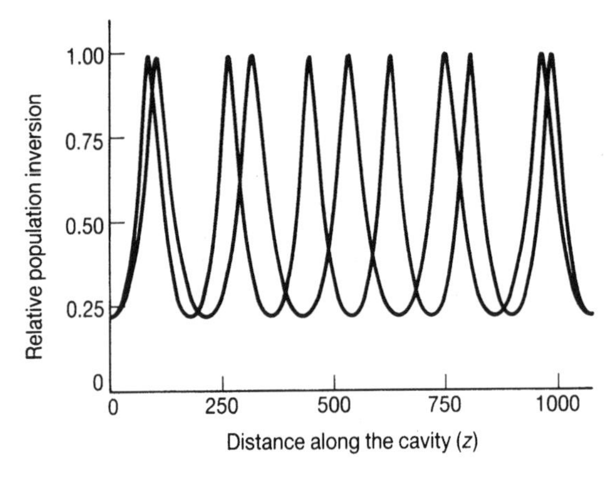

Figure 6.11 Standing wave fields patterns and the population inversion grating showing how adjacent modes can oscillate.

transition to oscillate simultaneously at more than one frequency. This latter point can be explained with reference to Figure 6.11. In this we have drawn the standing wave patterns for a qth mode and its neighbour the $q + 1$ mode. At the exact centre of the cavity the fields of the two modes are seen to be 180° out of phase and hence they do not compete for the gain. More than one mode of oscillation is possible in this cavity if a homogeneous broadened gain medium is placed at the centre of the cavity. However, placing the gain medium near one of the mirrors will force single-frequency operation. An example of this will be discussed in the next chapter in the section on CW dye lasers.

Using the spatial hole burning to force oscillation at a single mode is clearly one way of obtaining a single frequency and so highly coherent light. However, due to spatial hole burning all the gain is not being used and the potential output power is not being realized. Gas lasers do not experience spatial hole burning since the random motion of the atoms or molecules tends to remove any spatial variation of the gain. In liquid dye and solid-state lasers two different approaches can be used to remove the effect of spatial hole burning and hence increase the power in single-mode operation:

1. *Travelling wave (TW) laser*. By the addition of another mirror a linear laser can be configured in a ring cavity (see Figure 5.20) and, as before, the light can travel in both directions. However, if the light is forced to go in one direction only by using the optical equivalent of a diode then every point in the ring will experience the maximum and minimum excursions of the field since the wave will then be a travelling wave. Thus the population inversion will be used at all points in the gain medium and no spatial hole burning will be produced. Many ring geometries other than triangular are used and detailed discussion of these ring lasers will be given in the next chapter in the sections on dye and solid state lasers.
2. *Twisted-mode cavity* [Refs 2, 3]. The cavity shown in Figure 6.12 has a quarter-wave plate placed in front of each mirror with the gain medium (possibly a Nd-doped YAG crystal) in between. The quarter-wave plates and the laser rod have anti-reflection coatings to reduce the Fresnel reflection losses. The principal axes of the two plates are orientated with their fast axes perpendicular and lie at 45° to the plane of polarization determined by the Brewster angle plate B. Light travelling to the right is plane polarized when incident on P_1 and its field vectors can be resolved along the axes of the quarter-wave plate:

$$\mathscr{E}_{x1} \propto \sin(\omega t - kz) \text{ and } \mathscr{E}_{y1} \propto \sin(\omega t - kz) \qquad (6.34)$$

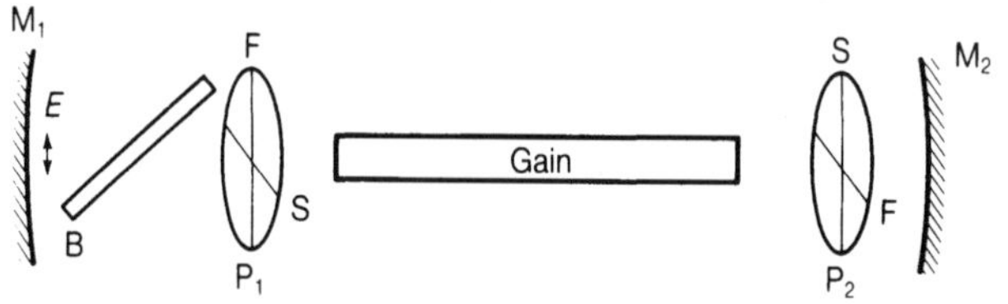

Figure 6.12 Twisted-mode cavity.

After passing through the plate P_1 the components are

$$\mathscr{E}_{x2} \propto \sin(\omega t - kz + \pi/2) \text{ and } \mathscr{E}_{y2} \propto \sin(\omega t - kz) \tag{6.35}$$

The light thus has become circularly polarized. Passage through the second plate P_2 induces a relative phase shift of $\pi/2$ in E_y and the polarization is returned to linear. After a reflection from mirror 2 and returning through P_2 the components become

$$\mathscr{E}_{x3} \propto \sin(\omega t - 2kd + kz + \pi/2) \text{ and } \mathscr{E}_{y3} \propto \sin(\omega t - 2kd + kz + \pi) \tag{6.36}$$

A further passage through P_1 and a reflection from mirror 1 returns the light to a plane polarized[2] state. The total of the running circularly polarized fields in the rod can be found by adding Eqs (6.35) and (6.36) to yield the components

$$\mathscr{E}_x \propto 2\cos(\omega t - kd)\cos k(d-z) \text{ and } \mathscr{E}_y \propto 2\cos(\omega t - kd)\sin k(d-z) \tag{6.37}$$

The resultant $\mathscr{E}$ field of this circularly polarized standing wave has the shape of a twisted ribbon with a spatial period of one wavelength. We could also consider it as two orthogonally plane polarized standing waves but displaced from each other by $\lambda/4$. The total light irradiance at any point, z, is proportional to

$$\mathscr{E}_x^2 + \mathscr{E}_y^2 \propto 4\cos^2(\omega t - kd) \tag{6.38}$$

which is independent of z and no spatial modulation of the population inversion along the gain medium occurs (i.e. no spatial hole burning).

6.4.4 Amplified spontaneous emission (the mirrorless laser)

There are some laser systems in which the gain is so high that a cavity is not necessary. In these cases spontaneous emission produced at either end of the gain medium will be amplified in travelling to the other end and emerge as a collimated beam [Ref. 4]. The spatial coherence and profile of the beam may not be very different from a beam generated in a cavity but its temporal coherence will be very much worse. In a cavity, amplification of the spontaneous emission can act to reduce the gain available for the cavity mode and this is an important factor in very high-gain solid-state lasers.

The mathematical analysis is complicated since we are required to determine the saturation of the gain at all frequencies across the transition bandwidth and not simply at a single frequency as was considered in Chapter 3. We can, however, begin by neglecting gain saturation and considering a homogeneously broadened transition. Figure 6.13 is a schematic diagram of a mirrorless laser and shows at a point z spontaneous emission being added to the growing incoherent light. The spontaneous emission added within the solid angle of the gain medium need only be

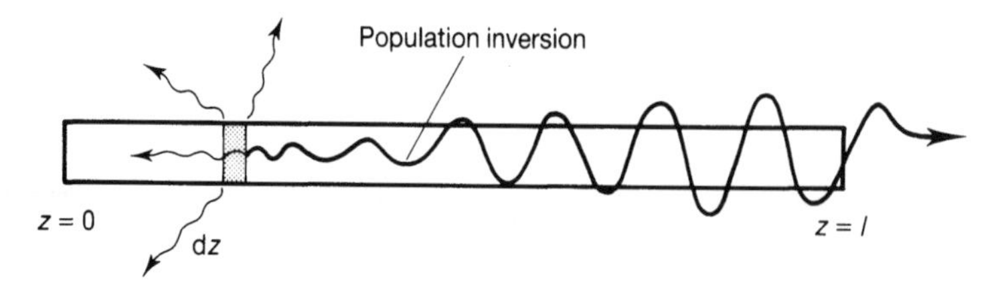

Figure 6.13 Schematic diagram of a mirrorless laser.

considered. The rate of growth of the irradiance $I^+(\nu, z)$ within a bandwidth of $\delta\nu$ at a frequency ν at a distance z from one end is given by

$$\frac{d}{dz}(I^+(\nu, z)\,\delta\nu) = \gamma_0(\nu)I^+(\nu, z)\,\delta\nu + h\nu A_{21}N_2 g(\nu)\,\delta\nu\,\frac{d\Omega}{4\pi} \tag{6.39}$$

The first term on the right-hand side of this equation represents the increase of the irradiance arriving at z due to amplification while the second gives that added incoherently to it by the spontaneous emission produced within the solid angle. Integrating, with the boundary condition that there is no injected radiation (i.e. $I^+(\nu, z = 0) = 0$, gives the irradiance at the end of the gain medium (i.e. at $z = l$):

$$I^+(\nu, l) = \frac{h\nu A_{21}N_2 g(\nu)}{\gamma_0(\nu)}\,[\exp(\gamma_0(\nu)l) - 1]\,\frac{d\Omega}{4\pi} \tag{6.40}$$

Substituting for the small signal gain coefficient gives

$$I^+(\nu, l) = \frac{8\pi h\nu^3}{c^2}\,\frac{N_2}{N_2 - (g_2/g_1)N_1}\,[\exp(\gamma_0(\nu)l) - 1]\,\frac{d\Omega}{4\pi} \tag{6.41}$$

Inspection of Eq. (6.41) shows that this amplified spontaneous emission will be spectrally narrower than if amplification were negligible, i.e. the optically thin case. This is due to the frequency variation contained in the exponential term of Eq. (6.41). This spectral narrowing as a function of increasing gain coefficient for a Lorentzian line shape is shown in Figure 6.14. The spectral narrowing is similar when the transition is inhomogeneously broadened. It is only when saturation of the gain is important that the two broadening types differ, since hole burning is important with inhomogeneous broadening.

Saturation of the gain will begin to occur when the irradiance at either end of the gain medium becomes of the order of the saturation irradiance. The exact analysis involves consideration of the gain saturation at each frequency and then integration over the whole gain bandwidth. The calculations are complex and for further information the reader is referred to Ref. 4. The general result of these calculations shows that after saturation occurs the spectral line narrowing slows and becomes constant, and in the case of an inhomogeneous broadened transition the spectral width will begin to increase again towards the spontaneous emission width.

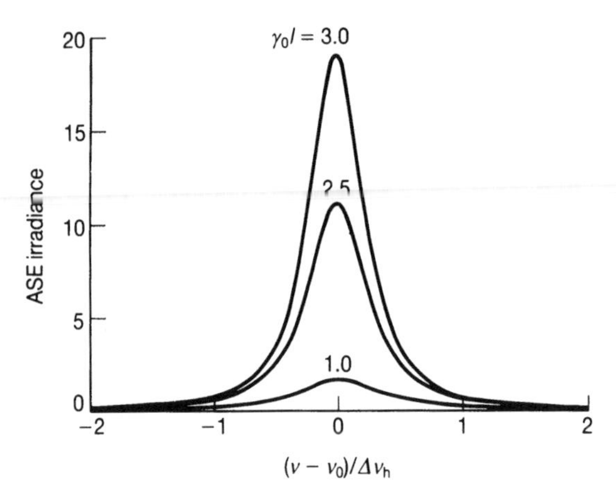

Figure 6.14 Spectral narrowing for ASE in an unsaturated homogeneously broadened transition for different values of unsaturated gain.

6.5 Output power of a laser

We now return to our conventional laser in which the two-way running waves extract energy from the population inversion and grow in irradiance. We may now calculate the power which can be coupled to the outside world through the output mirror, and in so doing we may investigate the efficiency of this process and determine the optimum value of this output coupling. From Eq. (6.33) the saturated gain coefficient can be written

$$\gamma(\omega, z) = \frac{\gamma_0(\omega)}{1 + [I_\omega^+(z) + I_\omega^-(z) + 2(I_\omega^+(z)I_\omega^-(z))^{1/2} \cos 2kz]\, \bar{g}(\omega)/I_s} \tag{6.42}$$

The irradiance will grow in both directions and hence decrease the available gain. This continues until an equilibrium is reached when the roundtrip increase in the irradiance is equal to the cavity loss. The differential equations for the positive- and negative-going single-frequency waves are

$$\frac{\mathrm{d}I_\omega^+(z)}{\mathrm{d}z} = [\gamma(\omega, z) - \alpha_T]\, I_\omega^+(z) \tag{6.43}$$

$$\frac{\mathrm{d}I_\omega^-(z)}{\mathrm{d}z} = -\,[\gamma(\omega, z) - \alpha_T]\, I_\omega^-(z) \tag{6.44}$$

where α_T is the intra-cavity loss coefficient averaged over the length of the gain medium and given by

$$\alpha_T = \frac{\log_e(1/R_1R_2) + \log_e[1/(1-L)]}{2l} \tag{6.45}$$

and we note that from Section 6.2 that $\alpha_T = \gamma_{th}$.

In Eq. (6.44) the negative sign accounts for the wave moving in the $-z$ direction. The exact solution of these differential equations, which gives the intra-cavity irradiance in terms of the gain and loss, is mathematically complex, but with appropriate approximations it is possible to obtain solutions for conditions which will suit certain lasers. These special cases will now be considered.

6.5.1 Uniform field low-gain (lightly coupled) laser

In looking for a solution to Eqs (6.43) and (6.44) we approximate by considering the laser to have a low gain per pass and so a low output coupling (perhaps less than 20 per cent). A schematic diagram of such a laser is shown in Figure 6.15. In addition, we assume that the spatial hole burning is absent and so the gain is constant along the length z. We are therefore considering a situation of *uniform gain* and *irradiance* in the active medium. Neglecting any spatial hole burning implies that we take the average of the cos kz term in Eq. (6.33) to be zero. This is the uniform field approximation. In addition, we assume that the irradiance does not change much over the length of the gain section, which implies that the spatial derivatives in Eqs (6.43) and (6.44) can be put equal to zero. Hence

$$\gamma(\omega) = \alpha_T \tag{6.46}$$

From Eq. (6.42)

$$\gamma(\omega) = \frac{\gamma_0(\omega)}{1 + [I_\omega^+ + I_\omega^-]\bar{g}(\omega)/I_s} \tag{6.47}$$

Thus

$$\alpha_T = \frac{\gamma_0(\omega)}{1 + [I_\omega^+ + I_\omega^-]\bar{g}(\omega)/I_s} \tag{6.48}$$

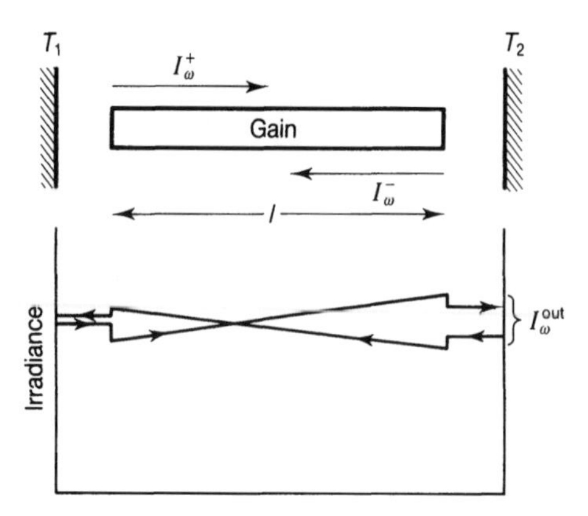

Figure 6.15 Schematic diagram of travelling wave irradiances and output in a weakly coupled laser.

In terms of the irradiances

$$I_\omega^+ + I_\omega^- = \frac{I_s}{\bar{g}(\omega)}\left(\frac{\gamma_0(\omega)2l}{\alpha_T 2l} - 1\right) \quad (6.49)$$

If output from the laser is taken from mirror 2, which has a power transmission coefficient T_2, then the output irradiance is given by

$$I_\omega^{out} = T_2 I_\omega^+ \quad (6.50)$$

where I_ω^+ is the irradiance at the lasing frequency, ω, which is propagating towards the output coupling mirror.

Since we are assuming little difference in the two irradiances we can write

$$I_\omega^- \cong I_\omega^+ \quad (6.51)$$

Thus from Eq. (6.49)

$$I_\omega^{out} = \frac{I_s}{\bar{g}(\omega)}\left(\frac{\gamma_0(\omega)2l}{\alpha_T 2l} - 1\right)\frac{T_2}{2} \quad (6.52)$$

We note that the output power is zero when either T_2 is zero or when the roundtrip gain equals the losses. This latter condition is, of course, the threshold condition.

From Eq. (6.45)

$$\alpha_T 2l = \log_e(1/R_1R_2) + \log_e(1/(1-L)) \quad (6.53)$$

$$\alpha_T 2l = \log_e(1/(1-T_1)) + \log_e(1/(1-T_2)) + \log_e(1/(1-L)) \quad (6.54)$$

We have assumed weak output coupling and low internal loss so we can use the approximation that $\log_e(1/(1-x)) \cong \log_e(1+x) \cong x$ for small x. Thus

$$\alpha_T 2l \cong T_1 + T_2 + L \quad (6.55)$$

and we can write Eq. (6.52) as

$$I_\omega^{out} = \frac{I_s}{\bar{g}(\omega)}\left[\frac{\gamma_0(\omega)2l}{T_2 + T_1 + L} - 1\right]\frac{T_2}{2} \quad (6.56)$$

As stated above we have two values of output coupling, T_2, at which the output irradiance is zero and there must then be a value for T_2 which will give maximum output irradiance. This is readly determined by differentiation of Eq. (6.56) with respect to T_2, equating to zero and solving for T_2. The result is (see Problem 6.6) the optimum output coupling.

$$T_{2,opt} = \sqrt{(L + T_1)\gamma_0(\omega)2l} - (L + T_1) \quad (6.57)$$

Substitution of this optimum output coupling into Eq. (6.56) gives the maximum output irradiance as

$$I_{\omega \max}^{\text{out}} = \frac{I_s}{2\bar{g}(\omega)} \left\{ \frac{\gamma_0(\omega)2l}{(T_1 + L) + \sqrt{(L + T_1)\gamma_0(\omega)2l} - (L + T_1)} - 1 \right\} \times \{\sqrt{(L + T_1)\gamma_0(\omega)2l} - (L + T_1)\}$$

which readily simplifies to

$$I_{\omega \max}^{\text{out}} = \frac{I_s}{\bar{g}(\omega)} [\sqrt{\gamma_0(\omega)l} - \sqrt{(L + T_1)/2}]^2 \tag{6.58}$$

Equation (6.57) implies that as $L + T_1$ decreases towards zero then the optimum output coupling will become zero. This seeming paradox is resolved by noting that as the losses decrease towards zero then, as Eq. (6.49) indicates, the intra-cavity power will approach infinity. In any real laser this does not occur since there is always some form of loss no matter how small. The output irradiance as a function of output coupling as given by Eq. (6.56) is plotted in Figure 6.16 for different values of losses $L + T_1$. The ordinate used here is $(I_\omega \bar{g}/I_s)$, referred to as the normalized output power. The resulting curves are a set of hyperbolas with a common oblique asymptote, which is the zero loss line. With loss in the cavity all the available power cannot appear at the output since some is being wasted by the internal optical scattering and absorption. When $L + T_1 = 0$ and output coupling $T_2 \Rightarrow 0$ we define the available output irradiance to be, using Eq. (6.58),

$$I_\omega^{\text{avail}} = \frac{I_s}{\bar{g}(\omega)} \gamma_0(\omega)l \tag{6.59}$$

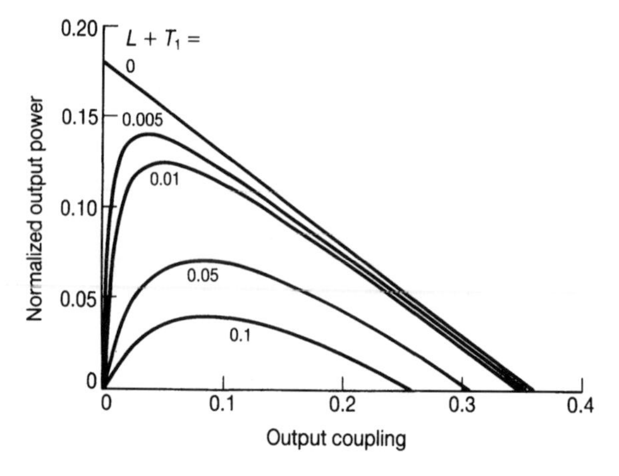

Figure 6.16 Weakly coupled laser output irradiance as a function of output coupling. The ordinate, the normalized output irradiance, is the output irradiance in units of the saturation irradiance. The small signal gain $\gamma_0 l$ as 0.18.

We define an *optimum power extraction efficiency* η_{opt} such that

$$\eta_{opt} = \frac{I_{\omega_{max}}^{out}}{I_{\omega}^{avail}} = \left[1 - \sqrt{\frac{L + T_1}{2\gamma_0(\omega)l}}\right]^2 \tag{6.60}$$

We see from this equation that η_{opt} is a function only of the loss-to-gain ratio and in Figure 6.17 we plot this optimum power efficiency as a function of gain-to-loss ratio. It shows the very large effect of small changes of loss relative to the gain on the efficiency. The loss has a greater effect on the output irradiance since it is constant with irradiance while the gain decreases as it saturates. We saw from Eq. (3.129) that maximum irradiance is extracted from the population inversion when the gain is driven down to zero by saturation. The laser oscillator provides a better way of doing this than the single-pass amplifier since all that is required is to reduce the internal losses to zero. It must be noted that we have assumed that the irradiance is constant radially and that the gain saturation is then uniform. Exact calculation of the output power from the irradiance requires a knowledge of the radial distribution of the irradiance.

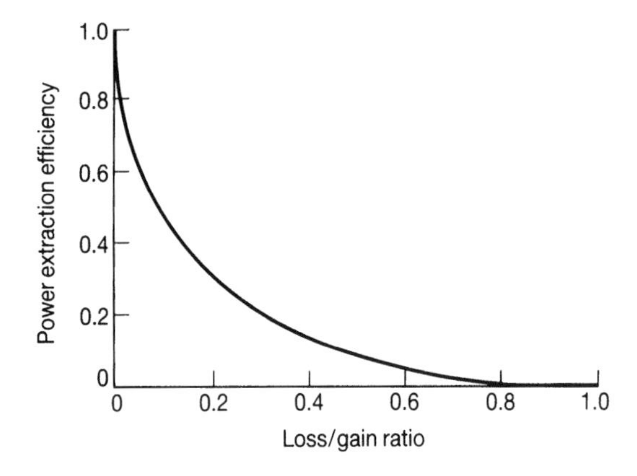

Figure 6.17 Plot of optimum power extraction efficiency against loss-to-gain ratio.

6.5.2 Lightly coupled laser with spatial hole burning

As before, we assume

$$I_{\omega}^{+} \cong I_{\omega}^{-} = I_{\omega}/2 \tag{6.61}$$

The total irradiance then becomes

$$I_{\omega} + I_{\omega} \cos 2kz \tag{6.62}$$

and the saturated gain coefficient becomes

$$\gamma(\omega, z) = \frac{\gamma_0(\omega)}{1 + (\bar{g}(\omega)/I_s)I_{\omega}[1 + \cos 2kz]} \tag{6.63}$$

If we use the boundary condition that at one of the mirrors ($z = 0$) the total irradiance is zero, Eq. (6.63) simplifies to

$$\gamma(\omega, z) = \frac{\gamma_0(\omega)}{1 + (\bar{g}(\omega)/I_s)I_\omega 2 \sin^2 kz} \tag{6.64}$$

The change in total irradiance as a function of z in the cavity becomes

$$\frac{dI_\omega}{dz} = \gamma(\omega, z)I_\omega = \frac{\gamma_0(\omega)I_\omega 2 \sin^2 kz}{1 + (\bar{g}(\omega)/I_s)I_\omega 2 \sin^2 kz} \tag{6.65}$$

The total gain in power of the field from the population inversion is

$$\int_0^l dI_\omega = \int_0^l \gamma(\omega, z)I_\omega, dz = \int_0^l \frac{\gamma_0(\omega)\ I_\omega 2 \sin^2 kz\, dz}{1 + (\bar{g}(\omega)/I_s)I_\omega\ 2 \sin^2 kz} \tag{6.66}$$

Since we are evaluating the change in the cavity irradiance and we are considering low gain then a valid approximation is to consider the irradiance, I_ω, in Eq. (6.66) to be constant as a function of z. The value of this integral for $kl \gg 1$ is obtained from tables [Ref. 5] and we find that

$$\int_0^l dI_\omega \cong \frac{lI_s\gamma_0(\omega)}{\bar{g}(\omega)2I_\omega} \left\{1 - \frac{1}{\sqrt{(1 + 2I_\omega\bar{g}(\omega)/I_s}}\right\} \tag{6.67}$$

The total power lost from the laser is given as before in terms of mirror transmittances and internal loss as

$$I_\omega^+ T_2 + I_\omega^- T_1 + (I_\omega^+ + I_\omega^-)L/2 = \frac{I_\omega}{2}(T_1 + T_2 + L) \tag{6.68}$$

We can now equate the loss (Eq. (6.68)) with the energy gain (Eq. (6.67)) to give

$$\frac{I_\omega}{2}(T_1 + T_2 + L) = \frac{\gamma_0 lI_s}{\bar{g}(\omega)} \left\{1 - \frac{1}{\sqrt{(1 + 2I_\omega\bar{g}(\omega)/I_s)}}\right\} \tag{6.69}$$

This equation may now be solved for the irradiance, I_ω. To simplify the notation we put

$$X = \frac{2\gamma_0 l}{T_1 + T_2 + L} \tag{6.70}$$

which we recognize as the gain-to-loss ratio. With some rearrangement we can write Eq. (6.69) as

$$\frac{1}{\sqrt{(1 + 2I_\omega\bar{g}(\omega)/I_s}} = 1 - \frac{I_\omega}{I_s X} \tag{6.71}$$

whence we then obtain a quadratic equation for $(I_\omega\bar{g}(\omega)/I_s$ thus:

$$\left(\frac{I_\omega\bar{g}(\omega)}{I_s}\right)^2\left(\frac{2}{X^2}\right) + \left(\frac{I_\omega\bar{g}(\omega)}{I_s}\right)\left(\frac{1 - 4X}{X^2}\right) + \frac{2(X - 1)}{X} = 0 \tag{6.72}$$

so that the normalized irradiance is given by

$$\frac{I_\omega \bar{g}(\omega)}{I_s} = X - \frac{1}{4} \pm \sqrt{\frac{X}{2} + \frac{1}{16}} \tag{6.73}$$

We take the negative sign since this ensures that the threshold condition is given by $X = 1$. The output irradiance is still given by Eq. (6.52) and this, along with Eq. (6.53) results in

$$I_\omega^{\text{out}} = \frac{T_2 I_s}{2\bar{g}(\omega)} \left\{ \frac{2\gamma_0 l}{T_1 + T_2 + L} - \frac{1}{4} - \sqrt{\frac{\gamma_0 l}{T_1 + T_2 + L} + \frac{1}{16}} \right. \tag{6.74}$$

Figure 6.18 compares the output power with (Eq. (6.74)) and without (Eq. (6.58)) spatial hole burning and shows the significant reduction in the irradiance of about 30 per cent when spatial hole burning is included.

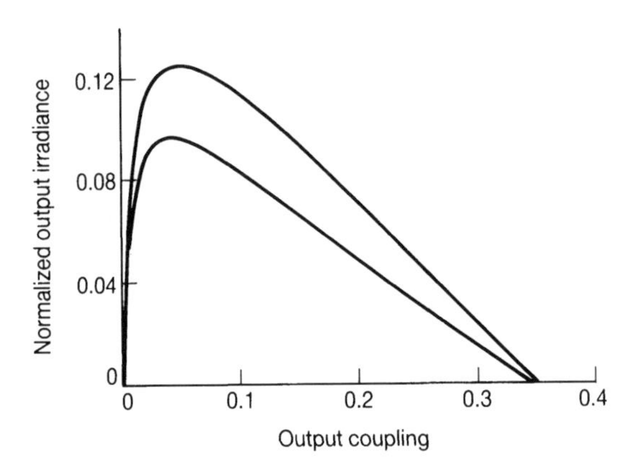

Figure 6.18 Plot of output irradiance as a function of output coupling for a lightly coupled laser showing the effect of spatial hole burning. The unsaturated gain of the medium is 0.18 (18 per cent) and the loss is 0.01 (1 per cent).

6.5.3 Large-output coupled laser

We now consider a more exact analysis known as the *Rigrod analysis* [Ref. 6] which allows for large variation in the two running fields. Thus there can be a large gain and therefore large output coupling. The following simplifications are made in order to make the mathematics tractable.

1. All the distributed intra-cavity losses are assumed small compared with the mirror losses and for mathematical convenience are lumped into a single loss at the high reflecting mirror.
2. As in the previous cases, we assume no radial variation of the irradiance and negligible beam divergence in the cavity.

3. Spatial hole burning is neglected.
4. The gain is assumed to be homogeneously broadened.

We still use Eq. (6.47) for the saturation of the gain by the two running fields at any plane z in the cavity, i.e.

$$\gamma(\omega) = \frac{\gamma_0(\omega)}{1 + [I_\omega^+ + I_\omega^-]\bar{g}(\omega)/I_s} \tag{6.75}$$

The two differential equations (6.43) and (6.44) are modified to exclude the distributed cavity loss. As in Section 6.5.2, the total gain of each running field over the length of the gain medium will be evaluated and the irradiances related by their boundary conditions as shown in Figure 6.19:

$$\frac{dI_\omega^+}{dz} = \gamma(\omega, z)I_\omega^+(z) \tag{6.76}$$

$$\frac{dI_\omega^-}{dz} = -\gamma(\omega, z)I_\omega^-(z) \tag{6.77}$$

We have included the negative sign in Eq. (6.77) since the wave is travelling in the negative z direction. These two coupled differential equations can be linked by writing the derivative of the product of the irradiances:

$$\frac{d}{dz}[I_\omega^+(z)I_\omega^-(z)] = I_\omega^+(z)\frac{dI_\omega^-(z)}{dz} + I_\omega^-(z)\frac{dI_\omega^+(z)}{dz}$$

$$= -\gamma(\omega, z)I_\omega^-(z)I_\omega^+(z) + \gamma(\omega, z)I_\omega^-(z)I_\omega^+(z) = 0 \tag{6.78}$$

Hence we have

$$I_\omega^-(z)I_\omega^+(z) = \text{a constant} \tag{6.79}$$

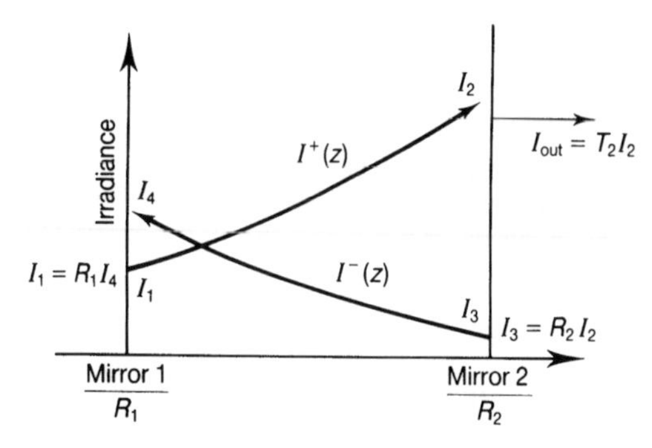

Figure 6.19 Boundary conditions for the left and right travelling irradiance in a strongly coupled laser.

To simplify the writing of the equations we will use the irradiance in units of the saturation parameter and write

$$I^+ = I_\omega^+(z)\bar{g}(\omega)/I_s \text{ and } I^- = I_\omega^-(z)\bar{g}(\omega)/I_s \tag{6.80}$$

We then write Eq. (6.79) as

$$I^+ I^- = C \tag{6.81}$$

The two differential equations (6.76) and (6.77) can now be decoupled and become

$$\frac{\mathrm{d}I^+}{\mathrm{d}z} = \frac{\gamma_0(\omega)I^+}{1 + I^+ + C/I^+} \tag{6.82}$$

$$\frac{\mathrm{d}I^-}{\mathrm{d}z} = -\frac{\gamma_0(\omega)I^-}{1 + I^- + C/I^-} \tag{6.83}$$

Figure 6.20 shows the boundary conditions for the two counter-propagating waves which can be used in the integration of Eqs (6.82) and (6.83). Integration of these equations gives

$$\int_{I_1}^{I_2} \left(1 + \frac{1}{I^+} + \frac{C}{(I^+)^2}\right) \mathrm{d}I^+ = \int_0^l \gamma_0(\omega)\,\mathrm{d}z \tag{6.84}$$

$$\int_{I_4}^{I_3} \left(1 + \frac{1}{I^-} + \frac{C}{(I^-)^2}\right) \mathrm{d}I^- = -\int_0^l \gamma_0(\omega)\,\mathrm{d}z \tag{6.85}$$

The integrals are readily evaluated to yield

$$\gamma_0 l = \ln(I_2/I_1) + I_2 - I_1 - C\left(\frac{1}{I_2} - \frac{1}{I_1}\right) \tag{6.86}$$

$$\gamma_0 l = \ln(I_4/I_3) + I_4 - I_3 - C\left(\frac{1}{I_4} - \frac{1}{I_3}\right) \tag{6.87}$$

In addition to these two relationships between the irradiances we also have from the boundary conditions at the mirrors:

$$\begin{aligned} &\text{At mirror 1:} \quad && I_1 = R_1 I_4 \\ &\text{At mirror 2:} \quad && I_3 = R_2 I_2 \end{aligned} \tag{6.88}$$

$$\text{and from Eq. (6.81)} \quad I_1 I_4 = I_2 I_3 = C \tag{6.89}$$

Adding Eqs (6.86) and (6.87) and using (6.89) to eliminate C, we obtain

$$2\gamma_0(\omega)l = \log_e\left(\frac{1}{R_1 R_2}\right) - 2(I_1 - I_2 + I_3 - I_4) \tag{6.90}$$

Using Eq. (6.88) together with the relation between the irradiances (Eq. 6.89) we can obtain an expression for I_2:

$$I_2 = \frac{R^{1/2}}{[R_1^{1/2} + R_2^{1/2}][1 - (R_1 R_2)^{1/2}]} [\gamma_0 l + \log_e (R_1 R_2)^{1/2}] \quad (6.91)$$

We can now evaluate the output irradiance as before by multiplying Eq. (6.91) by the output coupling, T_2, and using Eq. (6.80) we obtain for the output irradiance at the lasing frequency

$$I_\omega^{\text{out}} = \frac{I_s T_2 R_1^{1/2}}{\bar{g}(\omega)[R_1^{1/2} + R_2^{1/2}][1 - (R_1 R_2)^{1/2}]} [\gamma_0 l + \log_e (R_1 R_2)^{1/2}] \quad (6.92)$$

In Figure 6.20 the output irradiance as a function of output coupling as given by Eq. (6.92) is plotted for varying values of the gain. The curves show a very broad maximum, and so a very much larger tolerance on choosing the exact optimum output coupling than in the weakly coupled case.

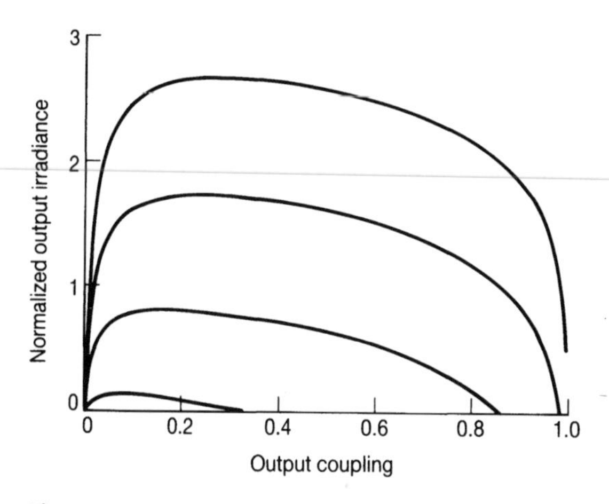

Figure 6.20 Output power as a function of output coupling for a high-gain laser. The unsaturated gain coefficients $\gamma_0 l$ are 3, 2, 1. The high reflectivity mirror is taken as 0.98.

6.5.4 Inhomogeneously broadened lasers

In many lasers the saturation of the gain does not take place uniformly over the linewidth but in a region centred on the laser frequency, and in these cases the gain saturation cannot be properly described as either homogeneous or inhomogeneous. We consider first the special case of a single-mode laser with inhomogeneous gain saturation. In this case we can use Eq. (3.139) for the saturated gain coefficient and modify the steady-state relationship for a lightly coupled laser in the uniform field approximation. Equating the saturated gain to the loss coefficients as given by Eq. (6.46) we obtain

$$\alpha_T = \frac{\gamma_0(\omega)}{[1 + [I_\omega^+ + I_\omega^-]/I_s]^{1/2}} \tag{6.93}$$

Rearrangement of this gives, as before, an equation for the gain-to-loss ratio, X, in terms of the one-way cavity irradiance:

$$X = \frac{\gamma_0(\omega)2l}{\alpha_T 2l} = \left[1 + \frac{2I_\omega}{I_s}\right]^{1/2} \tag{6.94}$$

Further rearrangement gives

$$I_\omega = \frac{I_s}{2}\,[X^2 - 1] \tag{6.95}$$

As in the homogeneously broadened case we obtain the output irradiance by multiplying Eq. (6.95) by the transmittance of the output coupling mirror. To compare with the homogeneously broadened case we rewrite Eq. (6.52) as

$$I_\omega = \frac{I_s}{2\bar{g}(\omega)}\,[X - 1] \tag{6.96}$$

These equations show the cavity radiation increasing linearly as the homogeneous broadened laser is pumped above threshold while for an inhomogeneous broadened laser it increases quadratically. This is a consequence of the fact that in the latter case the hole burnt in the gain curve broadens as it deepens, and hence can extract gain from an ever-increasing number of atoms.

Equation (6.95) applies to most types of inhomogeneous broadening. In the special case of Doppler broadening with the laser frequency tuned away from line centre the left and right travelling waves interact with different velocity groups of atoms, and in this case the factor 2 should be removed from Eq. (6.95). When the laser frequency is tuned to line centre the waves are then interacting with the same velocity group and the 2 remains, so predicting the Lamb dip discussed in Section 6.4.2.

6.5.5 Mixed broadening

In practice, gain saturation in real lasers lies somewhere between homogeneous and inhomogeneous broadening and the exact calculation must take this into account. The best way to illustrate this is to consider the most common case, that of Doppler broadening in a gas laser, since here analytic expressions are available for the lineshape function. The saturated gain in this case can be obtained by evaluating the integral in Eq. (3.138) and, as before, equating the cavity gain to the loss. This is an extremely tedious procedure but analytical expressions are possible. We define a broadening parameter as the ratio of homogeneous linewidth to Doppler linewidth:

$$\eta = \Delta\nu_h/\Delta\nu_D \tag{6.97}$$

The two broadening extremes are

$\eta = 0$: purely inhomogeneous and $\eta = \infty$: purely homogeneous

As Eqs (6.95) and (6.96) show, the two extremes of broadening are linear for homogeneous and quadratic for inhomogeneous. The mixed broadening will fall in between these two. For details of exact calculations the reader should consult see Refs 7 and 8.

6.5.6 Methods of measuring the gain and optimum output coupling

Measurements of the gain can be obtained by inserting known loss into the cavity until lasing ceases. The total intra-cavity loss will then equal the roundtrip gain. This could be done by having a selection of output coupling mirrors or optical flats of known transmission which can be put into the cavity. In many cases it is difficult to determine the exact threshold, and a better method is to be able to vary the loss continuously rather than by discrete values. A suitable technique for this is to pass a knife-edge across the intra cavity beam and monitor the decrease in output power. The transmission past the knife edge can be obtained by integration using Eq. (4.94) in cartesian coordinates.[3]

A widely used technique which continuously varies the effective reflectivity of the output coupling mirror is shown in Figure 6.21. An intra-cavity optical flat at the Brewster angle couples no power out of the cavity. However, when the flat is tilted power is coupled out from its two surfaces due to Fresnel reflection. Two methods have been used corresponding to the two axes about which the plate can be tilted. In one technique the angle ϕ is set to the Brewster angle and the plate is tilted about the axis AA′. In this case the tilted angle is referred to as the *skew angle*. This method suffers the disadvantage that the loss as a function of skew angle is difficult to calculate and depends on the number of Brewster angle surfaces in the cavity. In addition, the plane of polarization of the light changes as a function of the skew angle. A full Jones matrix analysis of this method is given in a paper by Taché [Ref. 9]. The second and more popular method involves setting the skew angle to zero and tilting the plate away from the Brewster angle, so varying the angle ϕ. By adding the multiple Fresnel reflections within the optical flat but neglecting interference effects its transmission can be shown to be given by

$$T = \frac{1 - R}{1 + R} \tag{6.98}$$

where R is the reflectance of one dielectric air interface given by Eq. (2.65).

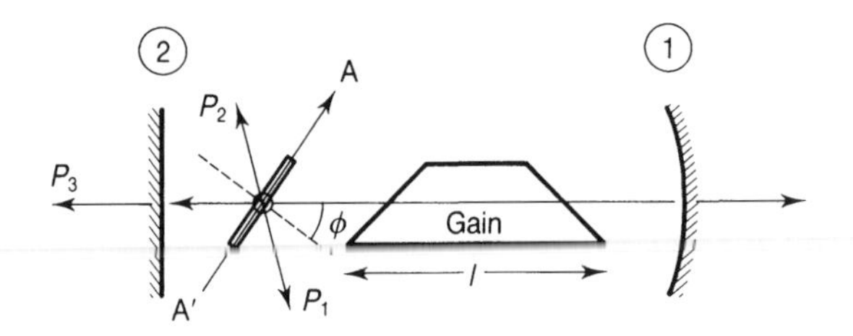

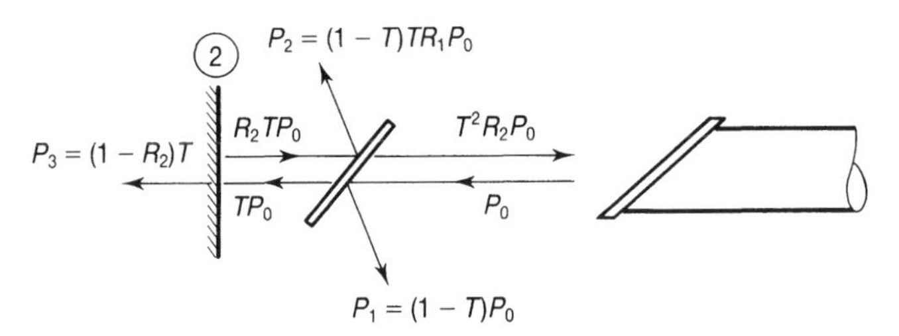

Figure 6.21 Intra-cavity rotatable reflector for the measurement of laser parameters.

With reference to Figure 6.21 it can be seen that the combination of tilted optical flat and the plane mirror can be regarded as a single mirror with an effective reflectance given by

$$R_e = T^2 R_2 \tag{6.99}$$

where R_2 is the reflectance of mirror 2. The total output coupling is then

$$T_{out} = 1 - T^2 R_2 \tag{6.100}$$

The total output coupled power is

$$P_T = P_1 + P_2 + P_3(1 - T^2 R_2)P_0 \tag{6.101}$$

where P_0 is the power incident on the plate mirror combination. This total output power can be obtained from any of the three outputs shown in Figure 6.21. The easiest one to monitor is P_3 since its direction is independent of the angle of tilt of the plate.

Figure 6.22 shows a result of applying this technique to a helium–neon laser operating on the 633 nm transition. The tube was 3 mm diameter and 600 mm long and had a number of bar magnets distributed along its length to reduce the amplified spontaneous emission from the 3.39 μm transition (see Section 7.2.3). The hemispherical cavity was adjusted to ensure TEM_{00} mode operation throughout the experiment and the reflectivity of the plane mirror was obtained from an accurate measurement of its transmission at 633 nm. The data in Figure 6.22 is seen to be

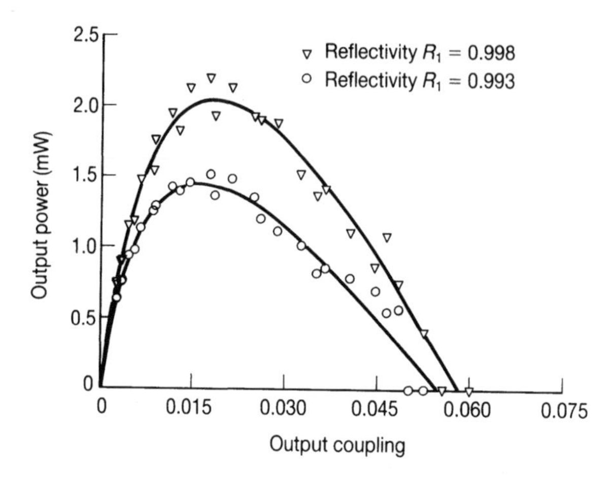

Figure 6.22 Result of intra-cavity reflector experiment with a helium–neon laser operating on the 633 nm transition. The data were curve fitted to Eq. (6.56).

an excellent fit to Eq. (6.56), showing that we can consider the multimode gas laser to be described by homogeneous broadening since there is significant overlap between the holes burnt in the gain curve. From a knowledge of the average spot size in the gain medium we can obtain an average cross-section area for the beam. This allows the saturation irradiance to be calculated from the curve fit.

6.6 Laser output frequency

6.6.1 Exact oscillation frequency

We have tacitly assumed until now that the oscillating frequencies of the active lasing cavity are identical to those of the passive cavity. However, this is not the case since allowance must be made for dispersion – the variation of the refractive index of the gain medium with frequency. Even in gases where the refractive index is not very different from unity these effects are small but still measurable.

In the electron oscillator model of Chapter 2 we saw how the refractive of a dilute gas differed from unity when the frequency of the light was in the region of an atomic absorption. Similar effects also occur for solids and liquids when refractive indices are much larger. For this discussion we are interested in how the refractive index varies as the frequency passes through the 'resonant' or absorption frequency of the transition. Thus we will neglect any 'background' refractive index and absorption which will be present and consider only the change of refractive index from unity. Equations (2.45) and (2.47), respectively, give the real and imaginary parts of the refractive index of the active medium and are rewritten as follows:

$$n_{\mathrm{r}}^2 = 1 + \frac{Ne^2}{2\varepsilon_0 \omega m}\left(\frac{(\omega_0 - \omega)}{(\omega_0 - \omega)^2 + (\Delta\omega_{\mathrm{L}}/2)^2}\right)$$

$$n_{\mathrm{i}} = \frac{Ne^2}{4\varepsilon_0 m\omega}\left(\frac{\Delta\omega/2}{(\omega_0 - \omega)^2 + (\Delta\omega_{\mathrm{L}}/2)^2}\right)$$

We remember from Eq. (2.39) that we associate the imaginary part of the refractive index with absorption. Using the same approximation that leads to Eq. (2.51) we can write the real part of the refractive index as

$$n_{\mathrm{r}} = 1 + \frac{\omega_0 - \omega}{\Delta\omega_{\mathrm{L}}}\frac{\lambda}{2\pi}\alpha \tag{6.102}$$

The equation is identical when we have gain instead of loss, since gain is the negative of absorption and the relationship between refractive index, and the gain becomes:

$$n_{\mathrm{r}} = 1 + \frac{\omega_0 - \omega}{\Delta\omega_{\mathrm{L}}}\frac{\lambda}{2\pi}(-\gamma(\omega)) \tag{6.103}$$

Using Eq. (5.60), which gives the family of passive cavity resonant frequencies for the mn transverse mode, together with Eq. (6.103) we can obtain an expression for the oscillating frequency ν_{osc} in hertz (we again assume that the gain length is the same as the cavity length):

$$\nu_{\mathrm{osc}} = \frac{c}{2d}\left(1 + \frac{\lambda}{2\pi}\frac{\nu_0 - \nu}{\Delta\nu_{\mathrm{L}}}(-\gamma(\nu))\right)^{-1}(q + (1 + m + n)\cos^{-1}\sqrt{g_1 g_2})$$

Assuming that

$$\frac{\lambda}{2\pi}\frac{\nu_0 - \nu}{\Delta\nu_{\mathrm{L}}}(-\gamma(\nu)) \ll 1$$

then

$$\nu_{\mathrm{osc}} = \nu_{nmq} + \nu_{mnq}\left(\frac{\lambda}{2\pi}\frac{\nu_0 - \nu_{\mathrm{osc}}}{\Delta\nu_{\mathrm{L}}}\gamma(\nu_{\mathrm{osc}})\right) \tag{6.104}$$

The oscillating frequency will usually be close to the passive cavity resonant frequency and we may put

$$\nu_{\mathrm{osc}} \simeq \nu_{\mathrm{mnq}} \tag{6.105}$$

and using

$$\lambda\nu_{mnq} = c$$

we obtain

$$\nu_{\mathrm{osc}} = \nu_{mnq} + \frac{c}{4\pi}\frac{\nu_0 - \nu_{mnq}}{\Delta\nu_{\mathrm{L}}}\gamma(\nu_{mnq}) \tag{6.106}$$

We see from this equation that (to a first order) the passive cavity frequencies are pulled in towards the centre of the transition by the active gain medium. This is illustrated graphically in Figure 6.23. The ordinate represents the q numbers of the cavity resonances while the abscissa gives the oscillating frequency. The straight line gives the phase shift for each axial mode number, q, of the passive cavity. The effect of the dispersion of the gain is shown by the dispersion-shaped curve drawn about the straight line. The exact oscillating frequency is given by the intercept between the mode number, q, and the dispersion curve. We can see that a mode at the centre of the transition is not changed in frequency from its passive cavity value. However, the two modes, one above and one below line centre, are both pulled in towards the

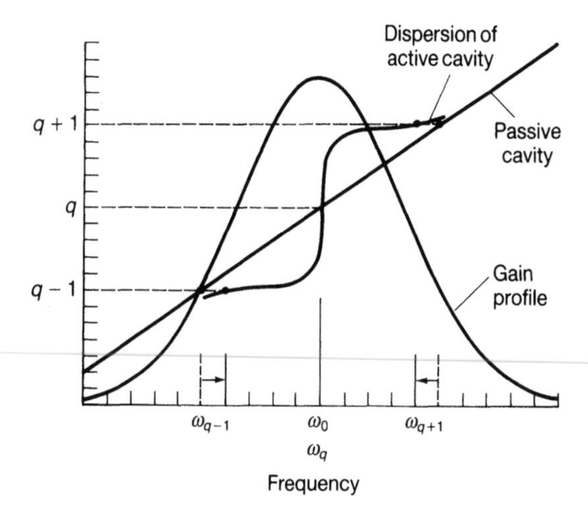

Figure 6.23 Diagram showing mode pulling by the dispersion of the active medium.

centre frequency. Since the gain of an oscillating laser is clamped at threshold the gain coefficient in Eq. (6.106) must be the threshold value which will also be equal to the loss. Thus lasers which can tolerate higher loss and those whose transitions have small linewidths will exhibit the largest mode pulling. A calculation illustrating this effect is given in Example 6.3.

Although this analysis has been carried out for a homogeneous broadened transition a similar result can be predicted for an inhomogeneous broadened transition or indeed for mixed broadening. In the Doppler-broadened case the exact dispersion function is obtained by an overlap integral similar to that required for the saturated gain calculation. The shape of the dispersion curve will then be determined by the shapes and depth of the holes burnt in the gain curve. A further complication occurs when more than one mode oscillates. In this case, the holes burnt in the gain curve produce an opposite effect on the dispersion and the frequencies are pushed away from line centre (mode pushing). Such analysis is complex and for further reading the student should consult reference 8.

Example 6.3 Estimation of the magnitude of mode pulling in a helium–neon laser
The threshold gain coefficient for a helium–neon laser operating at 633 nm is 0.5 per cent. The Doppler width is 1500 MHz and we can take the mode to be tuned at 500 MHz from line centre.

Rearrangement of Eq. (6.106) gives the difference between the empty cavity resonance frequency and the frequency of the oscillating mode:

$$\nu_{\text{osc}} - \nu_{mnq} = \frac{c}{2\pi} \frac{\nu_0 - \nu_{mnq}}{\Delta\nu_{\text{L}}} \gamma(\nu_{mnq})$$

In this case $\Delta\nu_{\text{L}} =$ Doppler width $= 1500$ MHz

$$= \frac{3 \times 10^8 \times 5 \times 10^8}{2\pi \times 1.5 \times 10^9} 0.005 = 100 \text{ kHz}$$

We can see from Example 6.3 that mode pulling shifts from the empty cavity resonant frequency are extremely small when compared with the optical frequency of 473 THz. Nevertheless, it can be measured using optical heterodyne techniques. In such an experiment the output of a helium–neon laser is simultaneously monitored using a scanning Fabry–Perot interferometer and a photodiode. The output from the photodiode produces a heterodyne (see Appendix 7) signal at frequencies equal to the mode separations and is recorded on a radiofrequency spectrum analyzer. The scanning Fabry–Perot directly records the optical spectrum.

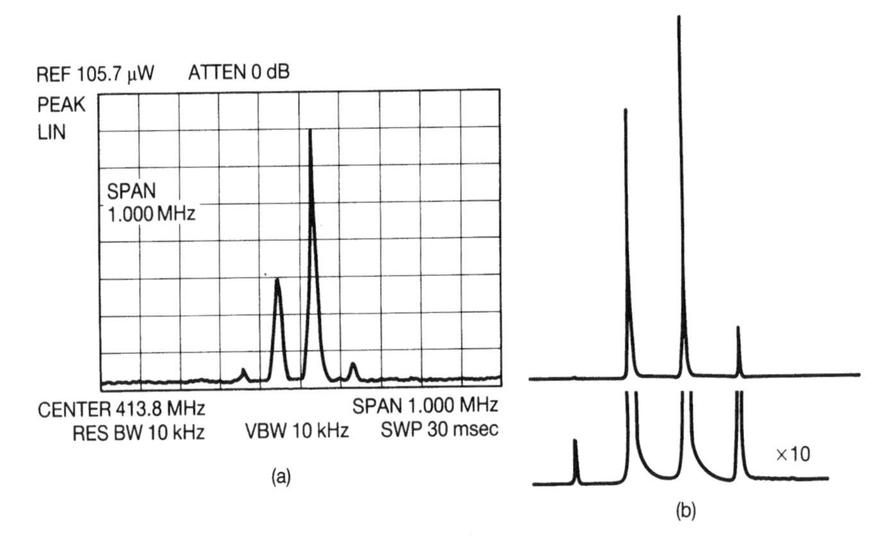

Figure 6.24 The spectrum of the modes asymmetrically located in the gain curve of the 633 nm transition in neon. (a) Heterodyne signal showing four beat frequencies. (b) The optical spectrum of this asymmetrical condition.

Figure 6.24(a) shows the heterodyne signal for the case when the four oscillating modes, shown in the optical spectrum of Figure 6.24(b), are situated asymmetrically about line centre. Four beat frequencies are detected since the separations between neighbouring modes are different on each side of line centre. On the other hand, when the modes are positioned symmetrically under the gain curve as shown in the spectrum of Figure 6.25(b) then, although the modes are pulled in frequency, this pulling is equal for modes on either side of line centre. The heterodyne signal then collapses into one beat frequency as shown in Figure 6.25(a).

In lasers with smaller Doppler widths and larger gain (e.g. the 3.39 μm helium–neon laser) much larger mode pulling can occur and in these cases the approximation leading to Eq. (6.106) may not be valid.

6.6.2 The laser linewidth

In Section 6.4 it was implied that, above threshold, the frequency of the laser oscillation would become infinitely narrow as is the case with any resonator in which the loss is balanced by the gain. However, in this model we omitted spontaneous emission. In the discussion of gain saturation it has been implied that the gain saturates at a value equal to the loss. The spontaneous emission can add to the cavity fields and ensures that the gain always saturates at a value slightly less than the loss. The result is that we now have a resonator with a very small loss and hence a small but finite bandwidth. Let us now develop an equation for this bandwidth.

The rate of change of the number of photons in a cavity mode, N_p, will equal the number added by stimulated emission in a roundtrip divided by the roundtrip time

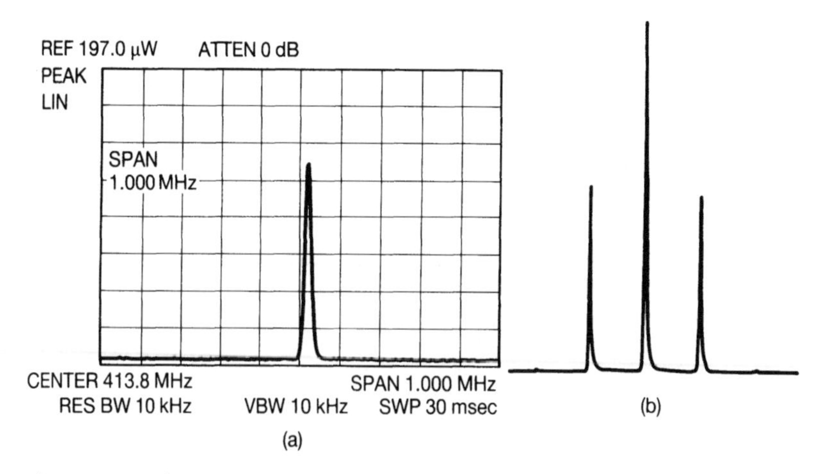

Figure 6.25 The spectrum of mode located symmetrically about line centre. (a) The heterodyne signal shows one beat frequency since the mode pulling is equal on both sides of the gain curve. (b) The optical spectrum for this symmetrical condition.

together with the rate of gain of photons added by way of spontaneous emission. This is Eq. (6.12) written here in a different but still valid way:

$$\frac{dN_p}{dt} = \frac{R_1 R_2 \, e^{2\gamma l} - 1}{2nd/c} N_p + c\sigma_{se} N_2 \tag{6.107}$$

We note that without the inclusion of the term for the rate of spontaneous emission we will obtain the threshold condition when γ is the threshold gain. Above threshold the gain must be clamped slightly below the loss. A steady state will be reached when the rate of change of cavity photons is zero, thus at equilibrium

$$(1 - R_1 R_2 \, e^{2\gamma l}) \frac{N_p}{2nd/c} = c\sigma_{se} N_2^s \tag{6.108}$$

In this equation the upper state population density, N_2^s, is the saturation value and γ is the saturated gain which is approximately equal to the threshold gain. The output power of the laser is given by the rate at which photons leak out of the cavity, that is,

$$P_0 = h\nu N_p / \tau_c \tag{6.109}$$

Using Eq. (5.56) for the cavity lifetime the number of photons in the cavity mode in the steady state is given by

$$N_p = \frac{P_0 2nd/c}{h\nu \, \log_e(1/R_1 R_2)} \tag{6.110}$$

Using this in Eq. (6.108) we obtain

$$c\sigma_{se} N_2^s = (1 - R_1 R_2 \, e^{2\gamma l}) \frac{P_0}{h\nu \, \log_e(1/R_1 R_2)} \tag{6.111}$$

The spectral width of thc lasing cavity resonance will be determined from the familiar Fabry–Perot function. The finesse of the resonator with internal loss is given by Eq. (5.52), therefore replacing loss by gain the lasing spectral width, $\Delta\nu_{osc}$ then becomes

$$\Delta\nu_{osc} = \frac{1 - e^{\gamma l}(R_1 R_2)^{1/2} c}{\pi [e^{\gamma l}(R_1 R_2)^{1/2}]^{1/2} 2nd} \tag{6.112}$$

The gain will be close to the threshold value, and since at threshold we have

$$e^{\gamma l}(R_1 R_2)^{1/2} \cong 1$$

We can then use the approximation

$$1 - e^{2\gamma l}(R_1 R_2) = (1 - e^{\gamma l}(R_1 R_2)^{1/2})(1 + e^{\gamma l}(R_1 R_2)^{1/2}) \\ \cong 2[1 - e^{\gamma l}(R_1 R_2)^{1/2}] \quad (6.113)$$

since we are concerned only with the departure of the term $e^{\gamma l}(R_1 R_2)$ from unity. Thus

$$\Delta\nu_{osc} \cong \frac{c}{4\pi nd}[1 - e^{2\gamma l}(R_1 R_2)] \quad (6.114)$$

Substitution from Eq. (6.111) for the term in brackets gives

$$\Delta\nu_{osc} = \frac{c^2}{4\pi nd}\sigma_{se} N_2^s \frac{h\nu}{P_0}\log_e(1/R_1 R_2) \quad (6.115)$$

We now write the threshold condition in its familiar form using the population inversion density given by Eq. (6.3). For simplicity we assume that the gain length is equal to the cavity length (i.e. $l = d$) and that the internal loss, L, is zero:

$$N_2^s \sigma_{se}\left(1 - \frac{g_2 N_1^s}{g_1 N_2^s}\right) = \frac{1}{2d}\log_e(1/R_1 R_2) \quad (6.116)$$

Substitution of Eq. (6.116) into (6.115) for $N_2^s \sigma_{se}$ and taking the refractive index to be unity gives

$$\Delta\nu_{osc} = \frac{1}{2\pi}\left[\frac{h\nu}{P_0}\right]\left[\frac{c}{2d}\log_e(1/R_1 R_2)\right]^2\left(1 - \frac{g_2 N_1^s}{g_1 N_2^s}\right)^{-1} \quad (6.117)$$

Finally, using expression (5.56) for the cavity lifetime we obtain

$$\Delta\nu_{osc} = \frac{1}{2\pi}\left[\frac{h\nu}{P_0}\right]\left[\frac{1}{\tau_c}\right]^2\left(1 - \frac{g_2 N_1^s}{g_1 N_2^s}\right)^{-1} \quad (6.118)$$

This is one form of an expression for the laser spectral width known as the *Schawlow–Townes limit* after the two Americans who first derived it in 1958 [Ref. 11]. It represents the ultimate or quantum limit for the laser spectral bandwidth. It must be stressed that in practice the spectral width of most lasers is not quantum limited but is determined by external agencies such as temperature changes and mechanical vibration which have the effect of altering the cavity length. These external effects are known as technical noise and if the Schawlow–Townes limit is to be approached, real lasers require some form of active frequency stabilization. Techniques for achieving this will be described in Section 6.6.4.

Another form of Eq. (6.118) may be obtained by assuming high-reflectivity mirrors and use of Eq. (5.57(a)) for the cavity lifetime in terms of the passive cavity bandwidth, thus:

$$\Delta\nu_{osc} = 2\pi\left[\frac{h\nu}{P_0}\right][\Delta\nu_{1/2}]^2\left(1 - \frac{g_2 N_1^s}{g_1 N_2^s}\right)^{-1} \quad (6.119)$$

Since the rate of spontaneous emission is clamped at the threshold value then, as the laser is pumped harder above threshold, the stimulated emission increases. The laser linewidth, being attributed to the contribution of the spontaneous emission to the radiation, will then be inversely proportional to the power, as Eq. (6.119) shows.

Example 6.4 Estimation of the minimum oscillating linewidths of helium–neon and semiconductor lasers

In this example we assume that the population of the lower laser state is much less than that of the upper state, and we put

$$\left[1 - \frac{g_2 N_1^s}{g_1 N_2^s}\right]^{-1} = \frac{g_1 N_2^s}{g_1 N_2^s - g_2 N_1^s} \cong 1$$

This approximation is not true for all lasers.

We consider a helium–neon laser at 633 nm with mirror reflectivities of 99.8 per cent and 99 per cent. The passive cavity bandwidth, for a 50 cm long laser, from Eq. (5.49) is

$$\Delta\nu_{1/2} = 5.75 \times 10^5 \text{ Hz}$$

Such a laser, with 5 mW output power, would have an oscillating linewidth of

$$\Delta\nu_{osc} = 2\pi \frac{3.13 \times 10^{-19}}{5 \times 10^{-3}} (5.75 \times 10^5)^2 \times 1 = 0.13 \times 10^{-3} \text{ Hz}$$

This is an extremely small linewidth which is not realized in practice due to other external factors such as noise and vibrations which alter the cavity length and so broaden the laser linewidth. Only in the ring laser gyroscope is it possible to detect this limit.[4]

We may now carry out a similar calculation for a typical semiconductor laser. Taking mirror reflectivities of 99.8 per cent and 30 per cent, a cavity length of approximately 300 μm and a material refractive index of 3.5, we obtain from Eq. (5.49)

$$\Delta\nu_{1/2} = 8.7 \times 10^{10} \text{ Hz}$$

For a 5 mW laser operating at 800 nm we have

$$\Delta\nu_{osc} = 2.4 \times 10^6 \text{ Hz}$$

Again, in practice, the laser linewidth is very much larger than this calculation indicates. In this case the main contribution to the linewidth is random fluctuations of current and modulation of the refractive index by the spontaneous emission.The linewidth can be reduced considerably by using a larger cavity, thus reducing $\Delta\nu_{1/2}$. For this the diode can be removed from its package and an anti-reflection coating put on the output facet. This device can then be placed in a suitably designed external cavity.

Figure 6.26 shows the spectrum obtained using a Fabry–Perot interferometer of a semiconductor laser and a helium–neon laser. The wavelength of the semiconductor laser is 670 nm and that of the helium–neon laser is 633 nm so they are close enough for the instrument resolution to be the same for both. The much narrower linewidth for the gas laser is clearly seen in the spectrum.

6.6.3 Techniques for single-frequency operation

We have seen that lasers whose gain media are inhomogeneously broadened will oscillate in as many modes as can be contained within the gain bandwidth. For homogeneous broadened gain media we have discussed how the spatial hole burning can permit multimode oscillation. Single-frequency output, and hence long coherence length, can be achieved from multimode lasers by having a filter at the output which can pass only one of the modes. This could be done using a tunable Fabry–Perot filter. Internal-mirror helium–neon lasers have two orthogonally polarized modes which can be easily separated using a a piece of Polaroid.

A more efficient approach is to force the laser into single-frequency oscillation. This can be achieved by making the cavity short enough for only one mode to be accommodated within the gain bandwidth. The cavity must, of course, still be long enough to provide gain in excess of cavity loss. Since reducing the cavity length has the effect of lowering the gain, the output power will be smaller. For example, helium–neon laser tubes of 150 mm length can give single-frequency output power of about 1 mW at 633 nm with the mode close to line centre.

For high-power single-frequency operation the loss is modified so that all modes are suppressed except the one closest to the centre of the gain profile. The simplest

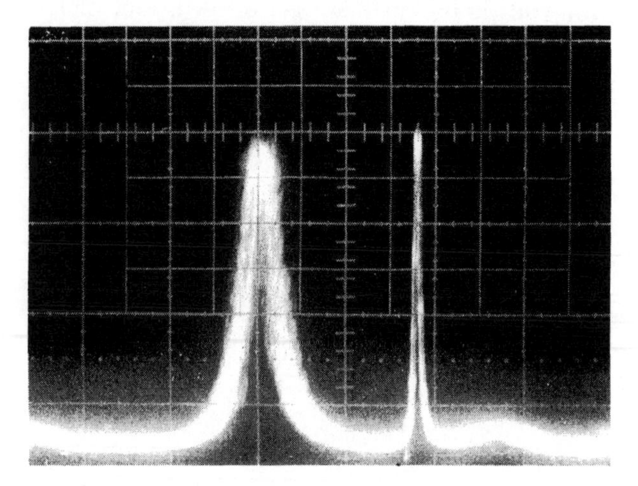

Figure 6.26 Optical spectrum taken using a scanning Fabry–Perot interferometer of a semiconductor laser at 670 nm and a helium–neon laser at 633 nm. The smaller laser linewidth of the helium–neon laser is clearly seen.

way of achieving this is to use a tilted etalon in the cavity as shown in Figure 6.27. The etalon is tilted so that reflections from either side of it are directed out of the cavity and do not set up additional resonances with the mirrors. The etalon is normally a solid material which transmits, with low loss, at the lasing wavelength and has coatings on each surface with reflectivities determined by the finesse required. Figure 6.28 shows the cavity loss which, instead of being constant as in Figure 6.2, is modulated by the transmission of the etalon. The comb of axial modes is shown along the bottom of the diagram. In this case we see how only one mode is going to be above threshold. Roughly, then, the resolution of the etalon must be smaller than the cavity mode spacing. Fast-frequency tuning can be achieved by altering the tilt angle by suspending the etalon on the shaft of a fast-response galvanometer or using piezoelectric length control in the case of air-spaced etalons.

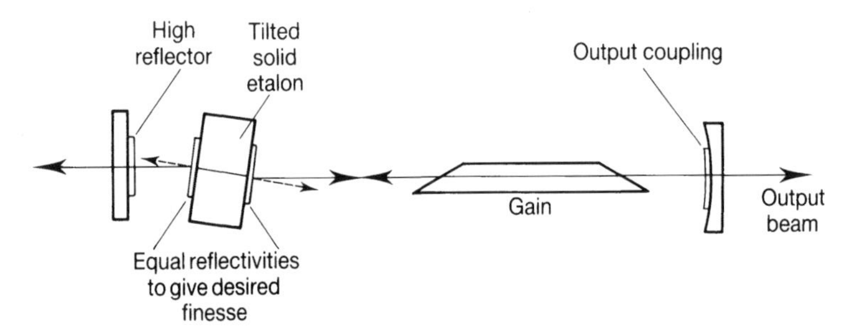

Figure 6.27 Cavity showing a tilted intra-cavity etalon to force single axial mode oscillation in a cavity in which multimode oscillation would otherwise occur.

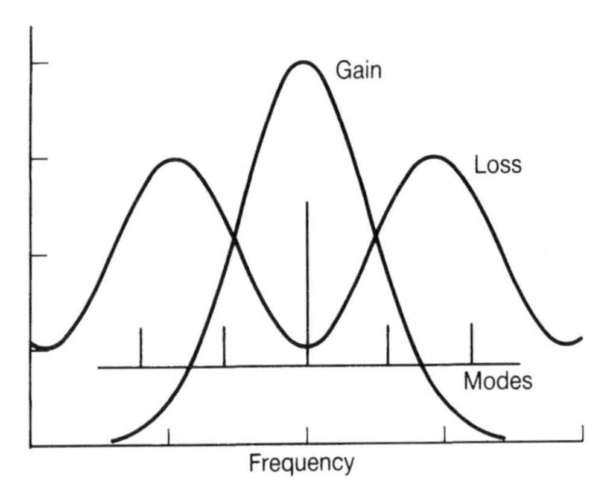

Figure 6.28 Diagram showing the etalon loss shown on top of the gain. The comb of axial modes is also shown and the diagram illustrates how only one mode can be above threshold.

Slower but stable frequency control can be achieved by changing the etalon temperature.

The tilting of the etalon causes additional loss due to transverse walk-off by the beam and the intra-cavity etalon can only be used for high-gain lasers such as argon–ion (Section 7.2.4) and dye lasers (Section 7.3.2). A lower walk-off loss can be achieved using the coupled cavity arrangement in which a third mirror is added, so producing another shorter cavity. Oscillation will occur at a frequency which is simultaneously resonant in both cavities. If the short cavity is arranged to have a free spectral range greater than the laser gain bandwidth then single-frequency oscillation can occur. The most successful of the three mirror cavities is the *Fox–Smith interferometer* as shown in Figure 6.29. In this diagram the short cavity is defined by mirrors 3, 2 and 4 while the long cavity is defined by mirrors 1 and 3. Mirror 2 is a low-loss coating of reflectivity R_2 on a transparent substrate and acts as a beam splitter. It is set at the Brewster angle to reduce the loss at its other surface. The condition that resonance will occur at a wavelength, λ, simultaneously in the two coupled cavities is

$$d_1 + d_2 = q_1 \frac{\lambda}{2} \text{ and } d_2 + d_3 = q_2 \frac{\lambda}{2} \tag{6.120}$$

where q_1 and q_2 are two large integers.

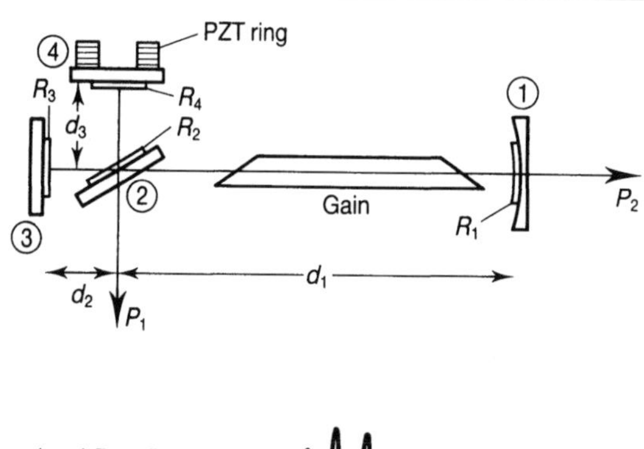

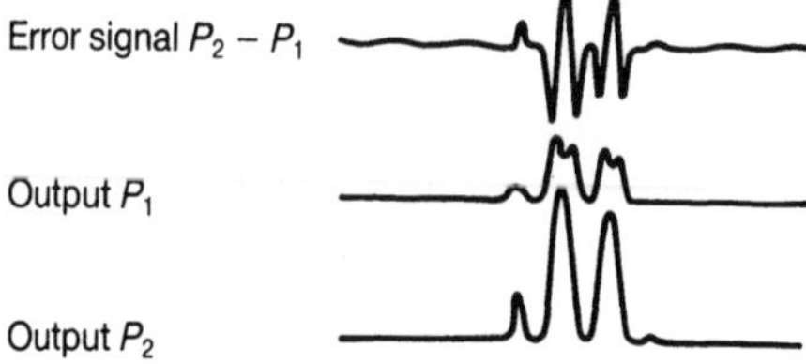

Figure 6.29 Diagram of Fox–Smith interferometer cavity for single-frequency oscillation in long laser cavities. Oscillograms are shown of the output powers P_1 and P_2 as the short cavity is detuned by applying a voltage to the PZT ring on mirror 4.

For a TEM_{00} mode to be a mode of both cavities simultaneously it is necessary, as discussed in Section 4.7, that wavefront curvatures should match mirror curvatures and so mirror 4 would normally be convex. However, in practice very small loss is incurred by the wavefront mismatch when a plane mirror is used. The reflectance of the short cavity can be determined by the same analysis as was done for the Fabry–Perot interferometer in Chapter 5. The equivalent reflectivity along the laser axis is given by

$$R_{FS} = \frac{T_2^2 R_2 (1 - A_{2s})^2}{(1 - R_2 (R_2 R_3)^{1/2})^2 + 4R_2 (R_3 R_4)^{1/2} \sin^2(\phi/2)} \tag{6.121}$$

where A_{2s} is the loss in the substrate of mirror 2, ϕ is the roundtrip phase change given by

$$\phi = \frac{2n\omega}{c}(d_2 + d_3) \tag{6.122}$$

and

$$T_2 = 1 - R_2 - A_2 \tag{6.123}$$

As for the conventional Fabry–Perot the separation of the reflectivity peaks or free spectral range is given by

$$\Delta\nu_{FS} = \frac{c}{2n(d_2 + d_3)} \tag{6.124}$$

their full width at half maximum is given by

$$\Delta\nu_{1/2\ FS} = \frac{c[1 - R_2(R_3 R_4)^{1/2}]}{2n(d_2 + d_3)\pi(R_2(R_3 R_4)^{1/2})^{1/2}} \tag{6.125}$$

and the finesse can be defined as before.

In Figure 6.30 the equivalent reflectivity and the finesse of the Fox–Smith interferometer are shown for different values of the beam splitter reflectivity, R_2. Absorption in the coating and substrate is assumed to be zero and the reflectivities of the other two mirrors taken to be 0.998. The choice of a value of R_2 will depend on the amount of suppression of the axial modes of the longer cavity required. As a rough guide we could make the FWHM (Full Width Half Maximum) of the reflectivity peak equal to the axial mode separation in the large cavity. In practice, we may only need to introduce as much loss as required to suppress neighbouring modes, as illustrated in Figure 6.28. The output from this laser can be either from the beam splitter, P_1 or through mirror 1, P_2. In the former case the output coupling can be determined by the value of R_2 as shown in Figure 6.30 but suffers from the disadvantage that a double-spot occurs unless the two cavities are in exact resonance. The two cavities can be actively locked together by obtaining an error signal (see next section) from the difference $P_2 - P_1$ as shown in Figure 6.29.

Single-frequency oscillation can also be forced onto a laser when the gain medium has a spatial grating burnt into it by the pumping mechanism in a similar way to

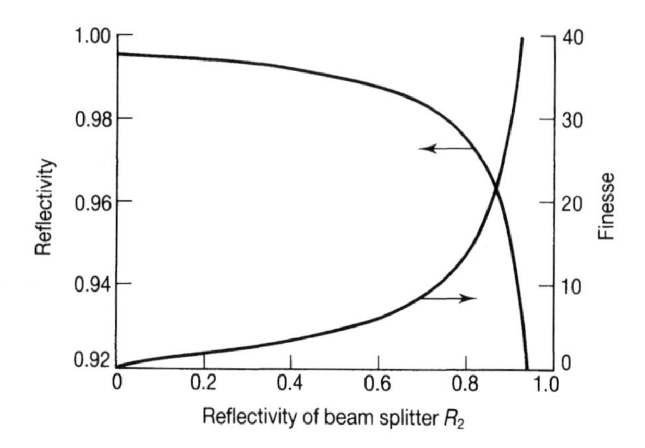

Figure 6.30 Effective reflectivity and finesse of the Fox–Smith interferometer as a function of the beam splitter reflectivity. $R_3 = R_4 = 0.998$ and $A_{2s} = A_2 = 0.001$.

spatial hole burning. This can be done by creating an interference pattern in the gain medium by the pumping source as occurs in optically pumped lasers (see Section 7.3). In semiconductor lasers a grating can be etched into the material forming the *distributed feedback laser* which will be discussed in Section 7.4.6.

6.6.4 Techniques for stabilizing the laser frequency

The output of a single-frequency laser is determined by the optical length of the cavity which will be affected by noise and vibration in the environment which alter the cavity length or the intra-cavity refractive index. These external sources of frequency instability are known as technical noise. Significant improvement in the frequency stability can be achieved by proper opto-mechanical design to reduce vibration of the cavity and the use of low expansivity materials to reduce thermal expansion. For major improvement in the frequency stability, active stabilization techniques using electronic servo-systems must be employed.

With active frequency stabilization it is necessary to have some form of frequency reference and from this a method of generating an error signal which can then be applied to a cavity length control element in such a way as to compensate for the frequency fluctuations. The reference frequency can be the resonance of a Fabry–Perot interferometer, or a spectral feature such as a molecular absorption or the gain curve of the laser itself. The error signal must have a zero crossing at the frequency at which it is desired to lock the laser.

Figure 6.31 illustrates how an error signal can be obtained from a resonant feature. Frequency modulation of the light on either side of the resonance will result in the power being modulated. If the frequency of the light is exactly at the peak of the resonance then no power modulation will occur. A phase shift in the modulated power signal results as the lasing frequency is scanned through the peak.

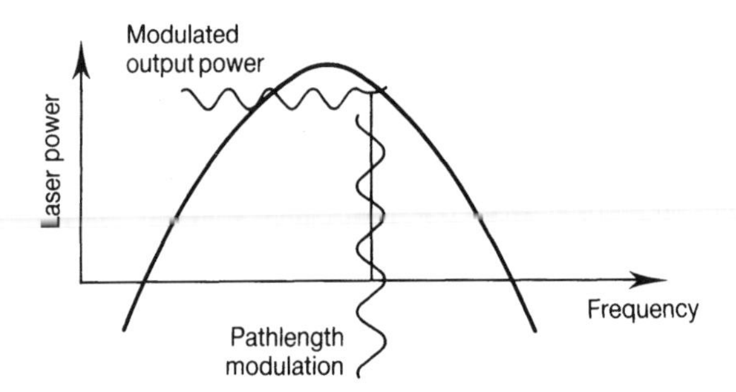

Figure 6.31 Method of obtaining an error signal from a resonant feature.

Phase-sensitive detection of the modulated power signal will produce a dispersion-like curve with a zero crossing at the centre of the resonance. If one of the laser cavity mirrors is fixed to a piezoelectric transducer so that the cavity length can be changed by the application of a voltage, then both frequency modulation and frequency change can be affected. Thus feeding back the error signal in anti-phase to the piezoelectric driver can lock the laser output frequency to the zero crossing point. It should be emphasized that this method will produce frequency-stabilized light having a small frequency modulation.

Locking to the centre of the gain curve has been successfully carried out by this technique. However, this does not produce a very sharp frequency discriminant due to the fact that the top of this curve is relatively flat even in a gas laser. The discriminant can be improved in a gas laser by using the Lamb dip (Section 6.4.2) as the resonant feature. However, a gain medium, especially a gas discharge, is a very noisy environment and a passive cell containing an atomic or molecular vapour having a spectral feature at some frequency within the laser gain bandwidth is often preferred. Absorption features like this are due to excitation of atoms or molecules in the cell from the ground state or a state close to the ground state which is thermally excited. The iodine isotopic molecule $^{127}I_2$ has a particular wealth of hyperfine[5] absorption features in the visible. In particular, it has seven which lie within the gain bandwidth of the 632.8 nm ^{20}Ne laser transition [Ref. 12]. The lower temperature and the larger mass of iodine compared to that of the neon in the helium–neon laser gas discharge means a lower Doppler width, but some means of resolving the absorption features is still required, since the homogeneous linewidth is significant.

Doppler-free spectra can be obtained using saturation spectroscopy. For this, light is made to traverse the cell in both directions interacting with the same gas molecules in both directions. When its frequency is at the centre of the transition a drop in the absorption will occur. This is exactly the reverse of the Lamb dip, and instead of a drop in power a small increase in the transmission will take place. A convenient way of achieving this is to insert the vapour cell inside the laser cavity and detect the small absorption peak by an increase in the output power from the laser. This absorption peak will occur on top of a background of the power versus frequency curve which follows the shape of the gain curve. It can be shown (see Appendix 6) that if sinusoidal frequency modulation is used the third harmonic signal of the output power is the third derivative of the feature and is a dispersion-shaped curve about zero. These zero crossing curves produced by scanning a single mode through the gain profile in a 633 nm helium–neon laser are shown in Figure 6.32 and can be used to lock the laser onto any one of the seven features. The frequency of this iodine-stabilized laser is very repeatable and realizes the definition of the metre in the laboratory to one part in 10^9. Other species with absorption features which have been used include methane at 3.39 μm again for the helium–neon laser and ^{85}Rb at 780.0 nm and 794.8 nm for semiconductor lasers.

A particularly simple technique which locks a short helium–neon laser with its modes symmetrically placed in the gain profile is the *mode balancing technique* originally devised by Balhorn *et al.* [Ref. 13] and subsequently refined by a number of workers. Figure 6.33 is a schematic diagram of such a stabilized laser. A short (~150 mm) helium–neon laser tube operating at 633 nm with internal mirrors (see Section 7.2.3) will oscillate with two axial modes either side of the gain curve. These modes are usually found to be orthogonally polarized. A signal from each of the orthogonally polarized modes can be obtained by splitting the light using a polarizing beam splitter and directing each onto a photodetector. If the modes are symmetrically placed in the gain curve then the signal in each photodiode is the same and the output from the differential amplifier is zero. In this system the cavity length cannot be altered with a piezoelectric transducer since the mirrors are sealed to the

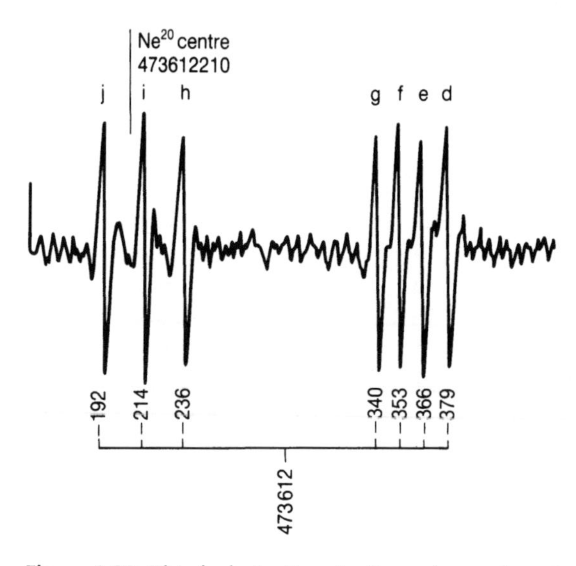

Figure 6.32 Third derivative iodine absorption features at 633 nm used to stabilize a helium–neon laser. The internationally accepted frequencies are given in hertz.

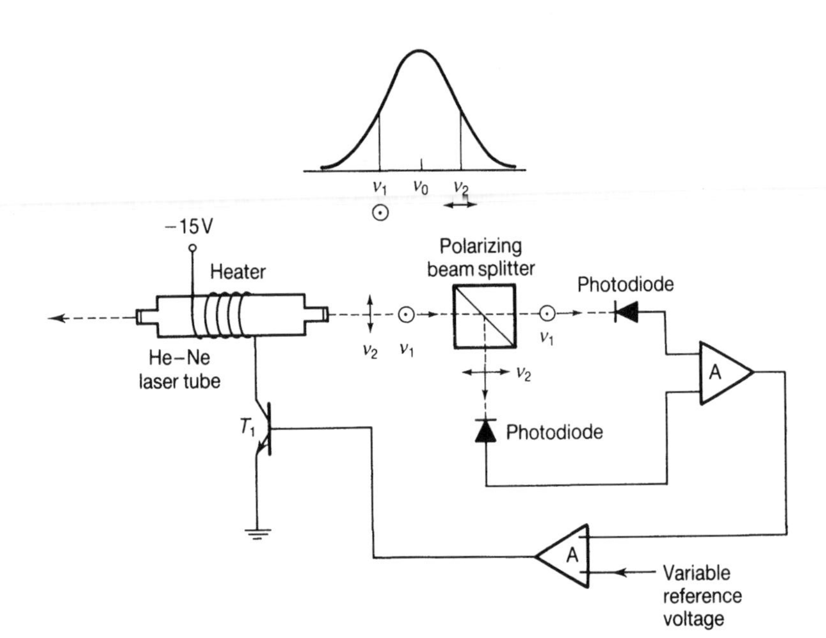

Figure 6.33 Schematic diagram of the mode-balancing technique for frequency stabilization of a short helium–neon laser operating at 633 nm with two orthogonally polarized modes.

tube, so the relatively slower technique of heating the tube is used. The driver transistor T_1 (Figure 6.33) is biased with a steady DC current with zero voltage applied from the differential amplifier. When the modes move due to some external effect then their amplitudes change and a signal appears from the differential amplifier. This signal is applied to the base of T_1 and controls the heater current in such a way as to drive the cavity resonant frequency back to its original unperturbed value.

This scheme has the advantage of simplicity and low cost as well as giving an output which is not frequency modulated. A disadvantage of the method is that perturbations in the electronics will cause the modes not to balance exactly in the same location in the gain curve each time the laser is turned on, although a reference voltage as shown in Figure 6.33 can be incorporated to trim the lasing frequency. Since it uses heat to control the laser cavity length its response time is slow and hence fast-frequency fluctuations are not removed.

It is often more convenient to stabilize the laser frequency to the resonance frequency of a Fabry–Perot interferometer. Passive Fabry–Perot cavities can be manufactured using a very low expansion glass ceramic material, zerodur,[6] as the mirror spacer. Such cavities, isolated from mechanical noise and with narrow resonances determined by high mirror reflectivities, provide excellent reference for laser stabilization even though they are not referenced to any atomic standard. An alternative to locking to the peak of the resonance is to lock on the side of a fringe.

This does not require frequency modulation and obtains an error signal by comparing the light transmitted on the side of the resonance (fringe) of the cavity with that directly from the laser.

In semiconductor lasers frequency control may be carried out by applying the error signal to the drive current and in so doing the frequency response of the feedback system can be made much higher – perhaps up to 100 MHz. However, in most other lasers, where piezoelectric transducers are used for frequency control, then cavity-length fluctuations which occur at frequencies beyond the bandwidth of the piezoelectric transducer (~ 8 kHz) cannot be corrected.

A wide bandwidth technique known as the *Pound–Drever scheme* [Ref. 14] has been adapted from use with microwaves and can achieve sub-hertz laser frequency stability. Figure 6.34 is a simplified schematic diagram of the optical arrangement. Plane polarized light from a single-frequency laser is passed through an optical isolator (Section 7.3.1.5) and then through a phase modulator. Modulation of the phase modulator at a radiofrequency of about 20 MHz produces a carrier and two sidebands (see Section 8.3.2). This light is then directed into a stable reference cavity and the reflected light detected. Figure 6.35(a) shows the carrier at the centre of the cavity resonance and the sidebands well removed from the resonance. This is arranged by designing the cavity bandwidth, perhaps a few hundred kilohertz, to be less than the modulation frequency. Figure 6.35(b) shows the phase changes which occur when the carrier is scanned through the resonance. Exact forms of this phase shift are discussed in Section 5.4.1 and shown graphically in Figure 5.12. When the carrier is at the centre of the cavity resonance no beat frequency between the carrier and the sidebands is detected since the heterodyne signals are each 180° out of

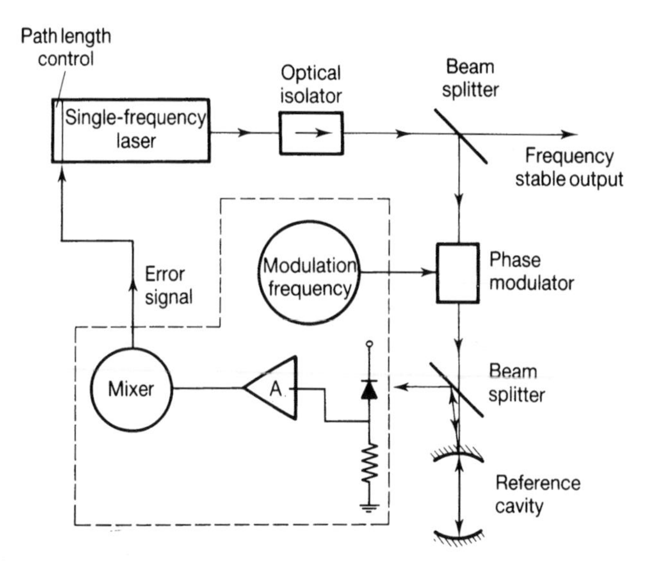

Figure 6.34 Pound–Drever scheme for laser frequency stabilization.

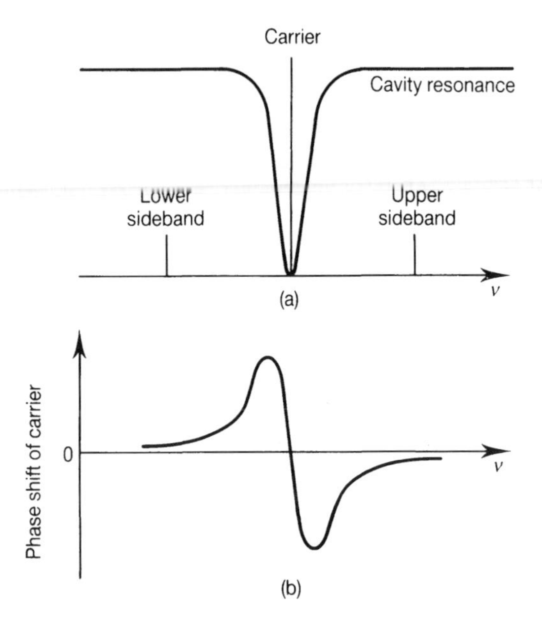

Figure 6.35 (a) Relationship of the carrier and the sidebands to the reference cavity resonance as observed in reflection. (b) Phase shift or dispersion curve for the cavity.

phase. If the laser frequency moves, then its phase will change and a beat at the modulation frequency will be detected. Its phase, however, will depend on the direction of movement of the laser frequency. The phase information can be detected in the mixer since the phase of the beat signal with respect to the signal on the phase modulator varies in the same way as that of the reflected carrier. Figure 6.36 shows the spectrum of the phase-modulated light and the error signal from the

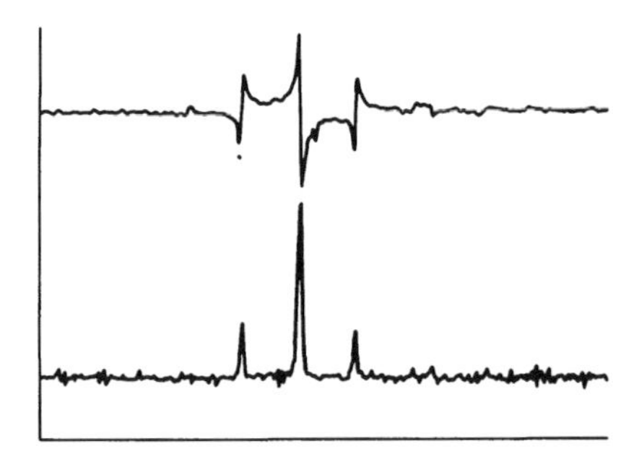

Figure 6.36 Lower trace shows a scanning Fabry–Perot spectrum of the phase-modulated light. It illustrates the carrier and the sidebands. The upper trace shows the error signal which comes from the mixer.

mixer. This error signal, as before, has a zero crossing and can be applied to alter the optical length of the laser cavity and pull the frequency back to the centre of the reference cavity. The bandwidth of this servoloop will now be determined by the phase-modulation frequency which can be tens of megahertz. We can only take advantage of this wide bandwidth if the laser frequency can be controlled at this rate, and this is not possible with piezoelectric transducers. Hence the laser must also have some form of electro-optic phase control, either internal or external to the cavity. Normally, these systems will have slow loops using piezoelectric transducers and a fast loop using an electro-optic device such as a phase modulator.

6.6.4.1 Methods of monitoring frequency stability

Coarse monitoring of the frequency stability of a laser can be performed by observing the movement of the spectrum on a scanning Fabry–Perot interferometer provided the interferometer has a better stability than the laser. For precise frequency-stability measurement the laser under test must be heterodyned with a reference laser having better stability. Figure 6.37 shows how the beams from two lasers can be combined onto a photodetector such as a photodiode. If the detector is a square-law detector then the current in the external circuit will have a DC component proportional to the total power incident on the detector and an AC term at the inter-laser beat frequency (see Appendix 7). It is the frequency excursions of this AC term which gives the frequency stability of the test laser. The following requirements must be satisfied in order to detect the heterodyne or beat signal:

1. The two beams of light must not be orthogonally polarized at the detector. Maximum signal will be achieved when they have the same polarization.
2. The electrical bandwidth of the detector and the associated electronics must be wide enough to accommodate the maximum beat frequency. This can be typically 1 GHz.
3. If all the light is focused onto the detector as shown in Figure 6.37 then the angle between the two beams, θ, must be small enough to produce one interference fringe at the detector since we are observing a moving fringe pattern. The loss of signal due to angular misalignment and difference in the spot sizes is given by Rowley [Ref. 15] as

$$\text{loss} = \exp[-\pi(w_1 w_2 \theta)^2/\lambda^2(w_1^2 + w_2^2)] \tag{6.126}$$

   where w_1 and w_2 are the two spot sizes. Using into this equation will show that angular alignment tolerances for visible light of better than milliradians is required.
4. The two beams should be concentric.
5. Optical feedback into the laser must be eliminated. Some optical isolation can be achieved using a Polaroid and a quarter-wave plate but best isolation is obtained with a Faraday isolator (see Section 7.3.1.5).

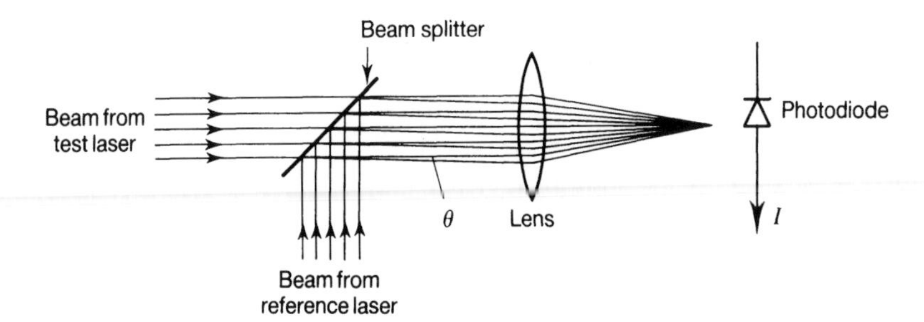

Figure 6.37 Optical arrangement for heterodyning two frequency-stable lasers.

A measure of the frequency fluctuations, which may have a steady drift superimposed, is given by the *Allan Variance*, σ, where

$$\langle \sigma^2(2, \tau) \rangle = \frac{1}{N} \sum_{i=1}^{N} \left(\frac{\bar{\nu}_i - \bar{\nu}_{i-1}}{2} \right)^2 \tag{6.127}$$

where $\bar{\nu}_i$ and $\bar{\nu}_{i-1}$ are the successive frequency averages during sampling times, τ, and N is the number of pairs of readings taken. The frequency stability of the laser can then be defined by

$$S(\tau) = \frac{\langle \sigma^2(2,\tau) \rangle^{1/2}}{\nu_0} \tag{6.128}$$

The laser stability is then specified by a $\mathrm{Log}_{10}\, S(\tau)$ versus $\log_{10} \tau$ plot as shown in Figure 6.38. An analysis by Rowley [Ref 15] identifies four regions of the graph

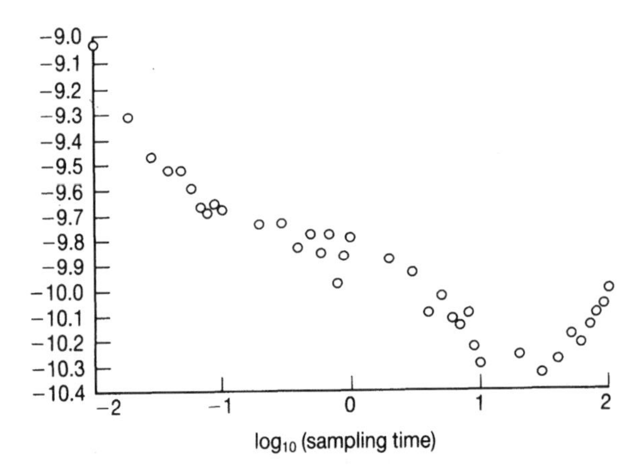

Figure 6.38 Log-log plot for the stability of a helium–neon laser stabilized to the peak of a zerodur reference cavity using the frequency dither technique.

which are associated with different forms of noise. Straight-line segments can be drawn in these regions and are analyzed as follows:

1. At low sampling times a slope of −1 is associated with the discrete frequency modulation necessary for the servosystem.
2. White noise such as that due to thermal energy has a slope of −1/2.
3. Flicker noise has a $1/f$ spectrum and zero gradient.
4. A steady drift has a slope of +1.

Figure 6.38 is the result of a stability measurement of a helium–neon laser stabilized to a zerodur reference cavity. The four straight-line sections can be drawn instead of a curve through the points and clearly show the fundamental limit for this laser-stabilization system.

6.6.5 Second harmonic generation

The technique of second harmonic generation (or frequency doubling) is a popular method of extending the normal operating frequency range of a laser. As its name implies, it enables the output frequency to be doubled (that is, the wavelength to be halved). Thus the output of the Nd:YAG laser at 1.06 μm can be frequency doubled to give 0.53 μm radiation. Frequency doubling relies on the presence of a nonlinear relationship between the electrical polarization of a material and the applied electric field. As we have seen in Chapter 2, the polarization arises from the displacement of the outer electrons in the presence of an applied field. Taking, for simplicity, a one-dimensional atom and writing the potential in which the electrons move as $V(x)$ where x is the measurement from the minimum energy position, then the equilibrium displacement, x_0, of an electron in a DC field $\mathscr{E}$ will be given by equating the total force on the electron to zero. This requirement can be expressed as

$$-\left(\frac{\mathrm{d}V(x)}{\mathrm{d}x}\right)_{x=x_0} - e\mathscr{E}_0 = 0$$

Writing the potential as a series expansion in x[7]

$$V(x) = A + Cx^2 + Dx^3 + \text{etc.} \tag{6.129}$$

the equilibrium displacement is then given by solving

$$3Dx_0^2 + 2Cx_0 + e\mathscr{E}_0 = 0 \tag{6.130}$$

If the first term in Eq. (6.130) is missing (i.e. if $D = 0$) then x_0 will have the same magnitude but the opposite sign when the direction of $\mathscr{E}_0$ is reversed. If D is non-zero, however, this is not so. A different magnitude of equilibrium displacements results when the field direction is reversed. It follows, therefore, that the magnitude of the polarization obtained will also depend on the direction of the field. Such a situation can only arise if $V(x) \neq V(-x)$. Since the electron potential must reflect the symmetry of the crystal structure, the requirement $V(x) \neq V(-x)$ will only hold in crystals where the structure lacks inversion symmetry.

We now express the polarization, $\mathscr{P}$, as a power series in the applied electric field $\mathscr{E}$, thus:

$$\mathscr{P} = \varepsilon_0(\varkappa\mathscr{E} + \varkappa_2\mathscr{E}^2 + \varkappa_3\mathscr{E}^3 + \ldots) \tag{6.131}$$

where $\varkappa$ is the polarizability (see Eq. (2.20)) and $\varkappa_2$ and $\varkappa_3$, etc. represent the terms arising from the presence of nonlinear terms in the electron potential. If the applied field is of the form $\mathscr{E} = \mathscr{E}_0 \sin \omega t$, then substitution into Eq. (6.131) gives

$$\begin{aligned}\mathscr{P} &= \varepsilon_0(\varkappa\mathscr{E}_0 \sin \omega t + \varkappa_2\mathscr{E}_0^2 \sin^2\omega t + \varkappa_3\mathscr{E}_0^3 \sin^3\omega t + \ldots)\\ &= \varepsilon_0\left(\varkappa\mathscr{E}_0 \sin \omega t + \frac{1}{2}\varkappa_2\mathscr{E}_0^2(1 - \cos 2\omega t) + \ldots\right)\end{aligned} \tag{6.132}$$

The polarization thus contains a term which oscillates at a frequency twice that of the applied field, and this can then serve as a source of radiation at frequency of 2ω. The efficiency of conversion should be proportional to $\varkappa_2\mathscr{E}_0^2$, and thus increase linearly with the irradiance of the input radiation. It is customary to use d to represent the quantity $\varepsilon_0\varkappa_2$, and we will follow this convention from now on.

So far, we have restricted the discussion to one dimension. In three dimensions the polarization at frequency 2ω is related to the field at frequency ω by a third-rank tensor:

$$\begin{aligned}\mathscr{P}_x^{2\omega} = {} & d_{111}\mathscr{E}_x\mathscr{E}_x + d_{122}\mathscr{E}_y\mathscr{E}_y + d_{133}\mathscr{E}_z\mathscr{E}_z\\ & + 2d_{123}\mathscr{E}_y\mathscr{E}_z + 2d_{131}\mathscr{E}_z\mathscr{E}_x + 2d_{112}\mathscr{E}_x\mathscr{E}_y\end{aligned} \tag{6.133}$$

Similar relations are needed for $\mathscr{P}_y^{2\omega}$ and $\mathscr{P}_z^{2\omega}$. Fortunately, crystal symmetry considerations often reduce the number of coefficients that need to be considered.[8] Table 6.1 gives values for the principal d coefficients for a number of nonlinear crystals.

Another factor which strongly influences the conversion efficiency is the requirement that both the input beam (at frequency ω) and the frequency doubled beam (at frequency 2ω) must travel with the same speed through the crystal. This ensures that the two beams retain the same phase relationship and thus at any point in the crystal the generated second harmonic radiation always adds coherently to the second harmonic radiation already present. In crystals where this requirement is not met the irradiance of the frequency-doubled radiation undergoes fluctuations as a function of distance traversed within the crystal.

The general form of this fluctuation can be obtained fairly simply. Suppose that the fundamental and second harmonic waves have wave factors which can be written as $\exp[\mathrm{j}(k_1z - \omega t)]$ and $\exp[\mathrm{j}(k_2z - 2\omega t)]$. The amplitude of the second harmonic after a length, L, of crystal is given by adding together the contributions from each element $\mathrm{d}z$ within the length L. Thus we may write

$$\mathscr{E}(2\omega, l) \propto \int_0^L \mathscr{E}^2(\omega, z)\,\mathrm{d}z$$

Table 6.1 Nonlinear optical coefficients for a number of crystals

Crystal	d_{ijk} *Value* *units of* $\frac{1}{9}\times 10^{-22}$ m^3 V^{-1} s^{-2}
$LiNbO_3$	$d_{131} = 4.4$
	$d_{122} = 2.3$
$NH_4H_2PO_4$ (ADP)	$d_{321} = 0.45$
	$d_{123} = 0.5$
KH_2PO_4 (KDP)	$d_{321} = 0.45$
	$d_{123} = 0.35$
KD_2PO_4 (KDP)	$d_{312} = 0.42$
	$d_{123} = 0.42$
Quartz	$d_{111} = 0.37$
CdS	$d_{333} = 29$
	$d_{311} = 30$
	$d_{312} = 33$
GaAs	$d_{123} = 72$

Assuming a low conversion efficiency so that the amplitude of the primary beam is not appreciably diminished after a length l, then

$$\mathscr{E}(2\omega, l) \propto \int_0^L \exp[2j(k_1 z - \omega t)]\, dz \tag{6.134}$$

Light from a point z within the crystal will take a time $(L - z)k_2/(2\omega)$ to reach the end, so that

$$\mathscr{E}(2\omega, l) \propto \int_0^L \exp 2j\left[k_1 z - \omega\left(t - \frac{L-z}{2\omega}k_2\right)\right] dz \tag{6.135}$$

The integration is readily carried out and, on squaring the result to obtain the irradiance, we have

$$|\mathscr{E}(2\omega, L)|^2 \propto \left(\frac{\sin(\Delta k L)}{\Delta k}\right)^2 \tag{6.136}$$

where $\Delta k = k_1 - k_2/2$. As a function of L the right-hand side of this equation oscillates between the values 0 and $[2/(2k_1 - k_2)]^2$, the maximum being reached after a length $\pi/(2k_1 - k_2)$. The distance corresponding to half that between maxima this function is known as the *coherence length*, l_c, where

$$l_c = \left|\frac{\pi}{2k_1 - k_2}\right|$$

$$= \left|\frac{\lambda_1}{4(n_1 - n_2)}\right| \tag{6.137}$$

where n_1 and n_2 are the refractive indices at frequency ω and 2ω, respectively, and λ_1 is the wavelength of the primary beam. Typical values for l_c are of the order of 10^{-5} m or so (see Example 6.5).

The actual efficiency of the conversion process, η_{SHG}, can be shown to be given by [Ref. 16][9]

$$\eta_{SHG} = 2\left(\frac{\mu_0}{\varepsilon_0}\right)^{3/2} \frac{\omega^2 d^2}{n_1^2 n_2} \left(\frac{\sin(\Delta k L)}{\Delta k}\right)^2 I_1 \tag{6.138}$$

where I_1 is the irradiance of the primary beam. The maximum efficiency will be reached when $L = l_c$, so that

$$\eta_{SHG}^{max} = 2\left(\frac{\mu_0}{\varepsilon_0}\right)^{3/2} \frac{\omega^2 d^2}{n_2^2 n_2} \left(\frac{2l_c}{\pi}\right)^2 I_1 \tag{6.139}$$

With a coherence length of the order of 10^{-5} m, conversion efficiencies are very low (see Example 6.5).

Example 6.5 Second harmonic generation efficiency

In KDP the refractive index (ordinary ray) at 1 μm is 1.4960 while at 0.5 μm it is 1.5149. Thus using Eq. (6.137) we obtain for the coherence length

$$l_c = \frac{10^{-6}}{4(1.5149 - 1.4960)}$$

$$= 1.3 \times 10^{-5} \text{ m}$$

From Table 6.1 a typical value for d is $0.42 \times \frac{1}{9} \times 10^{-22}$ or 5.25×10^{-24} m^3 V^{-1}s^{-2}, and assuming an irradiance of 10^{12} W m^{-2} Eq. (6.139) gives that the maximum second harmonic generation efficiency is

$$\eta_{SHG}^{max} = 2\left(\frac{4\pi \times 10^{-7}}{8.85 \times 10^{-12}}\right)^{3/2} \frac{1}{1.51}$$

$$\times \left(\frac{4\pi \times 3 \times 10^8}{10^{-6}} \frac{5.25 \times 10^{-24}}{1.51} \frac{1.3 \times 10^{-5}}{\pi}\right)^2 10^{12}$$

$$= 1.5 \times 10^{-7}$$

Second harmonic generation was first observed by Franken and his co-workers in 1961 [Ref. 17], who, by focusing the output from a ruby laser operating at 694.3 nm onto a quartz crystal obtained a small amount of radiation at 347.15 nm. Not surprisingly, in view of the result of the calculation in Example 6.5, the conversion efficiency was very low ($\approx 10^{-8}$).

Obviously, the longer we can make the coherence length, the longer we can make the crystal and hence the higher will be the resulting efficiency. Ideally we would like to have n_1 equal to n_2 ($=n$). When this is so Eq. (6.138) becomes simply

$$\eta_{\mathrm{SHG}} = 2\left(\frac{\mu_0}{\varepsilon_0}\right)^{3/2} \frac{\omega^2 d^2 L^2}{n^3} I_1 \tag{6.140}$$

Of course, we must remember that this expression is only valid provided the conversion efficiency is low so that the irradiance of the primary beam remains essentially unchanged. When this is not the case then the efficiency is given by

$$\eta_{\mathrm{SHG}} = \tanh^2 \left(2\left(\frac{\mu_0}{\varepsilon_0}\right)^{3/2} \frac{\omega^2 d^2 L^2}{n^3} I_1\right)^{1/2} \tag{6.141}$$

It is obvious that if high conversion efficiencies are to be achieved then some way has to be found of ensuring that n_2 is as close to n_1 as possible. This is most commonly done by using a technique called *phase* or *index matching* which makes use of the birefringent properties of the nonlinear medium. Figure 6.39 shows the variation in the refractive index experienced by both the ordinary and extraordinary rays as a function of the angle, θ, between the ray direction and the optic axis in a negative uniaxial crystal. The curves are drawn assuming that the extraordinary ray has twice the frequency of the ordinary ray. The ordinary ray refractive index is, of course, independent of θ while that for the extraordinary ray varies with θ (see Section 2.1.9). We see that along a particular direction (i.e. at an angle of θ_m to the optic axis in the diagram) the ordinary ray velocity (at frequency ω) will be equal to the extraordinary ray velocity (at frequency 2ω). In a uniaxial crystal the refractive index seen by the extraordinary ray depends on the angle θ between the optic axis according to (see Eq. 2.85))

$$\frac{1}{n_e^2(\theta)} = \frac{\cos^2\theta}{n_0^2} + \frac{\sin^2\theta}{n_e^2} \tag{6.142}$$

Provided $n_e^{2\omega} < n_0^{\omega}$, then it is possible to find an angle θ_m for which $n_e^{2\omega}(\theta_m) = n_0^{\omega}$. The angle θ_m may be determined by finding the angle where the ellipse of Eq. (6.142) for the radiation of frequency 2ω intersects with a circle of radius n_0^{ω}, in other words when

$$\frac{1}{(n_0^{\omega})^2} = \frac{\cos^2\theta_m}{(n_0^{2\omega})^2} + \frac{\sin^2\theta_m}{(n_e^{2\omega})^2}$$

or

$$\sin^2\theta_m = \frac{(n_0^{\omega})^{-2} - (n_0^{2\omega})^{-2}}{(n_e^{2\omega})^{-2} - (n_0^{2\omega})^{-2}} \tag{6.143}$$

The above discussion has been in terms of a negative uniaxial crystal. Phase matching is also possible in positive uniaxial crystals with slightly different conditions (see Problem 6.15).

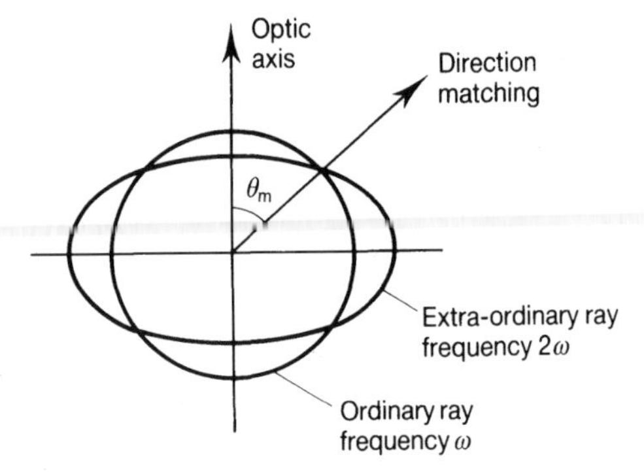

Figure 6.39 The index ellipsoids for the ordinary ray of frequency ω and the extra-ordinary ray of frequency 2ω in a uniaxial birefringent crystal showing the possibility of index matching along a direction at an angle θ_m from the optic axis.

Example 6.6 Phase matching angle in ADP

Consider the determination of the phase matching angle in ADP when the output from a Nd:YAG laser is being used. The appropriate refractive indices are[10]

$\lambda(\mu m)$	n_0	n_e
1.06	1.4943058	1.4603276
0.53	1.5132318	1.4712203

Inserting numerical values into Eq. (6.143) then gives

$$\sin^2\theta_m = \frac{0.44784 - 0.43671}{0.46200 - 0.43671} = 0.44$$

whence $\theta_m = 26^0$

The angular accuracy with which the beam must be launched into the crystal can be estimated by noting that when the phase match condition is not quite obeyed then the frequency doubling efficiency will be reduced by the factor F, where, from Eq. (6.138)

$$F = \left(\frac{\sin(\Delta k(\theta)L)}{\Delta k(\theta)}\right)^2$$

Now $\Delta k(\theta)$ $(=k_1(\theta)-k_2(\theta)/2)$ can be written as $(\omega/c)(n_o^\omega - n_e^{2\omega}(\theta))$, and we may use Eq. (6.142) to expand $n^{2\omega}(\theta)$ as a series in θ near to θ_m, the result being (see Problem 6.16)

$$\Delta k(\theta) \approx A(\theta_m - \theta)$$

where the factor A is given by

$$A = \frac{2\omega}{c}\sin(2\theta_m)\frac{(n_o^{2\omega})^{-2}-(n_e^{2\omega})^{-2}}{2(n_o^\omega)^{-3}}$$

The frequency-doubled output irradiance, $I_{2\omega}(\theta)$, should therefore depend on θ according to

$$I_{2\omega}(\theta) \propto \frac{\sin^2[A(\theta_m-\theta)]}{[A(\theta_m-\theta)]^2} \tag{6.144}$$

The function on the right-hand side of this equation has a maximum at $\theta=\theta_m$, and falls to zero when $A\,|\,\theta_m-\theta\,|=\pi$. Thus for efficient frequency doubling we require that $|\,\theta_m-\theta\,| \ll \pi/A$. This usually implies that (see Problem 6.15) angular alignment should be to within a hundredth of a degree or so. The predictions of Eq. (6.144) have been verified experimentally on a number of occasions [Ref. 19].

The simplest technique for second harmonic generation is to focus the output from the laser onto a suitably oriented (for phase matching purposes) nonlinear crystal. It must be remembered, however, that the output beam is likely to be Gaussian, while the above formulae were derived assuming a plane wave. The focused Gaussian beam will be characterized by the confocal parameter z_0. This represents the distance from the beam waist when the beam 'area' is double that at the waist. If $z_0 \gg L$, then the lateral spreading of the beam within the crystal may be ignored. By integrating the conversion efficiency over the radius of the beam it may readily be demonstrated that (see Problem 6.17)

$$\eta_{\text{SHG}}^{\text{G}} = 2\left(\frac{\mu_0}{\varepsilon_0}\right)^{3/2}\frac{\omega^2 d^2}{n_1^2 n_2}\left(\frac{\sin(\Delta kL)}{\Delta k}\right)^2\left(\frac{P_1}{\pi w_0^2}\right) \tag{6.145}$$

where P_1 is the total power in the beam of frequency ω. At first, Eq. (6.145) would seem to indicate that the conversion efficiency may be increased by reducing w_0. This is indeed true up to a point, but eventually z_0 (which is given by $\pi w_0^2 n/\lambda$) will become comparable to L, in which case we can no longer make the assumption that the beam does not spread appreciably over the length of the crystal. The approximate maximum efficiency is given by putting $2z_0 = L$ (a situation known as *confocal focusing*), giving

$$\eta_{\text{SHG}}^{\text{G,max}} \approx \frac{2}{\pi c}\left(\frac{\mu_0}{\varepsilon_0}\right)^{3/2}\frac{\omega^3 d^2 L}{n_1 n_2}\left(\frac{\sin(\Delta kL)}{\Delta k}\right)^2 P_1 \tag{6.146}$$

A more exact calculation gives a value which is some 20 per cent larger than this.

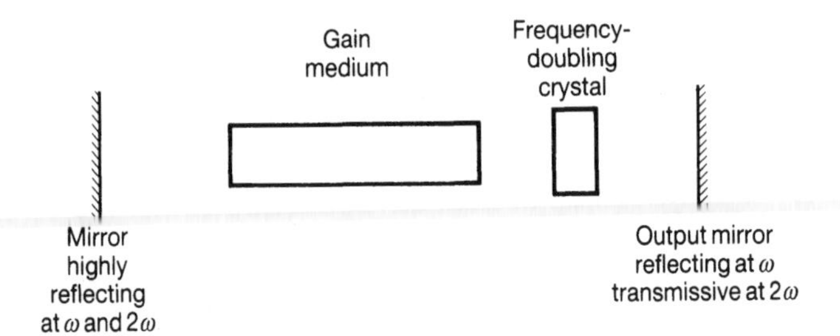

Figure 6.40 Arrangement for frequency doubling by placing the nonlinear crystal within the laser cavity.

Equation (6.146) indicates that the conversion efficiency for second harmonic generation is directly proportional to the pump irradiance. High conversion efficiencies usually require the high peak irradiances only available from pulsed lasers. CW operation is possible, however, by placing the nonlinear material inside the laser cavity where the available irradiance is many times that which emerges through the output mirrors. For example, the irradiance inside the cavity is some $(1 - R)^{-1}$ times its value outside. If R is close to unity then the enhancement can be very large, leading to high conversion efficiencies.

The basic layout is indicated in Figure 6.40. Ideally, one of the mirrors should be perfectly reflecting both at ω and 2ω, while the other (output) mirror should be perfectly reflecting at ω and perfectly transmitting at 2ω. As far as the laser is concerned, the frequency conversion can be regarded as a loss mechanism with a loss per pass given by Eq. (6.141). We know from the discussion in Section 6.5 that for a given laser with a particular pumping rate, etc. there is an optimum mirror transmission which will ensure maximum output power. If the loss per pass due to the frequency conversion can be made equivalent to that value, then this will ensure optimal output at the doubled frequency. As much power at frequency 2ω will then emerge as would emerge at frequency ω from the laser if it were without the frequency doubler but with optimal mirror coupling.

In practice there are are a number of difficulties. The crystals need to be of a very high optical purity to eliminate any additional cavity losses due to scattering within the crystal. In addition, absorption within the nonlinear medium either at ω or 2ω can cause temperature changes which may influence the refractive indices sufficiently to affect the index matching condition and so lower the conversion efficiency.

Problems

6.1 The active length of a laser has a peak gain coefficient of $0.4\ \mathrm{m}^{-1}$. If the mirrors have scatter loss of 0.5 per cent and transmission of 1 per cent and 15 per cent calculate the minimum active length for oscillation.

6.2 Show that the threshold condition for laser oscillation can be expressed in terms of the ratio τ_2/τ_c where τ_2 is the lifetime of the upper state and τ_c is the cavity lifetime.

6.3 A particular laser has an active medium with gain coefficient at line centre of 37 per cent per metre and a length of 30 cm. Its mirrors each have a scatter loss of 0.3 per cent and the high reflecting mirror has a transmission of 0.5 per cent. At what transmission of the output coupling mirror will laser oscillation cease?

6.4 Calculate the length of the gain medium to achieve threshold for a helium–neon laser at 633 nm if the mirror reflectivities are 99.8 per cent and 98 per cent. The degeneracy of the upper state is 3 and of the lower state is 5 and the Einstein A coefficient for the transition is $3 \times 10^6\ \mathrm{s}^{-1}$. Assume the population of the upper state to be $8 \times 10^{15}\ \mathrm{m}^{-3}$ and that of the lower state to be $10^{15}\ \mathrm{m}^{-3}$ and that the gas has a temperature of 100°C.

6.5 Show that the number of cavity modes at angular frequency, ω, per unit volume is given by

$$N_m/V = A_{21}/c\sigma_{se}(\omega)$$

6.6 Show that the optimum output coupling of a low-gain laser is given by Eq. (6.57).

6.7 Calculate the optimum output coupling of a low-gain homogeneously broadened laser which has an active length of 75 cm and a gain coefficient of 0.12 per cent per metre. Take the reflectivity of the high reflector to be 99.5 per cent and the internal single-pass loss to be 1.5 per cent. Determine the optimum output irradiance of this laser if the saturation parameter is $35\ \mathrm{W/cm}^2$.

6.8 Determine the output irradiance in Problem 6.7 if the laser were subject to spatial hole burning.

6.9 Starting with Eq. (6.95), derive an expression for the optimum output coupling for an inhomogeneously broadened transition.

6.10 A high-gain laser has a gain coefficient of $40\ \mathrm{m}^{-1}$, a gain length of 5 cm and a saturation irradiance of $700\ \mathrm{W/cm}^2$. If the intra-cavity single-pass loss is 2 per cent and the high reflector has a reflectivity of 99 per cent, calculate the output irradiance if the output coupling mirror has a reflectivity of 70 per cent. What would the output irradiance become if the single-pass internal losses were increased to 4 per cent?

6.11 Derive an expression for the gain bandwidth of a Doppler-broadened transition if the gain to loss ratio is three. Derive the same expression for a lorentzian broadened transition.

6.12 Determine the number of longitudinal modes in an argon–ion laser of length 70 cm if the lasing wavelength is 514 nm, the gas temperature is 7000 K and the gain-to-loss ratio is three.

6.13 Estimate the frequency pulling of the mode which is 100 MHz from line centre in a xenon laser operating at 3.51 μm. Take the roundtrip cavity loss to be 10 per cent and the gas temperature to be 500 K.

6.14 Estimate the oscillating line width of a diode-pumped Nd:YAG laser of output power 1 mW at 1.06 μm. Take the cavity length to be 10 cm and the rod length to be 5 cm with refractive index 1.86 and the mirror reflectivities 99.8 per cent and 98 per cent. Assume the degeneracies of the states to be equal and the population of the upper state to be $5.4 \times 10^{21}\ \mathrm{m}^{-3}$.

6.15 Show that for a positive uniaxial crystal phase matching may take place at an angle θ_m where

$$\sin^2\theta_m = \frac{(n_0^{\omega})^{-2} - (n_0^{2\omega})^{-2}}{(n_0^{2\omega})^{-2} - (n_e^{2\omega})^{-2}}$$

provided that $n_0^{2\omega} < n_e^{\omega}$.

6.16 Show that when the phase-matching condition for frequency doubling is not quite obeyed then the irradiance of the frequency doubled beam varies with angle θ as:

$$I_{2\omega}(\theta) \propto \frac{\sin^2[A(\theta_m - \theta)]}{[A(\theta_m - \theta)]^2}$$

where

$$A = \frac{2\omega}{c}\sin(2\theta_m)\frac{(n_0^{2\omega})^{-2} - (n_e^{2\omega})^{-2}}{2(n_0^{\omega})^{-3}}$$

Using the values contained in Example 6.6 to calculate a value for A for ADP and hence verify that for efficient frequency doubling angular alignment should be to within a hundredth of a degree or so. *Hint*: from Eq. (6.138) the frequency-doubling efficiency will be reduced by a factor F, where

$$F = \left(\frac{\sin(\Delta k(\theta)L)}{\Delta k(\theta)}\right)^2$$

and where

$$\Delta k(\theta) = k_1(\theta) - k_2(\theta)/2$$

$$= \frac{\omega}{c}(n_0^{\omega} - n_e^{2\omega}(\theta))$$

Equation (6.142) may be used to expand $n_e^{2\omega}(\theta)$ as a series in θ near to θ_m to give the required result.

6.17 Starting from Eq. (6.138), show that when a Gaussian beam is used for second harmonic generation then the conversion efficiency can be written as

$$\eta_{SHG} = 2\left(\frac{\mu_0}{\varepsilon_0}\right)^{3/2}\frac{\omega^2 d^2}{n_1^2 n_2}\left(\frac{\sin(\Delta kl)}{\Delta k}\right)^2\left(\frac{P_1}{\pi w_0^2}\right)$$

where P_1 is the total power in the beam of frequency ω and where it is assumed that $z_0 \gg 1$.

Notes

1. These reflectivities are chosen so that the cavity resonances are resolved in our graph.
2. This can be more formally proved by finding the eigenvalue of the Jones matrix of the cavity (see Ref. 10).
3. These integrals can be shown to be the error function and numerical solutions are obtained from tables of these functions.

4. A ring laser gyroscope senses rotation by the frequency difference between two counter-propagating modes in a ring laser. Since the modes occupy the same cavity and are thus equally affected by external vibrations of the cavity then the limit of sensitivity of the device will be determined by the smallest uncertainty in the laser frequencies of the two modes.
5. Transitions associated with vibrational rotational energy changes which additionally include nuclear spin.
6. Schott glass trade mark. This material is thermally cycled during manufacture and can have an expansion coefficient of almost zero at room temperature.
7. There is no term in x in this expansion because $V(x)$ has a minimum when $x = 0$.
8. Often the notation is simplified further by abbreviating the last two subscripts according to the following scheme: $(11) \equiv 1$, $(22) \equiv 2$, $(33) \equiv 3$, $(23) \equiv 4$, $(31) \equiv 5$, $(12) \equiv 6$.
9. This expression is still based on the assumption that no significant amounts of energy are lost from the incoming beam and that all attenuation processes are ignored.
10. See, for example, Ref. 18, although the values quoted here have been obtained by interpolation.

References

[1] W. E. Lamb Jr, 'Theory of an optical maser', *Phys. Rev.*, **134**, 1429, 1964.
[2] V. Evtuhov and A. E. Seigman, 'A twisted mode technique for obtaining axially uniform energy density in laser cavity', *Applied Optics*, **4**, 142, 1965.
[3] D. A. Draegert, 'Efficient single-longitudinal-mode Nd:YAG laser', *IEE Journal of Quantum Electronics*, **QE-8**, 235, 1972.
[4] L. W. Casperton, 'Threshold characteristics of mirrorless lasers', *Journal of Applied Physics*, **48**, 256, 1977.
[5] *Tables of Integrals, Series and Products*, Academic Press, New York, 1965, equation 3.615-1.
[6] W. W. Rigrod, 'Saturation effects in high gain lasers', *Journal of Applied Physics*, **36**, 2487 1965.
[7] L. W. Casperton, 'Laser power calculations: sources of error', *Applied Optics*, **19**, 422, 1980.
[8] D. C. Sinclair and W. E. Bell, *Gas Laser Technology*, Holt Rinehart and Winston, New York, 1969.
[9] J. P. Taché, 'Intracavity skew-Brewster-angle plates as a calibrated attenuator for gas lasers', *J. Phys. D.*, **19**, 943, 1986.
[10] R. J. Oram, I. D. Latimer, S. P. Spoor and S. Bocking, 'Longitudinal mode separation tuning in 633 nm helium–neon lasers using induced cavity birefringence', *J. Phys. D: Appl. Phys.*, **26**, 1993.
[11] A. L. Schawlow and C. H. Townes, 'Infrared and optical masers', *Physical Review*, **112**, 1940, 1958.
[12] A. J. Wallard, 'Frequency stabilization of the helium–neon laser by saturated absorption in iodine vapour', *J. Phys. E*, **5**, 926, 1972.

[13] R. Balhorn, H. Knunzmann and F. Lebowsky, 'Frequency stabilisation of internal mirror HeNe lasers', *Applied Optics*, **11**, 742, 1972.

[14] R. W. P. Drever, J. L. Hall, F. V. Kowalski, J. Hough, G. M. Ford, A. J. Munley and H. Ward, 'Laser phase and frequency stabilization using an optical resonator', *Applied Physics*, **B31**, 97, 1983.

[15] W. R. C. Rowley, 'Analysis of laser frequency stability by heterodyne measurement', National Physical Laboratory (London) Report *MOM 78*, 1986.

[16] A. Yariv, *Optical Electronics*, 4th edn, Saunders College Publishing, 1991, Section 8.2.

[17] P. A. Franken, A. E. Hill, C. W. Peters and G. Wienreich, 'Generation of optical harmonics', *Phys. Rev. Lett.*, **7**, 118–119, 1961.

[18] F. Zernike Jr, 'Refractive indices of of ammonium dihydrogen phosphate and potassium dihydrogen phosphate between 2000Å and 1.5 μm', *J. Opt. Soc. Am.*, **54**, 1215–1220, 1964.

[19] A. Ashkin, G. D. Boyd and J. M. Dziedzic, 'Observation of continuous second harmonic generation with gas masers', *Phys. Rev. Lett.*, **11**, 14–17, 1963.

7

Laser excitation and systems

7.1 Introduction

In this chapter we will apply the theory discussed in previous chapters to real lasers and also consider the means of achieving population inversion. Although population inversion and optical gain were understood since the early work of Einstein it was not until after the Second World War with advances in microwave technology that the ideas could be harnessed. In 1954 the MASER was developed by Townes and in 1959 Maimen achieved self-sustaining oscillations in the ruby laser. It is perhaps remarkable that the laser was not invented much earlier due to the large variety of systems (solids, liquids and gases) which can lase. Indeed, in some cases a very small amount of mirror reflectivity is required for self-sustained oscillation. Not every excited medium will exhibit population inversion and sometimes the exact conditions for population inversion are extremely critical. For example, in the helium–neon laser the gas mixture, pressure and discharge current are all critical parameters for achieving inversion. We will confine the discussion to lasers which are manufactured commercially, since a full description of all lasing systems could fill many volumes. For further reading the reader will be referred to books and review articles on specific lasers or types of lasers.

We remind ourselves of the equation for the gain coefficient from Chapter 3 (Eqs (3.104) and (3.109)):

$$\gamma(\omega) = \frac{A_{21}\lambda_0^2}{4n^2}\, g(\omega)\left[N_2 - \frac{g_2}{g_1} N_1\right] = \sigma_{se}(\omega)N \tag{7.1}$$

The stimulated emission coefficient, $\sigma_{se}(\omega)$, involves the parameters of the transition over which there is very little control. On the other hand, the population inversion density is strongly determined by the excitation mechanisms of the system and some control by the laser scientist is possible. For example, in the helium–neon laser we can vary the discharge parameters to obtain optimum population inversion in the gas. Generally, long wavelength transitions have large values of σ_{se} (note the λ_0^2 factor in Eq. (7.1)). However, this may be offset by smaller values of the spontaneous emission coefficient (A_{21}).

Transitions with a favourable lifetime ratio ($\tau_2 > \tau_1$) are the most common lasing system since population inversion will be achieved with similar pumping rates to both states and so no enhanced excitation of the upper state is required. However, population inversion and lasing is possible for transitions where $\tau_1 > \tau_2$. These are known as *cyclic lasers*. In these cases continuous wave (CW) oscillation is not possible and only a pulsed output can be obtained. We can generalize by distinguishing lasers according to whether they have a very short or a long upper state lifetime. States with long upper state lifetimes are normally easier to pump but have lower A coefficients and hence lower stimulated emission cross-sections. These lasers have the ability to store large amounts of energy in the population inversion. Systems with short upper state lifetimes achieve large gain by having a large stimulated emission cross-section. It must be emphasized that these are very broad generalizations and comparisons between the many different systems should be interpreted carefully.

In Figure 7.1 a schematic energy-level diagram of a typical laser is shown. The pumping is achieved by the input of energy, either directly to the upper laser state or by cascade transitions from some higher states marked 3. Ideally, the laser will produce one stimulated emission quantum for each input quantum of energy required to excite state 2. This is achieved when the stimulated emission rate out of

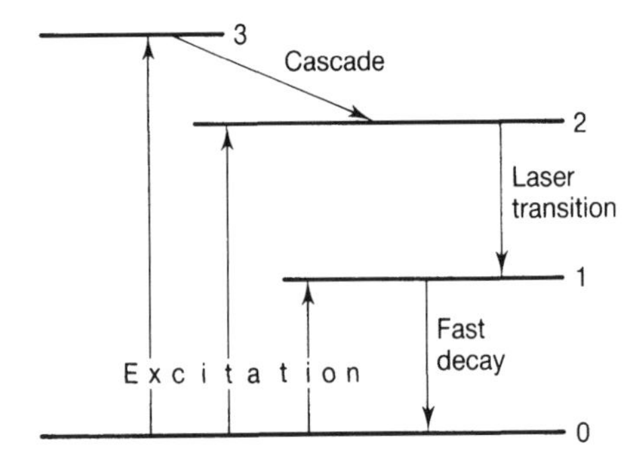

Figure 7.1 Schematic diagram of the energy levels of a typical laser system showing the excitation and de-excitation routes.

state 2 is much larger than any of the spontaneous transition rates. For this ideal laser the quantum efficiency is given by

$$\text{Ideal quantum efficiency} = \frac{E(2) - E(1)}{E(2)} = \frac{\nu_{21}}{\nu_{20}} \tag{7.2}$$

In practice, the input energy will inevitably excite states other than '2' and so Eq. (7.2) represents an ideal efficiency which is approached in only a few special cases. Many lasers have energy efficiencies which are several orders of magnitude less than this ideal, particularly if a significant amount of the excitation occurs due to cascade from higher states.

It is useful to classify lasers according to the phase of the active medium since usually this determines the means of excitation. This is shown in the following table:

Laser type	Pumping method
Gas lasers	Electrical discharge
Ionic crystalline (solid state)	Optical pumping
Liquid lasers	Optical pumping
Semiconductor lasers	Injection of electrons and holes

We now turn to a detailed consideration of the excitation and output parameters of lasers under the above headings.

7.2 Gas lasers

Of the three states of matter, the gas phase has the lowest atom density and, as a consequence, will have the smallest gain per unit volume. Therefore high-power gas lasers are physically large and usually require high voltage supplies. Excitation of the atoms or molecules in the gas is normally achieved by setting up an electrical discharge, although other methods of excitation can be used (for example, adiabatic expansion, exploding wires or even nuclear explosions). Commercial gas lasers use discharge excitation and our discussion will be limited to these.

7.2.1 Excitation mechanisms in ionized gases

An ionized gas or plasma is produced when an electric field, high enough to break down the gas and so create free electrons and positive ions, is applied and the gas becomes conductive. An ordinary fluorescent lamp is an everyday example. The electric field can be applied to the gas as a DC field or as pulse or high-frequency radiofrequency field. Once the gas becomes conductive the current is carried by the free electrons and the positive ions. We can in fact consider the ionized gas or plasma to consist of three interacting fluids namely; the electrons, the positive ions and the neutral atoms or molecules. The densities of these three species will be defined by n_e, n_i, and n, respectively, as follows:

Electron density n_e = number of electrons per cubic metre
Positive ion density n_i = number of positive ions per cubic metre
Neutral density n = number of neutral particles per cubic metre

Assuming that the neutral particles in the plasma obey the ideal gas law and have a Maxwellian energy distribution we can ascribe the gas with a temperature, T_g. The energy distribution of the charged particles, particularly the electrons, is a more complicated issue and requires the solution of the Boltzmann collision equation. To avoid this complexity it is normally assumed that the electrons and the ions have a Maxwellian energy distribution and thus can be assigned temperatures T_e and T_i. The determination of these temperatures and densities is a topic which has absorbed the energies of plasma physicists for many years, particularly those with an interest in nuclear fusion. In a plasma the overall system must be electrically neutral even though this may not be the case on the microscopic scale. Therefore we can write

$$n_e \approx n_i$$

On the microscopic scale the random motion and collisions of the particles will cause the charges to separate microscopically. It can be shown [Ref. 1] that the average charge 'separation' distance, known as the *Debye length*, is given by

$$\lambda_D = \sqrt{\frac{\varepsilon_0 k T_e}{n_e e^2}} \tag{7.3}$$

In addition to charge neutrality we can now define a plasma by:

$$\lambda_D \ll \text{diameter of the discharge tube}$$

The excitation and ionization of the gas and the subsequent emission of light is primarily a result of collisions between the electrons and the other particles. The electrons will gain energy from the applied electric field and lose energy by collisions. There are two main types of collision processes:

1. *Elastic collisions* in which there is no change in potential energy of the collision partners and total kinetic energy is conserved.
2. *Inelastic collisions* in which the potential energy of the system changes. It is these collisions which excite the atoms of the gas and a few examples pertinent to laser excitation will now be considered.

1. Excitation
The atom or molecule is raised to an excited state, the energy being provided by the kinetic energy of the electron. This is illustrated by a chemical equation in which A denotes an atom and the asterisk an excited atom:

$$e + \mathrm{A} \rightarrow \mathrm{A}^* + e$$

In most cases this excited atom will decay by the spontaneous emission of a photon or, if in a laser, by stimulated emission. However, in some cases the upper state lifetime is so long that the transition will not take place before another collision, or before the excited atom impacts with the wall of the tube. These excited states are referred to as *metastable.*

2. Ionization
In this process the positive ions and free electrons are produced which keep the gas conductive.

$$e + \mathrm{A} \rightarrow \mathrm{A}^+ + e + e$$
$$e + \mathrm{A} \rightarrow \mathrm{A}^{++} + 2e + e$$

The positive ions have excited states since their orbiting electrons can be excited in just the same way as neutral atoms. Their excitation can occur in the plasma in one collision or by a second electron collision with a species already excited. These are illustrated by the following reactions:

$$e + \mathrm{A} \rightarrow \mathrm{A}^{+*} + e + e$$
$$e + \mathrm{A}^+ \rightarrow \mathrm{A}^{+*} + e$$
$$e + \mathrm{A}^{\mathrm{m}} \rightarrow \mathrm{A}^{+*} + e + e$$

where m denotes a metastable state.

3. Dissociation
The above processes will also occur for molecules with the further complication that the molecule can fragment or dissociate after a collision. This reaction is written:

$$e + \mathrm{AB} \rightarrow \mathrm{A} + \mathrm{B} + e$$

The situation in a gas discharge is very complex since all the above processes (and many others) will be going on simultaneously, each with a different probability. It is therefore useful to define a parameter which is the probability that a particular process will occur as a result of a collision. The collision cross-section for a particular process is defined as *the number of collision events of the particular type per unit path length per atom per particle flux*. The cross-section has the units of area (m^2) and is a function of kinetic energy of the collision partners. The determination of its function of energy both theoretically and experimentally is an important branch of atomic and molecular physics, and many of the collision processes responsible for the population inversion in gas lasers have been extensively studied.

With this definition we can write the number of electron excitation events which create a particular state of the atom per unit volume per second as:

$$n_e v \sigma_x(E) n$$

where v is the electron velocity and $\sigma_x(E)$ the excitation collision cross-section of the state.

To calculate the excitation rate by electron collisions in a plasma we have to take into consideration the fact that the electrons are not mono-energetic and, as noted above, may be represented by a Maxwellian energy distribution function. The fractional number of electrons with kinetic energy in the differential interval E to $E + \mathrm{d}E$ is given by [Ref. 1]

$$\frac{\mathrm{d}n_e}{n_e} = F(E)\,\mathrm{d}E = \frac{2}{kT_e}\sqrt{\frac{E}{\pi k T_e}}\exp\left(-\frac{E}{kT_e}\right)\mathrm{d}E \tag{7.4}$$

The rate of excitation per m^3 by this energy group is given by

$$\mathrm{d}n_e(v\sigma_x(E)n) = n_e F(E) v \sigma(E) n\,\mathrm{d}E \tag{7.5}$$

Using

$$E = \frac{1}{2} m v^2$$

and integrating over the whole energy distribution remembering that the cross-section is zero below the excitation energy of the state, E_x, gives the total rate of excitation of the state per unit volume:

$$R = n_e n \int_{E_x}^{\infty} F(E)\sqrt{\frac{2E}{m}}\,\sigma_x(E)\,\mathrm{d}E \tag{7.6}$$

If the energy distribution function of Eq. (7.4) is used then Eq. (7.6) is normally written in a shorthand notation as

$$R = n_e n \langle \sigma(E) v \rangle \tag{7.7}$$

where the angled brackets indicate the average of the cross-section over a Maxwellian velocity distribution.

Figure 7.2 graphically illustrates this integration. Here we have plotted the electron energy distribution corresponding to an electron temperature of 10 ev along with the electron impact excitation cross-section as a function of energy for two different types of states. The sharply peaked function is characteristic of a forbidden transition from the ground state while the broad function is for an allowed transition. Equation (7.7) is seen to be the overlap of the electron energy distribution with the collision cross-section.

4. *Excitation transfer*[1]

Although electron collisions are the fundamental mechanism of transfer of energy from the applied field to the excited atomic states another important method of

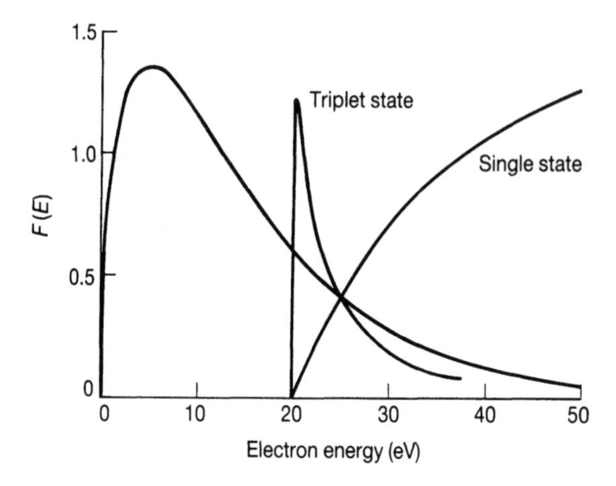

Figure 7.2 Electron energy distribution function (electron temperature 10 eV) along with a triplet and a singlet excitation function (variation of collision cross-section with energy).

energy transfer between the excited atoms exists. This mechanism can preferentially excite particular states and hence produce population inversion. It often happens in nature that two atoms or molecules can have states which have almost the same energy above the ground state – an accidental coincidence in energy. A collision between these atoms can result in energy transfer from one to the other:

$$A + B^* \rightarrow A^* + B \pm \Delta E \tag{7.8}$$

There will always be some energy defect ΔE and it can be positive (endothermic) or negative (exothermic). In either case it will appear as a kinetic energy change of the collision partners. The reaction can also proceed in the reverse direction and the rates, with atomic concentrations, being determined by the principle of detailed balance. In this the reaction rates are determined by the concentration of the active species and the rate of forward reaction equals the rate in the reverse direction. In processes of interest for laser excitation the atom B^* will normally be metastable and A^* will be an upper laser state and decay rapidly by spontaneous or stimulated emission. Thus the favoured direction of the reaction is as shown above. The cross-section for these processes is known to increase when

1. The magnitude of ΔE decreases
2. Decreasing relative velocity of the atoms which is dependent on the gas temperature and atomic masses
3. Spin is conserved in the collision – Wigner spin rule [Ref. 8].

5. Penning ionization
This is the excitation transfer process where an excited neutral atom collides with, and simultaneously ionizes, another neutral atom:

$$A^* + B \rightarrow A + B^{+*} + e \pm \Delta E \tag{7.9}$$

6. Charge transfer
The collision of an ion with a neutral atom can excite an upper state of the ion

$$A^+ + B \rightarrow A + B^{+*} \pm \Delta E \tag{7.10}$$

7. Trapping of resonance radiation
Radiation emitted by atoms near the centre of the plasma is likely to suffer reabsorption. This will re-excite the upper state and so effectively increase its lifetime. This occurs most frequently when the lower state of the transition is the atomic ground state, and in this case the radiation associated with the transition to ground state is known as *resonance radiation*. The trapping of this radiation is most effective at high pressure and in large-radius tubes and plays an important role in the population inversion mechanisms.

7.2.2 Characteristics of electrical discharges

Population inversion in gases is normally achieved by an electrical discharge through the gas contained in a narrow-bore tube. In this section we will summarize some of the important features of the electrical discharge. Figure 7.3 is a typical circuit which could be used to pass a DC current through a discharge used for a CW gas laser. In some cases the cathode is heated and supplies electrons by thermionic emission while in others it is cold. The distinction between the two is illustrated if we plot the variation of the voltage between cathode and anode as shown in Figure 7.4. We see from this figure that the cold cathode has a very much larger voltage drop at the cathode than does the hot cathode. The mechanisms at the cathode are complex but it is sufficient to state that this large voltage drop at the cathode (cathode fall) accelerates the positive ions to produce electrons by positive ion bombardment. The

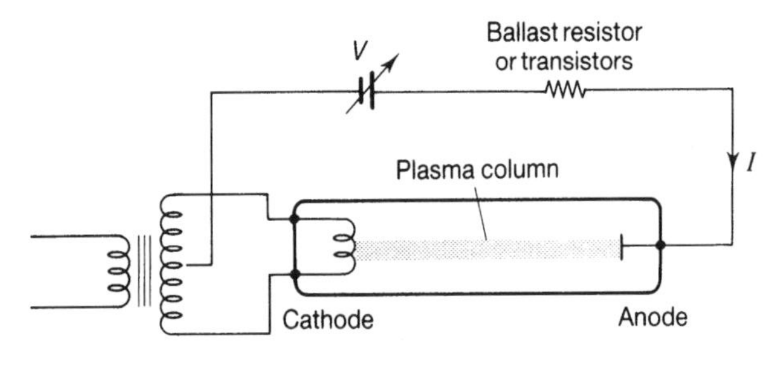

Figure 7.3 DC circuit for a discharge tube.

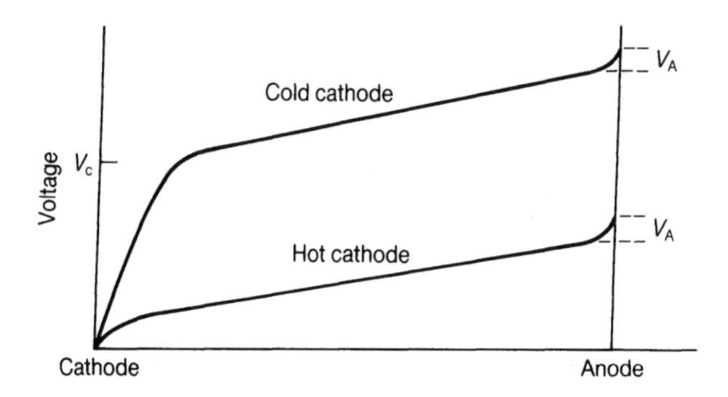

Figure 7.4 Variation of voltage between cathode and anode for a hot and a cold cathode discharge.

supply of electrons at the cathode must balance those lost at the anode and so maintain current continuity in the circuit. The hot cathode, on the other hand, provides a copious supply of electrons and does not require such a large voltage drop.

Both hot and cold cathode discharges have negative current voltage characteristics giving the circuit a negative dynamic impedance and the series-positive resistor is necessary to prevent relaxation oscillations of the circuit. Often a number of transistors are placed in series to provide amplitude stabilization of the discharge current and thus the gain. The cold cathode is usually made of a hollow cylinder of aluminium while the hot cathode can be a heated tungsten filament or indirectly heated metal oxide such as barium oxide. The hot cathode must not come into contact with any oxidizing gas and requires a heater circuit. The cold cathode, on the other hand, is restricted to low discharge currents otherwise sputtering (removal of metal atoms by ion bombardment) occurs. These sputtered atoms can migrate to optical surfaces and cause absorption of the laser light as well as burying the gas atoms and decreasing the pressure in the plasma tube. A small voltage drop at the anode exists which accelerates electrons, resulting in heat and, as a consequence, with high current discharges anodes must be cooled.

The excitation and the population inversion takes place in the region designated as the *positive* (or *plasma*) column. From Figure 7.4 we note the field is uniform there, that is, dV/dx is constant. Thus from Gauss's law the positive column is neutral, thus satisfying one of our definitions of a plasma discussed in Section 7.2.1. The positive column is not a necessary part of the discharge but serves as a conducting region between the cathode and anode regions of the discharge. In the plasma electrons are gaining energy from the field and losing energy by collisions. Electrons are being produced in these collisions by ionization but, at the same time, are diffusing to the walls where they are lost by recombination with the positive ions. A theory for the plasma, known as the *Schottky theory*, equates the rate of ionization, evaluated using the integral of Eq. (7.7) with the excitation cross-section replaced by the ionization cross-section, to the diffusive loss obtained from known measured diffusion coefficients [Ref. 1]. The assumption made in this theory is that the ion mean-free path is less than the tube radius. The results of this theory are presented in Figure 7.5, which shows the electron temperature to be a decreasing function of the product of *pressure* and *tube diameter*. Without recourse to the complexity of the theory we can explain this result as follows. Increasing the pressure or the tube diameter will result in a decrease in the rate of diffusion of electrons to the tube walls and so a lower ionization rate and hence a lower electron temperature is required to sustain the plasma; additionally, at higher pressures the ionization rate is higher. The discharge current has no effect on the electron temperature but changes the electron density in direct proportion. A further result of this diffusion theory is that the electron density varies across the tube diameter as a zero-order Bessel function, resulting in a radial gain variation which follows the same function. In a low-pressure discharge through narrow-bore tubes the ion mean-free path can be larger than the bore of the tube and the electrons can fall freely without collisions to the tube walls. This is known as the *Tonks–Langmuir regime* and although the magnitudes are different the electron temperature is again found to be a decreasing function of the product *pressure* × *diameter*.

7.2.3 The helium–neon lasers

The first report of lasing in a gas was that of Javan *et al.* [Ref. 2] who excited a mixture of helium and neon using a radiofrequency discharge and obtained continuous wave (CW) oscillation at 1.1523 μm. After this development a period of intense activity from 1961 to 1965 resulted in the discovery of thousands of lasing

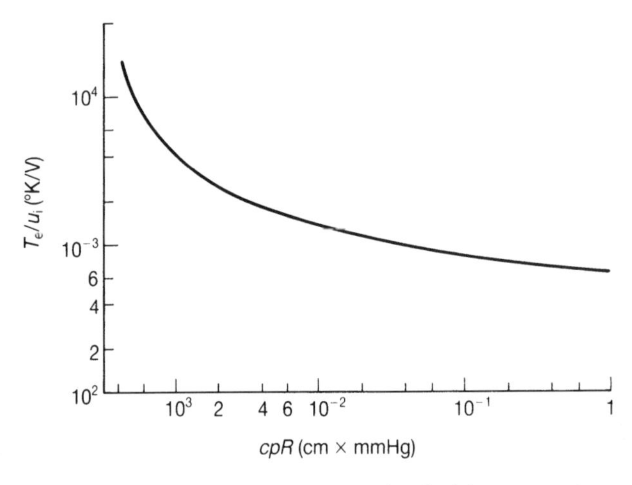

Figure 7.5 Electron temperature divided by ionization potential U_i as a function of pD product from the Schottky diffusion theory. The value of c is determined by the gas. For helium it is 0.0039 and for neon 0.0059.

transitions in both molecular and atomic gases. The helium–neon laser system is one of a group in which population inversion can be produced between excited states of the neutral atom, in this case neon. Bennett [Ref. 3] lists over 180 transitions of neon in which laser action has been observed. Of course, the majority of these have little commercial interest either because they are too weak or are not a useful wavelength. The energy-level system for these first laser lines and the subsequent ones which have found commercial interest is shown in Figure 7.6. The energy coincidence between two metastable states of helium and two states of neon is shown in Figure 7.7. In a glow discharge in the gas mixture electrons excite the metastable helium states 2^1S and 2^3S which, under the optimum discharge conditions, can make collisions with neon atoms in the ground state and transfer excitation to the $3s_2$ and $2s_2$ states of neon.[2] The excitation transfer processes can be written thus:

$$\text{He}(2\,^1\text{S})\uparrow\downarrow + \text{Ne}(1\,^1\text{S})\uparrow\downarrow \rightarrow \text{He}(1\,^1\text{S})\uparrow\downarrow + \text{Ne}(3\text{s}_2)\uparrow\downarrow - 47\ \text{meV} \tag{7.11}$$

$$\text{He}(2\,^3\text{S})\uparrow\uparrow + \text{Ne}(1\,^1\text{S})\uparrow\downarrow \rightarrow \text{He}(1\,^1\text{S})\uparrow\downarrow + \text{Ne}(2\text{s}_2)\uparrow\downarrow + 39\ \text{meV} \tag{7.12}$$

The vertical arrows indicate the resultant total spin of the atom, i.e. two parallel arrows indicate a triplet state and two anti-parallel arrows a singlet state. In a collision designated by Eq. (7.11) the system prefers to take the largest energy jump and is endothermic but conserves spin. In the collision of Eq. (7.12) the smallest energy defect is preferred, the reaction is exothermic but here spin is not conserved. Thus since the $3s_2$ and the $2s_2$ states are preferentially excited (or pumped) they can, under the right circumstances, achieve more population than states below them. These are therefore the upper laser states, while the lower laser states are those to which these upper states make strong transitions (high A coefficients). Some of the important transitions are shown in Figure 7.6 while Table 7.1 lists the transition data.

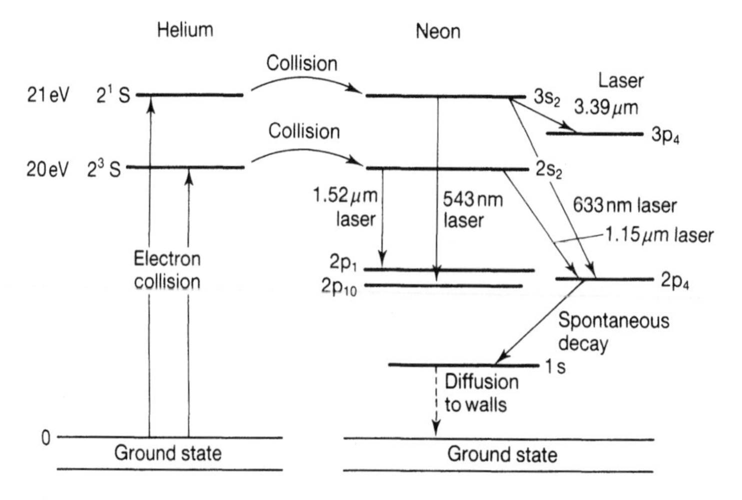

Figure 7.6 Energy level system of neon transitions of the helium–neon laser.

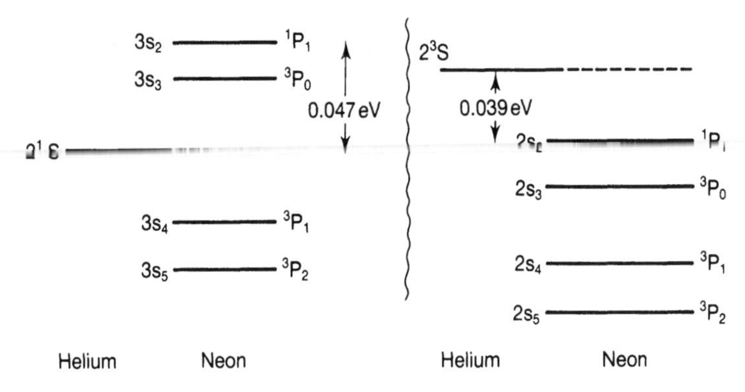

Figure 7.7 Energy level coincidence of the metastable states of helium and the $3s_2$ and $2s_2$ states of neon. Paschen and Russell–Saunders notations for the neon states are used.

Table 7.1 Data for the neon laser transitions of the helium–neon system

Transition	Wavelength (nm)	$A_{21}(10^6\ s^{-1})$	$\sigma_{SE}(10^{-22}\ m^2)$
$3s_2 \rightarrow 2p_1$	730.5	0.48	7.3
$2p_2$	640.1	0.6	6.14
$2p_3$	635.2	0.7	7.0
$2p_4$	632.8	6.56	65.0
$2p_5$	629.4	1.35	13.0
$2p_6$	611.8	1.28	11.4
$2p_7$	604.7	0.68	6.96
$2p_8$	593.9	0.56	4.58
$2p_{10}$	543.4	0.59	3.6
$3s_2 \rightarrow 3p_4$	3391.3	2.87	4400
$2s_2 \rightarrow 2p_1$	1523.1	0.802	110
$2p_2$	1176.7	4.089	259
$2p_3$	1160.2	0.801	48
$2p_4$	1152.3	6.537	388
$2p_5$	1140.9	2.301	133
$2p_6$	1084.4	7.543	372
$2p_7$	1062.1	0.816	38
$2p_8$	1029.5	0.726	31
$2p_{10}$	889.5	1.708	35

The stimulated emission cross-section is evaluated using a gas temperature of 400 K to determine the Doppler line shape function value at line centre

The lower states for the lasing transitions are the 2p and 3p groups of states and these are excited by electron collisions with both neon ground state atoms and neon atoms in the metastable 1s states. In addition, the lower states are excited by cascade from higher excited neon states. In particular, the 2p states can be populated by transitions from states excited by excitation transfer from the helium metastable atoms. The radiative lifetimes of the $3s_2$ and $2s_2$ levels are considerably lengthened due to radiation trapping of the strong vacuum ultra-violet transitions to the ground state while the transitions to the ground state from the 3p and 2p states are forbidden and are depopulated by the fast decay to the metastable 1s manifold. These 1s atoms are de-excited by diffusion and subsequent collisions with the tube walls. These mechanisms of excitation and decay are summarized in Figure 7.8, and it is possible from a knowledge of the cross-sections, spontaneous emission coefficients and metastable diffusion coefficients to set up rate equations which allow the population inversion to be calculated. However, by using some general arguments about the mechanisms it is possible to explain the experimentally observed variation of the gain with discharge parameters, and useful scaling laws result which can be used for laser design.

The densities of the helium and neon metastable atoms in the discharge will determine the optimum operating conditions and the ultimate gain. Let us consider the effects of pressure and tube diameter on these atom densities. Most experimental studies have been performed on the laser transitions at 3.39 μm, 633 nm, which

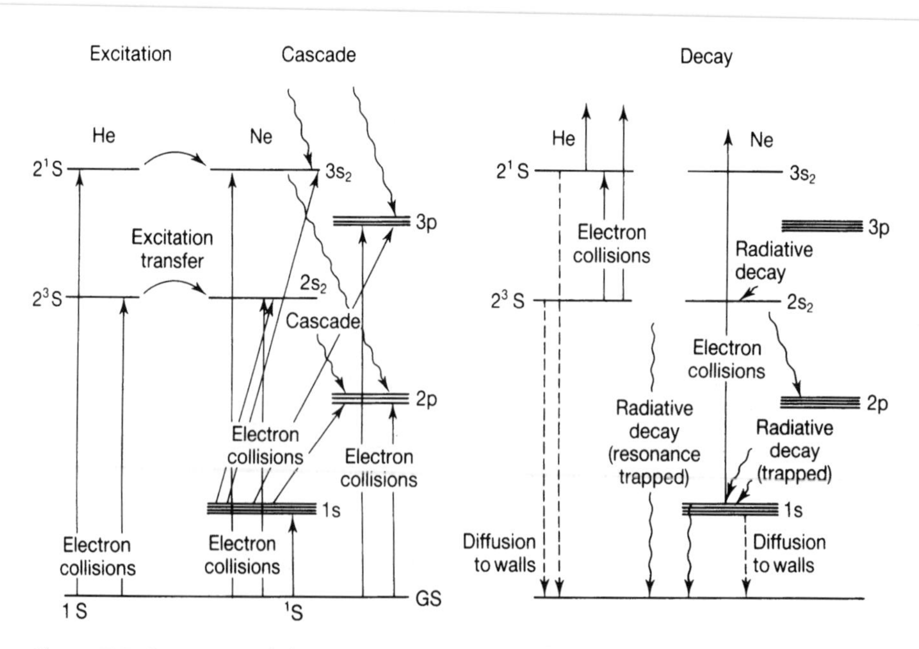

Figure 7.8 Summary of the excitation and decay mechanisms in helium–neon laser discharges.

share the same upper state, 1.15 μm and more recently [Ref. 6] on the very weak 543 nm transition. For the 633 nm line the largest gain is obtained when the pressure × diameter product is 3.6 torr mm. This gives, from Figure 7.5, an electron temperature of about 100 000 K (10 ev). Increasing tube diameter will decrease the rate of diffusion to the walls for the metastable atoms but, at the same time, a reduction in pressure will be required to keep the electron temperature at 100 000 K. This reduction in pressure causes reduced gain simply because there are fewer atoms available for excitation. The net effect is that the gain decreases with diameter and we have a scaling law which is often used in gas lasers, namely that *gain is inversely proportional to tube diameter*. A further complication is the dominance of the 3.39 μm transition. Its stimulated emission cross-section is almost 70 times that of the 633 nm transition. In long-gain sections (> 50 cm) its amplified spontaneous emission will deplete the $3s_2$ population and so reduces the gain for all the $3s_2 \rightarrow 2p_n$ transitions. This problem can be overcome in three ways:

1. Ensure that the mirror reflectivities are appropriate to the lasing wavelength required and use a Littrow prism with a coating (see Figure 7.11(c)).
2. Use some intra-cavity absorber for 3.39 μm such as a methane cell.
3. Apply an inhomogeneous magnetic field by placing small bar magnets along the length of the tube. This causes a larger broadening as a fraction of the wavelength for the infrared than for the visible transition.

The electron density is directly proportional to the discharge current density and so one might expect the gain to increase in direct proportion to the current. This most certainly is not so for these lasers, which are limited to being low-power devices. The explanation is that increasing the current in the discharge increases the density of 1s neon metastables, and this in turn saturates the population of the lower laser states by reabsorption of the radiation from 2p → 1s transitions and by electron excitation. In addition, the helium atoms in the metastable states, which are so effective in exciting the upper laser states of neon, will be destroyed by electron impact by exciting higher helium states, ionization or de-excitation to the ground state. This increases with electron density and so current.

A simple analysis gives an expression for gain variation with current and has been applied with success to the 633 nm and 3.3 μm transitions. For these we assume that the excitation of the $3s_2$ state, the upper laser state, is due only to the excitation transfer from the helium metastable 2^1S atoms and so is proportional to the density of the latter. The excitation of helium atoms is due to electron collisions and so is proportional to current, while their de-excitation is caused by excitation transfer to the neon, diffusion to the walls and electron collisions. The diffusion and excitation transfer is independent of current so the rate equation for the metastable helium atom density, M, can be written:

$$\frac{\mathrm{d}M}{\mathrm{d}t} = K_1 I - M(K_2 + K_3 I) \qquad (7.13)$$

where K_1, K_2 and K_3 are appropriate constants, and I the discharge current density.

At equilibrium the time derivative is zero and the metastable density as a function of current density is given by

$$M = \frac{K_1 I}{K_2 + K_3 I} \tag{7.14}$$

The rate equation for the density of $3s_2$ state neon atoms, N_2, is

$$\frac{dN_2}{dt} = MN_0K_4 - \frac{N_2}{\tau_2} \tag{7.15}$$

where N_0 is the density of neon atoms in the ground state and K_4 is a constant. At equilibrium we have for N_2

$$N_2 = \tau_2 N_0 K_4 M \tag{7.16}$$

Using Eq. (7.14) for M and lumping the constants together gives for N_2 into K_5

$$N_2 = \frac{K_5 I}{K_2 + K_3 I} \tag{7.17}$$

The density of the lower laser state, N_1, can be described by using the rate equation

$$\frac{dN_1}{dt} = K_6 I - \frac{N_1}{\tau_1} \tag{7.18}$$

At equilibrium we get

$$N_1 = K_7 I \tag{7.19}$$

Now the gain coefficient is proportional to the population difference $N_2 - N_1$, so introducing further constants, A, B and C. We can write the form of the gain with current density as:

$$\gamma = \frac{AI}{1 + BI} - CI \tag{7.20}$$

This simplified model therefore shows that maximum gain occurs at a discharge current which depends on the transition involved and the tube bore diameter. Experimental evidence for this simplified description has been provided for the 3.39 μm transition by White and Gordon [Ref. 4], for the 633 nm line by Spoor [Ref. 5] and for the 543 nm line by Gray [Ref. 6]. Data for the 1.15 μm line are not available.

Further evidence for the decrease in gain as the electron density (and hence current density) is increased comes from the radial variation of the gain. It has been noted in Section 7.2.2 that the radial variation of electron density follows a zero-order Bessel function and so we would expect the gain similarly to follow this functional form. This is broadly true, however at the centre of the tube the electron density can go beyond optimum as determined by Eq. (7.20) while it is less than optimum nearer to the tube wall. This is especially noticeable at higher pressures when the electron density is higher for a given current. Since the gain follows the electron density it is found that at higher pressures the gain can experience a dip in the centre of the tube. Measurements of the radial variation of the gain presented by Spoor and Latimer [Ref. 7] given in Figure 7.9 clearly show this effect. One consequence of this higher gain towards the tube walls is to enhance the higher order transverse modes shown in Figure 4.21. Optimum conditions for some neon transitions are shown in Table 7.2.

By far the largest number of helium–neon lasers manufactured are short (~ 12 cm) tubes operating at 633 nm. A schematic diagram of a typical commercial tube is shown in Figure 7.10. The mirrors are sealed to the end of the tube by a high-temperature glass-to-metal seal and are aligned before sealing. The tube is connected to a clean high-vacuum system where it is pumped and baked to remove impurities. After filling with the desired gas mixture the tube is sealed off from the vacuum system. Tubes manufactured in this way have lifetimes in excess of 10 000 hours. These tubes normally emit about 1 mW of CW optical power in the TEM_{00} mode with a spot size of about 0.3 mm. Their gain bandwidth is such that two longitudinal modes equally spaced about line centre will normally oscillate with orthogonal polarization, thus reducing gain competition between them. If the cavity length changes due to a temperature change during warm-up then the modes will move through the gain curve and the laser can change from two- to one-mode operation with the single mode near to line centre (see Section 6.6.4). With single-mode

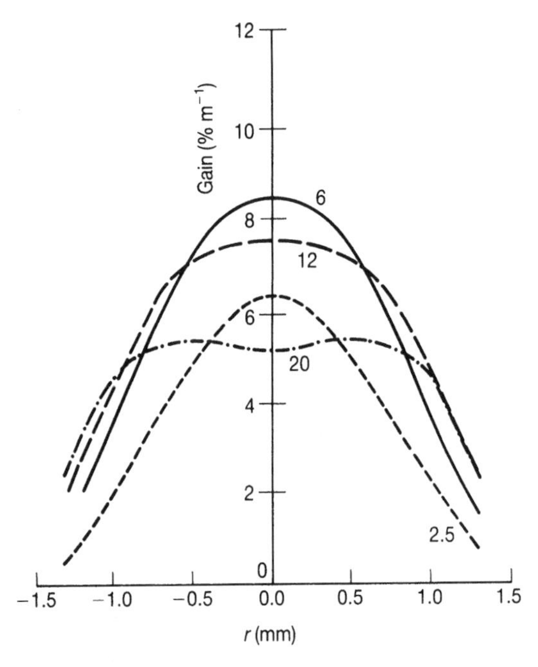

Figure 7.9 The radial distribution of gain at 633 nm with 0.2 torr Ne and 3.3 torr ^{3}He. Current ------ 2.5, ——— 6.0, – – – – 12.0, – · – · – · 20.0 mA (Data taken from Ref. 7).

Table 7.2 Optimum gain conditions for some neon laser transitions in the helium–neon laser

	633 nm	543 nm	1.15 μm
Gain (%/m) $D = 1$ mm	36	0.9	
Press. × Diam. Torr × mm	3.6–4.0	1.6	14–17
T_e (ev)	10–11		6
n_e (m^{-3})	10^{17}		10^{17}
Current $D = 1$ mm	3–5 mA	6 mA	~6
He/Ne ratio	5:1	5:1	10:1

The gain is increased by about 15 per cent using He^3 in place of He^4 due to the increased collisional rate with the lighter isotope. Parameters for the visible transitions are given with the 3.39 μm transition suppressed.

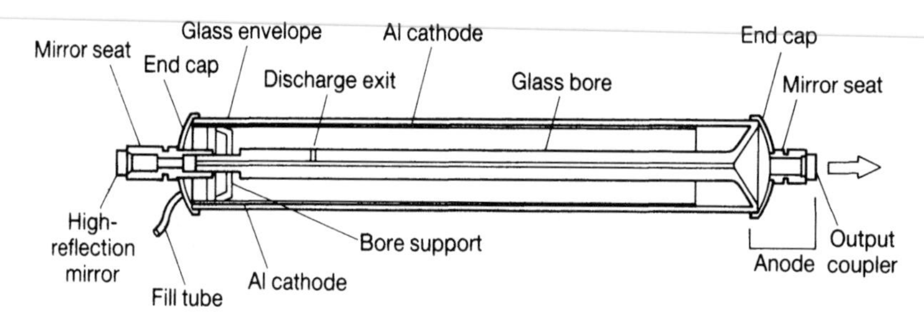

Figure 7.10 Schematic diagram of a commercial internal mirror helium–neon laser tube. (Diagram reproduced by kind permission of Penwell Publishing.)

operation the light output is plane polarized. To produce light which is plane polarized in both modes a flat of fused silica is placed internally at the Brewster angle during manufacture. In recent years improvement in mirror technology by ion sputtering deposition techniques as described in Section 5.4 has allowed the development of helium–neon lasers operating on the very low-gain $3s_2 \rightarrow 2p_{10}$ transition at 543 nm. This green wavelength offers an alternative to the 633 nm red line in applications such as alignment, interferometry and as a beam pointer for infrared lasers used in medicine. The gain at 543 nm is about 30 times smaller than at 633 nm so very low-loss reflectors are required which also have low reflectivity at 633 nm and

3.39 μm. In spite of this, tubes are available which are a little longer than the 633 nm tubes and which emit 1 mW output power.

Although 633 nm lasers with internal mirrors and outputs of up to 20 mW can be produced commercially it is often necessary to use an external cavity and a tube sealed with windows at the Brewster angle giving output at the low-loss *p*-polarization. A schematic diagram of such a laser is shown in Figure 7.11 in which a Littrow[3] prism is used to tune between the wavelengths of the $3s_2 \rightarrow 2p_n$ transitions. If oscillation at 633 nm only is required then the Littrow prism is replaced by a high reflecting plane mirror since only the 3.39 μm transition requires suppressing. Such lasers are capable of up to 75 mW in the TEM_{00} mode at 633 nm. Although

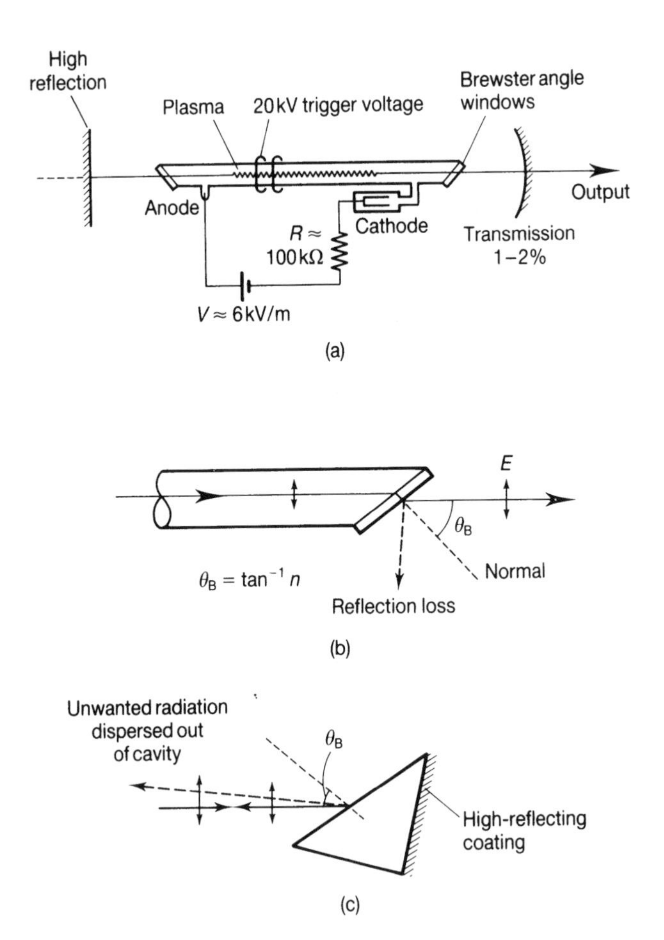

Figure 7.11 Schematic diagram for a high-power tunable visible helium–neon laser. The Littrow prism is used to tune the different wavelengths of the $3s_2 \rightarrow 2p$ transitions. For operation at 633 nm only the Littrow prism can be replaced by a plane mirror with high reflectivity at 633 nm.

the discussion has centred on the visible transitions, commercial lasers are available operating on the 1.15 μm and 1.52 μm transitions. The former normally can be made to lase by changing mirror reflectivity from 633 nm to 1.15 μm even when the tube fill pressure is not optimized for the infrared transition. Short internal mirror tubes are used for the 1.5 μm transition, which is of interest since fused silica optical fiber used for communications shows minimum attenuation at this wavelength.

7.2.4 Noble gas ion lasers

In 1964 laser action in excited states of the argon ion (i.e. in the ArII spectrum) was discovered simultaneously by two groups, one in the USA and the other in France [Ref. 8]. To excite states of the ion, high-discharge current densities, which are a few hundred times that used in the helium–neon laser, are required. Electron densities in such discharges are in the region of 10^{19} m^{-3} and were achieved initially in pulsed discharges in argon with continuous-wave oscillation being achieved very soon afterwards. The other noble gas ions were soon made to lase and later lasing in higher ionized species of these ions was observed producing wavelengths from the infrared to the ultraviolet. Ion lasers can produce very high CW power and this along with the range of wavelengths available through the visible and ultraviolet have made these lasers an extremely useful industrial and research tool.

The excitation of the laser states is by electron collisions in the plasma and is similar for all the noble gases. We can confine our discussion to argon with a simple modification to the other gases by altering the principal quantum number of the outer electrons.[4] The electronic structure of the outer electrons of the ground state of atomic argon is $3s^23p^6$ ($^1S_0$) and, when excited, one of the six 3p electrons is promoted to a higher orbit. The ground state of the ion Ar^+ is therefore $3s^23p^5(^2P)$ while the ground state of the ion Ar^{++} is $3s^2\,3p^4(^3P)$. The excited states of the singly charged ion are obtained by adding an electron (known as the *running electron*) in an excited orbital to the ground state of the doubly charged ion. Thus we are describing the coupling of a running electron to a core which in this case is the ^3P core. We can also add the running electron to an excited state core and for Ar^{++} these are ^1D and ^1S states, which, we note, are the first two excited states of the atom having atomic number 2 less than argon in the periodic table, oxygen. The strongest laser transitions occur from states associated with the ^3P core and by far the most important are transitions in the 4p → 4s manifold of states. We can use the rules of $L-S$ coupling to determine the state designations as follows:

^3P core has total spin = 1 and total angular momentum $L = 1$
Adding an electron with spin $\pm\frac{1}{2}$ gives total spin $= \frac{3}{2}$ or $\frac{1}{2}$
That is, the resulting states are either quartets or doublets (multiplicity $= 2S + 1$).
Adding a 4p electron with $l = 1$ to the core will give D, P, or S states.
The doublet states are: $^2D_{5/2,\,3/2}$; $^2P_{3/2,\,1/2}$; $^2S_{1/2}$
The quartet states are: $^4D_{7/2,\,5/2,\,3/2,\,1/2}$; $^4P_{5/2,\,3/2,\,1/2}$; $^4S_{3/2}$

Similarly, for the 4s running electron with the ^3P core the states are:
$^2P_{3/2,\,1/2}$ and $^4P_{5/2,\,3/2,\,1/2}$
The preferred selection rules for $L-S$ coupling are:
Change in core angular momentum = 0; change in total spin = 0
$\Delta J = \Delta L = +1$ (Δ = *upper* − *lower*).

These preferred selection rules give the strongest lines in spontaneous emission which usually implies strong laser emission. The selection rules $\Delta J = 0, +1$ with $J = 0 \rightarrow J = 0$ forbidden are obeyed for all laser lines while some strong transitions violate the spin conservation rule, namely the strong Ar^+ laser line at 514 nm which is a quartet-to-doublet transition.

Figure 7.12 is an energy-level diagram of Ar^+ showing the important blue/green 4p → 4s lasing transitions, while Table 7.3 lists a selection of strong CW ion laser lines observed in the noble gases taken from the comprehensive list of wavelengths and transition assignments of Davis and King [Ref. 9]. Excitation is through electron impact with ground state and metastable atoms and ions in the discharge as well as by recombination from higher ionized species and cascade processes from higher excited states. The possible processes are numerous but they can, in general, be considered as single-step, double-step, triple-step, etc. mechanisms. These can be shown to be proportional to current, current squared and current cubed,

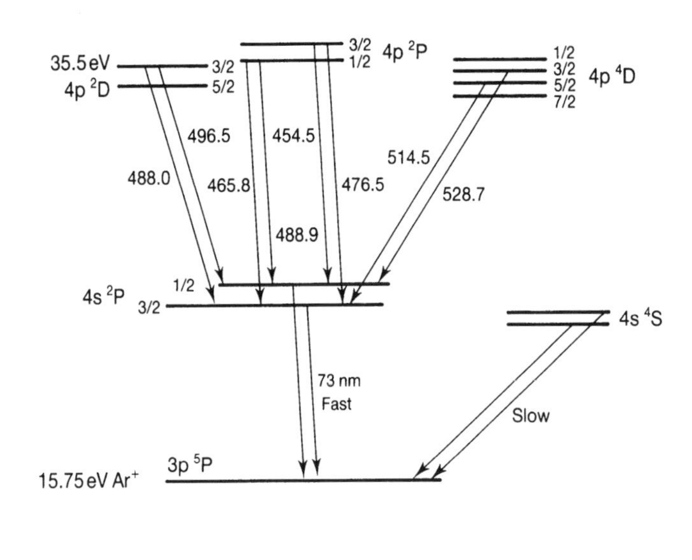

Figure 7.12 Energy-level diagram of Ar^+ showing the laser lines associated with the 4p → 4s transitions in nm.

Table 7.3 Some strong ion laser transitions

λ (nm) In air	Transition assignment		Ion species
332.4	$(^3P)3p_2P \rightarrow$	$(^3P)3s^2P$	Ne^+
337.8	$(^3P)3p^2P_{1/2} \rightarrow$	$(^3P)3s^2P$	Ne^+
339.3	$(^3P)3p^2P \rightarrow$	$(^3P)3s^2P$	Ne^+
371.3	$(^3P)3p^2D_{5/2} \rightarrow$	$(^3P)3s^2P_{3/2}$	Ne^+
291.3	$(^3P)4p^2D_{5/2} \rightarrow$	$(^3P)4s^2P_{3/2}$	Ar^{+++}
292.6	$(^3P)4p^2D_{3/2} \rightarrow$	$(^3P)3s^2P_{1/2}$	Ar^{+++}
333.6	$(^2D)4p^3F_4 \rightarrow$	$(^2D)4s^3D_3$	Ar^{++}
334.5	$(^2D)4p^3F_3 \rightarrow$	$(^2D)4s^3P_2$	Ar^{++}
335.9	$(^2D)4p^3F_2 \rightarrow$	$(^2D)4s^3D_1$	Ar^{++}
351.1	$(^4S)4p^3P_2 \rightarrow$	$(^4S)4s^3S_1$	Ar^{++}
363.8	$(^2D)4p^1F_3 \rightarrow$	$(^2D)4s^1D_2$	Ar^{++}
454.5	$(^3P)4p^2P_{3/2} \rightarrow$	$(^3P)4s^2P_{3/2}$	Ar^+
457.9	$(^3P)4p^2S_{1/2} \rightarrow$	$(^3P)4s^2P_{1/2}$	Ar^+
465.8	$(^3P)4p^2P_{1/2} \rightarrow$	$(^3P)4s^2P_{3/2}$	Ar^+
472.7	$(^3P)4p^2D_{3/2} \rightarrow$	$(^3P)4s^2P_{3/2}$	Ar^+
476.5	$(^3P)4p^2P_{3/2} \rightarrow$	$(^3P)4s^2P_{1/2}$	Ar^+
488.0	$(^3P)4p^2D_{5/2} \rightarrow$	$(^3P)4s^2P_{3/2}$	Ar^+
496.5	$(^3P)4p^2D_{3/2} \rightarrow$	$(^3P)4s^2P_{1/2}$	Ar^+
501.7	$(^1D)4p^2F_{5/2} \rightarrow$	$(^1D)3d^2D_{3/2}$	Ar^+
514.5	$(^3P)4p^4D_{5/2} \rightarrow$	$(^3P)4s^2P_{3/2}$	Ar^+
528.7	$(^3P)4p^2D_{3/2} \rightarrow$	$(^3P)4s^2P_{1/2}$	Ar^+
350.7	$(^4S)5p^3P_2 \rightarrow$	$(^4S)5s^3S_1$	Kr^{++}
356.4	$(^4S)5p^3P_1 \rightarrow$	$(^3P)3s^3S_1$	Kr^{++}
406.7	$(^2D)5p^1F_3 \rightarrow$	$(^2D)5s^1D_2$	Kr^{++}
413.1	$(^4S)5p^5P_2 \rightarrow$	$(^4S)5s^3S_1$	Kr^{++}
476.2	$(^3P)5p^2D_{3/2} \rightarrow$	$(^3P)5s^2P_{1/2}$	Kr^+
482.5	$(^3P)5p^4S_{3/2} \rightarrow$	$(^3P)5s^2P_{1/2}$	Kr^+
520.8	$(^3P)5p^4P_{3/2} \rightarrow$	$(^3P)5s^4P_{3/2}$	Kr^+
530.9	$(^3P)5p^4P_{5/2} \rightarrow$	$(^3P)5s^4P_{3/2}$	Kr^+
568.2	$(^3P)5p^4D_{5/2} \rightarrow$	$(^3P)5s^2P_{3/2}$	Kr^+
647.1	$(^3P)5p^4P_{5/2} \rightarrow$	$(^3P)5s^2P_{3/2}$	Kr^+
676.4	$(^3P)5p^4P_{1/2} \rightarrow$	$(^3P)5s^2P_{1/2}$	Kr^+
752.5	$(^3P)5p^4P_{3/2} \rightarrow$	$(^3P)5s^2P_{1/2}$	Kr^+
793.1	$(^1D)5p^2F_{7/2} \rightarrow$	$(^3P)4d^2D_{5/2}$	Kr^+
799.3	$(^3P)5p^4P_{3/2} \rightarrow$	$(^3P)4d^4D_{1/2}$	Kr^+
345.4	$(^2D)6p^1D_2 \rightarrow$	$(^2D)6s^1D_2$	Xe^{++}
378.1	$(^4S)6p^3P_2 \rightarrow$	$(^4S)6s^3S_1$	Xe^{++}

Table 7.3 *(continued)*

λ (nm) In air	Transition assignment	Ion species
460.3	$(^3P)6p\,^4D_{3/2} \rightarrow (^3P)6s\,^4P_{3/2}$	Xe^+
495.4		Xe^{+++}
515.9		Xe^{+++}
526.0		Xe^{+++}
535.3		Xe^{+++}
539.4		Xe^{+++}
541.9	$(^3P)6p\,^4D_{3/2} \rightarrow (^3P)6s\,^4P_{3/2}$	Xe^+
595.5		Xe^{+++}
597.1	$(^1D)6p\,^2P_{3/2} \rightarrow (^1D)6s\,^2D_{3/2}$	Xe^+
798.8	$(^3P)6p\,^4P_{1/2} \rightarrow (^3P)6s\,^4P_{1/2}$	Xe^+
871.6	$(^3P)6p\,^4D_{3/2} \rightarrow (^3P)5d\,^2P_{3/2}$	Xe^+
969.8	$(^3P)6p\,^4D_{3/2} \rightarrow (^3P)5d\,^4P_{5/2}$	Xe^+

respectively. Since similar arguments apply to the excitation of the upper and lower laser states we can write the population difference or gain as a polynomial thus:

$$\gamma = AI + BI^2 + CI^3 \tag{7.21}$$

where A, B and C are constants which in some cases can be negative.

Although electron collisions are efficient in exciting the 4p states of Ar^+, it is the favourable lifetime ratio between the 4p and 4s states which is of primary importance in achieving population inversion. The upper state lifetimes are in the region of 10 ns while those of the lower states are smaller by a factor of 20. The $4s\,^2P$ states are is depopulated by a very rapid vacuum ultraviolet transition to the ground state of Ar^+ (we remember from Eq. (3.27) that the transition probability is proportional to the cube of the frequency). The $4s\,^4P$ states cannot be lower laser states since their lifetimes are in the region of 50 ns (their transitions to the ion ground state violate the spin selection rule). Thus the lower laser state is depopulated by a rapid resonance transition to the ion ground state. Fast resonance transitions also occur in neutral atoms but the levels involved are rarely lower laser states.

Excited ionic states suffer much less trapping of this resonance radiation than do neutral atom states and hence maintain a short lifetime up to much higher pressures. There are two reasons for this. First, ionized states have a lower density of ground state ions. Second, the Doppler width is much higher since the discharges have, by necessity, high currents and so high temperatures. In addition, and perhaps of more importance, the excited ions can increase their velocity and so temperature by acceleration in the electric field of the discharge. Ion lasers require very high current densities to achieve population inversion between their excited states and often an axial magnetic field is used to increase the current density. Once above threshold the gain, and so the output power, increases rapidly with current as indicated in Eq. (7.21). The gain does eventually reach a maximum with current at which point it has been observed that laser transitions in the doubly charged ion reach threshold. At high currents the population of ions in the ground state increases and resonance trapping becomes important. This increases the density of the lower laser states and thus reduces the population inversion.

Even the most efficient ion lasers have comparatively low-power efficiencies, and the argon–ion blue/green laser has an efficiency of about 0.1 per cent. This is because there are many processes taking place other than the ones discussed above which do not contribute to the population inversion. The excess energy is dissipated as heat and, since their invention, much effort has been put into design and development of tubes which can efficiently dissipate this heat. Pulsed discharges have been used to give high current with low average power dissipation. These discharges have been particularly useful in exciting high states of ionization, exciting new laser lines and understanding the details of the excitation mechanisms. Figure 7.13 illustrates the variation of gain of the 488 nm Ar^+ transition during the current pulse. A large transient spike occurs due to population of the ion by simultaneous excitation and ionization (the single-step mechanism). The gain then turns to loss when the ground state ion density has built up sufficiently to trap the resonance radiation from the lower laser state. In pulse B the gain has returned since plasma heating increases the Doppler width and removes the resonance trapping. Pulse B is known as the quasi-CW pulse since its duration increases with increased current pulse duration and it follows the shape of the current pulse. In pulse B excitation occurs in two steps. First, the ion ground state is excited after which the upper laser states are excited.

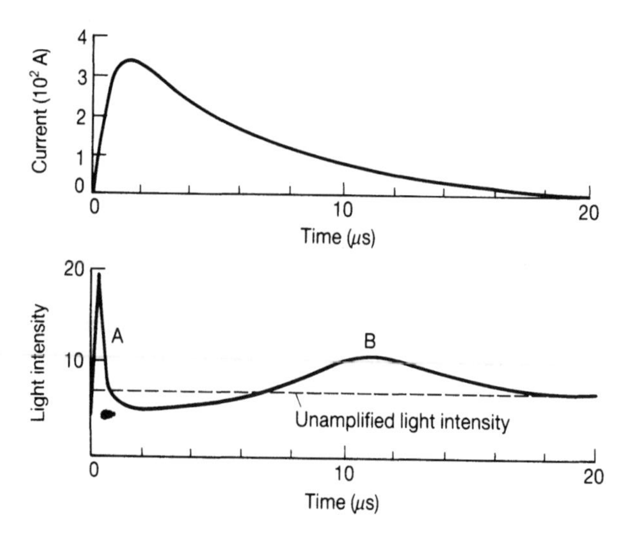

Figure 7.13 Gain and exciting current dependence in a pulsed argon ion amplifier at 488.0 nm and pressure of 15 m torr. (Data reproduced by kind permission of T. A. King (Ref. 9).)

This is called the double-step mechanism. Commercial argon ion pulsed lasers were available during the 1960s but only pulsed xenon ion lasers are now produced since they can produce much higher peak power.

Over thirty years of development in ion laser technology has significantly improved the tube lifetimes and output power, particularly for the doubly ionized lines in the ultraviolet. At present commercial ion lasers can produce 20 W CW visible power in the blue/green with argon and 5 W in the red with krypton. Two basic designs to cope with the problems of high current and high power dissipation have been adopted by manufacturers. The most common is to use a tube of high thermal conductivity ceramic, such as beryllium oxide, to confine the discharge. The heat is absorbed in a coaxial jacket with water cooling capable of dissipating $\sim 300 \text{ W cm}^{-1}$. The more expensive alternative is a segmented metal disk structure. This has an improved lifetime and has been adopted by a few manufacturers. Figure 7.14 shows a schematic diagram of such a tube while Figure 7.15 illustrates the complete laser system. Many different disk structures and materials have been

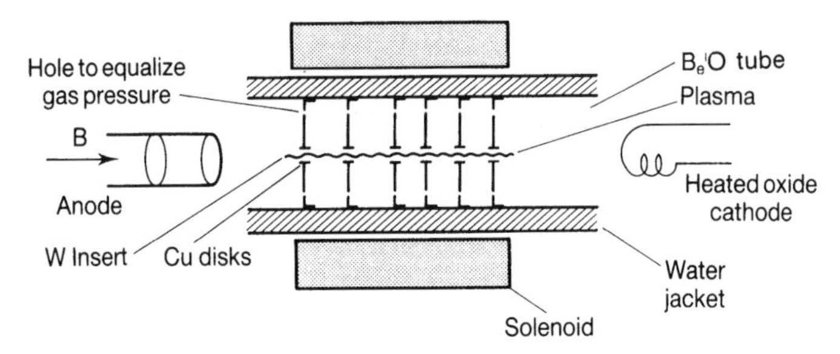

Figure 7.14 Schematic diagram of segmented metal disk structure ion laser tube.

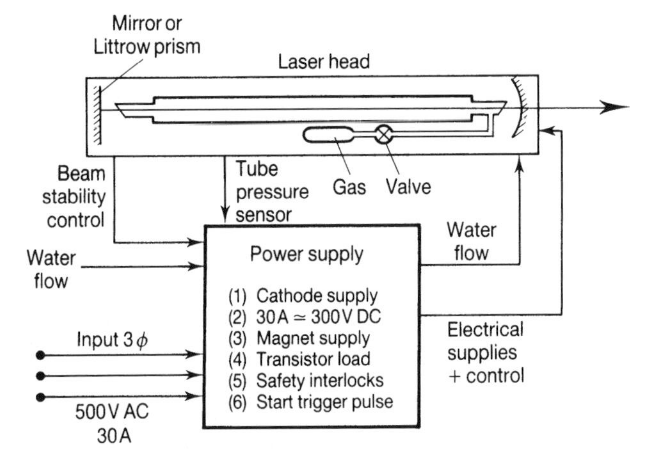

Figure 7.15 Schematic diagram of ion laser system.

studied, and the most successful uses tungsten inserts pressed into the centre of copper disks. The disks are electrically insulated from each other and have a 2–3 mm hole through the tungsten insert to confine the plasma. The tungsten is resistant both to the very high temperatures involved (~ 2000 K) and also to material removal by ion impact (sputtering). The copper disks conduct the heat radially out to the water cooling jacket. Holes on the periphery of the copper disks serve to equalize the gas pressure along the length of the discharge since gas is driven out of the bore by gas pumping caused by neutral atoms forced towards to the anode by electron collisions and flow of positive ions to the cathode. An axial magnetic field of about 10^{-1} T is used to increase the electron density for a given current density. This is particularly important for the ultraviolet lines, and also helps to stabilize the discharge. Ion bombardment results in the burial of gas atoms in the walls and a subsequent drop in pressure from the initial optimum fill. Consequently a gas reservoir is incorporated into the laser head and used periodically to replenish the lost gas.

An alternative to these very expensive and complex systems is the air-cooled ion laser. The system cost can be as low as 20 per cent of the water-cooled version but the output power is usually less than a few hundred milliwatts and is normally available with argon only. The air-cooled lasers have lower discharge currents, no magnetic solenoid (since this would restrict the air cooling) and no gas refill. They can be operated with a single electrical phase input while the water-cooled lasers usually require a three-phase supply.

As shown in Figure 7.12, gain can occur on a number of argon ion transitions and these can lase simultaneously. If a Littrow prism is used as shown in Figure 7.11 then oscillation on any one of the lines can be selected by rotation of the prism. The output power of the individual lines from a water-cooled ion laser using a Littrow prism is shown in Figure 7.16. Figure 7.17 shows the longitudinal mode

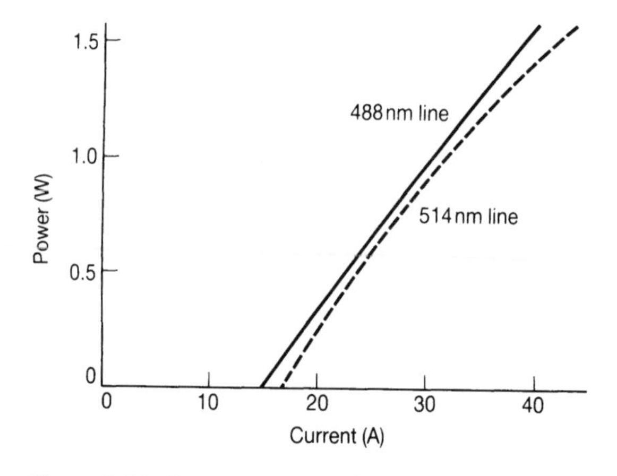

Figure 7.16 Output power plotted against discharge current for some of the blue/green lines of the singly ionized argon ion.

spectrum for the 488 nm and 514 nm lines from an air-cooled Ar^+ laser obtained with a confocal Fabry–Perot interferometer. The air-cooled laser has a Littrow prism and the power in each line has been adjusted to 10 mW in the TEM_{00} mode. The lower-gain 514 nm transition has fewer modes under the gain curve than the much higher-gain 488 nm transition. However, the output power of the two lines is the same since the saturation irradiance (Eq. (3.122)) for the 514 nm line is much larger. The gain bandwidth for these lasers is much greater than that of the neutral gas lasers since the Doppler widths are much larger due to the higher gas temperatures, typically 7000 K. Single-frequency operation of ion lasers can be readily achieved using a tilted plane intra-cavity etalon or the Fox–Smith interferometer described in Section 6.6.3. The Doppler widths are ~5–10 GHz and the homogeneous linewidths are about 500 MHz, with the result that up to 80 per cent conversion from multi- to single-frequency operation can be achieved.

7.2.5 Metal vapour ion lasers

An important class of ion laser exists in which the excitation is by Penning and charge transfer collisions (Eqs (7.9) and (7.10), respectively). The important lasers in this group are those having a metal vapour excited in a discharge with a noble gas as the buffer. Many of the metallic elements in the periodic table can be made to lase in this way in both pulsed and DC discharges. The CW helium–cadmium laser has attracted most commercial interest due to its ability to produce CW short wavelength laser emission at 441.6 nm and 325 nm using discharges technically a little more complex than those used for helum–neon lasers. Our discussion will concentrate on the helium–cadmium laser, and the reader is referred to various review articles [Refs 8, 9] for details of other metal vapour ion lasers.

The energy-level diagram for ionized cadmium in Figure 7.18 shows the excitation routes for the lasing transitions in Cd^+ in a heated glow discharge of helium and cadmium vapour. The Penning reaction (Eq. (7.9)) between the helium metastable atoms and atomic cadmium is mainly responsible for the excitation of the upper states of the 441.6 nm and the 325 nm transitions. The charge transfer reaction (Eq. (7.10)) excites the upper states of the higher-lying levels. The 441 nm and the 325 nm

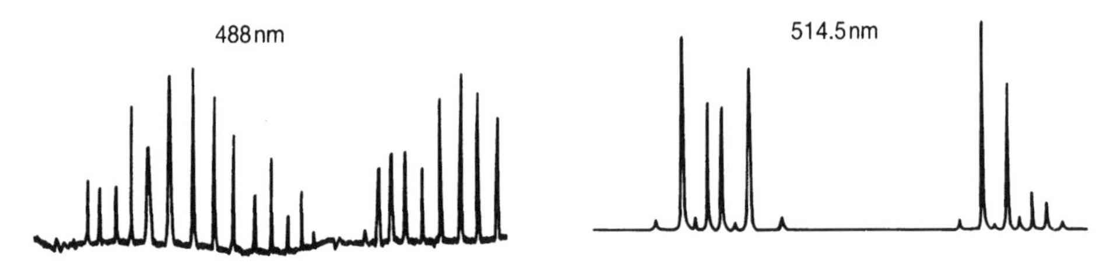

Figure 7.17 Longitudinal mode spectrum of 488 nm and 514 nm argon ion laser lines from an air cooled ion laser with a mode spacing of 438 MHz. The output power for each line has been adjusted to 10 mW. The gain bandwidth for the 514 nm line is 4 GHz which for the 488 nm line is 6.5 GHz. Notice the change in the mode structure in each scan of the spectrum due to the noisy cavity.

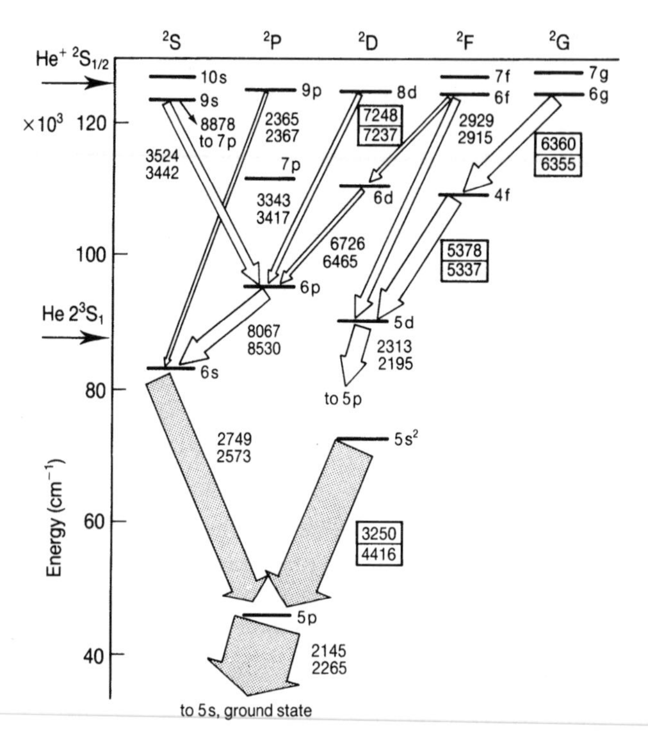

Figure 7.18 Energy diagram of the Cd^+ ion showing the excitation routes for some laser transitions. (Data taken from Ref. 10.)

transitions are helped by having very favourable lifetime ratios: the upper and lower state lifetimes are ~700 ns and ~4 ns, respectively. Preventing the metal vapour from condensing on the laser optics and maintaining a uniform distribution of vapour along the length of the plasma are the main technical problems in a discharge which is otherwise similar to that of helium–neon lasers. Very strong cataphoresis or vapour pumping towards the cathode end of the discharge occurs, and this has been used to produce a uniform flow of cadmium vapour towards the cathode. One typical design is shown in Figure 7.19 in which a hot cathode discharge tube has a hot source of cadmium at the anode end and a condenser at the cathode end. Between these reservoirs and the optics (mirrors or windows) there is a layer of helium plasma which protects them. An oven around the tube maintains a constant temperature in the active region. A further refinement of this design is to have a cathode in the middle and two anodes, one at each end of the tube, and in effect have two discharges. Typical operating parameters quoted in the literature [8, 9] for such a laser are source temperature 220°C; vapour pressure of cadmium 0.002 torr; pressure of helium 3.4 torr; tube current 110 mA; tube diameter 2.4 mm and length 1.43 m giving output power at 441 nm of 220 mW and at 325 nm of 20 mW.

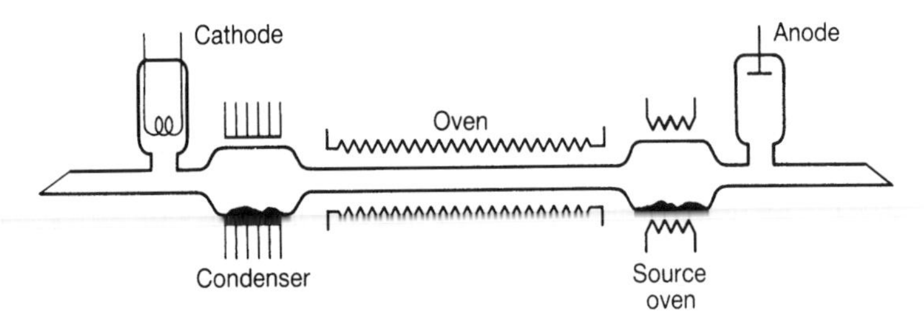

Figure 7.19 Design for discharge tube configuration used in catephoric flow He–Cd$^+$ laser.

7.2.6 Cyclic metal vapour lasers (MVL)

Population inversion can be achieved between levels which do not have a favourable lifetime ratio, that is, when the lifetime of the lower state is in fact longer than that of the upper state. However, in these cases we require a much greater excitation rate to the upper state than to the lower state and the inversion is only possible for a very short time.

An energy level model for an ideal cyclic laser is shown in Figure 7.20. The upper laser level is optically connected to the ground state by a fast transition ($A_{21} \sim 10^8\ \mathrm{s}^{-1}$) and to the lower laser state by a slower transition ($A_{20} \sim 10^4$–$10^8\ \mathrm{s}^{-1}$). Transitions from the lower state to the ground state are forbidden and it can only relax slowly by nonradiative collisions with the walls, electrons or other atoms. Under the proper discharge conditions the electron excitation rate to the upper state can be larger than that to the lower state. This may result from, for example, a difference in the electron collision cross-section as illustrated in Figure 7.2. The resonance radiation $2 \rightarrow 0$ is trapped and hence the decay of the upper state is only through $2 \rightarrow 1$ transitions. Population inversion will then exist for a short time during the discharge current pulse until the lower level fills due to the $2 \rightarrow 1$ stimulated and spontaneous transitions. It is therefore necessary to have the rise time of the electrical pulse faster than A_{21}^{-1} so that no population of state 2 is lost by spontaneous emission before oscillation threshold is reached. The exciting pulse should be no longer than the

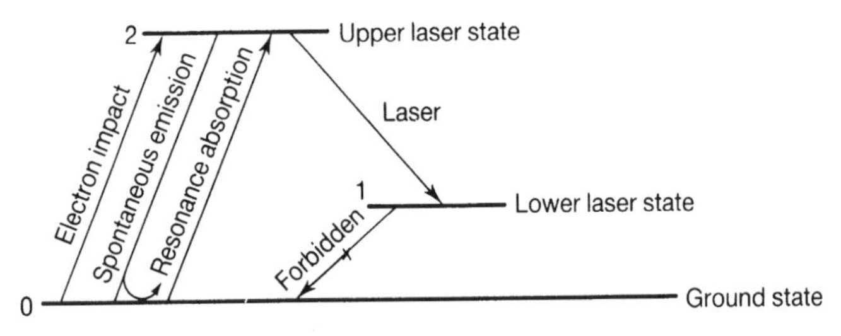

Figure 7.20 Schematic energy-level diagram of an ideal cyclic laser.

self-terminating laser pulse and the pulse repetition rate is limited by the relaxation rate of the lower state.

A large number of metal vapours and indeed noble gases can exhibit cyclic laser action but the most successful are transitions in copper and gold vapour. Two very simplified energy-level diagrams for atomic copper and gold are shown together for comparison in Figure 7.21. They both resemble the ideal cyclic laser discussed above except that in both cases a pair of laser lines are emitted. Figure 7.22 shows a schematic diagram of the laser configuration and its driving circuit. Pulse duration is rarely more than 50 ns and in this time light will be able to make only a few transits through the gain medium. It is not necessary therefore to confine the light using high-finesse stable cavities and output couplings of 50 per cent are typical. The output from such lasers is more like amplified spontaneous emission with the gain becoming saturated after a few transits of the gain medium. The beam is not a diffraction limited Gaussian, as obtained from a stable resonator, but is spatially narrowed with an angular divergence determined by the diameter-to-length ratio of the gain medium. The linewidth is narrower than the Doppler width but never smaller than the Fourier transform limit (that is, the reciprocal of the stimulated emission pulse width).

Since there is no requirement to confine the beam to a TEM_{00} mode in these lasers then the bore diameter can be increased to typically 20 mm, thus increasing the

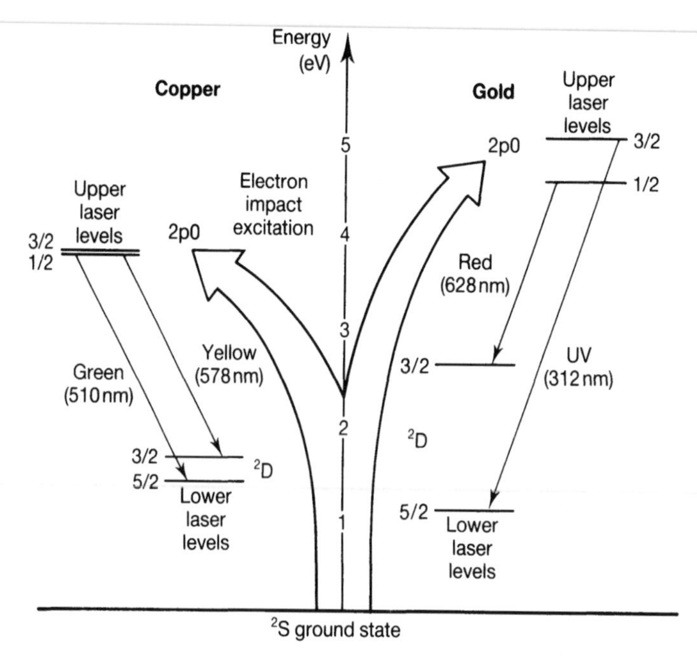

Figure 7.21 Energy-level diagram for cyclic metal vapour lasers copper and gold.

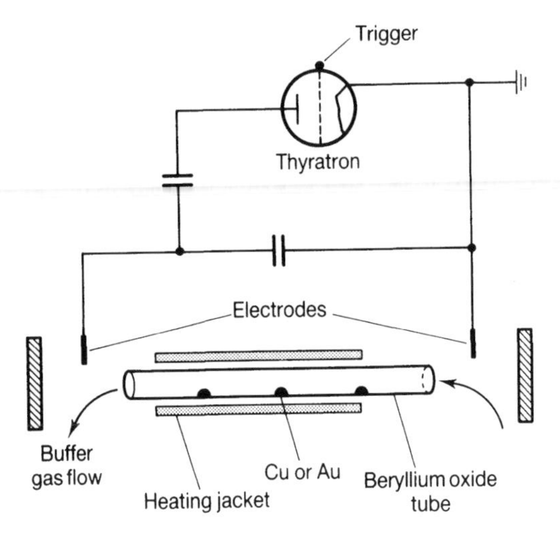

Figure 7.22 Schematic diagram of cyclic metal vapour laser configuration.

output power. The discharge tube used is a ceramic material, such as beryllium oxide, capable of withstanding temperatures in excess of 1000°C. Pieces of the metal placed along the length of the tube are vaporized by the heat input from the discharge in the helium or neon buffer gas and insulation around the bore allows the temperature to rise without any external heat input. The high repetition rate (3–20 kHz) gives the beam a continuous appearance and in copper vapour an average power in the train of pulses of 50 W can be achieved. Efficiencies of copper vapour lasers are in the region of 1 per cent, making them a serious competitor to the green argon–ion laser in situations where average power only is important rather than high-quality beam or narrow-linewidth single-frequency operation. A typical application is in the pumping of dye lasers, which will be discussed in Section 7.3.2.

7.2.7 Molecular gas lasers

As discussed in Section 1.3, molecules have three degrees of freedom and consequently energy associated with each. The energy levels of molecules were discussed in Chapter 1 and here we extend this to include the resulting spectra and lasing transitions. Section 1.4 dealt with diatomic molecules but the basic ideas can be extended to molecules with more than two atoms provided allowance is made for the extra degrees of vibrational freedom involved. Molecular gas lasers can produce gain for infrared vibrational–rotational transitions (VR) and for visible/ultraviolet electronic transitions.

It is conventional to write Eq. (1.40), which gives the energy associated with the rotation of the molecule, using units of energy in cm^{-1} and a constant B_v such that

$$E(J) = B_v J(J+1)\ \text{cm}^{-1} \tag{7.22}$$

The constant B_v contains the moment of inertia of the molecule which will change with the vibrational state and the electronic state. The spacing between the rotational states is small compared with the average kinetic energy of the gas molecules ($\sim kT$) and collisions are effective in producing a thermal population of the rotational states. Applying Maxwell–Boltzmann statistics to the distribution of rotational states in a given vibrational manifold we have for the fractional number of molecules in a rotational state J:

$$\frac{N_{v,J}}{\sum_{J=0}^{J=\infty} N_{v,J}} = \frac{N_{v,J}}{N_v} = \frac{(2J+1)\exp\{-100hcB_v J(J+1)/kT\}}{\sum_{J=0}^{J=\infty} (2J+1)\exp\{-100hcB_v J(J+1)/kT\}} \tag{7.23}$$

where we note that $(2J+1)$ is the degeneracy of the quantum state J and the exponential terms are the Boltzmann factors. The factor 100 in these equations will allow for B_v being in units of cm^{-1}. With little error we can replace the summation in the denominator by an integral since J is quite high at the gas temperatures involved. It is left as a problem to show that Eq. (7.23) after integration becomes

$$\frac{N_{v,J}}{N_v} = \frac{100hcB_v}{kT}(2J+1)\ \exp\left[-\frac{100hcB_v J(J+1)}{kT}\right] \tag{7.24}$$

Figure 7.23 shows the distribution of rotational states at 400 K in the upper laser vibrational state of the CO_2 molecule with $B_v = 0.37\ cm^{-1}$. The maximum number

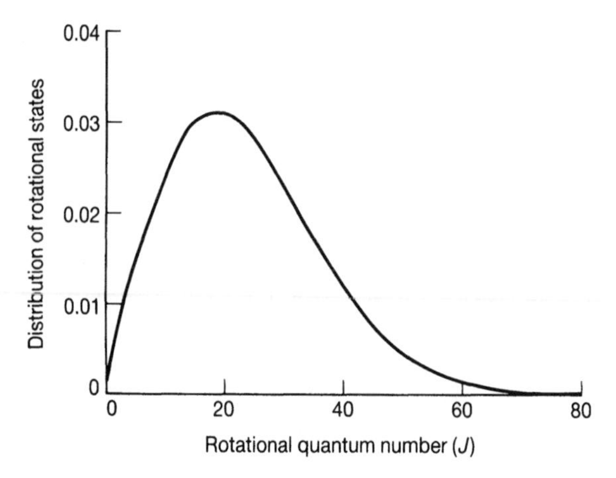

Figure 7.23 The distribution of rotational states for a vibrational state of CO_2 with $B_v = 0.37\ cm^{-1}$.

of molecules occurs for the rotational state J_{max} and is found by differentiating the right-hand side of Eq. (7.24) with respect to J and equating to zero to obtain:

$$J_{max} = \sqrt{\frac{kT}{200B_v hc}} - \frac{1}{2} \quad (7.25)$$

Transitions between two vibrational manifolds with a simultaneous change in rotational quantum number result in the emission of far infrared radiation. These are called *vibrational–rotational* (VR) transitions and take place according to the following selection rules:

$$\Delta v = \pm 1 \quad (7.26)$$

$$\Delta J = \pm 1,\ 0;\ \text{with } J = 0 \to J = 0 \text{ forbidden} \quad (7.27)$$

The energies involved in the rotational changes give the spectral band its fine structure and the three alternatives in Eq. (7.27) define different branches of the band. Using the convention that J refers to the rotational quantum number of the lower state we define:

P branch: J value of lower state is one larger than the upper: $\Delta J = -1$
Q branch: J value is unchanged: $\Delta J = 0$
R branch: J value of lower state is one less than the upper: $\Delta J = +1$

With this convention it can be easily shown that the R branch transitions are always of shorter wavelength than the P branch. An absorption spectrum of the 10.4 μm band of CO_2 is illustrated in Figure 7.24, the rotational lines being designated as P(J) or R(J).

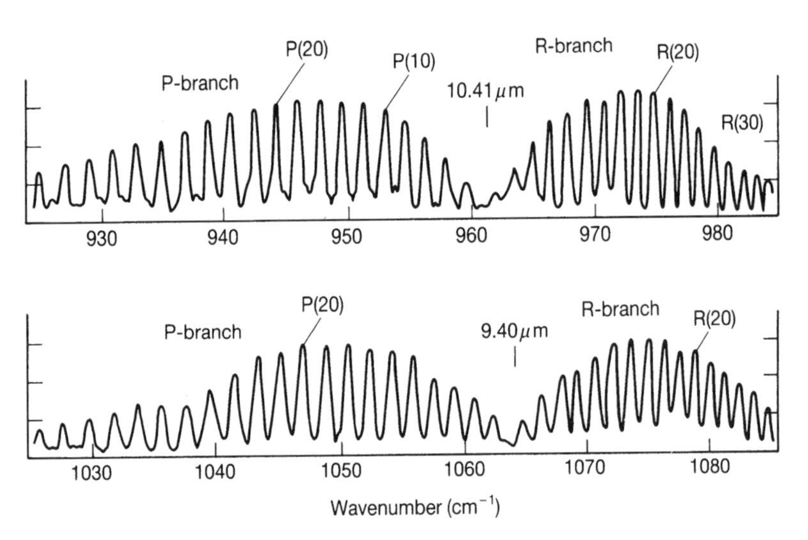

Figure 7.24 Absorption spectra for CO_2 in the 9.4 μm and 10.6 μm vibrational bands. (Data taken from Ref. 11.)

We may now now carry out a calculation to compare the optical gain for P and R branch for a VR transition. Using the expression for the gain coefficients Eq. (7.11), we have:

$$\gamma(\nu) = \sigma_{se}(\nu)\left(N_2 - \frac{g_2}{g_1} N_1\right) = \sigma_{se}(\nu) g_2 \left(\frac{N_2}{g_2} - \frac{N_1}{g_1}\right)$$

It is normal in molecular spectroscopy to express the upper vibrational and rotational quantum numbers with a prime and the lower ones with a double prime. Thus we have

$$g_2 = 2J' + 1 \text{ and } g_1 = 2J'' + 1$$

Using Eq. (7.24) for N_2 and N_1 together with the assumption that the upper and lower rotational constants B_v are the same for both vibrational manifolds the gain becomes

$$\gamma(\nu) = \sigma_{se}(\nu)(2J' + 1)\frac{100hcB_v}{kT}\left[N_{v'} \exp\left(-\frac{100hcB_v}{kT}(2J' + 1)\right) - N_{v''} \exp\left(-\frac{100hcB_v}{kT}(2J'' + 1)\right)\right] \quad (7.28)$$

For the P branch we have that $J'' = J$; $J' = J - 1$ and the gain for the P branch becomes

$$\gamma_P(\nu) = \sigma_{se}(\nu)(2J - 1)\frac{100hcB_v}{kT}\left[N_{v'} \exp\left(-\frac{100hcB_v}{kT}J(J - 1)\right) - N_{v''} \exp\left(-\frac{100hcB_v}{kT}J(J + 1)\right)\right] \quad (7.29)$$

Similarly, for the R branch $J'' = J$: $J' = J + 1$ and the gain in the R branch becomes

$$\gamma_R(\nu) = \sigma_{se}(\nu)(2J + 3)\frac{100hcB_v}{kT}\left[N_{v'} \exp\left(-\frac{100hcB_v}{kT}(J + 1)(J + 2)\right) - N_{v''} \exp\left(-\frac{100hcB_v}{kT}J(J + 1)\right)\right] \quad (7.30)$$

After some simplification the ratio of the gains in the two branches is

$$\frac{\gamma_P}{\gamma_R} = \frac{2J - 1}{2J + 3}\left[\frac{\dfrac{N_{v'}}{N_{v''}} \exp\left(\dfrac{100hcB_v}{kT} 2J\right) - 1}{\dfrac{N_{v'}}{N_{v''}} \exp\left(-\dfrac{100hcB_v}{kT} 2J + 2)\right) - 1}\right] \quad (7.31)$$

Careful inspection of equation (7.31) shows the gain in the P branch to be always larger than in the R branch for one particular value of J, and so in one vibrational manifold the P-branch laser lines will dominate.

7.2.6.1 *The carbon dioxide laser*

Excitation of carbon dioxide in an electrical discharge is achieved using a number of different configurations and very efficiently produces high-power laser output both pulsed and CW in the infrared. The molecule is triatomic and differs from a diatomic molecule in that it has four vibrational degrees of freedom of which two are frequency degenerate. The three modes of vibration of the carbon dioxide molecule shown in Figure 7.25 may be each assigned a vibrational quantum number as follows:

v_1 = vibrational quantum number of symmetric stretch mode. The carbon atom remains stationary and the two oxygen atoms move in opposite directions along the line of symmetry.
v_2 = vibrational quantum number of the bending mode. All the atoms move in a plane perpendicular to the axis of symmetry. This mode is twofold degenerate since vibrations can occur in the plane and perpendicular to the plane of the diagram.
v_3 = vibrational quantum number of the asymmetric stretch mode. All the atoms move along a line of symmetry, the carbon atom moving in the opposite direction to the two oxygen atoms.

The molecule can vibrate in more than one mode at the same time and so it can have more than one quantum of vibrational energy. By analogy with Eq. (1.36) for the diatomic molecule the total vibrational energy is

$$E_v = 100hc(\bar{\nu}_1(v_1 + \tfrac{1}{2}) + \bar{\nu}_2(v_2 + \tfrac{1}{2}) + \bar{\nu}_3(v_3 + \tfrac{1}{2})) \tag{7.32}$$

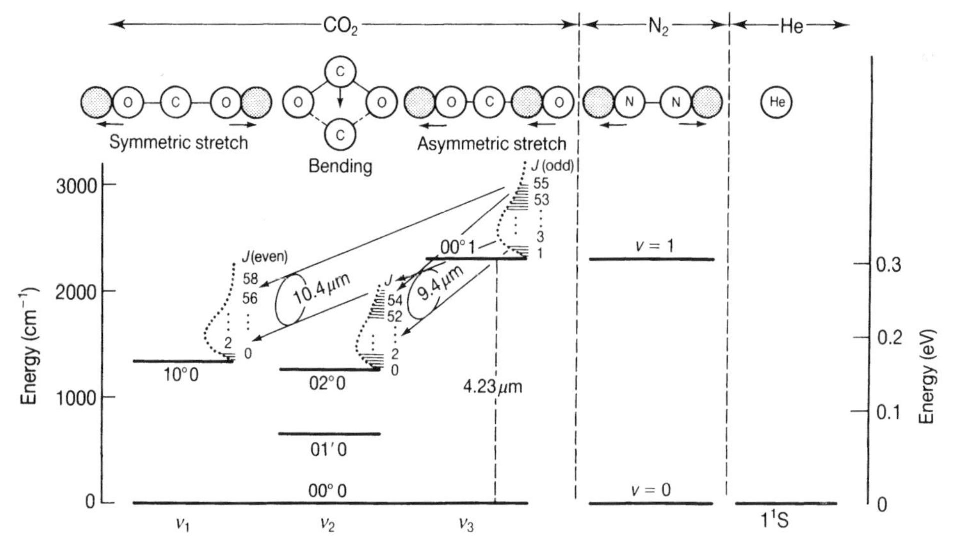

Figure 7.25 Energy-level diagram, showing the modes of vibration of the CO_2 molecule for the 10.4 μm and 9.4 μm bands of the CO_2 laser.

where we use the conventional units of wave numbers in cm^{-1} for the vibrational 'frequencies':

$$\bar{\nu}_1 = 1351\ cm^{-1},\ \bar{\nu}_2 = 667.3\ cm^{-1},\ \bar{\nu}_3 = 2396.4\ cm^{-1}$$

The vibrational states are represented with the notation indicating the number of vibrational quanta of each mode. Thus

$$(v_1 v_2^l v_3)$$

The superscript l on v_2 indicates angular momentum in the two degenerate states of the bending mode of vibration causing a very small energy splitting for the different values of l. For the purposes of our discussion of the CO_2 laser this can be overlooked.

Table 7.4 lists some of the vibrational transitions in carbon dioxide which have demonstrated gain. All these transitions have low gain except those at 10.4 μm and 9.4 μm. The energy levels pertinent to these two transitions together with the energy level of the ground electronic state of nitrogen are all shown in Figure 7.25. There is an accidental degeneracy of the states (10^00) and (02^00) since $\bar{\nu}_1$ is close to $2\bar{\nu}_2$ and their energies are measured, respectively, to be 1388.3 cm^{-1} and 1285.5 cm^{-1}. These values, however, are significantly smaller than those calculated from the vibrational frequencies given above. The two states are in *Fermi resonance*, causing a mixing of their wavefunctions. Each actual level is then a mixture of the two and the levels can no longer be unambiguously designated as 10^00 and 02^00. In terms of energy the former level is pushed up and the lower level pushed down so that the energy spacing between them increases, as indicated above. This Fermi mixing is responsible for the existence for the 9.4 μm band since it violates the $\Delta v = \pm 1$ rule. Notice that transitions from the upper laser vibrational state 00^01 has even rotational levels missing while for the lower states 10^00 and 02^00 odd rotational states are missing.[5]

Table 7.4 Vibrational lasing transitions in CO_2

Vibration transition		
Upper state	Lower state	Wavelength (μm)
$0\ 0^0\ 1$	$0\ 2^0\ 0$	9.4
$0\ 0^0\ 1$	$1\ 0^0\ 0$	10.4
$0\ 1^1\ 1$	$0\ 3^1\ 0$	11.0
$0\ 1^1\ 1$	$1\ 1^1\ 0$	11.0
$1\ 4^0\ 0$	$0\ 5^1\ 0$	13.2
$2\ 1^1\ 0$	$1\ 2^1\ 0$	13.5
$1\ 4^0\ 0$	$1\ 3^1\ 0$	16.6
$0\ 3^1\ 1$	$0\ 2^2\ 1$	17.0
$2\ 4^0\ 0$	$2\ 3^1\ 0$	17.4

The transition probabilities for VR transitions are many orders of magnitude smaller than for allowed electronic transitions. With the exception of the strong 4.3 μm resonance transition, A coefficients for individual rotational lines fall in the range 0.2–2 s^{-1}, resulting in very long state lifetimes. The upper laser state 00^01 has an A coefficient of 200 s^{-1} but with resonant trapping this is reduced to 10 s^{-1}. The lifetime of this upper state is in fact longer than the lower state, but this is hardly important with these long lifetimes since spontaneous emission plays little part in the relaxation processes. One might intuitively think that small A coefficients would imply small stimulated emission cross-sections. However, the stimulated emission cross-section which depends on the cube of the wavelength is significantly increased at the long wavelengths of these VR transitions. At the centre of the P(20) line at 10.6 μm it is about 1.78×10^{-20} m^2 and for the 9.4 μm line 0.7×10^{-20} m^2.

The excitation and relaxation mechanisms in the CO_2 laser are extremely complex so we will simply outline the main features. Although inversion can be obtained with discharges in pure CO_2 the addition of nitrogen N_2 to the discharge makes such a difference to the gain that excitation transfer must be of major importance. The vibrational energy spacing in N_2 is 18 cm^{-1} less than the energy of the 00^01 vibrational state of CO_2 and in fact this spacing is an excellent match up to $v = 4$ and within kT up to the $v = 8$ states of N_2. Since N_2 is a homonuclear diatomic molecule and has no dipole moment in any of its vibrational states, then VR transitions are forbidden and the vibrational states are metastable. This ensures that the transfer process proceeds more rapidly from N_2 to CO_2 than in the reverse direction. Electron collisions in the plasma are responsible for excitation of the CO_2 and the N_2. We can write these processes thus:

$$e + N_2(v = 0) \rightarrow N_2(v = 1, 2, \ldots, 8) + e \qquad (7.33)$$

$$CO_2(00^00) + N_2(v = 1, 2, \ldots, 8) \rightarrow CO_2(00^01) + N_2(v = 0, 1, \ldots, 7) \qquad (7.34)$$

$$e + CO_2(00^00) \rightarrow CO_2(00^01) + e \text{ upper laser state} \qquad (7.35)$$

$$\rightarrow CO_2(10^00) + e \text{ lower laser state} \qquad (7.36)$$

$$\rightarrow CO_2(02^00) + e \text{ lower laser state} \qquad (7.37)$$

$$\rightarrow CO_2(01^10) + e \qquad (7.38)$$

Cross-sections for these processes have been extensively studied using electron beam energy-loss techniques. The results of Boness and Schultz [Ref. 12] for the excitation of some of the vibrational states of CO_2 are shown in Figure 7.26 while Figure 7.27 shows excitation functions due to Schultz [Ref. 13] for the cross-sections of the first vibrational states of nitrogen. These cross-section data show that if the electron temperature in the discharge is in the region of 2 ev then very efficient excitation of the vibrational states of nitrogen can occur without exciting the lower laser states in CO_2. A higher electron temperature than this can lead to significant dissociation of the CO_2, which requires 2.8 ev. These excited nitrogen molecules will

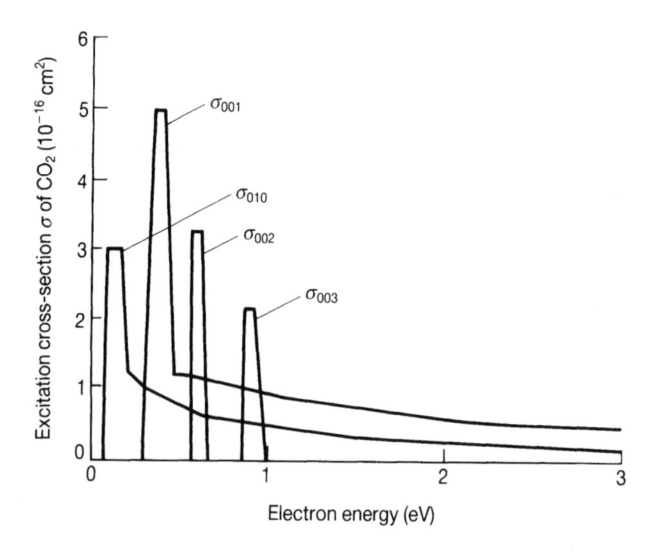

Figure 7.26 Excitation functions for electron impact excitation of CO_2.

transfer virtually all their energy to the 00^01 state of CO_2, so producing very efficient pumping of the upper state.

The spontaneous emission processes are not rapid, so efficient nonradiative de-excitation mechanisms are required for the lower laser states. Since these are dependent on molecule–molecule collisions higher pressure helps with the added bonus keeping down the electron temperature. Within each vibrational mode the levels are strongly coupled by collisions due to the near-resonant $v \rightarrow v \pm 1$ energy transfer. For example, the upper laser state can be de-excited by the resonant collision:

$$2\,CO_2\,(00^01) \Leftrightarrow CO_2(00^02) + CO_2(00^00) + 25\ \text{cm}^{-1} \tag{7.39}$$

The lower laser states, being mixed, can relax through the $01\,^10$ state, i.e. the vibrations v_1 are coupled to the vibrations v_2, and any excess energy is removed as kinetic energy ($V \rightarrow T$ collisions):

$$CO_2(10^00,\ 02^00) + CO_2(00^00) \Leftrightarrow 2CO_2(01\,^10) \pm \Delta E \tag{7.40}$$

In this process ΔE will be determined by whether the mixed state is at 1388 cm^{-1} or at 1285 cm^{-1}. The final relaxation process is the limiting mechanism which is the $V \rightarrow T$ transfer taking the CO_2 molecule back to the ground state:

$$CO_2(01\,^10) + CO_2(00^00) \rightarrow CO_2(00^00) + CO_2(00^00) + 667\ \text{cm}^{-1} \tag{7.41}$$

Helium is added to the discharge and increases the rate for this process by a factor of 20 due to its large thermal conductivity as well as stabilizing the plasma and controlling the electron temperature.

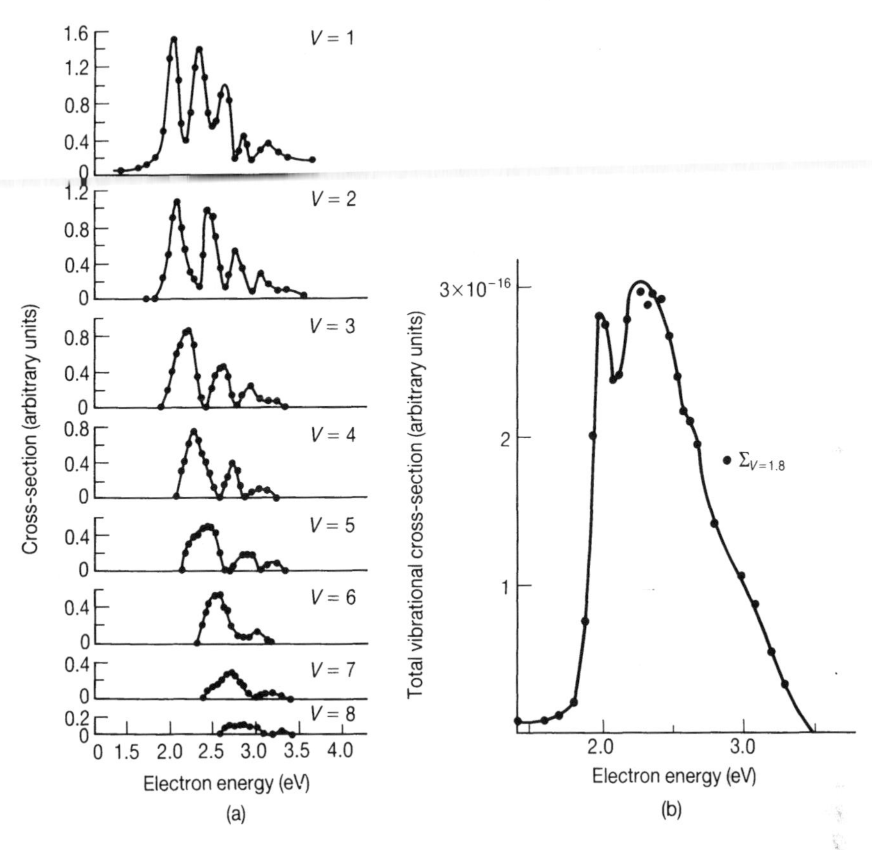

Figure 7.27 Excitation functions for (a) the excitation of the first eight vibrational states of nitrogen; (b) the sum of these eight cross-sections to give the total vibrational cross-section.

The simplest design of CO_2 laser, referred to as a conventional CO_2 laser, is shown in Figure 7.28. The gain medium is a water-cooled tube containing a mixture of helium, nitrogen and carbon dioxide with a typical ratio of 8 : 1 : 1. A continuous flow of the gas mixture ensures a fresh supply of gas, since CO_2 is dissociated in the discharge, and also helps to cool the gas. It is important to cool the gas for efficient removal of the kinetic energy produced in process (7.41). Reducing the gas temperature keeps down the number of rotational states and narrows the gain bandwidth. The optimum gain varies inversely as the bore diameter, D, with an optimum pD product of about 3 torr-cm where p is the partial pressure of CO_2. High-saturation irradiance, typically 100 W m^{-2} coupled with the high quantum efficiency and the efficient creation of the population inversion make the CO_2 laser a very efficient and high-power device. This conventional design can give output power in the region of 70 W m^{-1} in lengths up to 3 m decreasing to 50 W m^{-1} in lengths up to 10 m.

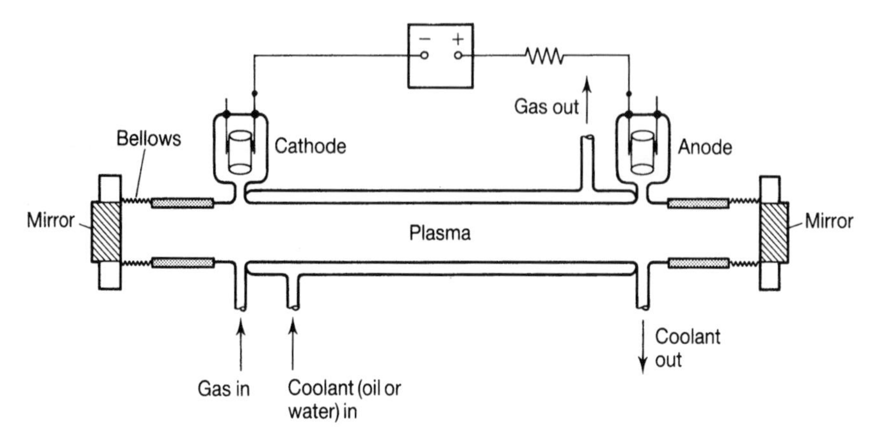

Figure 7.28 Schematic diagram of a conventional CO_2 laser.

A hemispherical cavity is normally used with the curved mirror having a gold high-reflecting coating on a copper or molybdenum substrate. The output coupler can be a flat of germanium, zinc selenide or other infrared-transmitting material. A multilayer dielectric coating can be used to optimize the output coupling, with an anti-reflection coating on the output surface, since the materials have high refractive indices and so high Fresnel reflectance. At output power levels greater than 100 W the optics, mirrors and external optics require water cooling. Large conventional flow carbon dioxide lasers are constructed by folding the tubes and using a 90° prism at each fold. Industrial lasers with outputs up to 1 kW are available using this design. For lower-power systems (~ 10–50 W) sealed-off tubes are used but lifetimes are then limited to a few thousands of hours due to the dissociation of CO_2.

Without any wavelength tuning element, such as a grating, in the cavity the P branch of the 10.4 μm band dominates. At total gas pressures of about 10 torr the homogeneous linewidth is about 50 MHz and the Doppler width of each rotational line is 60 MHz. The rotational levels experience strong thermal mixing and lasing will occur on the rotational line which has the largest gain-to-loss ratio. If TEM_{00} mode oscillation is forced using an aperture then single-frequency oscillation occurs. If higher-order transverse mode oscillation is permitted with a larger aperture then a few frequencies occur at modes which have small spatial overlap. The spacing between the rotational lines is about 5×10^{10} Hz so it is possible with long cavities to find a mirror spacing that could permit oscillation on two rotational lines simultaneously. This can occur at low pressures (< 1 torr) where the homogeneous linewidths are less than 10 MHz and Lamb dips have been observed [Ref. 14].

Using a grating in the Littrow configuration external to the output coupling mirror the cavity can be tuned for oscillation on specific rotational lines. The grating is tuned by adjusting its angle and position so that first-order diffracted light is reflected into the cavity and the zero order is the laser output. In this way the R branch and the other vibrational transitions given in Table 7.4 can be made to lase. The CO_2 laser can thus be used to study the spectrum of carbon dioxide in emission and obtain precise values of the transition frequencies required for international length standards. Lists of the transition frequencies for different isotope combinations of CO_2 are available [Ref. 11].

Other configurations used for carbon dioxide lasers are as follows:

1. Convection cooled laser. The cooling of the gas in the conventional laser relies on diffusion for cooling of the gas. If the gas is allowed to flow through the tube faster than the characteristic time for diffusive cooling then the heat is removed by convection. The electrical energy input can be increased and significantly more power extracted from the system. In this way, output power in the multikilowatt region has been obtained. The flow, which is at almost supersonic velocities, can be transverse or parallel to the optical cavity and the gas is cooled by a heat exchanger and recycled, often with catalytic conversion of the carbon monoxide back to carbon dioxide.
2. Gas dynamic lasers are an alternative to discharge excitation in which there is a direct conversion of heat energy into coherent optical energy. The gas mixture of nitrogen and carbon dioxide at a pressure of several tens of atmospheres is heated in a high-pressure container to 1500–2000 K. The gas is expanded through gas nozzles to a pressure of 50 torr and the temperature drops to about 300–400 K. The nitrogen vibrational states do not decay but store their energy for transfer to the CO_2 molecule. The lower laser states decay is rapid since it occurs by VT transfer. Output powers up to 60 kW have been achieved with this device.
3. A ceramic waveguide can be used to confine both the discharge and the light, enabling a reduction in the bore diameter and a consequent increase in the gain. The diffraction loss is low since the cavity mode is not a TEM mode but a waveguide mode with the light being confined by wall reflections. Excitation of the gas is normally achieved using a radiofrequency discharge and the mirrors are attached to the end of the waveguide. Very compact sealed-off waveguide lasers with powers up to 50 W are available.
4. The Herriott cell cavity, as described in Section 4.3, can be used to extend the gain volume. Radiofrequency excitation of the gas between two concentric cylindrical electrodes produces an annular gain region. Referring back to Figure 4.8, we see that a particular input ray can make a number of roundtrips of the annular cavity and be re-entrant. The particular ray can be defined by suitable apertures and the roundtrip determined by the mirror spacing and curvatures as described in Section 4.3.
5. Another approach to the problem of heat removal, and hence the depopulation of the lower laser states, is to use the heat capacity of the medium itself in short-pulsed operation. The heat capacity will be proportional to the pressure and the pulse duration, which is determined by the relaxation times of the excited molecules, decreases with increasing pressure. Thus the peak power will be

proportional to the pressure. At elevated pressures the conventional uniform glow discharge develops narrow constricted arcs and requires very large voltages. The approach adopted is to excite the gas transversely to the optical cavity. This reduces the distance between the electrodes to tens of millimetres and results in stable glow discharges. The helium, nitrogen and carbon dioxide gas pressure of these systems can be up to atmospheric pressure and hence they are known as TEA (*T*ransverse *E*xcited *A*tmospheric) lasers. To obtain stable glow discharges and submicrosecond rise times some pre-ionization of the gas before the main discharge is necessary and this is achieved using ultraviolet ionization or electron beams. In electron beam pre-ionization a 100–300 kev beam is fired through a thin metal foil from a vacuum region into the lasing gas. The field across the gas accelerates the resulting ions and electrons and excites the upper vibrational states of carbon dioxide. The discharge is not self-sustaining and so avoids arc formation. However, the system is large and requires a vacuum chamber for the electron gun and the ultraviolet pre-ionized self-sustaining discharge is preferred.

A spark gap, which can emit ultraviolet radiation into the discharge and is fired before the main gap, can be used to pre-ionize the gas, but more uniform discharges are produced using the corona which occurs at the surface of a dielectric. A schematic diagram of such a system is shown in Figure 7.29. The optical cavity is perpendicular to the page and can be either stable or unstable. The contours of the electrodes are designed with a profile[6] to produce a uniform electric field (no edge effects) at their surfaces. The cathode extension is close to the anode so that when the high voltage is switched on a strong electric field is produced across the dielectric, and surface charges are produced as shown in Figure 7.29. These fields produce free charges which develop into a corona surface discharge. Ultraviolet photons from this corona produce a very uniform ionization of the gas along the length of the cavity (but not necessarily across the gap) and subsequently a uniform glow discharge between the electrodes results. This uniform

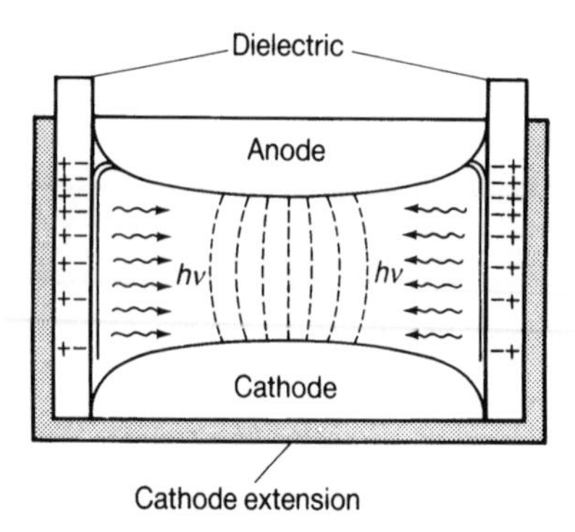

Figure 7.29 Schematic diagram of a section through a TEA CO_2 laser which uses a dielectric to produce a corona discharge for ultraviolet pre-ionization. The optical cavity is out of the plane of the paper.

discharge having a good balance between the currents in the corona and the main discharge is a necessary requirement for a good output beam profile. Figure 7.30 shows a typical current pulse from such a system, highlighting the distinction between the corona and main discharges. The laser pulse has to have a duration of about 20 ns and a peak power in the region of 1 MW.

7.2.6.2 The nitrogen laser

In addition to VR transitions, population inversion can also be produced between electronic-excited states of molecules in which the lasing transitions involve a change in the electron configuration and the vibrational quantum number. These transitions occur in the visible and ultra-violet and, as for VR transitions, any rotational energy change broadens the gain curve. An important example is the nitrogen laser which is a very useful source of pulsed radiation in the ultraviolet at 337 nm. Figure 7.31 shows a partial energy-level diagram for the nitrogen molecule. Each potential energy curve corresponds to a different configuration of the electron cloud and within these curves the horizontal lines represent the vibrational energy levels. The most probable electronic transitions (both up and down) are determined by the *Franck–Condon principle*. This states that electronic transitions take place along a vertical line on the energy-level diagram, that is, a constant internuclear separation. The classical explanation for this is simply that during the transition the light electrons can quickly readjust themselves while the heavier and more sluggish nucleii hardly change their positions. The energy-level diagram (Figure 7.31) shows that the vertical transition from the ground state can excite the $v = 0$; $C\,^3\Pi_u$ state with a higher probability than the lower $B\,^3\Pi_g$ state. Once electronically excited, the nucleii will have time to change by vibration and a vertical transition can take place to a lower state. The lifetime of the $C\,^3\Pi_u$ state is 10 ns while the lower $B\,^3\Pi_g$ has a lifetime of 40 μs and, as for the other cyclic lasers discussed in Section 7.2.6, the transient inversion is achieved by the much larger excitation rate to the upper state. These lasers are excited using the TEA configuration discussed for CO_2 lasers and require a very fast rise time current pulse, typically less than 10 ns. A cavity is

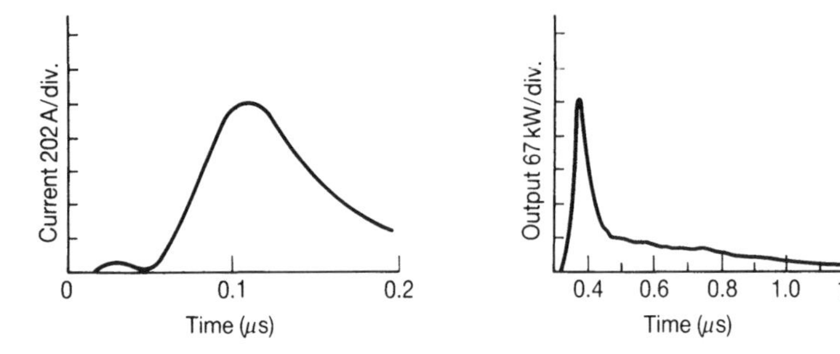

Figure 7.30 Current and laser pulse from a dielectric corona preionization pulsed discharge CO_2 laser. (Data reproduced by kind permission of W. J. Witteman (Ref. 11).)

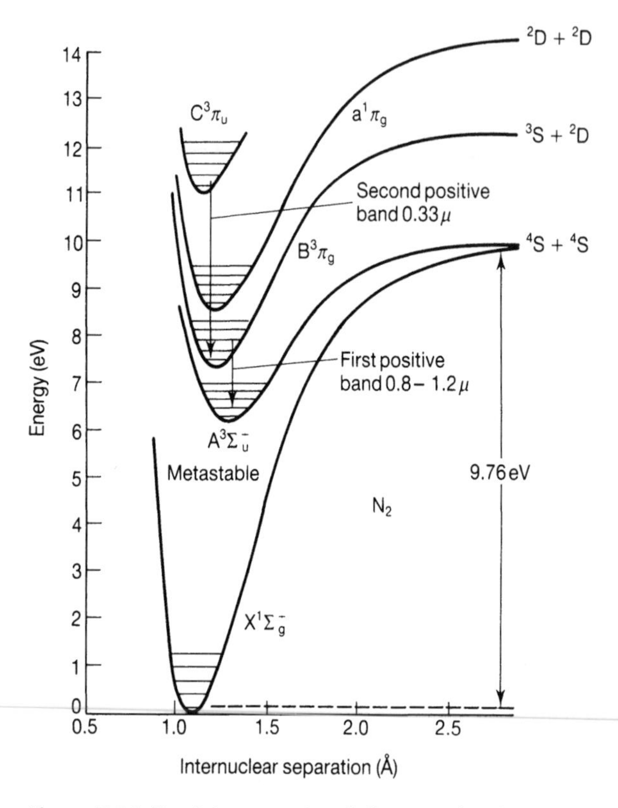

Figure 7.31 Partial energy level diagram for the nitrogen laser showing the two common laser transitions.

superfluous since the photons will only make a few roundtrips during the pulse. However, a high reflecting back mirror is used to gather the amplified light into a single-output beam. Output powers at 337 nm in excess of 1 MW lasting a few tens of ns can be achieved.

7.2.6.3 Excimer lasers

The term 'excimer' refers to a molecule which is bound in an excited electronic state but dissociated in the ground state. Figure 7.32 shows a schematic potential energy diagram of such a molecule indicating the vertical transitions between its electronic states. In this we have an ideal condition for producing population inversion since the lower laser state is unstable and, once the molecule finds itself there, it dissociates in about 10^{-13} s. Excimer molecules which have been studied [Ref. 16] for lasing include noble-gas dimers, metal-vapour dimers, rare-gas metal dimers and rare-gas halogen dimers. Rare-gas halogen dimers are the most important and our discussion will centre on these.

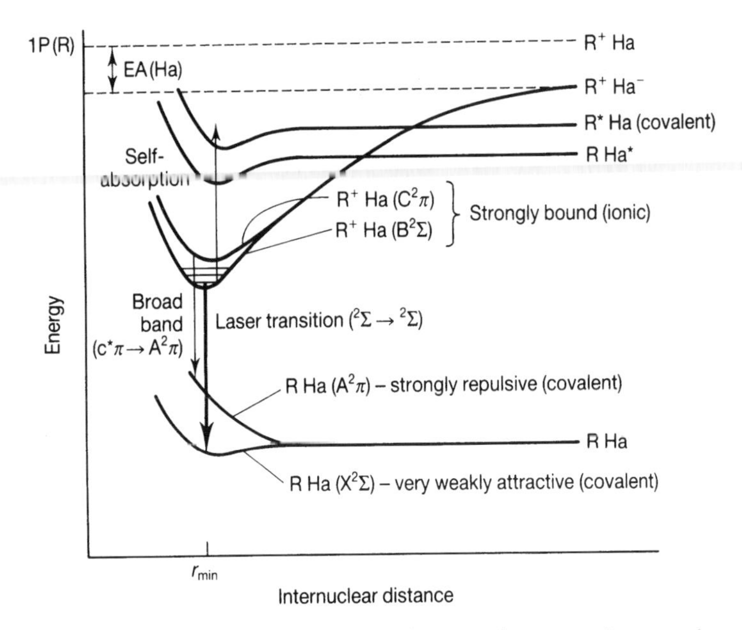

Figure 7.32 Potential energy-level diagram for an excimer molecule RHa (rare gas + halide).

The rare gases are normally thought of as being chemically inert, but if one of the outer electrons is excited it becomes more loosely bound and the atom will behave chemically as an alkali metal. The excited rare-gas atom will have a very much reduced ionization potential ($\sim$4–5 ev) and can be ionized when in the proximity of an electronegative halogen atom. The halogens readily form negative ions with an electron affinity of a few electronvolts. We define the electron affinity of a negative ion as the energy required to remove the attached electron. In Figure 7.32 the states of the free atoms which correlate to make the molecular states are shown on the right side of the diagram. The B state correlates with the separated ion pair $R^+(^2P) + Ha^-(^1S)$ where R represents a rare gas atom and Ha the halogen atom. It has a total energy of I.P. (R) – E.A. (Ha), where I.P. (R) is the ionization potential of the rare gas and E.A. (Ha) is the electron affinity of the halogen. This pair form a strong ionic bond indicated on the diagram by a minimum in the potential energy curve. On the other hand a weak covalent bond can be formed which correlates with either atom in an excited state. The energy of this ionic bonded state can then be estimated by assuming that the potential energy depression is due to the coulomb attraction between the ions.

$$\text{Energy of the B state} = \text{I.P. (R)} - \text{E.A. (Ha)} - \frac{e^2}{4\pi\varepsilon_0 r_{\min}} \tag{7.42}$$

where $r_{\min}$ is the bond length at the minimum in the PE curve.

Although electron beam excitation is used, particularly for very high-power systems, it is more convenient to excite these excimer-excited states using fast-pulsed discharges in high-pressure mixtures of rare gases and halogens by collisional reactions. A schematic diagram of the discharge arrangement and the power supply is shown in Figure 7.33. The TEA discharge is pre-ionized by ultraviolet radiation from the spark gap and the free electrons so produced gain energy from the main field and excite and ionize the noble gas atoms. The slow electrons in the energy distribution form negative ions with the halogen atoms by attachment. The bound upper state of the excimer can then be created by collisions between the positive and negative ions. As the ions approach each other (i.e. come in from the right in Figure 7.32) the potential energy decreases (Eq. (7.42)) and at some inter-nuclear separation the potential energy curve crosses that of the covalent state of the excited rare gas atom and the halogen. If this crossing occurs at relatively large internuclear separation then the probability of removing the electron from the negative ion is small and the system can move down the bound ionic state potential energy curve. A third body is necessary to remove the energy and allow the system to drop into the allowed vibrational states. This third body can be either another noble gas or a halogen atom. On the other hand, a collision between a metastable rare-gas atom and a halogen atom has almost unity probability of making the electron jump from the former to the latter and so taking the system on to the bound ionic potential energy curve. This reaction is called 'the harpooning reaction' and again requires a third body to remove the excess energy. Vibrational relaxation in the upper state allows most of the energy of excited states to be extracted in stimulated emission at one wavelength and the relatively flat repulsive lower state results in a tuning range of about 2 nm. In Table 7.5 the parameters of lasing excimer transitions are given and

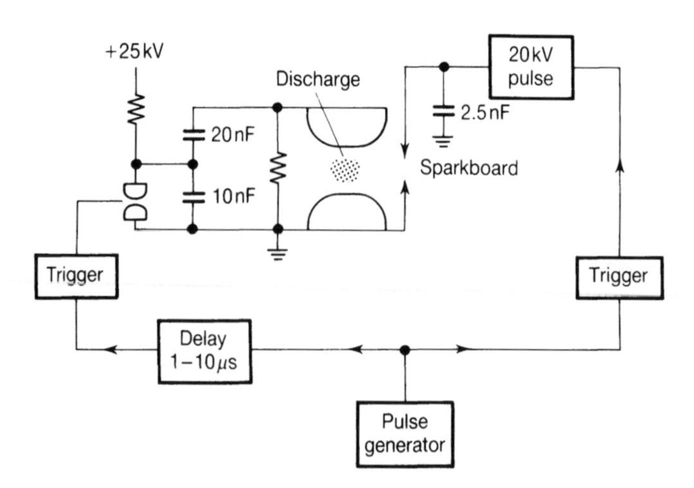

Figure 7.33 Schematic diagram of excimer discharge pumped laser.

Table 7.5 Rare gas–halogen lasing wavelengths, calculated lifetimes and stimulated emission cross-sections

	λ(nm)	τ(ns)	$\sigma_{SE}(10^{-16}\ cm^2)$
XeBr	282	12	2.2
XeCl	308	11	4.5
XeF	351	12–19	5.0
XeF (C → A)	490	~100	0.05
KrCl	222	–	–
KrF	249	6.5–9	2.5
ArCl	175	–	–
ArF	193	4.2	2.9

All transitions are $B \rightarrow X$ except for XeF, which also lases on the $C \rightarrow A$ transition.

we see that the wavelengths are shorter for lighter noble gases (higher ionization potential) and heavier halogens (lower electron affinity).

Excimer lasers can be up to 1 per cent efficient providing 30 ns ultraviolet pulses at rates up to 1 kHz. Average quasi-CW powers in excess of 100 W are achievable. Many excimer lasers use unstable cavities to improve beam quality, with high power being achieved with an oscillator amplifier combination. Stable cavity master oscillators with grazing incidence gratings for tuning and slave oscillators can produce Fourier transform-limited linewidths which are tunable over about 0.6 nm.

7.3 Optically pumped lasers

Optical pumping is a very important technique for producing population inversion and is applicable to solids, liquids and gases. In this section we will consider only its application to solid-state and liquid lasers. Light from either a coherent or incoherent source is directed, as efficiently as possible, into the gain medium. In systems which are pumped using a laser in the TEM_{00} mode it is possible, by careful cavity design, to overlap the lasing mode with the focused beam of the pump light. If the photon energy of the pump radiation corresponds to the energy of the excited state then almost total absorption of the pump radiation will create a high density of excited states. Subsequent fast decay of these states excites the upper laser state and emission at a longer wavelength than the pumping light results. In spontaneous emission this process is called fluorescence and the difference in wavelength between the pumping and emission is known as the *Stokes shift*. In solids packets of energy much smaller than the energy of visible photons can be released as quantized lattice vibrations known as phonons and this energy eventually is dissipated as heat.

Another important point is that this phonon energy cannot be absorbed and re-excite the state it came from. In liquids these packets of energy are in the form of rotational and vibrational energy of the molecules. As a method of inverting the population, optical pumping is often much simpler than using a gas discharge, and can provide alternative methods with different sources of exciting the same material. It often offers the alternative between obtaining pulsed and CW operation of a laser.

Since optical absorption can be very wavelength selective, especially if particular laser wavelengths are used, it is possible to attempt a theoretical model of the system using rate equations and so estimate the population inversion. Although the atom's energy level scheme is generally very complex we can often simplify it and consider only an N-fold system where N represents the number of levels involved in the excitation and lasing. It is not possible to achieve a steady-state (CW) population inversion with a two-level system since the pump transition and the lasing transition are at the same wavelength. Atoms pumped up will be induced to radiate by this very pump radiation. The best we can do is to distribute half the atoms in the upper state and only achieve the threshold of population inversion. At this point the medium is said to be saturated, and any further increase in the pumping radiation irradiance results in no further absorption (see Example 3.6). Population inversion can, however, be achieved in the three-level systems. This has a third level above the upper laser state and pumping into this state populates the upper laser state via fast radiationless transitions. This means that the pumping radiation will have a shorter wavelength than the laser light and the energy difference will appear as phonons and so heat.

A further and very important refinement is the four-level system in which the lower laser state is not the ground state. The schematic diagram (Figure 7.34) depicts the three- and four-level systems. If energy level 1 is so close to the ground state that it is significantly populated by the thermal energy of the system then we can consider the system to be a three-level one. The relative populations of levels 0 and 1 are determined by Boltzmann statistics:

$$\frac{N_1}{N_0} = \exp\left[-(E_1 - E_0)/kT\right] \tag{7.43}$$

Hence for a four-level system $E_1 - E_0 \gg kT$

and for a three-level system $\quad E_1 - E_0 \ll kT \quad$ (7.44)

Efficient pumping using a broadband source requires the pump levels (state 3) to be broad and the energy difference $(E_3 - E_2)$ to be much larger than kT so that the the states are not thermally mixed.

We can use a rate-equation approach to compare the threshold pumping required for the three- and four-level systems. The equations are modified to include non-radiative decay rates S_{32} and S_{10}, as well as pumping rate per atom, W_p, from the ground state to state 3. We will assume that the rate S_{32} is larger than any of the other decay rates of state 3 and consequently its population is zero, i.e. state 3 is

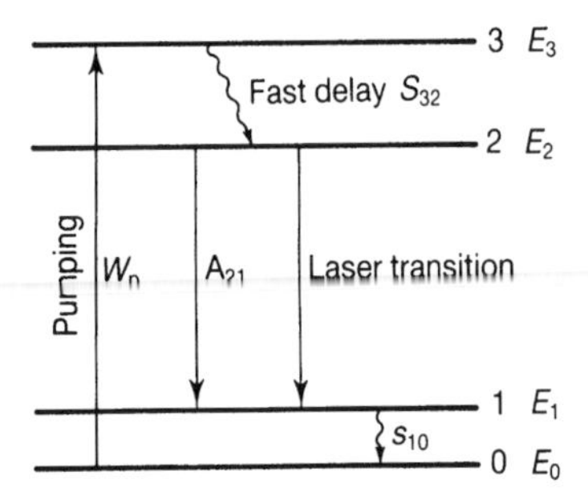

Figure 7.34 Schematic representation of energy levels for three- and four-level systems.

a necessary intermediate state which 'holds' the excitation for a very brief time. The rate equations can be written as before (Section 6.3):

$$\frac{dN_0}{dt} = -W_p N_0 + S_{10} N_1 \tag{7.45}$$

$$\frac{dN_1}{dt} = -S_{10} N_1 + (A_{21} + S_2) N_2 + \sigma_{se} \frac{c}{n} n_p \left(N_2 - \frac{g_2}{g_1} N_1 \right) \tag{7.46}$$

$$\frac{dN_2}{dt} = W_p N_0 - (A_{21} + S_2) N_2 - \sigma_{se} \frac{c}{n} n_p \left(N_2 - \frac{g_2}{g_1} N_1 \right) \tag{7.47}$$

Also we have from the conservation law that the total number of atoms, N_T, is given by

$$N_T = N_0 + N_1 + N_2 \tag{7.48}$$

We can expect that, for an effective lasing transition, the spontaneous decay rate A_{21} should dominate over the nonradiative phonon-assisted decay, S_2. It may be that state 2 can decay by other spontaneous transitions, and if this is the case the term A_{21} in these equations should be the total spontaneous decay rate. To determine the steady-state population densities we put the time derivatives equal to zero. To find the population inversion produced by the pumping we take the lasing photon density, n_p, to be zero. Solving Eqs (7.45) to (7.48) assuming equal degeneracies we then obtain for the steady-state population densities

$$N_0 = \frac{S_{10} A_{21}}{S_{10} A_{21} + S_{10} W_p + A_{21} W_p} N_T \tag{7.49}$$

$$N_1 = \frac{A_{21} W_p}{S_{10} A_{21} + S_{10} W_p + A_{21} W_p} N_T \tag{7.50}$$

$$N_2 = \frac{S_{10} W_p}{S_{10} A_{21} + S_{10} W_p + A_{21} W_p} N_T \tag{7.51}$$

The steady-state (or quasi-CW) population inversion becomes

$$N_2 - N_1 = \frac{W_p(S_{10} - A_{21})N_T}{S_{10}A_{21} + S_{10}W_p + A_{21}W_p} \tag{7.52}$$

This equation shows that, as discussed previously, we can only achieve population inversion if

$$S_{10} > A_{21} \tag{7.53}$$

In many solid-state lasers the phonon-assisted relaxation rate of the lower state is much larger than the spontaneous transition rate, A_{21}, and Eq. (7.52) can be approximated by

$$N_2 - N_1 \cong \frac{W_p S_{10} N_T}{S_{10}A_{21} + S_{10}W_p} = \frac{W_p N_T}{A_{21} + W_p} \tag{7.54}$$

Let us now turn to the three-level system in which energy level 1 is so close to the ground state that we can consider it as one state. The rate equations for the densities N_1 and N_2 are all that are required and are determined from two rate equations:

$$\frac{dN_1}{dt} = -W_pN_1 + (A_{21} + S_{21})N_2 + \sigma_{se}\frac{c}{n}n_p\left(N_2 - \frac{g_2}{g_1}N_1\right) \tag{7.55}$$

$$\frac{dN_2}{dt} = W_pN_1 - (A_{21} + S_{21})N_2 - \sigma_{se}\frac{c}{n}n_p\left(N_2 - \frac{g_2}{g_1}N_1\right) \tag{7.56}$$

This time the conservation law gives $N_T = N_1 + N_2$ (7.57)

Making the same assumptions as for the four-level analysis we can show that the steady-state atom densities are as follows:

$$N_1 = \frac{A_{21}}{W_p + A_{21}}N_T \tag{7.58}$$

$$N_2 = \frac{W_p}{W_p + A_{21}}N_T \tag{7.59}$$

The population inversion then becomes

$$N_2 - N_1 = \frac{W_p - A_{21}}{W_p + A_{21}}N_T \tag{7.60}$$

We see immediately from Eq. (7.60) that population inversion is only possible when the pumping rate is larger than the spontaneous decay rate of the laser transition, i.e. $W_p > A_{21}$. There was no such requirement for four-level systems since inversion occurs immediately there is any pumping and, as Eq. (7.54) shows, for small pumping rates the inversion increases almost linearly. Figure 7.35 illustrates these differences graphically between the three- and four-level systems showing the much lower pumping threshold required for the four-level systems.

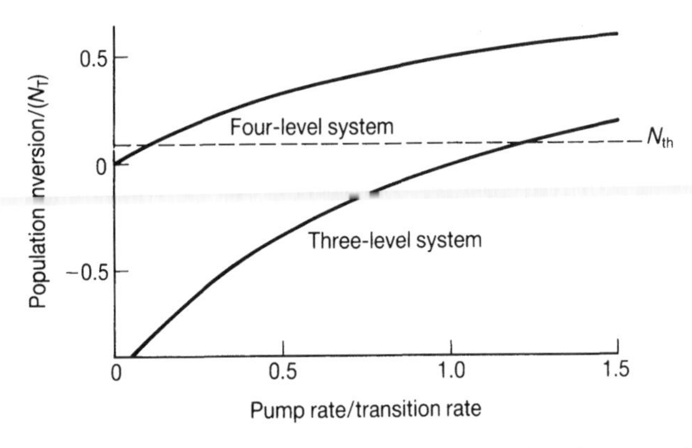

Figure 7.35 Comparison of the pumping for three- and four-level systems.

Equations (7.54) and (7.60) can be used to compare the pumping power required per unit volume to achieve lasing threshold for the two systems with the broad assumption that the atomic parameters are roughly similar. In both cases the pumping power per unit volume of active medium is given by:

$$\frac{P}{V} = (\text{pump rate/atom})(\text{total no. of atoms/m}^3)(\text{energy of pump transition})$$

$$\frac{P}{V} = W_p N_0 h\nu_{30} \tag{7.61}$$

At threshold we require $N_1 = N_2$ and using Eq. (7.57) for the three-level system we obtain

$$N_0 = N_1 = N_2 = \tfrac{1}{2} N_T \tag{7.62}$$

This simply means that half of the total atoms are in the upper laser state while the other half are in the lower state (which in this case is the ground state). Hence for a three-level system the threshold pumping power per unit volume becomes

$$\left(\frac{P_{th}}{V}\right)_{3\text{-level}} = W_p \frac{N_T}{2} h\nu_{30}$$

Using the requirement noted above that for the three-level system to arrive at threshold we require $W_p = A_{21}$, then

$$\left(\frac{P_{th}}{V}\right)_{3\text{-level}} = A_{21} \frac{N_T}{2} h\nu_{30} \tag{7.63}$$

We can produce a similar equation for the four-level system by putting the inversion density in Eq. (7.54) equal to the threshold inversion density to give

$$N_{th} = \frac{W_p}{W_p + A_{21}} N_T$$

Rearranging this gives an expression for W_p

$$W_p = \frac{A_{21}N_{th}}{N_T + N_{th}} \tag{7.64}$$

Substituting Eq. (7.64) into (7.61) gives

$$\left(\frac{P_{th}}{V}\right)_{4\text{-level}} = \frac{A_{21}N_{th}}{N_T + N_{th}} N_0 h\nu_{30} \tag{7.65}$$

In systems with high concentrations of active atoms, N_T, and with large stimulated emission cross-sections, then pumping rates much less than the spontaneous emission rate are satisfactory and we can put $W_p \ll A_{21}$ and $S_{10} \gg W_p$. This gives from Eq. (7.49)

$$N_0 \cong \frac{A_{21}}{A_{21} + W_p} N_T \cong \frac{A_{21}}{A_{21}} N_T = N_T \tag{7.66}$$

and for the population inversion (Eq. (7.54))

$$N = N_2 - N_1 = \frac{W_p N_T}{A_{21} + W_p} \cong \frac{W_p}{A_{21}} N_T$$

Returning to Eq. (7.65), we can put the ground state atom density equal to the total atom density and neglect the threshold inversion density compared with the total atom density and obtain

$$\left(\frac{P_{th}}{V}\right)_{4\text{-level}} = A_{21}N_{th}h\nu_{30} \tag{7.67}$$

We can compare the threshold pumping rates for the two systems through Eqs (7.63) and (7.67) by assuming the A_{21} coefficients and the pumping energies to be equal for the two cases. The ratio of the pumping power required for the two systems becomes

$$\frac{\left(\frac{P_{th}}{V}\right)_{4\text{-level}}}{\left(\frac{P_{th}}{V}\right)_{3\text{-level}}} = \frac{2N_{th}\nu_{30}}{N_T\nu_{31}} \tag{7.68}$$

Example 7.1

Compare two very important ionic crystalline lasers, namely the *ruby laser*, which is a three-level system, and the *neodymium in YAG laser*, which is a four-level system. The relevant parameters are as follows:

Ruby laser

Active ion concentration: 1.6×10^{25} m^{-3}

Spontaneous emission coefficient: 330 s^{-1}

Stimulated emission cross-section: 2.5×10^{-24} m^2
Pumping band frequency: 6.25×10^{14} Hz
Volume of active medium: 19×10^{-7} m^3
Mirror reflectivities: 100 per cent and 70 per cent
Using Eq. (7.63) gives: $P_{th} \cong 2138$ W
Using Eq. (6.3) the threshold inversion density is

$$N_{th} = \frac{\log_e(R_1R_2)^{-1}}{2l\sigma_{se}} = \frac{\log_e(0.7)^{-1}}{6.5 \times 10^{-23} \times 0.2} = 7.10 \times 10^{23} \text{ m}^3$$

Nd:YAG laser
Active ion concentration: 1.38×10^{26} m^3
Total radiative decay rate: 2×10^3 s^{-1}
Stimulated emission cross-section 6.5×10^{-23} m^{-3}
Pump frequency 4×10^{14} Hz
Length of gain medium: 0.1 m
Volume of gain medium: 19×10^{-7} m^{-3}
Mirror reflectivities: 100 per cent and 70 per cent
From Eq. (7.67) we get for the pump power required: $P \cong 27$ W
The threshold inversion density using Eq. (6.3) is 2.7×10^{22} m^{-3}

Clearly, significantly less pumping power is required to achieve threshold for the four-level laser than for the three-level system.

7.3.1 Solid-state lasers

The active medium of solid-state lasers consists of a crystalline or amorphous material containing a paramagnetic active lasing ion. The active ion can be substituted into the crystal lattice during growth or embedded in a suitable amorphous material such as glass. This active medium is normally in the form of a rod, slab or fiber and is optically pumped. The concentration of active ions can be controlled at the crystal growth stage and small variations in the ionic energy levels, and so the emitted wavelengths, occur between different host media. The range of active ions and host crystals together with many different glass compositions have given laser engineers a wide selection of pumping transitions, lasing wavelengths and material properties.

Solid-state laser host materials may be grouped into two types: crystalline solids and glasses. For crystalline hosts the additive ion should match the valency and size of the host ion it replaces and this, along with interactions between the ion and its host, restricts the number of possible material combinations. Good host crystals have the following properties:

1. Optically uniform with few scattering centres and little variation in the refractive index

2. Magnitudes of expansivity and elasticity moduli which ensure low stress when subjected to thermal loading
3. The lattice sites which readily accept the dopant ions and local crystal fields induce narrow energy levels for the lasing transitions and broad levels for the pumping levels. Long upper state lifetimes are desirable and stimulated emission cross-sections of at least 10^{-24} m^2, are required.
4. Good optical-quality crystals should be easily grown with uniform distribution of the dopant ion.

The compositions of crystalline hosts include simple and mixed oxides, fluorides and more complex formulations. We will review those hosts which have found wide use and the reader is referred to Refs 18 and 19 for a very complete discussion of solid-state laser materials.

- *Sapphire*: This simple oxide (Al_2O_3) is host for transition metals which can readily substitute the Al site. These crystals are mechanically hard and have high thermal conductivity.
- *Garnets*: Complex oxides having many of the desirable properties for laser hosts, i.e. hardness, optically isotropic and good thermal conductivity. They are normally used with trivalent rare earth ions. Some examples are:
  Yttrium aluminium garnet (YAG): $Y_3Al_5O_{12}$. This is by far the most important host and will be discussed in relation to the Nd : YAG laser.
  Gadolinium gallium garnet (GGG): $Gd_3Ga_5O_{12}$
  Gadolinium scandium aluminium garnet (GSGG): $Gd_3Sc_2Al_3O_{12}$
- *Aluminate*: The complex crystal yttrium ortho-aluminate ($YAlO_3$) termed Yalo or YAP. This crystal can be grown faster than YAG but more attention to melt purity is required. In contrast to YAG, the crystal is anisotropic and so the orientation of the lasing axis can be chosen for particular performance.
- *Chysoberyl*: Another complex oxide ($BeAl_2O_4$) which is a host crystal for the transition group elements especially Cr, producing a crystal called alexandrite which is an important tunable solid-state laser.
- *Fluorides*: One which has become important is *yttrium aluminium fluoride* ($YLiF_4$) termed YLF.

Other crystals which are used with trivalent rare earths include *yttrium aluminium borate* (YAB) and *lanthanum beryllate* (BEL).

Glasses provide an alternative host medium with their own advantages and disadvantages over crystals. Glass, being easier to fabricate, permits casting in large sizes and various shapes such as long rods, plates or optical fibers. Their optical quality is better than that of crystalline materials and they can be doped with a higher concentration of active ion and give a more uniform distribution. However, they have lower thermal conductivities which can cause induced birefringence and optical distortion at high energies. An important feature is the much larger spectral broadening in glass hosts. Since the fields surrounding the impurity ions are not well defined and vary throughout the material the broadening can be described as inhomogeneous. The larger linewidths improve the efficiency of broadband pumping with flashtubes but reduce the stimulated emission cross-sections. The net effect is that the pumping thresholds are higher and ASE is smaller with glass than with crystals. Generally, glass is preferred for pulsed operation and high-energy storage while crystals are preferred for CW operation.

The spectroscopy of metal ions in crystal lattice was discussed in Section 1.3.4 and here we give a brief summary of it. The active ions used are the paramagnetic ions of the iron group transition metals and the lanthanide rare earths. The actinides are similar to the rare earths but only U^{3+} in a CaF_2 host has been shown to lase. The outer shell electronic structure of the rare earths from cerium (58) to Lutetium (71) is

$$4d^{10}4f^{n}5s^{2}5p^{6}6s^{2}$$

where n increases from 2 with cerium to 14 with lutetium. The exceptions are gadolinium (64) and lutetium itself which add a 5d electron instead of a 4f electron.

When a divalent ion is formed the outer 6s electrons are removed and removal of one of the 4f electrons (unless the atom has a 5d electron) will produce a trivalent ion. The rare earth ions are formed with a closed outer shell similar to xenon – $4d^{10}5s^{2}5p^{6}$. As long as the 4f shell is not completely filled then electrons in the 4f levels can be raised by the absorption of light into their empty levels within the $4f^{n}$ manifold. Since these levels are shielded by the outer closed shells they are little affected by the surrounding electrostatic fields. The result is a small Stark broadening giving the 4f–4f transitions relatively narrow bandwidth. On the other hand, electrons excited into the 5d orbitals give broad energy levels which act as good pump bands for flashlamp pumping.

Iron transition group metals have an electronic configuration:

$$1s^{2}2s^{2}2p^{6}3s^{2}3p^{6}3d^{n}4s^{2}$$

The divalent ion is produced by the loss of two 4s electrons and the trivalent ion by the loss of an additional 3d electron. Chromium is an exception having only one 4s electron. Ions of this group which have been shown to lase in suitable hosts are:

$$Ti^{3+}(3d^{1});\ V^{3+}(3d^{3});\ Cr^{3+}(3d^{3});\ Co^{2+}(3d^{7});\ Ni^{2+}(3d^{8})$$

The most important crystals are *ruby* ($Cr^{3+}:Al_2O_3$), *alexandrite* ($Cr^{3+}:BeAl_2O_4$) and *Ti:sapphire* ($Ti^{3+}:Al_2O_4$).

In contrast to the rare earths the outer (3d) electrons are more strongly affected by the local crystal fields, and greater Stark broadening of the levels occurs. This helps the pump bands for flashlamp pumping but also means broader laser transitions. In some cases, such as alexandrite and Ti:sapphire, further broadening of the transitions occurs due to lattice vibrational assistance and these are known as vibronic lasers. These lasers allow tuning of the laser light over a relatively large wavelength ranges of up to 200 nm.

7.3.1.1 The ruby laser

Ruby was the first material in which lasing action was demonstrated. It is a three-level optically pumped system, consequently it has been replaced in most applications by the more efficient rare earth Nd : YAG laser. It will be discussed in some detail here not only out of historical reverence but also because it is a good example of a three-level system. Chromium is an important impurity in tunable solid-state lasers to be described in a later section. The ruby crystal contains 0.05 per cent by weight of Cr_2O_3 doped into the Al_2O_3 where the chromium ions replace the aluminium ions. Good-quality crystals can be grown by the Czochralski method whereby the solid crystal is slowly pulled from a liquid melt with the growth initiated on a high-quality seed crystal. The crystals have a hexagonal unit cell with the axis of crystal symmetry being the *c*-axis. The crystal is uniaxial with the ordinary ray having its *E*-vector perpendicular to the *c* (optic) axis and the extraordinary ray having its *E*-vector parallel to the *c*-axis. The energy level diagram for Cr^{3+} in Al_2O_3 is shown in Figure 7.36 and the absorption spectrum in Figure 7.37. The $^4T_1$ and $^4T_2$ (spectroscopic notation from group theory) bands are formed by splitting of the ^4F free ion state by the crystal fields. Strong absorption is in the blue and green. Similarly, the ^2G free ion state is split into three states (^2E; $^2T_1$; $^2T_2$) which are very much narrower. The ^2E state is metastable[7] with a lifetime of about 3 ms and is split into two states, the lower one being the upper laser state. We thus have the requirements of a successful three-level laser, i.e broad pump bands for flashlamp pumping and narrow laser levels with long lifetime upper laser state.

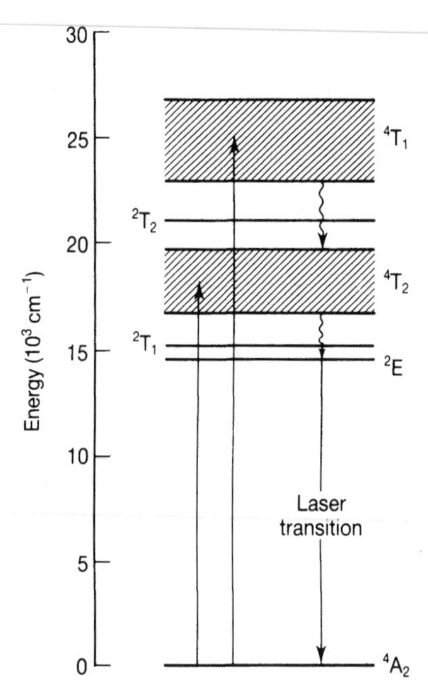

Figure 7.36 Energy-level diagram for ruby (Cr^{3+} : Al_2O_3). Spectroscopic notation is not conventional and the wavy lines indicate nonradiative transitions.

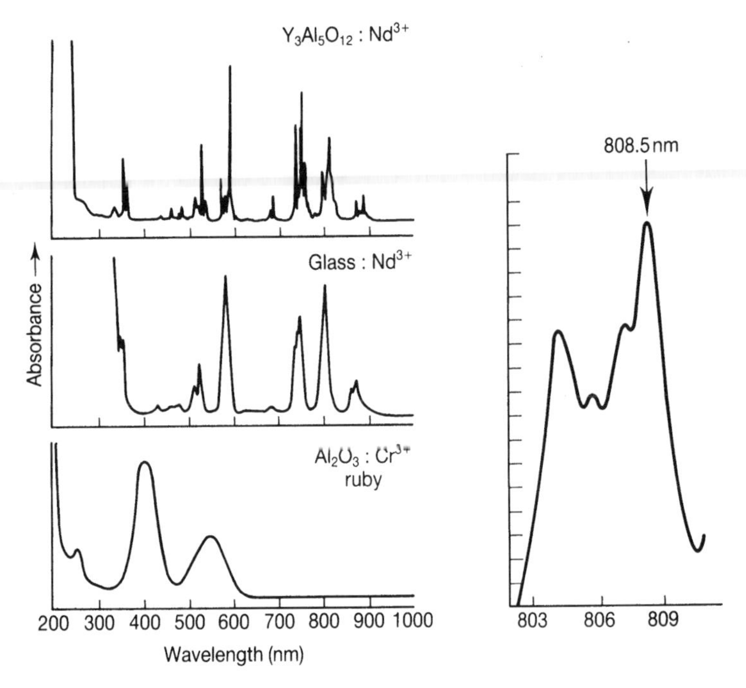

Figure 7.37 Absorption spectra for ruby, Nd in YAG and in phosphate glass. Also shown on the right is the expanded spectrum of Nd:YAG at 808 nm. (Data on left reproduced by kind permission of M. J. Weber (Ref. 18).)

7.3.1.2 Neodymium lasers

Neodymium doped into YAG crystal is a four-level system and is the most widely used solid-state laser. Its low pumping threshold, high energy storage capability and relatively narrow linewidth make it useful in numerous areas. Good-quality crystals which have about 1 per cent of sites occupied by Nd^{3+} can be grown by the Czochraski method and, although they take longer to grow than ruby, can produce rods up to 6 mm diameter and 100 mm in length. Since the effective radii of the two ions differ by about 3 per cent then significantly greater impurity concentrations create strained crystals which reduces the optical quality and increases the linewidths.

The energy-level diagram for Nd^{3+} in YAG is illustrated in Figure 7.38 and shows, on an expanded scale, the fine structure Stark splitting due to the electrostatic fields in the crystal. There are eight transitions in the wavelength region around 1.06 μm between the manifolds $^4F_{3/2} \rightarrow {}^4I_{11/2}$ as shown in the emission spectrum of Figure 7.39. With suitable intra-cavity etalons all can be made to lase. At room temperatures under normal operating conditions, lasing occurs at 1.064 μm, while, at lower temperatures, 1.061 μm has the lower threshold. The laser transition at 1.06 μm is therefore an overlap of the two transitions at 1.06414 μm and 1.0646 μm. Due to broadening (Stark effect and lattice vibrations), the gain curve

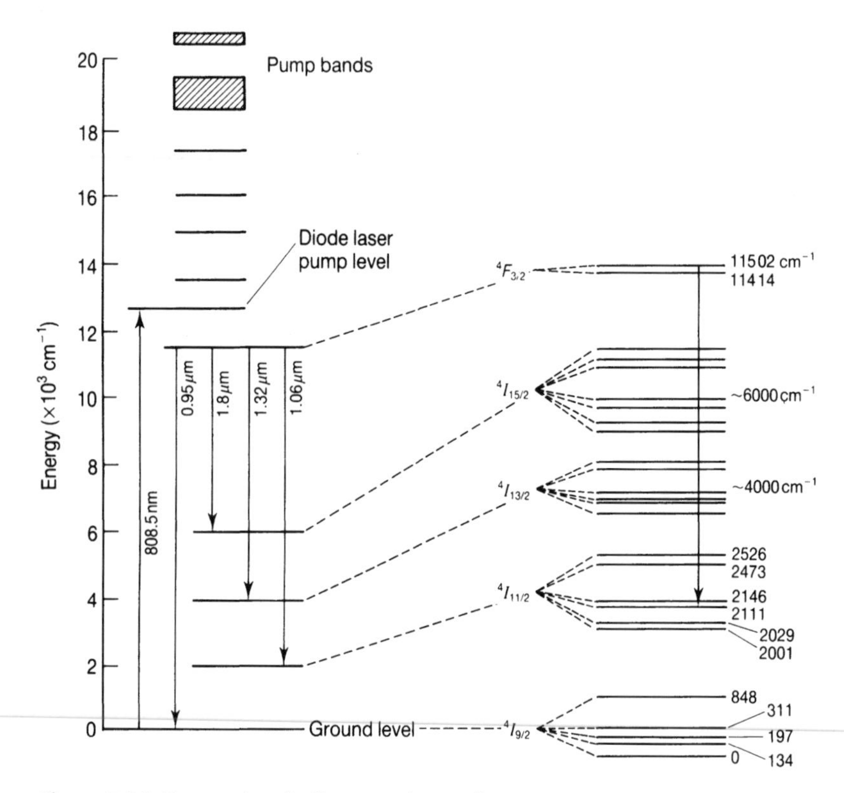

Figure 7.38 Energy-level diagram for Nd:YAG showing the important laser transitions and pumping levels.

is asymmetrical, being slightly weighted to the shorter wavelength as shown in Figure 7.40. Although different transitions contribute to the gain curve the thermal mixing and lattice vibrations make the transition homogeneously broadened.

The absorption spectrum for Nd : YAG, shown in Figure 7.37, displays much less broadband absorption than ruby. As well as the higher energy pump bands resulting from excitation of electrons to 5d orbits, there are three regions of absorption; namely 600 nm, 750 nm and 810 nm. Thus for efficient pumping it is important to ensure that the emission spectrum of the pumping source has a close overlap with the absorption spectrum of the rod. Flashtubes emitting mainly in the visible can emit at wavelengths which do not overlap the pumping transitions thereby wasting energy while semiconductor lasers, which can have their wavelength tuned about 810 nm by altering temperature and current, have emerged as extremely efficient and compact pumping sources. Figure 7.38(b) shows on an enlarged wavelength scale the absorption feature at 808.5 nm used for diode laser pumping of Nd : YAG.

Figure 7.38 illustrates other lasing transitions at 1.32 μm, 0.946 μm and 1.8 μm. Of these the transition at 0.946 μm has the lower gain since it behaves as a three-level system.

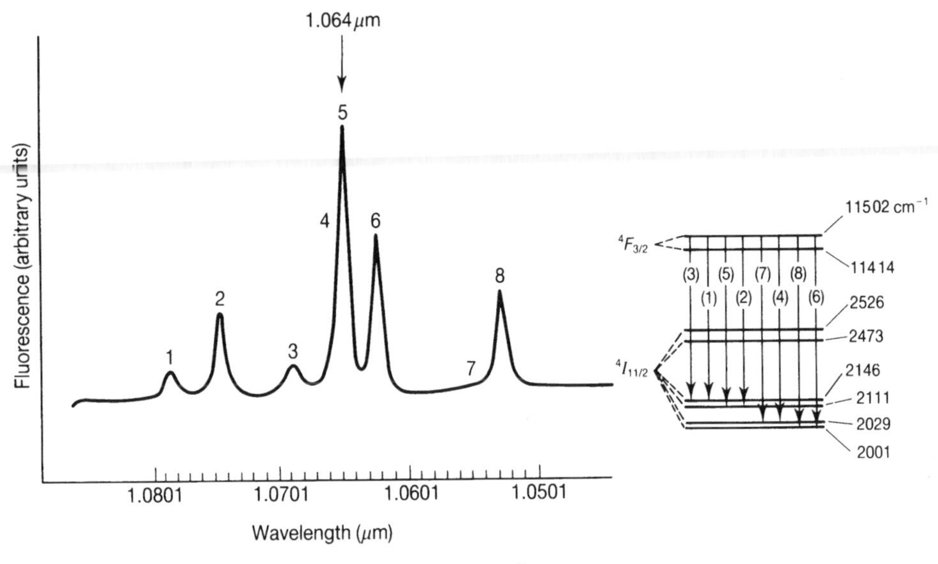

Figure 7.39 The fluorescence spectrum of Nd^{3+} : YAG at 300 K due to the transitions $^4F_{3/2} \rightarrow {}^4I_{11/2}$ in the region of 1.06 μm.

In contrast to crystalline hosts, both silicate and phosphate glass produce media of better optical quality, more uniform distribution of the neodymium and higher concentration of up to 6 per cent by weight of neodymium. The energy-level scheme for the lasing transitions is similar to the crystal but the energies are shifted slightly and the line broadening is much larger due to the large number of inequivalent sites. However, the material still behaves as a four-level system but with an inhomogeneously broadened transition. The fluorescent spectrum at 1.06 μm, shown in Figure 7.41, compares the doped crystal with doped glass and shows in the latter how the broadened spectrum includes all the eight lines at 1.06 μm and, in addition, transitions at 1.12 μm. The gain bandwidth is significantly larger and is weighted to the short wavelength side. The parameters associated with neodymium lasers at 1.06 μm are summarized in Table 7.6.

7.3.1.3 Other active ions and fluorescence sensitization

The efficiency of many solid-state lasers can be increased by co-doping the medium with other ions which can absorb the pump radiation more efficiently and transfer their energy to the upper laser state of the active ion. This transfer may occur radiatively or by phonon assistance. The sensitizer ion must not absorb at the laser wavelength and should have absorption bands which complement those of the active lasing ion.

The doping concentration must be high enough for the excitation transfer to occur during the lifetime of the activator. Chromium added to a host crystal along with neodymium uses the broadband absorption of the chromium in the blue/green for

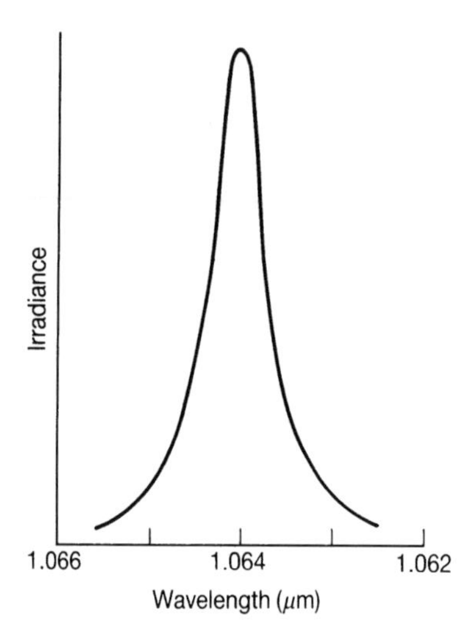

Figure 7.40 The fluorescence emission curve for the lasing transition at 1.064 μm for Nd : YAG.

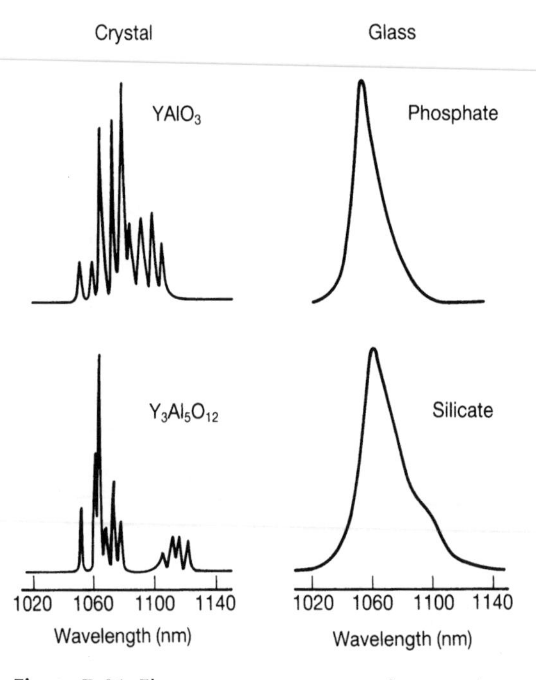

Figure 7.41 Fluorescent spectrum for neodymium in YALO,YAG and silicate and phosphate glass at 300 K. (Data reproduced by kind permission of M. J. Weber (Ref. 18).)

Table 7.6 Parameters for neodymium lasers and for the ruby laser

			Nd : Glass		
	Nd : YAG	Nd : YLF	Phosphate	Silicate	Ruby
Peak wavelength (nm)	1064	1053 (ordin.) 1047 (extra.)	1054	1062	694.3 692.9
σ_{SE} ($\times 10^{24}$ m^2)	65	30	4	2.8	2.5
Fluorescent lifetime (μs)	230	520	300	330	3000
Relaxation time $^4I_{11/2} \rightarrow {}^4I_{9/2}$	30 ns				Three-level
Index of refraction	182	$n_0 = 1.4481$ $n_e = 1.4704$	1.53	1.55	1.763 1.755
Fluorescence linewidth (nm)	0.45		19	26	0.53
Thermal conductivity (W/m/K)	13	6	1.0	1.3	42
Specific heat (J/kg/K)	590		710	920	760
Expansion coeff. ($\times 10^6$/K)	7.7	*a* axis 13 *c* axis 8	9.8	10.3	5.8

absorption of the pumping flashlamp radiation and couples the energy into the upper laser state of neodymium by broadband emission at 700–800 nm. The most successful host crystal for this is GSGG giving a doubling of the slope efficiency, defined in terms of output energy to input energy, over Nd : YAG.

Many of the rare earths can be made to lase very effectively using appropriate sensitizers. Of these, holmium and erbium ions have emerged as lasing ions giving wavelengths close to water absorption bands and hence finding applications in surgery. The Cr : Tm : Ho : YAG laser uses chromium as the sensitizer and transfers the excitation to the holmium via the thulium and lases at 2.1 μm at room temperature. Diode pumping using an absorption line in thulium for pumping also produces laser operation in holmium in the Tm : Ho : YAG crystal. Erbium which is sensitized with ytterbium or holmium and lases at 1.54 μm and 2.94 μm has been pumped using diode lasers.

7.3.1.4 Tunable solid-state lasers

When the photon emission is closely coupled to the emission of lattice vibrational quanta (phonons) the emitted light can be tuned over a wide wavelength range. The total energy of the lasing transition is fixed but can be partitioned between the phonons and the photons in a continuous way. These tunable solid-state lasers are known as *vibronic lasers*. Due to the special nature of their electronic structure the transition metal ions, discussed above, form the main vibronic lasers. The only rare earth used is the Ce^{3+} : YLF which is tunable between 300 and 400 nm when pumped with an excimer laser. Cerium lasers have high stimulated emission cross-sections ($\sim 8 \times 10^{-24}$ m^2) and short upper state lifetimes ($\sim$40 ns) and behave more like dye lasers (Section 7.3.2). Most vibronic lasers contain Cr^{3+} as the active ion and of these, alexandrite has received considerable commercial development. The energy level for the Cr^{3+} in alexandrite host is shown in Figure 7.42, and is, of course, very similar to that for ruby (Figure 7.36). It can lase as a three-level system analogous to lasing in ruby at a wavelength of 680 nm with a stimulated emission cross-section of 3×10^{-23} m^2. As a vibronic laser it is a four-level system with the upper laser level, $^4T_2$, thermally mixed with a storage level ^2E (the upper laser level for the 680 nm transition). The lower vibronic state causes the lasing transition to be broadened. At elevated temperatures both the upper and lower states increase their population densities, the former from the ^2E storage level and the latter from the ground state. The net effect is an increase in the gain and a shift to longer wavelength. The upper state lifetime determines the energy storage and, since this decreases as the temperature of the crystal is elevated, so the efficiency of *Q*-switching (see Chapter 8) decreases. Hence a compromise between gain and storage time is made with the design operating temperature.

Other materials for vibronic lasers are Co : MgF_2 (1.5–2.3 μm); Cr : GSGG (740–840 nm) and Ti : Al_2O_3 (660–1100 nm). Titanium sapphire has generated considerable interest since it has a large Stokes shift, good thermal properties and a high stimulated emission cross-section of 4×10^{-23} m^2. Titanium sapphire is pumped in the green with an argon–ion laser or a frequency doubled Nd : YAG laser as well as with flashtubes and appears to be a rival to dye lasers for tuned output in the red/infra-red.

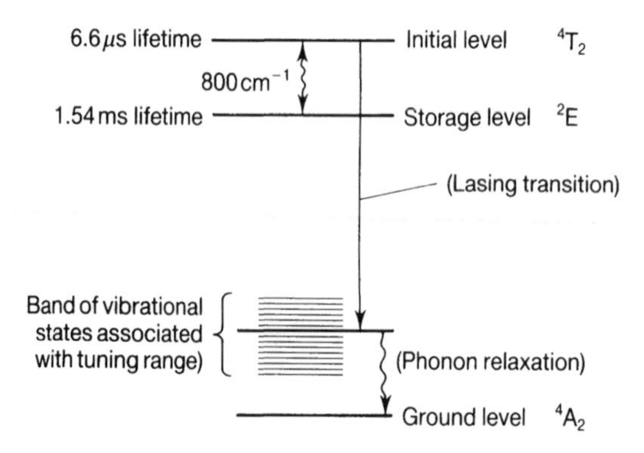

Figure 7.42 Energy-level diagram for chromium ions in alexandrite.

7.3.1.5 Real engineering systems

Any pump source must convert electrical energy as efficiently as possible into radiation at wavelengths in the absorption bands of the laser medium. Optical pump sources for solid-state lasers include noble gas and metal vapour discharge lamps, filament lamps, lasers and nonelectrical sources such as the sun. Until recent developments in high-power semiconductor lasers, the main pump sources have been noble gas flashlamps and filament lamps, the latter being used for CW operation only.

Flashlamps are either helical or linear quartz tubes with gas sealed in at pressures of between 300–700 torr. The electrodes are normally tungsten with the negative cathode having its work function lowered by impregnating it with thorium or barium and strontium. The electrodes are hard sealed to the quartz using a graded seal or an iridium solder via copper. Forced air cooling is used but for high-power application liquid cooling of the whole tube is necessary. At low currents the output spectrum of the noble gas-filled lamps is a combination of line and continuous spectra. At high power the line spectrum is broadened and the lamp behaves as a black body with an effective brightness temperature typically in excess of 5000 K. For pulsed operation xenon-filled lamps are used but, for pumping of Nd : YAG, krypton-filled tubes give an output spectrum which is a better match to the absorption bands, and these are preferred for CW lasers. The pulsed flashtube is driven by a charged capacitor which is discharged by triggering the flashtube with a 20 kV inductively coupled trigger pulse. A series inductor is used to slow the rise time of the pulse and for flat long pulses (~ 5 ms) a pulse-forming network is required. Light is coupled to the rod by placing the rod and flashtube in a pump cavity as shown in Figure 7.43. A common arrangement to optimize optical coupling between the flashtube and the rod is to shape the cavity as an ellipsoid of revolution with the rod at one focus and the tube at the other. From the geometry of ellipses rays from one focus pass through the other. Water cooling directly to the flashtube and the rod is necessary at high power levels, especially for CW operation. For details of flashtube and laser design the reader is referred to Ref. 19.

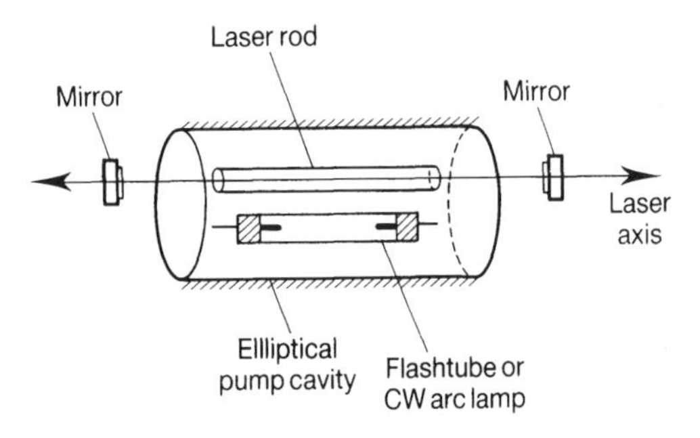

Figure 7.43 Diagram of pump cavity for a solid-state laser.

The advent of high-power semiconductor lasers, which can be tuned by altering temperature to emit at wavelengths that correspond closely to the absorption bands of the lasing medium, has allowed the development of very efficient compact solid-state lasers. With flashtube pumping energy is wasted, not only by poor spectral overlap but also by poor spatial overlap of the pump light and the lasing mode, thus heating the rod more than necessary. This in turn can produce thermal induced birefringence and optical loss. The overall efficiency (electrical to optical) for flash-lamp pumped lasers can be less than 1 per cent. With diode laser pumping this efficiency can be increased to around 30 per cent with the added bonus of miniaturization and no water cooling. This reduced thermal loading can allow the development of ultra-stable cavities and so lasers with extremely narrow linewidths. Neodymium lasers are pumped with radiation at 808 nm from a GaAs/GaAlAs semiconductor laser. The best spatial overlap between the pump and lasing light can be achieved by longitudinal or end pumping as shown in Figure 7.44. Light is pumped through one or both mirrors and the focusing optics is designed to achieve spatial overlap between the pump beam and the cavity mode. The pumping mirror has maximum reflectivity at the lasing wavelength and high transmission at 808 nm. More compact devices can be made with a monolithic cavity design in which the mirror coatings and curvatures are on the ends of the rod. Such a laser has pump thresholds of about 1 mW and can produce a single-frequency diffraction-limited beam. It effectively acts to improve the beam quality and spectrum of the diode laser as well as changing the output wavelength. To produce very high power (10 W to several kilowatts) transverse pumping geometry is necessary since many diode pump lasers are required. Side pumping of laser rods suffers from poorer matching of pumping radiation to the laser mode and a better approach is to pump a slab. The laser light is confined by total internal reflection and takes a zigzag path from one end of the thin slab to the other and thereby utilizes the full transverse extent of the gain medium. The flat edge of the slab allows it to be closely coupled to the output

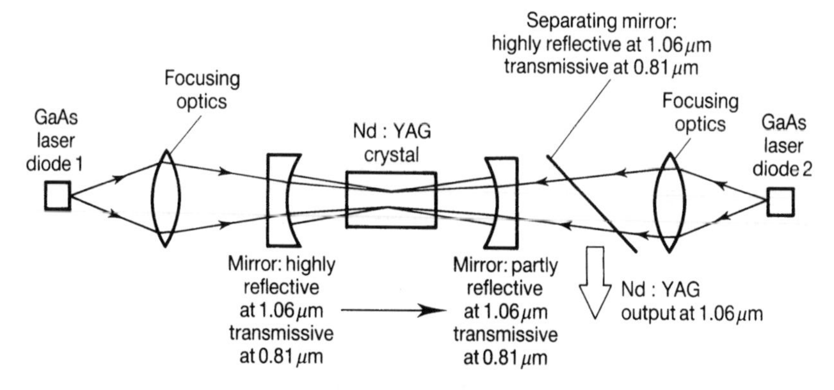

Figure 7.44 Diagram of end-pumped diode pumped solid-state laser. In this case the laser is pumped from each end.

surface of a diode array and efficient cooling can be effected by water-cooled heat sinks. Optical fibers can be used to couple light from a bank of laser diodes into the slab matched to the bounce separation of the zigzag path in the slab.

Solid-state crystal lasers are described by homogeneous broadening and so would be expected to produce high-power single-frequency output. Actively stabilized monolithic diode-pumped solid-state lasers can produce linewidths approaching the Schawlow–Townes limit due to their low heat dissipation and integral structure. However, at output power higher than about 1 mW spatial hole burning causes broadening of the linewidth and multifrequency operation. As discussed in Section 6.4.3, a unidirectional ring configuration has been shown to provide the most successful method of eliminating spatial hole burning. Oscillation in one direction only is achieved by having lower optical loss in the desired direction of propagation using a nonrecriprocal loss element or 'optical diode'. The cross-section diagram of such a device is shown in Figure 7.45. A stack of permanent magnets produces a very strong magnetic field (~ 0.2 T) in a magneto-optic material, so rotating the plane of polarization of light passing through it by the Faraday effect. The Faraday effect is reciprocal, meaning that light travelling in the opposite direction will be rotated in the opposite sense. The quartz crystal plate is a nonreciprocal element which rotates the plane of polarization by the same amount for light passing in either direction through it. This is the *chiral* effect. Hence in one direction the plane of polarization of the light is rotated by a small angle but in the other it is not affected. Loss will then occur at other Brewster surfaces in the cavity, such as the laser rod, for the direction of propagation in which the polarization rotation occurs. In a diode-pumped solid-state laser unidirectional operation can be accomplished by applying the magnetic field to the laser rod and using its own Faraday effect to rotate the plane of polarization of the light circulating in the ring cavity. To compensate for the Faraday rotation in one direction (i.e. perform the same function as the quartz plate) the ring is designed to operate in an 'out-of-plane' geometry. The schematic diagram of such a device (Figure 7.46), shows a solid block of Nd : YAG cut to give the desired out-of-plane ring cavity by five reflections and also the magnetic field perpendicular to the crystal provided by permanent magnets. This

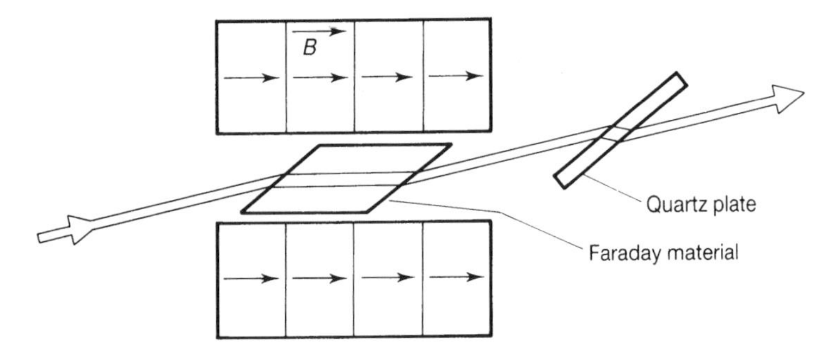

Figure 7.45 Cross-section diagram of an optical diode.

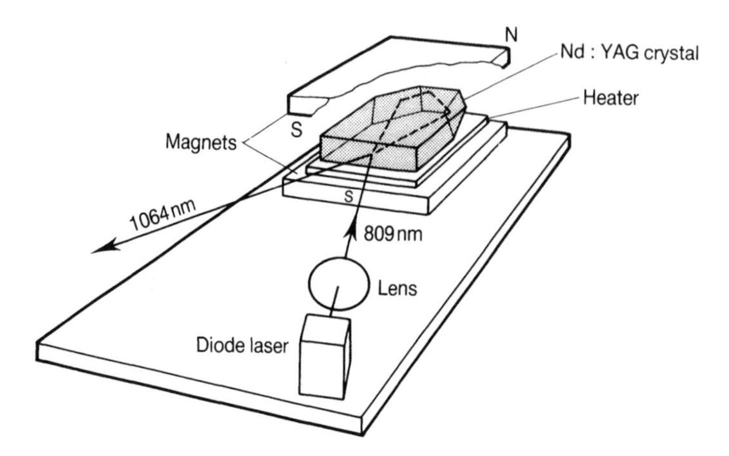

Figure 7.46 Schematic diagram of a monolithic nonplanar ring oscillator.

design has resulted in stable single-frequency operation and a laser linewidth of less than 3 Hz [Ref. 20].

In recent years glass optical fibers doped with rare earth materials have been used for optical amplifiers and for oscillators. As described in Section 2.1.7, optical fibers will transmit light over long distances by total internal reflection at the core cladding interface. The optical energy is contained in the core as modes of propagation (Appendix 3). Equation (2.72) gives the critical core radius for single-mode propagation and it is this single-mode fiber which is normally used for amplifiers and lasers. Fiber lasers are pumped with another laser due to the small diameter of single-mode fiber ($\sim 5\ \mu$m, see Example 2.1) and both argon–ion and diode lasers have been used for this. The fiber gain media can be made much longer than the normal rod gain media discussed above and hence very much less doping, typically 100 ppm, is required. In addition, the small cross-section of the core produces very high pumping irradiance and, as a result, low thresholds are achieved with these lasers. Neodymium-doped glass fiber lasers pumped with laser diodes at 810 nm and lasing at 1.088 μm have been widely studied having a pumping threshold of a few milliwatts and slope efficiencies of 30 per cent. In Figure 7.47(a) the pump light is focused into the fiber through one of the mirrors while in Figure 7.47(b) a wavelength selective coupler is used. Interest in these lasers for amplifier sections in long-haul telecommunications systems has centred on the rare earth erbium since it exhibits gain at 1.54 μm, which is the wavelength region for low loss in optical fibers. Erbium-doped fiber lasers can be pumped by diode lasers at 670 nm, 820 nm, 980 nm and 1.48 μm. If the gain bandwidth is large (as, for example, in erbium it is about 25 nm) then wavelength selection and tuning can be achieved by replacing the output mirror with a lens to expand the beam and a grating. The reader is referred to Ref. 22 for further information on fiber lasers.

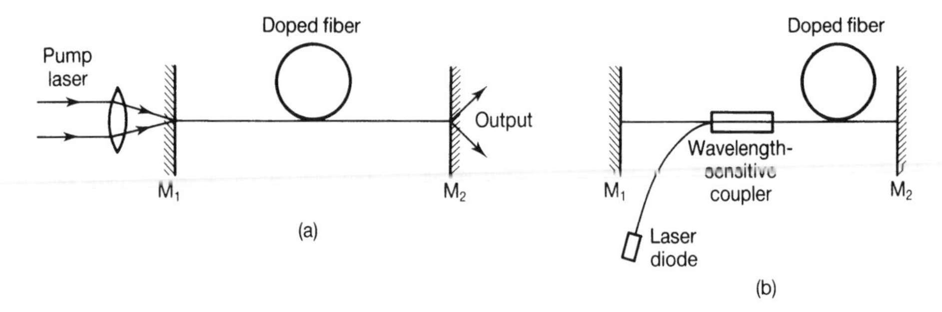

Figure 7.47 (a) End-pumped fiber laser. (b) Pumping by a wavelength-selective fiber coupler.

7.3.2 Dye lasers

Lasing in the liquid state is readily achieved by optically pumping organic dye molecules which are dissolved in a suitable solvent such as water, alcohol or glycol. These liquid lasers offer several advantages: ease of cooling by circulation, wide selection and tunability of wavelengths, less thermal strain than solids and ability to quickly change the gain medium. Disadvantages are: care must be exercised when handling these liquids since in some cases they are toxic and spillage stains are often very difficult to remove. In some cases, especially when pumped by ultraviolet, they have short lifetimes of the order of a few hours. The most important feature of dye lasers is their ability to tune continuously over a relatively large spectral range, typically 50 nm. The dye molecules have large complex structures with molecular weights of about 500 and many rotational and vibrational degrees of freedom. Thus each electron-excited state of the molecule will have an effective continuum of rotational and vibrational states associated with it. The energy-level diagram of such a molecule is shown in Figure 7.48 and is seen to be grouped into singlet (S) and triplet (T) manifolds of states. The singlet states have a total spin of zero while the triplets have total spin of 1 and the strong selection rule for radiative transitions of no spin change applies.

Figure 7.49 shows the absorption and fluorescence curves for the most common laser dye, rhodamine 6G. Broadband absorption of pump photons takes the molecule from the lowest state of S_0 to an excited state of S_1. We see that as a function of increasing energy (decreasing wavelength), the absorption curve in Figure 7.49 has a sharp rising edge and a much slower tail characteristic of excitation from a ground state to a quasi-continuum of states. Rearrangement of the molecule occurs in a few femtoseconds after which it relaxes to the lowest level of S_1, the excess energy appearing as heat. The transition back to S_0 takes place with a time constant in the region of 4 ns with the emission of broadband radiation the spectral shape of which is the mirror image of the absorption. It is in this tail of the fluorescence, where the overlap with the absorption is least, that lasing can occur. Finally, the molecule returns nonradiatively to the ground state in a few picoseconds.

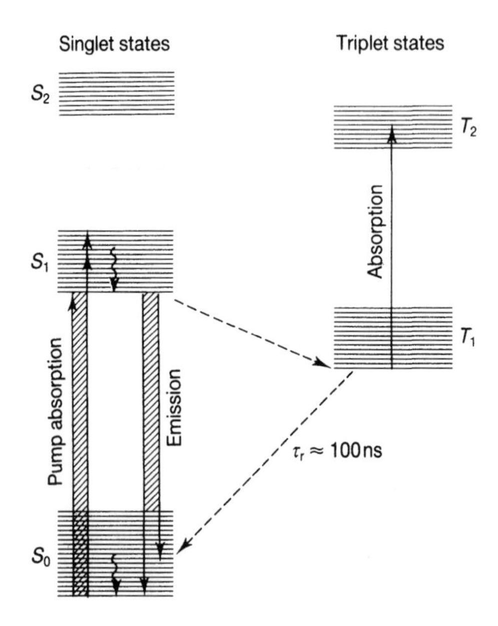

Figure 7.48 Energy-level diagram of an organic dye laser molecule.

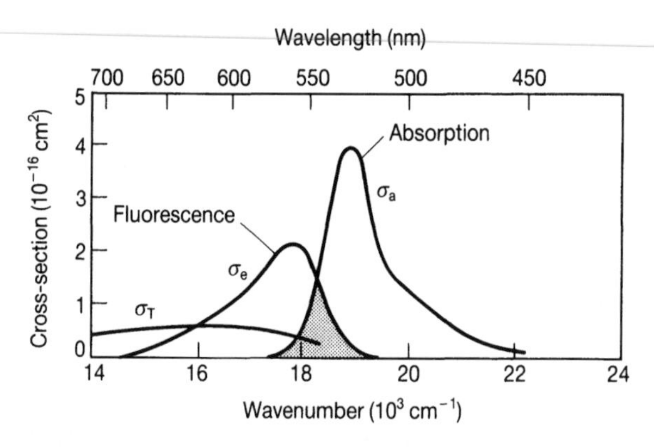

Figure 7.49 Absorption and emission spectra of a typical lasing dye, rhodamine 6G. Gain occurs under the fluorescent curve except in the region which overlaps with the absorption shown shaded. The absorption cross-section for the $S_0 \rightarrow S_1$ transition is σ_a, the emission cross-section is σ_e and the triplet absorption cross-section $T_1 \rightarrow T_2$ is σ_T.

We therefore have a four-level system similar to the solid-state lasers discussed in the previous section with the difference that the upper state lifetime is very much shorter and so has a larger stimulated emission cross-section (of the order of $2 \times 10^{-20}\,m^2$).

Molecules can be removed from the upper state by absorption of lasing or pump photons but a more important mechanism is their transfer to the triplet state T_1 with a time constant of about 50 ns. Since this T_1 state has a lifetime, determined by collisions, of about 100 ns then, under a steady state, the density of triplet molecules will build up and become similar to the S_1 molecules. Unfortunately, the absorption spectrum for the allowed transitions $T_1 \rightarrow T_2$ overlaps the laser emission spectrum and the triplet molecules behave like an intra-cavity absorber for the laser light. Fortunately, the higher singlet absorption spectrum $S_1 \rightarrow S_2$ overlaps more closely with the $S_0 \rightarrow S_1$ spectrum and so does not absorb the laser light. This triplet absorption has a significant effect on the gain and in some cases can drive the laser below threshold. Dye lasers are operated in three modes of operation: short pulsed, steady state or long pulsed and continuous wave.

7.3.2.1 Short-pulse laser-pumped dye lasers

In these lasers the pulse duration is shorter than the triplet state lifetime, typically 100 ns, so that the triplet states do not have time to build up to any appreciable density during the pulse. These devices do not reach equilibrium and, since the excitation duration is short, the number of roundtrips is small and high-gain short cavities are required. Figure 7.50 is a schematic diagram of a fast-pulsed dye laser. These lasers are pumped with another high-power pulsed laser giving the necessary pulse duration, rise time and wavelength for absorption by the dye. Fast laser pump sources include excimer for the ultraviolet, nitrogen for the near-ultraviolet, metal vapour and *Q*-switched frequency-doubled Nd:YAG for the visible.

The pumping light is focused transversely into the active volume contained in a dye cell so that a long narrow pumped volume results. This is achieved using a cylindrical lens for circular beams and a spherical lens for rectangular beams. The dye concentration is adjusted so that the excitation occurs within the Rayleigh range of the focused spot which will occur close to the wall of the cell. Amplified spontaneous emission (ASE) decreases the gain and appears as a background of high divergence

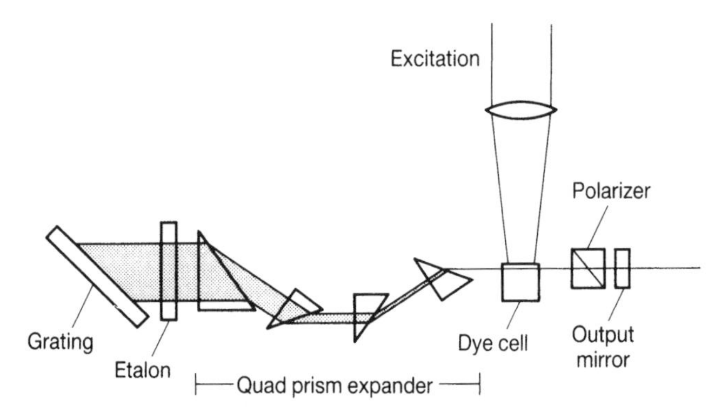

Figure 7.50 Schematic diagram of a short-pulse wavelength tunable dye laser using prisms to expand the intra-cavity beam.

and spectrally broadened radiation and limits the output power of these lasers. Arranging that the dye cell does not have parallel sides reduces parasitic oscillations within the cell.

A design aim for these lasers is to produce short narrow linewidth high-power pulses which can be wavelength tuned and so wavelength selection is an important feature. The laser shown in Figure 7.50 uses an intra-cavity etalon and a grating used in the Littrow arrangement such that maximum reflection back into the gain medium occurs when the grating condition is satisfied, and the angular setting of the grating with respect to the axis of the cavity selects the lasing wavelength. A notable feature of this cavity is the beam expander prisms which expand the beam in the one dimension perpendicular to the grating rulings so improving the resolution as well as reducing the power density incident on the grating. For extremely high output power a dye amplifier section can be combined with the oscillator and pumped with appropriate delay and ASE reduced by frequency and spatial filtering of the light. These amplifiers can be simple single-pass sections, multiple sections or regenerative using a cavity allowing a number of passes so that the irradiance will build up to levels which saturate the gain and efficiently extract the energy. In the latter case careful mode matching to the regenerative amplifier cavity is required.

7.3.2.2 Long-pulsed flashtube-pumped lasers

The second type of dye laser has the pump pulse lengthened to a few microseconds so that the triplet state population may be considered to be in equilibrium with the singlet state population. The addition of triplet quenchers such as oxygen, COT (C_8H_8), to the liquid can reduce the population of triplet states. Pulsed dye lasers of this type are pumped by very intense flashlamp sources and it is this application which has motivated developments in flashtube technology. Linear flashtubes similar to those used for solid-state lasers are used but with improved cooling, since a similar energy must now be dissipated in a very much shorter time (i.e. up to 100 J in 1–10 μs). An early design resembled that shown in Figure 7.43 for solid-state lasers. The rod is replaced by a tube or cuvette containing the dye which can be circulated, cooled and filtered. Wavelength tuning is achieved as before by using a Littrow grating (Figure 7.50).

Higher-energy devices use transverse excitation of the flowing dye as shown in Figure 7.51. The transit time of the active dye is shorter than longitudinal flow as used in a cuvette and the constriction increases the liquid flow rate. Turbulent flow of the liquid aids cooling by convection. A significant advance in flashlamp pumped dye lasers occurred with the development of the coaxial flashtube as shown in Figure 7.52(b). The flashlamp discharge region is a thin annulus surrounding the dye cell which leads to very efficient coupling of the pump light into the dye. A further advantage is the lower impedance of the discharge compared with the conventional filamentary flashtube. This enables much higher peak excitation to be obtained and single-pulse energies of up to 400 J for 10 μs have been achieved with these lasers.

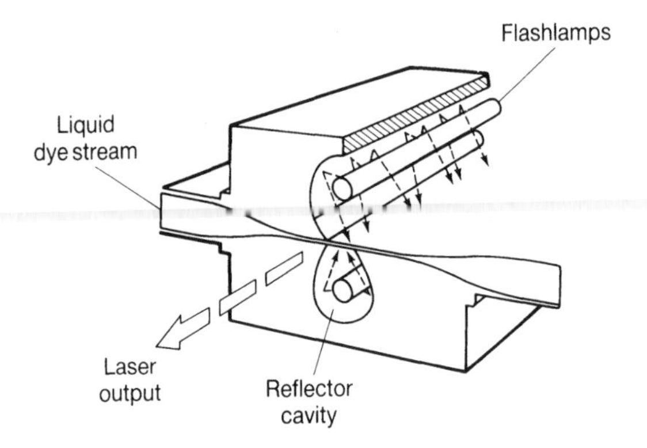

Figure 7.51 Flashlamp-excited dye laser with transverse flow.

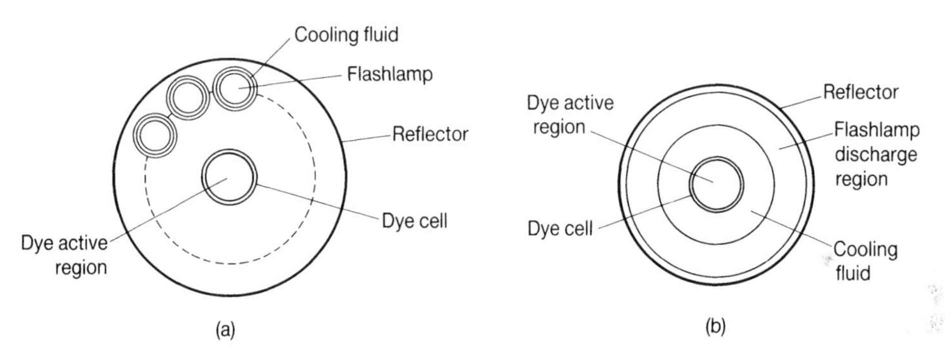

Figure 7.52 Cross-section diagram flashlamp-pumped dye laser. (a) Multiple lamp linear excitation configuration. (b) Coaxial flashlamp.

7.2.3.2 Continuous wave (CW) dye lasers

The development of continuous wave dye lasers has produced very narrow linewidth tunable lasers as well as mode-locked lasers with femtosecond pulse duration (see Section 8.4). For CW operation it is necessary to remove, as rapidly as possible, the accumulation of triplet-excited molecules in the active volume of the dye. This may be achieved by utilizing a very fast flow system with a small active region. Typically the active region is 30 μm in diameter and the liquid flow rates about 10 m/s, so the dye molecules only spend about 3 μs in the active region. This time is appreciably less than the triplet state lifetime. For successful operation the pump radiation must be focused efficiently into this extremely small active region and the dye resonator cavity must be designed so that this small region is within its mode volume. These requirements are met with the system depicted in

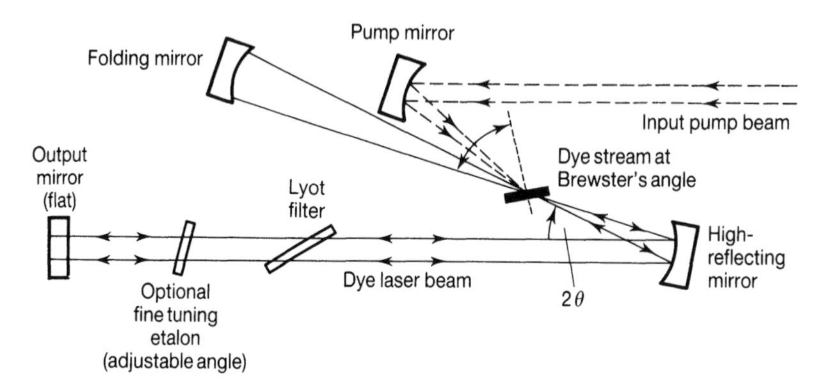

Figure 7.53 Three-mirror folded dye laser cavity using ion laser pump source.

Figure 7.53. The liquid is forced through a rectangular nozzle producing a jet of dye approximately 3 mm × 0.1 mm in cross-section. The jet has streamline flow characteristics and hence gives a good, optically uniform medium with smooth surfaces. The liquid is caught by an open tube and returned to a reservoir from where it is pumped through the jet via a filter to remove accumulated air bubbles and dirt. The fast flow serves to cool the liquid efficiently. Light is focused from a source which has a good Gaussian beam so that a tightly diffraction limited focused spot is achieved. This can produce pump irradiance in the region of 10^9 W/m^2. To date, the best lasers for this are the argon and krypton ion lasers.

Two mirror stable resonators which have such small waist sizes are either marginally stable or require mirrors with very short radii of curvature with small separation. The marginally stable cavities are difficult to align and the short cavities have no space for the wavelength tuning elements so important in dye lasers. The three-mirror cavity of Figure 7.53 produces a small beam waist where the active region in the jet is located and a long region of about 50 cm where the beam is almost parallel and where the wavelength selective optics can be located. By careful design of the resonator and the pump mirror the cavity waist and the pump beam waist can be made almost equal. The tilted folding mirror has two focal lengths; one in the tangential plane f_T – in Figure 7.53, the plane of the paper; and the other, f_s in the sagittal plane – perpendicular to the paper. These are given by:

$$f_T = f \cos \theta \tag{7.69}$$

$$f_s = f/\cos \theta \tag{7.70}$$

where 2θ is the folding angle of the resonator. This causes astigmatism and the resonator stability condition will not be the same in the two planes. The jet, although thin, being placed where the cavity is very sensitive to any optical element also causes astigmatism of the beam. Since the jet is set at the Brewster angle its optical thickness is different in the sagittal and tangential planes. This means that the beam expansion in one plane is different from the other plane. Fortunately, the two sources of astigmatism have opposite signs and a folding angle can be found such that they cancel. This angle can be calculated from:

$$f\,(\sin\theta)\,(\tan\theta) = t(n^2 - 1)\,\frac{\sqrt{n^3 + 1}}{n^4} \tag{7.71}$$

where t is the thickness of the jet and n its refractive index.

Typical operating conditions for a rhodamine 6G CW dye laser which is pumped using 514 nm or all lines from an argon ion laser are as follows:

High-viscosity solvent ethelyene glycol with concentration adjusted for 85 per cent absorption of the pump radiation
Polished stainless steel jet 3×1 mm
Folding angle $2\theta = 4^\circ$
Argon ion laser pump power at 514 nm = 3 W
Pump mirror radius of curvature = 50 mm
Folding mirror radius of curvature = 75 mm
High reflector radius of curvature = 50 mm
Output coupling of plane mirror = 15 per cent
Output power of dye laser = 1 W

The gain is homogeneously broadened and, without any wavelength selector in the cavity, lasing will occur at the peak gain wavelength which is higher than the peak of the fluorescent curve. The oscillation may jump rapidly between a number of cavity modes. Wavelength selection can be effected using a prism or a Lyot filter. The latter has low loss and gives better dispersion and so is used in all commercial CW dye lasers. It consists of three quartz birefringent plates with their optic axis in the plane of each plate and all aligned parallel. The thickness of each plate is integrally related to the others, for example 1 : 4 : 16 is commonly used [Ref. 23]. The plates are held at the Brewster angle in the cavity and can be rotated about an axis perpendicular to their surface. The wavelength selectivity of the filter is a consequence of the rotation of the plane of polarization of the light by the plates and the polarization dependence of their reflectivity at the Brewster angle. At a particular wavelength the light will pass through the plates without any change in low cavity loss *p*-polarization state and so suffer no reflection loss. The use of multiple plates integrally related narrows the bandwidth and restricts the oscillation to a few cavity modes as shown in Figure 7.54. Rotation of the plates about their perpendicular axis changes the optical thickness of the plates and thus wavelength tunes the laser.

Analysis of Figure 7.54 reveals that the longitudinal mode spacing is three times that expected from the free spectral range of the cavity. Since the gain is homogeneously broadened then oscillation at a single frequency would be expected. However, spatial hole burning allows frequencies to lase which are 90° out of phase at the jet where amplification occurs. The condition that two optical frequencies, ω_1 and ω_2, will oscillate, is then that the distance from the high reflector to the jet takes them 90° out of phase since they are in phase at the high reflector, this point

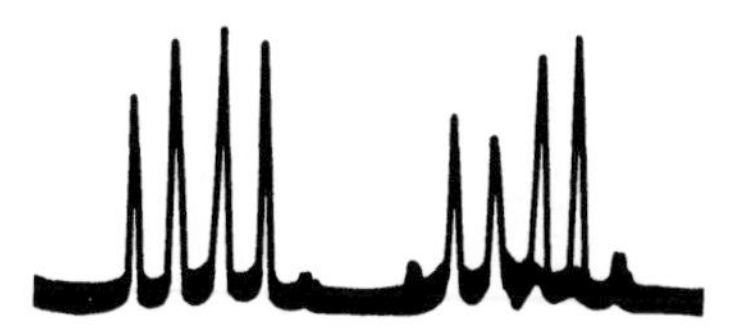

Figure 7.54 Two Fabry–Perot (8 GHz FSR) scans of the longitudinal modes of a CW dye standing wave laser similar to Figure 7.53 but without the intra-cavity etalon. The mode spacing is calculated to be 1.12 GHz for a cavity length of 0.565 m.

being a node in the standing wave field. Since the distance of this mirror to the jet is its radius of curvature R, we can write

$$\frac{\omega_1 R}{c} - \frac{\omega_2 R}{c} = \frac{\pi}{2}$$

$$\omega_1 - \omega_2 = \frac{\pi c}{2R}$$

The frequency difference in hertz then becomes:

$$\Delta\nu = \frac{c}{4R_2} \qquad (7.72)$$

The laser used to produce the spectrum in Figure 7.54 had a high reflector with $R = 75$ mm giving a predicted mode separation of 1.12 GHz as observed. The cavity had a total length of 0.565 m and a free spectral range of 265 MHz, showing that two longitudinal modes are missing from the spectrum.

Further narrowing of the gain bandwidth can be achieved with one or possibly two intra-cavity etalons but the output power is limited due to spatial hole burning and, as for solid-state lasers, this can be overcome by using a ring configuration. Figure 7.55 is a schematic diagram of such a laser used by Johnston *et al.* [Ref. 24], who have reported a very complete parameter study of this laser through the spectrum from 400 to 900 nm. The cavity is derived from the standing wave cavity of Figure 7.53 by tilting the plane mirror and introducing another high reflector which in this case is curved in order to produce another waist in the cavity for astigmatism correction. The cavity has a Faraday effect 'optical diode' as described in Section 7.3.1.5 to force oscillation in one direction. A Lyot filter allows broad-band tuning of the lasing wavelength. The galvanometer-tipped thin etalon and the PZT-scanned thick solid etalon are used for single-frequency scanning. The vertex-mounted Brewster plate and the PZT-mounted folding mirror constitute the cavity length scanning covering about 30 GHz. Active stabilization of the laser frequency is achieved by obtaining an error-correction signal (see Section 6.6.4) from an atomic absorption spectrum or from a reference cavity. The signal in this case is from a reference cavity and is applied to the folding mirror PZT and Brewster plate

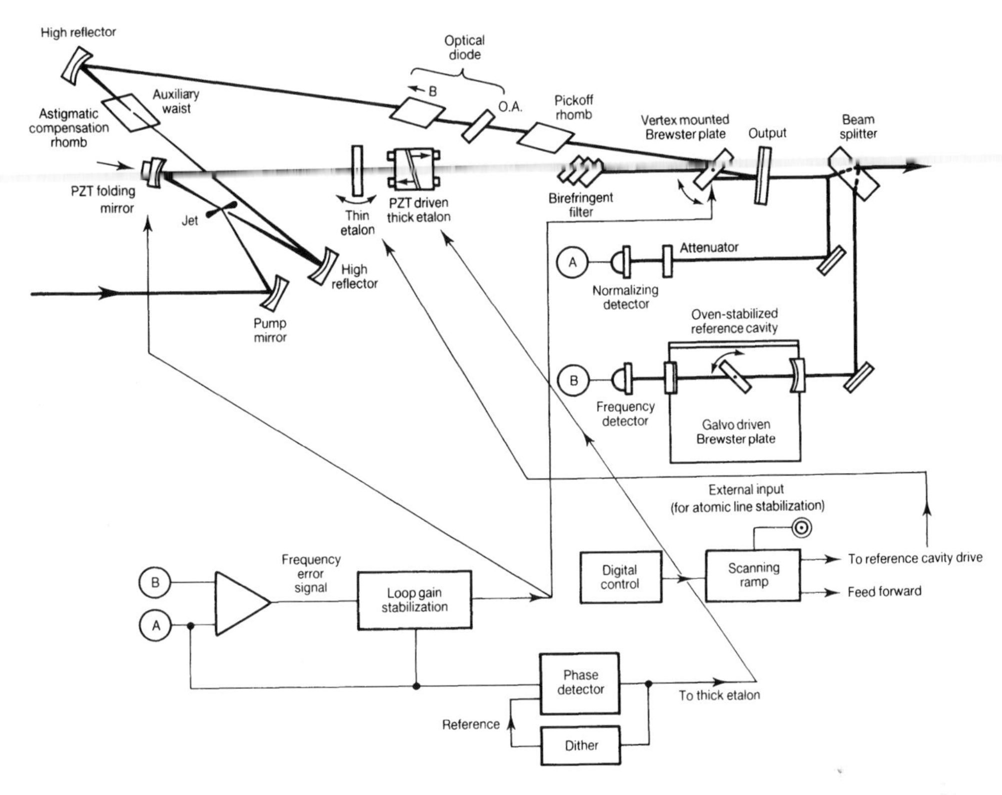

Figure 7.55 Schematic diagram of single-frequency CW dye laser.

from the servocontrol amplifier. Frequency stability of better than 1 MHz is commercially obtained with these lasers with sub-hertz stability obtained with the Pound–Drever stabilization scheme (Section 6.6.4). The CW ring dye laser is a source, tunable through the visible spectrum, of stable single-frequency high-power light. This device plays an important role in absorption spectroscopy and, in particular, Doppler free spectra using saturated absorption.

7.4 The semiconductor laser

The semiconductor laser occupies a rather special position among the various types of laser. In contrast to most others it is very efficient (≈ 30 per cent), has very simple power supply requirements (i.e. a few volts with currents from 10 mA up to several amps) and is physically very small. During the initial development stage, however, there were several problems – operating lifetimes were quite short and output powers small. In addition, the first wavelengths that were available were in the near-infrared ($\approx 900\ \mu$m). Since then most of these problems have been overcome. Lifetimes and powers have become very respectable and the range of wavelengths available has now been extended both down into the red region of the spectrum and further out into the infrared.

7.4.1 The homojunction semiconductor laser

At its simplest a semiconductor laser consists of a semiconductor p–n junction, as shown in Figure 7.56, operated in forward bias. Because the materials on either side of the junction are formed by doping the same basic starting material, such junctions are termed *homojunctions*. Lasing takes place in the plane of the junction when a sufficiently high current flows. Usually there are no external mirrors, the high gain of the laser medium ensures that it is possible to rely on the natural reflectance of a cleaved end face of the semiconductor crystal even though this only has a value of about 0.32 (see Example 7.2). Lasing in a direction perpendicular to the required direction is prevented by roughening the appropriate end faces.

Example 7.2 Reflectance at a GaAs/air interface

Using the Fresnel equations at normal incidence (i.e. Eqs (2.56) and (2.60) with $\theta_i = \theta_t = 0$) gives a reflectance of

$$R = \left(\frac{n_2 - n_1}{n_2 + n_1}\right)^2$$

The refractive index of typical semiconductor laser materials is quite high, for example in GaAs $n = 3.6$, so that

$$R = \left(\frac{3.6 - 1}{3.6 + 1}\right)^2$$

$$= 0.32$$

Under forward bias the current flows are similar to those in light-emitting diodes as described in Chapter 3. Electrons in the conduction band of the n-type material are injected across the depletion region and into the p-material while holes are injected in the opposite direction from the p to n material. These excess carriers then diffuse away from the junction, recombing as they do so. According to the discussion in Section 3.3.4, a situation of population inversion can be obtained between a level E_2 in the conduction band and a level E_1 in the valence band provided the quasi-Fermi levels are such that (Eq. (3.153))

$$E_2 - E_1 < E_{Fc} - E_{Fv}$$

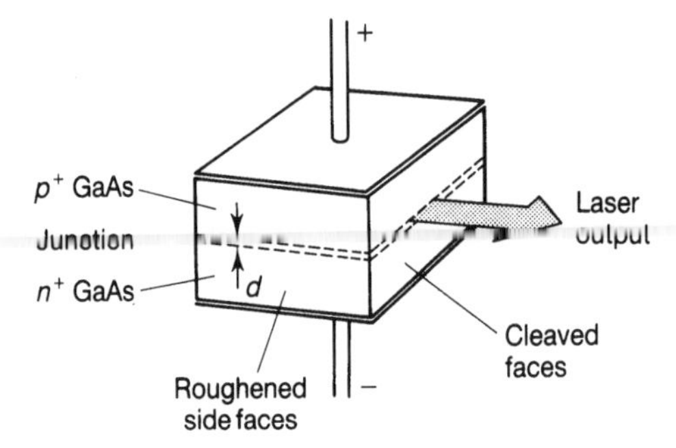

Figure 7.56 Schematic construction of a semiconductor homojunction diode laser based on GaAs. Lasing takes place in the plane of the junction, the length being typically a few hundred microns.

Figure 3.15 shows the relative gain curves predicted for different quasi-Fermi levels. We note from this that the gain starts from zero at the bandgap frequency (i.e. when $\hbar\omega = E_g$) increases to a maximum and becomes negative when $\hbar\omega > E_{Fc} - E_{Fv}$. However, to deal quantitatively with a particular situation is somewhat involved. The relationship between the electron carrier density and the quasi-Fermi level E_{Fc} may be derived by using the function $F_c^e(E)$ in place of $F(E)$ in Eq. (3.36), thus:

$$n = \int_{E_c}^{E_T} D(E)F_c^e(E)\,\mathrm{d}E \tag{7.73}$$

The density of states function, $D(E)$, is given by Eq (1.59). The integral must then be evaluated numerically[8] for a particular value of E_{Fc}. Similar calculations can be done for the valence band. Once the values of E_{Fc} and E_{Fv} are known as functions of n and p, then they can be substituted into Eq. (3.151) to give the gain for a particular carrier concentration as a function of photon frequency (or energy). The curves will be similar to those shown in Figure 3.15. Of particular interest is value of the peak gain. Figure 7.57 shows this as a function of carrier concentration in GaAs (the results are taken from Ref. 25). Carrier densities of about 2×10^{24} m^3 are readily achieved leading to gains as high as 10^4 m^{-1} or so. Such high gains imply that the length of the gain medium need only be quite small, indeed commercial semiconductor lasers are typically some 250 μm long.

The involved nature of the above calculation arises because the carriers are distributed over a band of energy levels, the overall energy spread being about $2kT$ (Section 3.1.6). At a sufficiently low temperature, however, we may assume the carriers are effectively in 'single' energy levels. This enables us to use Eq. (3.106) directly. Thus putting $N_1 = 0$,[9] $A_{21} = 1/\tau_{21}$ and $g(\omega_0) = 1/\Delta\omega$,[10] the gain can be written[11]

$$\gamma(\omega_0) = \frac{N_2\pi^2c^2}{\Delta\omega\omega^2n^2\tau_{21}} \tag{7.74}$$

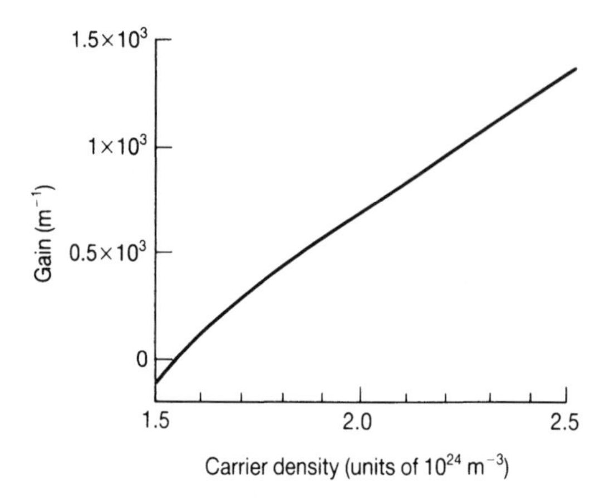

Figure 7.57 The calculated peak gain in GaAs as a function of the carrier density.

or, writing N_2 in terms of $\gamma(\omega_0)$,

$$N_2 = 8\pi\gamma(\omega_0)\Delta\nu n^2/\lambda_0^2 \tag{7.75}$$

We now consider the current densities required to achieve threshold lasing conditions. The situation in the homojunction laser is complicated by the fact that the carriers injected into the active region diffuse away from the junction, with the result that the carrier concentration (and hence gain) varies within the gain region. The carrier flow is governed by the so-called *continuity equation*, which, for the excess electron population ($\Delta n(x)$), can be written [Ref. 26]

$$\frac{d^2\Delta n(x)}{dx^2} - \frac{\Delta n(x)}{\tau_e D_e} = 0 \tag{7.76}$$

where τ_e and D_e are the electron lifetime (i.e. τ_{21}) and diffusion coefficient. The equation is readily solved to yield

$$\Delta n(x) = \Delta n(0)\exp(-x/L_e) \tag{7.77}$$

where L_e ($=\sqrt{D_e\tau_e}$) is known as the *diffusion length*. This exponential decay in excess carrier population with distance from the junction obviously causes problems from the point of view of the present calculations.

To simplify the analysis we make the (rather drastic!) approximation that the injected carriers are uniformly distributed over a length equal to the diffusion length. (In fact, L_e is the *average* distance an electron diffuses away from the junction before recombining, see Problem 7.16). The situation for hole diffusion is similar to that for the electrons. Thus the thickness, *d*, of the region over which we expect population inversion is then, approximately, $L_e + L_h$.[12]

In GaAs the value for D_e is 0.0015 $m^2 s^{-1}$, assuming a lifetime of 4×10^{-9} s, which gives $L_e = 2.4$ μm. The value for D_h is some twenty times smaller than D_e so we may ignore the contribution from L_h. Thus d is approximately 2 μm, and comparable to the wavelength of the laser light itself. Unless some form of confinement of the radiation is provided the light will spill out into the surrounding highly absorbent regions. A small amount of confinement is provided by the fact that the refractive index of the material is slightly increased by the presence of excess electrons. The increase is typically between 0.1 per cent and 1 per cent and creates a rather weak optical waveguide (see Section 2.1.7).

Another guiding process is provided by the gain itself. A mode which has too much of its field outside the gain region may not be able to achieve threshold. Whatever the guiding mechanism, we may assume that the mode field will extend for some distance outside the gain region (Figure 7.58). We designate the fraction of the total mode power that is within the gain region as Γ_m, and make allowance for the fact that not all the mode field 'sees' gain by reducing the effective gain within the gain region by the factor Γ_m.

If the current density (i .e. current per unit area of the semiconductor) is J A m^{-2} the number of electrons being injected into unit volume of the active volume per second is $J/(ed)$. The electron flow per unit volume required to maintain an electron density of N_2 is then given by

$$\frac{J}{ed} = \frac{N_2}{\tau_e}$$

so that

$$J_{th} = \frac{N_2^{th} ed}{\tau_e} \tag{7.78}$$

Calculations of J_{th} for a GaAs laser are given in Example 7.3. The values obtained when using Figure 7.57 to determine N_2^{th} are in reasonable agreement with those

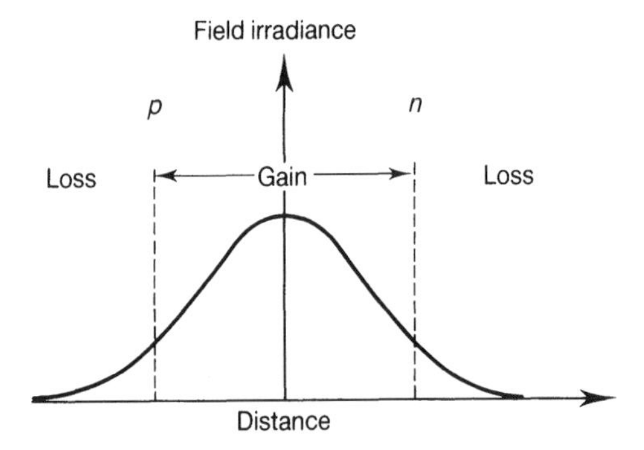

Figure 7.58 Illustration of the spreading of the mode field in a semiconductor laser out of the gain region.

determined experimentally at room temperature. As expected, the use of Eq. (7.75) to determine N_2^{th} results in much smaller values (by about an order of magnitude) for J_{th} which only agree with the measured values at low temperatures (i.e. 77 K or below).

Example 7.3 Threshold current densities in a GaAs laser

We use the following data for a GaAs laser diode, emission wavelength 0.84 μm; linewidth, 30 nm; refractive index 3.6; laser length 250 μm; active layer thickness 2 μm; mode confinement factor 0.75; electron lifetime 4×10^{-9} s and lumped absorption coefficient 3000 m^{-1}.

We start by calculating the threshold gain γ_{th}. From the result of Example 7.2 we have that $R_1 = R_2 = 0.32$. The threshold gain is then given by Eq. (6.4) as

$$\gamma_{th} = 3000 - \frac{1}{250 \times 10^{-6}} \log_e(0.32)$$

$$= 3000 + 4557$$

$$= 7557 \text{ m}^{-1}$$

Because of the confinement factor Γ_m, however, the actual gain required will be a factor $1/\Gamma_m$ larger than this, i.e. 10 076 m^{-1}. From Figure 7.57 the carrier density needed to give a gain equal in magnitude to these losses is $\approx 2.3 \times 10^{24} \text{ m}^{-3}$. Equation (7.78) then gives

$$J_{th} = \frac{1.6 \times 10^{-19} 2 \times 10^{-6} 2.3 \times 10^{24}}{4 \times 10^{-9}}$$

$$= 1.8 \times 10^8 \text{ A m}^{-2} \ (= 180 \text{ A mm}^{-2})$$

If the whole junction area is $250 \times 500 \ \mu\text{m}^2$, then the threshold current will be 22.5 A. These values are reasonably close to what is observed in practice for diodes at room temperature.

We may now repeat the calculation using Eq. (7.75) instead of Figure 7.57 to determine N_2^{th}. The laser linewidth of 30 nm converts into a frequency bandwidth of 1.3×10^{13} Hz and we take $\tau_{21} = \tau_e$ so that

$$N_2^{th} = 8\pi \times 10\,076 \times 1.3 \times 10^{13} 4 \times 10^{-9} 3.6^2/(0.84 \times 10^{-6})^2$$
$$= 2.4 \times 10^{23} \text{ m}^{-3}$$

We may note that this is an order of magnitude smaller than that calculated above, the corresponding threshold current density being

$$J_{th} = 1.9 \times 10^7 \text{ A m}^{-2} (= 19 \text{ A mm}^{-2})$$

As indicated in the text, values of this order are only seen in diodes at low temperature.

As the injection current increases above threshold, laser oscillations build up in the usual way, clamping the population inversion at the threshold value. The power available from stimulated emission can then be written

$$P = A(J - J_{th})\frac{\eta_i h\nu}{e}$$

where A is the junction area. Some of this power will be dissipated inside the laser cavity and the rest coupled out through the crystal faces. These two components are proportional to α_L and $(1/2l)\log_e(1/R_1R_2)$, respectively, so that the output power, P_0, can be written

$$P_0 = A(J - J_{th})\frac{\eta_i h\nu}{e}\frac{\log_e(R)}{[\log_e(R) - l\alpha_L]} \tag{7.79}$$

We define an external differential efficiency η^d_{ext} as the ratio of the incremental increase in photon output for a corresponding increase in electron injection. Since the photon output is $P_0/h\nu$ and the electron injection rate is $(A/e)(J - J_{th})$, then we have from Eq. (7.79)

$$\eta^d_{ext} = \eta_i \frac{\log_e(R)}{\log_e(R) - l\alpha_L} \tag{7.80}$$

Equation (7.80) enables the determination of η_i from the experimentally determined dependence of η^d_{ext} on l. In GaAs it is found that η_i usually lies within the range 0.7–1.

If the forward bias voltage applied across the junction is V_f, the power input is V_fAJ and the efficiency of the laser in converting electrical input to laser output is given by

$$\eta = \frac{P_0}{V_fAJ}$$

$$= \eta_i\left(\frac{J - J_{th}}{J}\right)\left(\frac{h\nu}{eV_f}\right)\left(\frac{\log_e(R)}{\log_e(R) - l\alpha_L}\right) \tag{7.81}$$

All the terms on the right-hand side of this equation (provided $J \gg J_{th}$) have values that are not very much less than unity,[13] which leads to power efficiencies as high as 50 per cent.

Equation (7.81) indicates that the output from a semiconductor laser should be proportional to $J - J_{th}$. Thus as a function of current supplied to the diode we expect that no laser output is obtained until threshold is reached, and that thereafter the output should rise linearly. In fact, below threshold the device functions rather like an LED and a small amount of incoherent radiation is emitted, resulting in an output-drive current characteristic as indicated in Figure 7.59.

As the temperature is lowered the spread of electron energies in the conduction band reduces and the gain resulting from a given electron density increases, thus lowering the threshold current. Figure 7.60 illustrates this behaviour. The variation

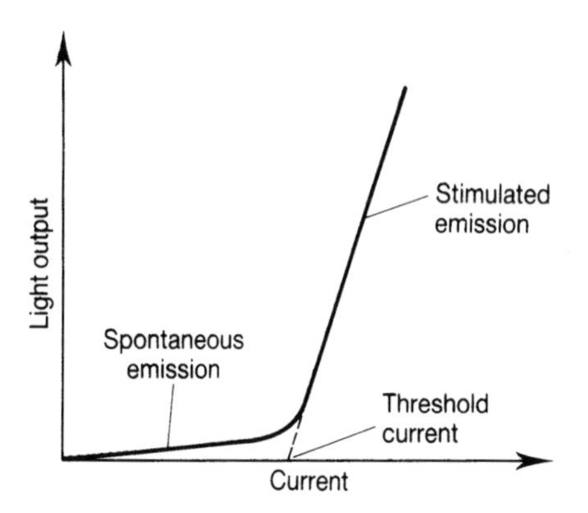

Figure 7.59 Light output-current characteristic for a semiconductor laser.

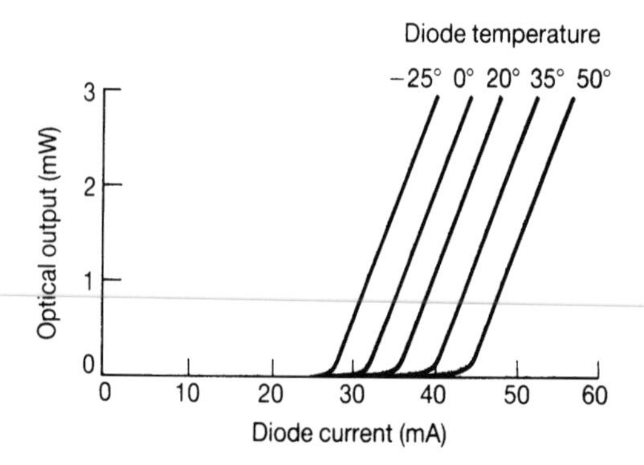

Figure 7.60 The output-current characteristics of a GaAlAs laser diode at different temperatures.

in threshold current density with temperature is often described in terms of the semi-empirical expression

$$J_{\mathrm{th}} = J_0 \exp(T/T_0) \tag{7.82}$$

where J_0 and T_0 are constants for a particular diode.

7.4.2 Heterojunction structures

The high threshold currents for the simple p–n junction type of semiconductor lasers described above usually preclude their use in anything but pulsed modes of operation. CW operation at room temperature requires that the threshold currents be substantially reduced, otherwise there are severe heating problems. We see from Eq. (7.78) that the threshold current is proportional to d, the width of the active volume, and initial attempts to reduce J_{th} were centred on devising structures that reduce d. We saw in the previous section that the magnitude of d is determined mainly by carrier diffusion away from the junction. The early 1970s saw the development of ***heterojunction*** structures to restrict this diffusion and thereby reduce the threshold current density. A double heterostructure design is shown in Figure 7.61(a). A layer of GaAs is sandwiched between two layers of the ternary compound $Ga_{1-x}Al_xAs$ which has both a larger bandgap than GaAs and also a lower refractive index. The resulting energy level structure is shown in Figure 7.61(b). Here we see that when electrons are injected from the *N*-GaAlAs[14] layer into the *p*-GaAs layer they are prevented from freely diffusing into the *P*-GaAlAs layer by the presence of a potential barrier.

Similarly, holes flowing from the *P*-GaAlAs into the *p*-GaAs to make up for those holes used up in the recombination processes cannot cross into the *N*-GaAlAs again because of a potential barrier. Thus the GaAs layer now acts as the active region and this can be made as narrow as 0.2 μm or so. This technique of restricting the free movement of the carriers is known as *carrier confinement*.

Another factor which comes into play arises from the refractive index differences between the active layer of GaAs and the two surrounding layers of GaAlAs which are much larger than that resulting from carrier density effects in the homojunction structure. According to the discussion in Section 2.1.7, the amount of mode power within the guide core increases as the refractive index difference between core and cladding increases. The optical mode power is thus much more tightly confined to the active layer than in the homojunction structure. The result is an increased value for Γ_m and hence a reduction in J_{th}. This technique for reducing J_{th} is known as *optical confinement*. By these means threshold current densities of the order of 10^7 A m^{-2} can be realized and CW operation at room temperature becomes possible.

Another advantage of the heterojunction structure is that any part of the mode radiation that does travel in the GaAlAs layers finds itself in a wider bandgap material than that corresponding to its own wavelength, thus there can be no resonant absorption. This helps to reduce the value of α_L and again increases the diode efficiency (via Eq. (7.78)).

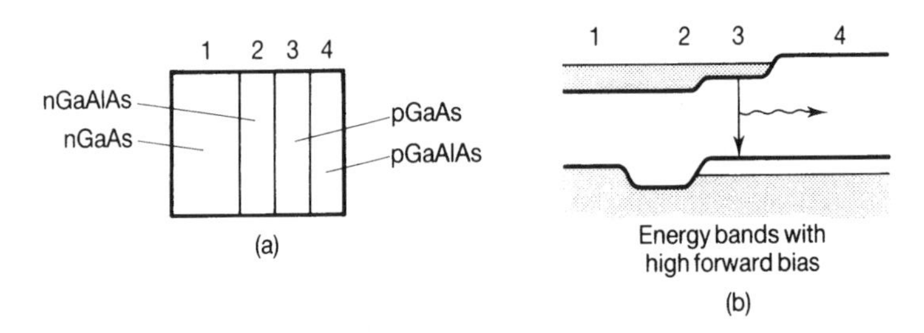

Figure 7.61 (a) Structure of a double heterostructure semiconductor laser. (b) The resulting energy band structure.

The output from a semiconductor laser is usually predominantly plane polarized with the electric field vector lying in the plane of the junction. This can be explained by assuming that the gain region approximates to a planar dielectric waveguide. Reference to Eqs (2.64) and (2.65) and also Figure 2.9 shows that the phase change on reflection for TM polarized light is greater than for TE polarized light. Thus solutions of Eq. (2.66) (Figure A.3.2) give rise to ray angles within the guide that are larger for TE than for TM polarization. Now modes that are nearer to cut-off (i.e. whose ray angles are closer to the critical angle) tend to have an evanescent field which extends further into the cladding and, in consequence, experiences greater loss. Thus the TE_0 mode should experience greater gain than the TM_0 mode and consequently oscillate preferentially.

As it is normally manufactured, the laser chip is symmetrical and emits radiation equally from both ends. Usually only the radiation emitted from one end is allowed to emerge from the encapsulated laser. In some designs the radiation from the other end falls onto a photodetector (Figure 7.62) thus providing a useful reference signal proportional to the laser output. This can be used, for example, in an electronic feedback circuit to stabilize the main laser output power.

7.4.3 Gain and index guided lasers

In the double heterostructure lasers described in the previous section lasing can take place anywhere within the junction plane. In practice, lasing usually restricts itself to narrow filaments which unfortunately can move about within the plane depending on the current flow. This behaviour can give rise to 'kinks' in the output characteristics as the filaments jump from one position to another. The problem can be solved and also lower threshold currents obtained if some form of lateral confinement in both the current and photon flux is introduced in the junction plane. There are two main structures used at present to do this which are known as *gain guiding* and *index guiding* structures.

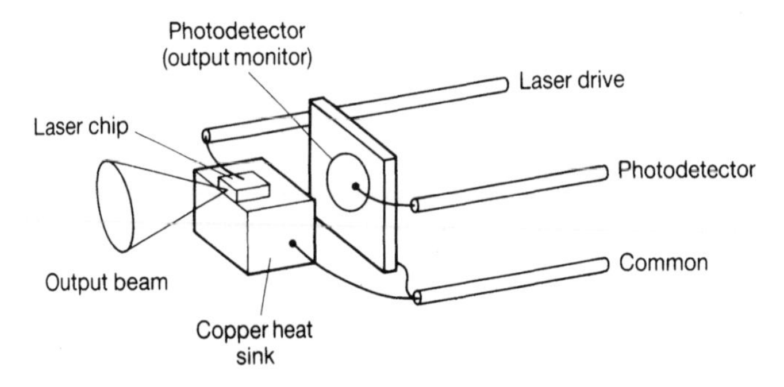

Figure 7.62 Laser diode packaging including an output power monitor. A detector is positioned to intercept the power output from the rear facet of the laser chip.

In the former the current flow along the junction plane is restricted into a narrow stripe which may be only a few microns wide. This can be achieved by using several different geometries, two of which are illustrated in Figure 7.63. In Figure 7.63(a), for example, the stripe has been defined by proton bombardment of the adjacent regions to form highly resistive material, whereas in Figure 7.63(b) a mesa-like structure has been formed by etching; an oxide mask prevents shorting of the junction during metallization to form the contacts. In both cases the assumption is that the current does not spread out too far laterally before it reaches the active layer. Such structures enable threshold currents of several tens of milliamps to be achieved.

We may note that no attempt is made at optical confinement in these structures, although, as we have seen before, the current flow (via a change in refractive index) does lead to a small waveguiding effect. Stripe geometry structures have enabled threshold current densities to be reduced to some 10^5 A m^{-2}. Such lasers are capable of producing output powers of some 10 mW with currents of less than 50 mA.

In index guided structures an optical waveguide is created to enable guidance to take place parallel to the junction plane. The most obvious technique is to create some type of *buried heterostructure* as shown schematically in Figure 7.64. Here the lasing region is surrounded on all four sides by material of lower refractive index. This structure then forms an optical waveguide not too dissimilar to the planar optical waveguides discussed in Section 2.1.7.

Depending on its size and the refractive index change at the boundaries such a structure can, in general, support a number of different transverse modes similar in principle to the laser TEM$_{mn}$ modes discussed in Section 4.8. It is usually desirable, for a number of reasons, for the laser to operate on its lowest-order mode. The required waveguide structure is then known as a *single-mode waveguide*.

Although structures similar to those indicated in Figure 7.64 have been made there are some problems which relate, basically, to the size of the buried layer. To appreciate these we consider a simple planar dielectric waveguide with core thickness d and core and cladding refractive indices of n_1 and n_2, respectively (Figure 2.10). For the waveguide to be single mode we require that the parameter V be less than $\pi/2$. Since

$$V = \frac{\pi d}{\lambda_0}(n_1^2 - n_2^2)^{1/2}$$

we require

$$\frac{\pi d}{\lambda_0}(n_1^2 - n_2^2)^{1/2} < \pi/2 \tag{7.83}$$

or

$$d^2 < \frac{\lambda_0^2}{4(n_1 + n_2)\Delta n} \tag{7.84}$$

where $\Delta n = n_1 - n_2$.

Now for the simple buried heterostructure of Figure 7.64 the index differences along the junction plane that can be achieved fairly easily lie within the range

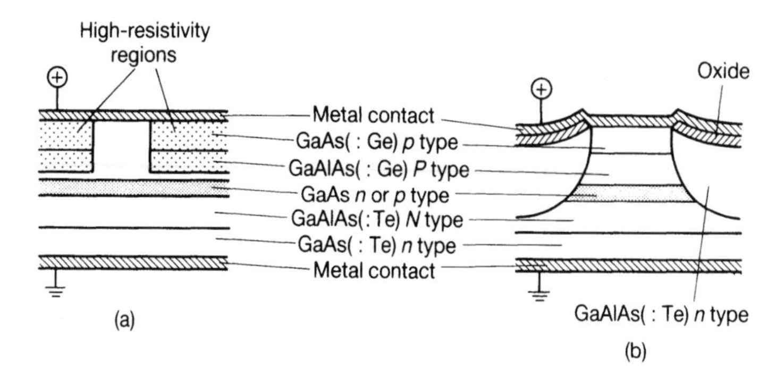

Figure 7.63 Schematic end view of two types of stripe geometry laser diode.

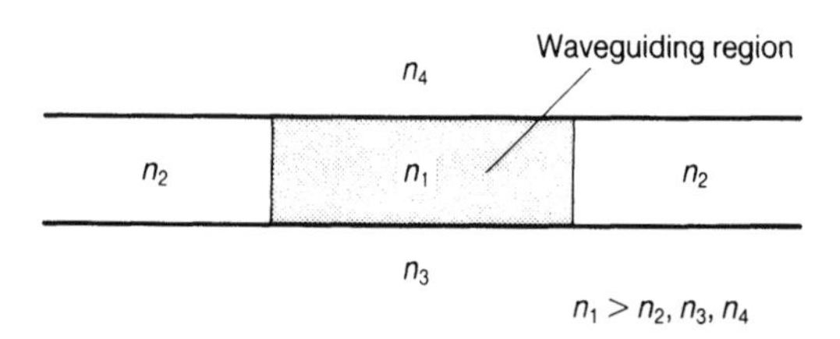

Figure 7.64 End view of a buried heterostructure which could be used as an optical waveguide structure in a semiconductor laser.

0.01–0.1. Putting $\Delta n = 0.01$ and $n_1 + n_2 = 7$ (for GaAs $n \approx 3.5$) in Eq. (7.84) gives that $d \approx 2\lambda_0$. This indicates that using a simple buried heterostructure single transverse modes are only possible when the waveguides are one or two microns wide. This restriction results in large beam divergence and low power.[15]

In the early 1980s more sophisticated structures were evolved to provide adequate mode control as well as larger mode cross-sectional areas. To obtain values of d that are several times larger than λ we need Δn to be rather smaller than 0.01. However, Δn has to be at least several times 0.001, since the current flow itself induces refractive index changes of this order. The necessary tight control can be provided by a structure of the type indicated in Figure 7.65. Here a change in the thickness of the cladding layer above the waveguide creates an effective refractive difference of, say, 5×10^{-3}. A 'real-life' structure for an index guided visible laser diode is shown in Figure 7.66.

In general, gain guided lasers are easier to fabricate than index guided lasers but their poor optical confinement limits beam quality and makes stable single longitudinal mode operation difficult. The fact that the beam is spread over a larger area can be an advantage, however, when it comes to handling high power outputs. Index guided lasers have smaller threshold currents but limited output power ($\approx$200 mW or so), otherwise the power densities over the output facet become sufficient to cause damage.

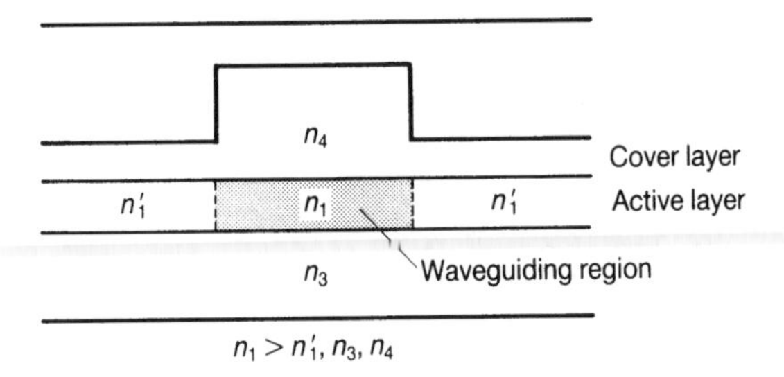

Figure 7.65 Schematic diagram of a structure which behaves like a buried heterostructure but which also provides very tight control over the refractive index differences. The varying thickness of the covering layer over the guiding layer gives rise to changes in the apparent refractive index of the layer underneath so that waveguiding action can take place.

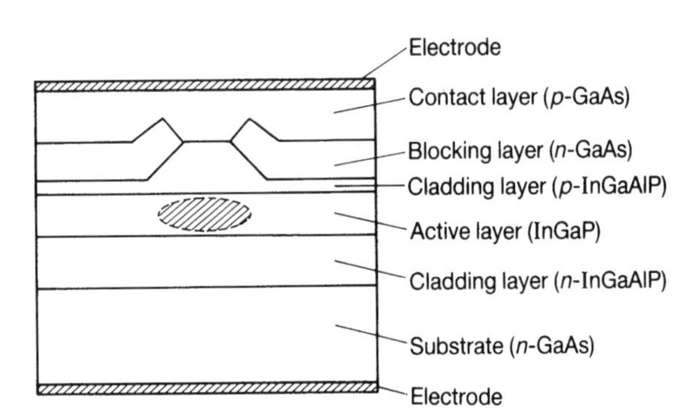

Figure 7.66 Index guiding structure used in a laser diode based on InGaAsP and emitting at about 670 nm.

7.4.4 Beam divergence

In the lasers we are discussing here the output emerges from an area whose dimensions are comparable with (and may even be less than) the wavelength of light. This small emitting area results in a comparatively large beam divergence (one of the less admirable properties of semiconductor lasers!). To obtain some feel for the beam divergence we assume that it is similar to the divergence of a plane parallel beam passing through a rectangular aperture ('Fraunhofer diffraction'). This is a standard problem in optics [Ref. 27] and turns out to be given by the product of the pattern given by two (orthogonal) single-slit patterns. The angular irradiance distribution from a single slit of width a is given by (Figure 7.67).

$$\mathrm{sinc}^2\left(\frac{\pi a \sin(\theta)}{\lambda}\right)$$

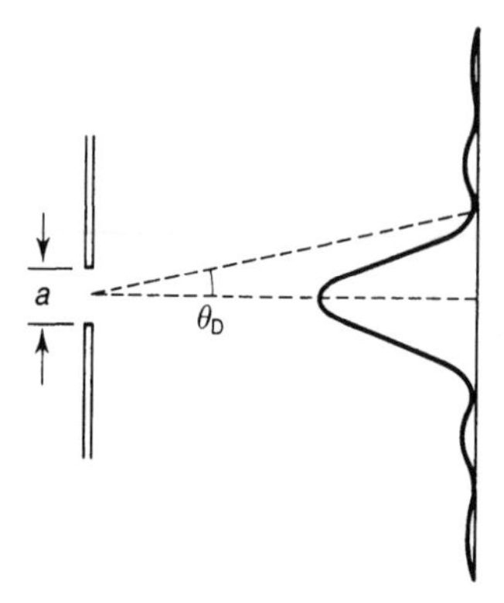

Figure 7.67 Diffraction of light from a narrow slit.

We have met the sinc function before (see Figure 2.22). Most of the diffracted beam energy will be contained within a central maximum which falls off to zero when $\sin(\theta) = \pm\lambda/a$. This corresponds to a total divergence angle $2\theta_D$ where

$$\theta_D = \sin^{-1}(\lambda/a)$$

Thus for a slit where the width is equal to twice the wavelength then $\theta_D = \sin^{-1}(0.5) = 30^\circ$.

The lasing area usually extends further along the plane of the junction than it does at right angles to it (i.e. Figure 7.65) so that the beam is expected to spread out more in a direction perpendicular (θ_{perp}) to the junction plane than in a direction parallel to it (θ_{par}) as indicated in Figure 7.68. Typical values for θ_{perp} might be as large as 30° or so.

Another observed feature is that the beams in these two orthogonal directions do not appear to originate from the same point on the axis of the laser, a property known as *astigmatism*. Index guided lasers exhibit lower astigmatism than do gain guided lasers.

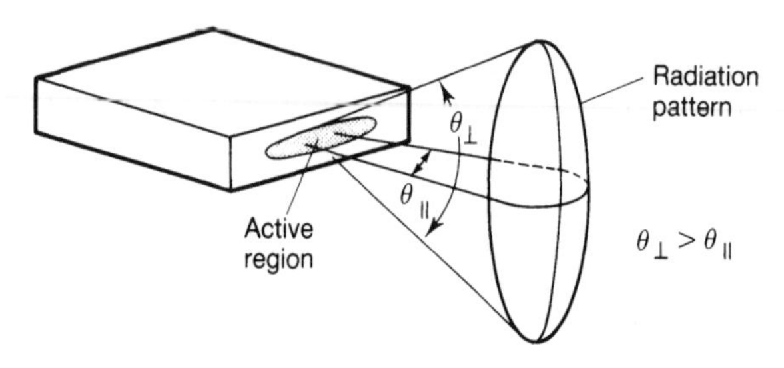

Figure 7.68 Angular divergence of a semiconductor laser beam.

7.4.5 Spectral output

Ignoring the complications of transverse mode structure the longitudinal mode cavity resonances occur at wavelengths λ_q, where

$$q\lambda_q = 2n(\lambda_q)l \tag{7.85}$$

Here l is the laser cavity length, $n(\lambda)$ the wavelength-dependent refractive index and q is an integer. Since q is very large we may regard it as a continuous variable, so that a change in q (Δq) is related to a change in λ ($\Delta\lambda$) by

$$\frac{\Delta q}{\Delta\lambda} \cong \frac{d}{d\lambda}\left(\frac{2n(\lambda)l}{\lambda}\right)$$

$$= 2l\left(\frac{1}{\lambda}\frac{dn(\lambda)}{d\lambda} - \frac{n(\lambda)}{\lambda^2}\right)$$

Dropping the explicit dependence of n on λ

$$\Delta\lambda = -\frac{\Delta q}{2l}\frac{\lambda^2}{n - \lambda(dn/d\lambda)}$$

The absolute value of the wavelength change between adjacent modes, $\Delta\lambda_m$ (i.e. $|\Delta\lambda|$ when $\Delta q = 1$), is then given by

$$\Delta\lambda_m = \frac{\lambda^2}{2l}\frac{1}{n - \lambda(dn/d\lambda)}$$

$$= \frac{\lambda^2}{2n_{eff}l} \tag{7.86}$$

where[16]

$$n_{eff} = n - \lambda\frac{dn}{d\lambda}$$

In semiconductor materials such as GaAs n_{eff} is appreciably larger than n (see Problem 7.17).

Example 7.4 Mode spacing in GaAs semiconductor lasers

Taking a value for n_{eff} in GaAs at 0.85 μm to be 4, and assuming a cavity length of 300 μm, then the longitudinal mode spacing can be calculated from Eq. (7.86) to be

$$\Delta\lambda_m = \frac{(0.85 \times 10^{-6})^2}{2 \times 4 \times 300 \times 10^{-6}}$$

$$= 0.3 \text{ nm}$$

A typical spectral width of a GaAs LED is 30 nm so that below threshold a GaAs laser would be expected to show about 100 longitudinal modes.

When the current is below threshold the output from the laser derives from spontaneous emission and consequently is incoherent and exhibits a spectral width similar to an LED. However, because the light is generated within a Fabry–Perot cavity a longitudinal mode structure is superimposed on the broad LED type spectrum (see Figure 7.69(a) and Example 7.4). Typically the number of modes seen is of the order of 100.

As the diode current is increased through threshold the expected 'ideal' behaviour for a homogeneous laser would be for the mode nearest the peak of the gain curve to achieve threshold first, thereafter clamping the gain so that no other modes could oscillate. In practice, unless special care is taken, more than one mode often

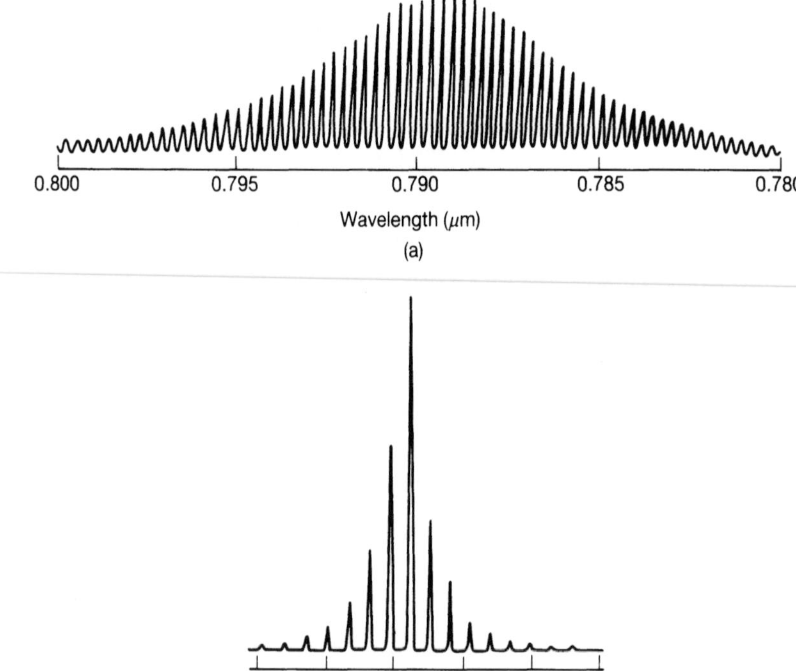

Figure 7.69 Emission spectrum of a GaAlAs laser diode both just below (a) and just above (b) threshold. Below threshold a large number of Fabry–Perot cavity resonances can be seen extending across a wide LED type spectrum. Above threshold only a few modes close to the peak of the gain curve oscillate. For the particular laser shown here the threshold current was 37 mA while spectra (a) and (b) were taken with currents of 35 mA and 39 mA, respectively.

oscillates (Figure 7.69(b)). For currents substantially above threshold usually one mode does predominate, although this is not always so.

The temperature behaviour of the output wavelength of a (single mode) semiconductor laser is shown in Figure 7.70. Basically we see a linear trend with a number of discontinuous 'jumps' superimposed. The steady change derives from the temperature behaviour of $\Delta\lambda_m$, thus from Eq. (7.85) we have

$$\frac{d\lambda_q}{dT} = \frac{2}{q}\left(l\,\frac{dn}{dT} + n\,\frac{dl}{dT}\right)$$

Usually the term on the right-hand side of this equation is dominated by the variation of refractive index with temperature. Assuming that this can be written

$$n = n_0(1 + \beta_n T)$$

then we have

$$\frac{d\lambda_q}{dT} = \lambda_q \beta_n$$

In practice β_n is a positive quantity, so that the mode wavelength increases with increasing temperature.

The discontinuities in Figure 7.70 are caused by 'mode hopping'. Both the comb of resonant mode frequencies and the gain curve[17] move in the same 'direction' (towards lower frequencies, i.e. higher wavelengths) as the temperature increases. However, the gain curve moves faster than the resonant frequencies so that a particular mode will not always remain the one closest to the peak of the gain curve.

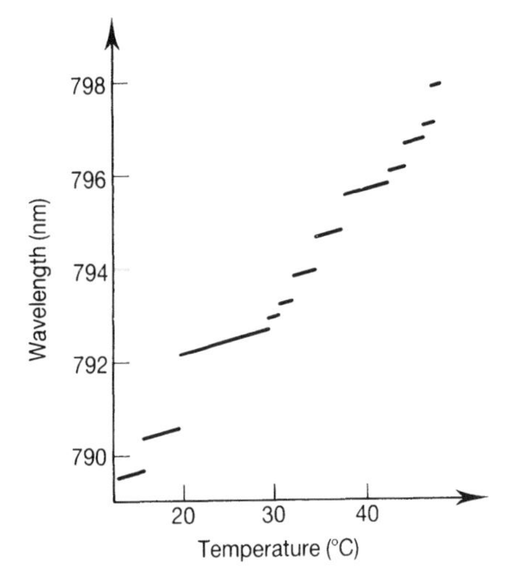

Figure 7.70 Variation in the output wavelength with temperature of a single-mode GaAlAs laser diode.

The shift in the resonant frequency allows the laser output to be tuned in wavelength either by varying the temperature of the diode directly or by changing the diode current. The occurrence of mode hopping, however, limits the useful tuning range. One way of eliminating mode hopping is discussed in the next section.

7.4.6 Distributed feedback structures

Although the cavity mirrors are usually provided for by using the cleaved semiconductor ends it is possible to make more efficient structures, one of these being the corrugated structure shown in Figure 7.71. This is called a *Bragg reflector.*[18] Suppose a light wave is propagating down a guide with such a corrugation on the surface. We may imagine that as the wave passes each corrugation some fraction of the beam is reflected back down the guide. If the corrugations have a periodicity of Λ, then the reflected portions of the wave will be in phase provided that $2\Lambda n = m\lambda_0$ where m is an integer. We expect strong reflections, therefore, when

$$\lambda_{0m} = \frac{2\Lambda n}{m}$$

Since m is usually of the order of unity, the wavelength separation between adjacent resonances is *much* larger than for the Fabry–Perot resonances for the cleaved end mirrors. Consequently only a single mode will ever oscillate. A formal treatment of the reflection from a series of such corrugations is somewhat involved [Ref. 28] and we must content ourselves here with a presentation of the results.

The reflectance from such a structure is given by

$$R_{\mathrm{B}} = \left| \frac{\varkappa}{\alpha/2 + \mathrm{j}\ \Delta\beta + S \coth(SL)} \right|^2 \tag{7.87}$$

where $\varkappa$ is a number representing the coupling between the corrugations and the wave, α the absorption coefficient, $S^2 = \varkappa^2 + (\alpha + \mathrm{j}\ \Delta\beta)^2$, and $\Delta\beta$ the deviation of the propagation constant from the Bragg condition. In the presence of gain then α must be replaced by the negative of the gain coefficient $(-\gamma)$. A curve of the resultant reflectance from a grating where $\varkappa L = 2$ and $\alpha = 0$ as a function of $L\ \Delta\beta$ is shown in Figure 7.72. As expected, we get a strong reflectance at the Bragg condition $(\Delta\beta = 0)$.

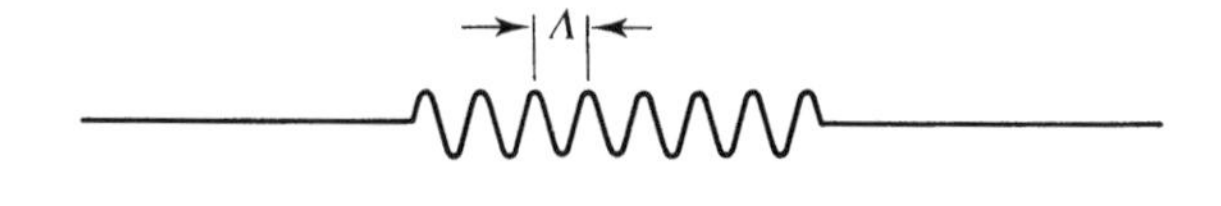

Figure 7.71 A corrugated portion of waveguide which can act as a mirror for certain wavelengths.

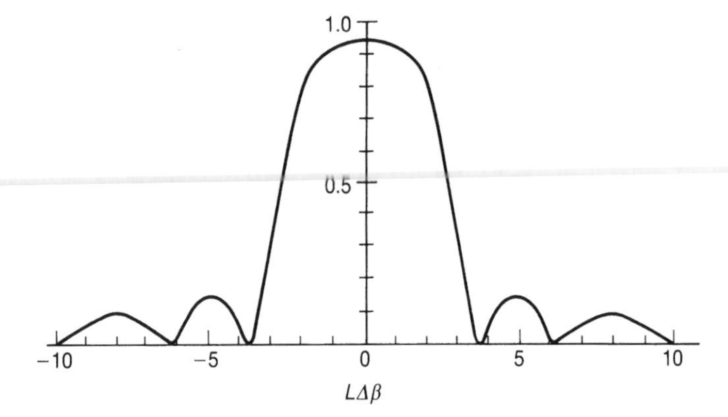

Figure 7.72 A plot of the reflectance from a Bragg-type corrugated waveguide according to Eq. (7.87) with $\varkappa L = 2$ and $\alpha = 0$.

Although lasers with Bragg-type reflectors at the end of the active cavity have been fabricated, a more usual scheme is to have the corrugations extend throughout the gain region itself as shown in Figure 7.73 to form a *distributed feedback* laser.[19] When the condition for achieving threshold in such a laser is examined in detail it turns out to be that the denominator for the expression for the transmission coefficient (which is identical to that for the reflection coefficient) should be zero.[20] The form of the denominator of Eq. (7.87), in particular its dependence on gain, means that threshold gains and resonant frequencies must be computed numerically for different circumstances. However, the general disposition of the resonant frequencies can be obtained by considering the positions of the zeros of the curve in Figure 7.72. Thus, rather surprisingly, oscillation is *not* obtained when $\Delta\beta = 0$, but at frequencies equally distributed on either side. The two nearest to the centre ($\Delta\beta = 0$) have the lowest thresholds.

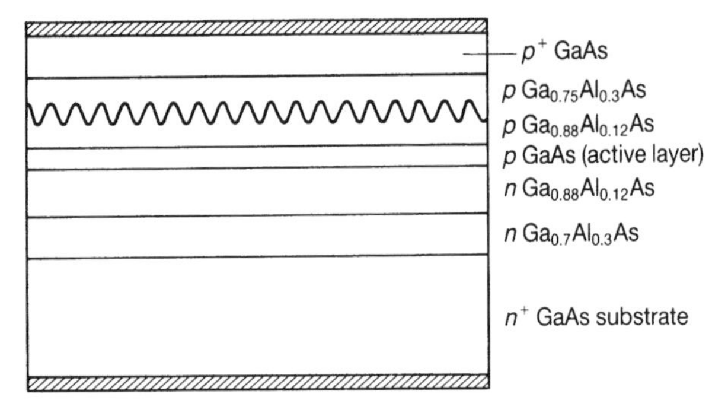

Figure 7.73 Schematic cross-section of a double heterojunction distributed feedback laser structure based on GaAs/GaAlAs.

In its simplest form then the distributed feedback laser is likely to operate multi-mode. Single-mode operation *can* be obtained, however, either by introducing a uniform section of length $\lambda_0 n/4$ in the centre or by removing a section of this length (Figure 7.74). This has the effect of allowing oscillation at the Bragg frequency. Such devices are available commercially and known as '$\lambda/4$ shifted' distributed feedback lasers.

7.4.7 Quantum well structures

In quantum well structures the thickness of the lasing region is made *very* small. To appreciate the effect of this on the electron energy levels we consider a slab of semiconductor material of dimensions L_1, L_2 and L_3, where $L_3 \ll L_1, L_2$. A solution of the Schrödinger equation shows that the energy levels can be written (see Eq. (1.48a))

$$E_{n_1 n_2 n_3} = \frac{h^2}{8m_e^*}\left[\frac{n_1^2}{L_1^2} + \frac{n_2^2}{L_2^2} + \frac{n_3^2}{L_3^2}\right] \tag{7.88}$$

where n_1, n_2 and n_3 are positive integers greater than zero.

Since L_3 is so small compared to the other two dimensions, states can be grouped into sub-bands where $n_3 = 1$, $n_3 = 2$, etc. (Figure 7.75). The lower states within the band will have a higher energy than 'normal' (i.e. where the L values are all comparable). For the electrons this energy shift has a magnitude of $h^2/(8m_e^* L_3^2)$. Taking $L_3 = 2 \times 10^{-9}$ m and $m_e^* = m_e$ gives a value of about 0.1 eV. A similar shift will be observed for the hole states in the valency band. Thus the effective energy gap of the semiconductor is increased, leading to a shorter wavelength of emission. Experimental investigations of such quantum size effects are described in Ref. 29.

Another interesting feature is that the electrons are less spread out in energy terms within the conduction band than normal, leading to reduced threshold currents. To see this we first consider the density of states function. For an electron in a two-dimensional potential well this is given by $D(E) = 4\pi m/h^2$ (see Problem 1.20), and is independent of energy. The number of electrons between E and $E + \mathrm{d}E$ will then

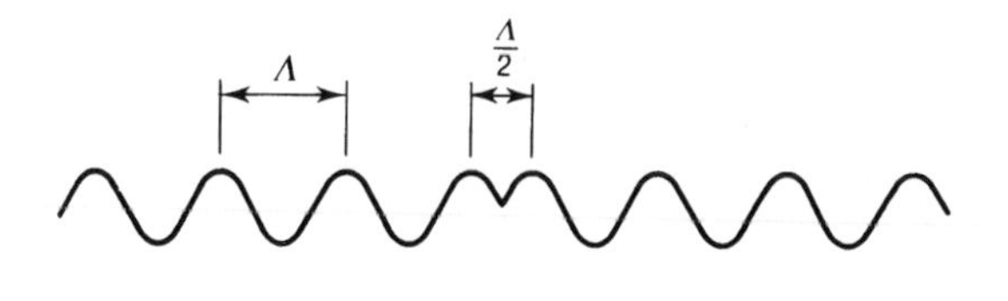

Figure 7.74 The central region of a '$\lambda/4$ shifted' distributed feedback laser with a missing section of length $\Lambda/4$ at the centre of the laser. This structure ensures that lasing can take place at wavelength corresponding to the Bragg condition.

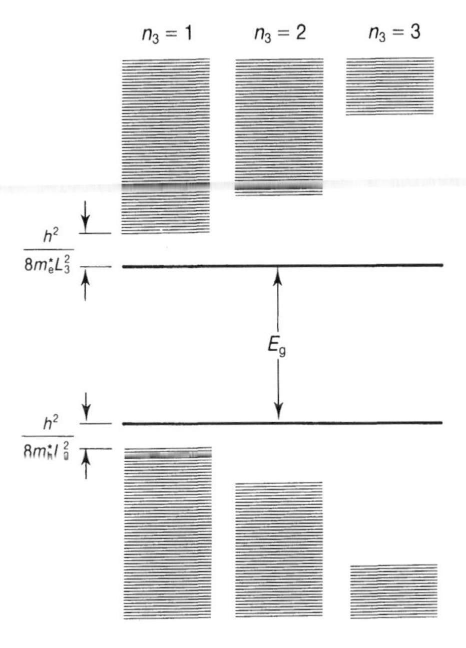

Figure 7.75 The energy-level structure of a quantum well laser. The sub-bands corresponding to $n_3 = 1,2,3$ etc. are shown separated for convenience, in reality they are superimposed on each other. Note the increase in the effective energy gap by an amount

$$\frac{h^2}{8L_3^2}\left[\frac{1}{m_e^*}+\frac{1}{m_h^*}\right]$$

simply vary with energy as the Fermi function itself, that is, as $\approx \exp(-E/kT)$. The 'halfwidth' for this distribution is $kT \log_e 2$, or 0.69 kT. This contrasts with the half-width for the three-dimensional case of 1.8 kT (Section 3.1.6).

One of the problems with the single-quantum well structure described above is that, because of the extreme narrowness of the gain region, optical confinement is poor. This causes higher losses and negates the potential advantage of low threshold currents. One way out of this difficulty is to use a multiple quantum well structure (Figure 7.76) with alternating high and low bandgap materials. The high bandgap barrier layers must be thick enough to prevent electrons 'tunnelling' through. This multiple structure allows a much higher optical confinement factor. Quantum well lasers have been constructed with threshold currents as low as a few milliamps.

7.4.8 Laser diode arrays

The output power from a semiconductor laser diode may be increased by using one-dimensional arrays of single-mode lasers as indicated in Figure 7.77. Such an array

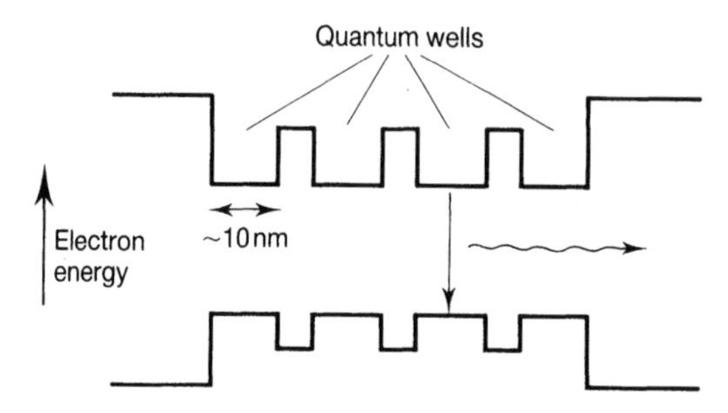

Figure 7.76 Multiple quantum well structure.

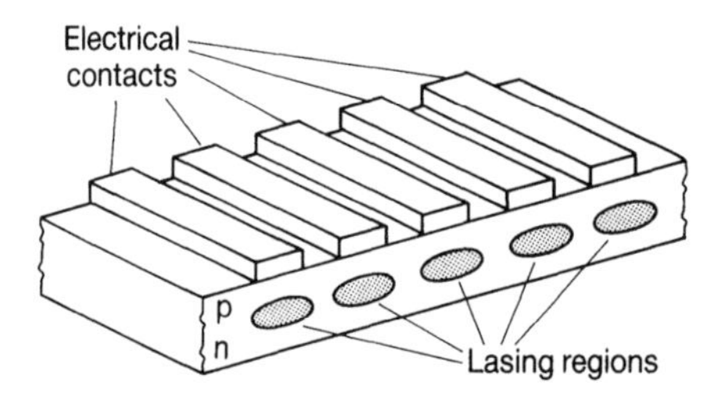

Figure 7.77 A linear array of lasing elements formed on a single semiconductor substrate.

is called a *phased* array since the electric fields associated with the different elements interact with each other, resulting in definite phase relationships between the elements. Usually the phase difference between adjacent elements is 180° (Figure 7.78). This is because this gives a zero resultant field at a point halfway between the active regions where there is likely to be absorption rather than gain, thus minimizing the overall loss. Unfortunately, this phase relationship also results in a far-field power distribution in the plane of the junction which has an angular distribution consisting of two main lobes (Figure 7.79). If the phase relationship between elements were 0° then a single-lobed distribution would result.

High-power arrays of as many as 200 individual laser stripes producing power outputs of several tens of watts can be fabricated on a 10 mm wide bar. Such arrays can also be stacked vertically to produce higher powers but efficient heat removal needs careful consideration and limits the maximum powers that can be obtained.

7.4.9 Materials and manufacturing techniques

So far we have made no specific mention of the manufacturing techniques and the constraints placed on the materials used for semiconductor lasers. Virtually all such lasers are fabricated upon substrates of binary compounds such as GaAs. Although these materials are easy to make their lattice constants, refractive indices and band-

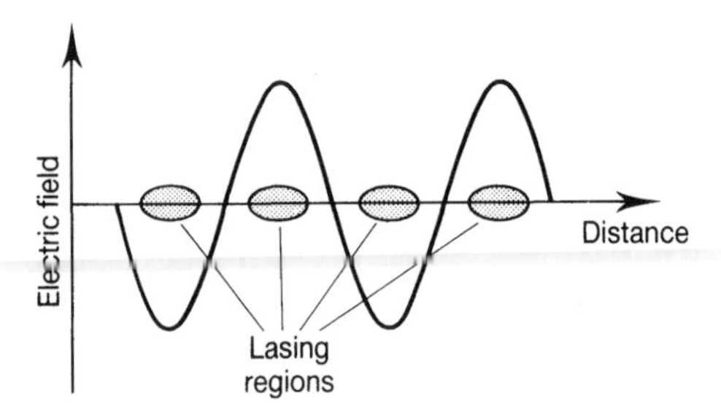

Figure 7.78 The electric field distribution resulting from the linear array of lasing elements shown in Figure 7.77. The field is zero halfway between the elements where there will be absorption rather than gain.

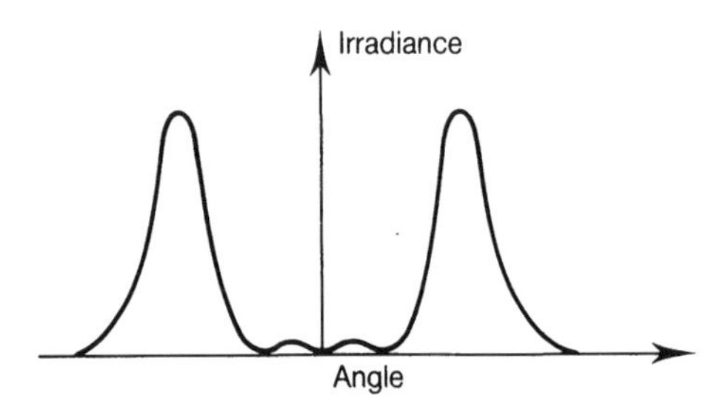

Figure 7.79 The multi-lobed angular distribution in the plane of the junctions which results from the electric field distribution of Figure 7.78.

gaps are fixed. Ternary compounds, where one element partially substitutes for another, are much more flexible in this respect. An example of this is $Ga_{1-x}Al_xAs$. The value of x here can be varied between 0 and 1, resulting in bandgaps which are intermediate between those GaAs ($x = 0$) and AlAs ($x = 1$).

A variety of different techniques may be used to build up the required layered structures, for example liquid phase, vapour phase or molecular beam epitaxial methods [Ref. 30]. Table 7.7 shows some of the semiconductor materials used to produce semiconductor lasers together with their emission wavelengths.

Table 7.7

Compound	Wavelength range (nm)
GaAs	904
$Ga_{1-x}Al_xAs$	620–895 ($0 < x < 0.45$)
AlGaInP	630–680
$In_{1-x}Ga_xAs_yP_{1-y}$	1100–1650

When layers of semiconductor materials are grown on top of each other good-quality strain free growth is not possible unless the lattice constants of the layers match each other quite closely (usually to within 0.1 per cent). This requirement obviously restricts the layer compositions that can be used. Figure 7.80 shows the lattice constant variation with bandgap for the alloy $In_{1-x}Ga_xAs_yP_{1-y}$. By a suitable choice of x and y, lattice matching to both GaAs and InP is possible. For example, two important wavelengths for optical fiber communication systems are 1.3 μm and 1.55 μm. Bandgaps corresponding to these wavelengths as well as a lattice match to InP can be obtained with $In_{0.73}Ga_{0.27}As_{0.58}P_{0.42}$ and $In_{0.58}Ga_{0.42}As_{0.9}P_{0.1}$, respectively.

The lattice-matching restraint makes it difficult to cover certain wavelength regions, an example being that between 0.9 μm and 1.1 μm. A recent development which has helped in this respect has been the finding that very thin layers (below a few tens of nanometers) can accommodate a lattice mismatch of more than 1 per cent. These layers are called *strained lattice* layers,[21] and the technique has enabled the fabrication of InGaAs lasers emitting at 980 nm with GaAlAs confinement layers and GaAs substrates [Ref. 31]. Strained layers are also used in quantum well structures to produce active layers which need not be precisely lattice matched to the surrounding layers. In 1991 the first room-temperature green diode laser (525 nm wavelength) was made using strained quantum wells of ZnSe [Ref. 32].

The wavelength range between 3 μm and 30 μm can be covered by a range of materials based on ternary compounds involving lead. The resulting lasers are called *lead salt* lasers and are mostly used in infrared spectroscopic applications. Typical

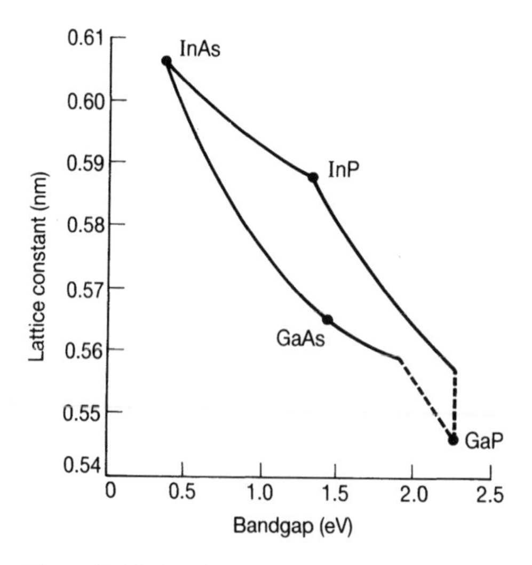

Figure 7.80 Lattice constant versus energy gap for the system InGaAsP. The solid lines correspond to direct bandgaps, the dashed lines to indirect bandgaps.

materials are $PbS_{1-x}Se_x$, $Pb_{1-x}Sn_xTe$ and $Pb_{1-x}Sn_xSe$. Coarse tuning of the diode output wavelength is carried out by varying the temperature between about 15 K and 100 K. This gives a typical tuning range of some 0.01 eV, which at 10 μm corresponds to a wavelength range of 1 μm. For fine tuning the diode current is varied which indirectly changes the device temperature although it also has the disadvantage of altering the output power.

Problems

7.1 Calculate the ideal quantum efficiencies of a helium–neon laser operating on the 633 nm and the 3.39 μm transitions.

7.2 Show that the average of a constant excitation cross-section, $\langle \sigma v \rangle$, over a Maxwellian energy distribution is given by

$$\sigma\sqrt{4kT_e/m\pi}$$

7.3 A low-pressure (0.01 mm Hg) gas is excited by an electron beam having an energy of 100 eV and a beam current density of 10^5 A/m^2. If the excitation cross-section for the upper and lower states are respectively 5×10^{-23} m^2 and 1×10^{-23} m^2 and if the upper state lifetime is 10 ns and the lower state lifetime is 1 ns calculate the population inversion density achieved. Take the Einstein A coefficient for the transition to be 1×10^7 s^{-1} and the wavelength 514 nm.

7.4 The helium–neon transition 543 nm has a gain of 1%/m. Using the data in Table 7.1 determine the threshold population inversion density if each mirror has a loss of 300 ppm, the single-pass diffraction loss is 200 ppm and the gain length is 25 cm, taking the gas temperature to be 100°C.

7.5 Determine the gain bandwidth of the laser in Problem 7.4 and hence how many longitudinal modes could oscillate.

7.6 A helium–neon laser with a discharge of 2 m in length has mirrors which have reflectivities of 30% at 3.39 μm and 99.8% and 98% at 633 nm. Using the data in Table 7.1 estimate the single-pass gain at 633 nm when the infrared line is at the threshold of oscillation.

7.7 Using the data given in this chapter and the theory of Section 6.5.1 design a helium–neon laser to give an output of 10 mW at 633 nm. Use a hemispherical cavity in your design.

7.8 Use the data given in Appendix 8 on laser safety to determine the optical density of eye protectors for a 100 mW laser for the wavelengths 633 nm and 1.15 μm. Take the spot size in both cases to be 0.5 mm.

7.9 The 488 nm and 514 nm transitions in Ar^+ share the same lower state. Estimate the ratio of their upper state populations if the gain at 488 nm is 3.0 dB/m and at 514 nm is 2.1 dB/m.

7.10 Calculate the Doppler width of the 10.6 μm transition in carbon dioxide if the gas temperature is 100°C.

7.11 A typical carbon dioxide laser has a gain coefficient for the P20 10.6 μm transition of 0.6 m^{-1}. If the transition is pressure broadened at the operating carbon dioxide pressure of 30 torr to 200 MHz calculate the fraction of the carbon dioxide molecules contributing to the upper laser state population taking the A coefficient to be 0.34 s^{-1}.

7.12 The gas ratio in a He : CO_2 : N_2 discharge is 8 : 1 : 1 and the total pressure is 10 torr. There is an electron density of 10^{17} m^{-3} and an electron temperature of 1 eV. If every vibrationally excited nitrogen molecule produces a vibrationally excited carbon dioxide molecule in the $00^{\circ}1$ state estimate the gain of the P20 transition neglecting the population of the lower laser state.

7.13 Calculate the temperature of a black body required to pump the ruby laser in Example 7.1 to threshold.

7.14 Estimate the threshold pumping power of a longitudinally diode-pumped Nd : YAG laser at 1.06 μm which has mirror reflectivities of 99.5 per cent and 98 per cent and a single-pass internal loss of 2 per cent. The absorption cross-section of the pump radiation is 5×10^{-23} m^3 and the density of neodymium ions in the crystal is 1.4×10^{26} m^3. Take the pump spot size and the cavity spot size to be 100 μm

7.15 Using the data in Figure 7.49 determine the concentration of rhodamine 6G for 80 per cent absorption in a liquid jet of 0.2 mm thickness.

7.16 Assuming that Eq. (7.76) describes the diffusion and recombination of minority carriers when they are injected into a semiconductor, show that when the semiconductor is very long (or 'semi-infinite') then the carrier density as a function of distance is given by Eq. (7.77). Show also that the average distance travelled by the carriers before recombining is simply the diffusion length.

7.17 From the data of Figure 7.69(a) given that the length of the laser cavity is 250 μm deduce a value for n_{eff}. Assuming that the refractive index of the semiconductor at the emission wavelength is about 3.6, estimate a value for $dn/d\lambda$.

7.18 A quantum well structure is made from GaAs and emission obtained at 700 nm. The emission from bulk devices of GaAs is at about 880 nm. Estimate the thickness of the quantum layer. You may assume that the effective masses for electrons and holes in GaAs are $0.068m_e$ and $0.56m_e$, respectively.

7.19 Section 7.4.7 discussed the use of 'two-dimensional' structures (quantum wells) to reduce semiconductor laser threshold currents. In this context is there an advantage in using one-dimensional structures ('quantum wires') [Ref. 32].

Notes

1. In many texts this is called a collision of the second kind since the excited atom is produced initially by electron collisions.

2. Paschen notation is normally used to designate the states of neon. It has no formal quantum-mechanical significance.

3. Some manufacturers have used a Lyot filter to tune between the different wavelengths. See Section 7.3.2 on dye lasers for a discussion of the Lyot filter.

4. The outer shell of eight electrons of a noble gas has the designation ns^2np^6 ($^1S_0$) where $n = 2, 3, 4, 5$ for neon to xenon, respectively.

5. This applies to molecules with identical oxygen atoms. Molecules with different isotopes have no rotational lines missing.

6. This is normally called the Rogowski profile.

7. The term 'metastable' is normally reserved for levels which do not decay by spontaneous emission. However, in this context the ^{2}E states are traditionally called metastable even though photons are emitted.
8. The procedure used to evaluate the integral in Section 3.1.6 cannot be used here since the quasi-Fermi level is close to or inside the conduction bend.
9. If the hole density is sufficiently large then the electron density may be neglected.
10. At the level of approximation being used here either inhomogeneous or homogeneous broadening (i.e. Eqs (3.87) and (3.89)) give $g(\omega_0) \approx 1/\Delta\omega$.
11. The student should note that from now on in this section electron densities in the conduction and valence band are referred to by using the upper case (i.e. N_2 and N_1). This is to avoid confusion with the use of n for refractive index.
12. The width of the depletion region itself has here been ignored.
13. With regard to the third term we may note that when a p–n junction is operating under heavy forward bias then the potential barrier between the p- and n-type regions is reduced to quite small values. With no bias applied this potential barrier, in the case of heavily doped materials, is almost equal to the bandgap. Thus $eV_f \approx E_g \approx h\nu$.
14. N and P are used here to represent the larger bandgap materials.
15. One of the major reasons for semiconductor laser failure is damage to the output facet which results when the power density is too high.
16. Note that this effective refractive index is not the same as that introduced in Chapter 3 to describe the wave group velocity.
17. The gain curve changes with temperature because the energy gap of the semiconductor changes, E_g decreasing with increasing temperature.
18. This nomenclature derives from the similarity of the situation with the reflection of X-rays from a periodic crystalline lattice which was first treated theoretically by Bragg.
19. If the grating is formed directly on the active layer then defects introduced by the grating fabrication process can introduce unwanted nonradiative recombination processes. Consequently it is separated from it by a layer of lower refractive index. The interaction then takes place via the evanescent tail of the optical field.
20. Since we have gain, the term $\alpha/2$ must be replaced by $\gamma/2$.
21. The deposition of several strained lattice layers on top of each other produces a strained layer superlattice which has several potentially interesting properties [Ref. 30].

References

[1] B. E. Cherrington, *Gaseous Electronics and Gas Lasers*, Pergamon Press, Oxford, 1979.

[2] A. Javan, W. R. Bennett, Jr and D. R. Herriott, 'Population inversion and continuous optical maser oscillation in a gas discharge containing a helium–neon mixture', *Phy. Rev.*, **6**, 106, 1961.

[3] W. R. Bennett, Jr, *Atomic Gas Laser Transition Data – a critical evaluation*, Plenum Press, New York, 1979.

[4] A. D. White and E. I. Gordon, 'Excitation mechanisms and current dependance of population inversion in He–Ne lasers', *Applied Physics Letters*, **3**, 197, 1963.

[5] S. P. Spoor, *The determination of the radial distribution of optical gain at 633 nm in small bore helium neon discharges*, PhD thesis, Newcastle-upon-Tyne Polytechnic, 1985.

[6] B. S. Gray, *An investigation of the 543 nm ($3s_2$–$2p_4$) neon laser transition*, PhD. thesis, Newcastle-upon-Tyne Polytechnic, 1991.

[7] S. P. Spoor and I. D. Latimer, 'An accurate determination of the radial variation of gain at 633 nm in small bore helium neon discharges', *J. Phys. D : Applied Physics*, **17**, 1984.

[8] W. B. Bridges, *Methods of Experimental Physics*, Vol. 15: *Quantum Electronics*, Part 2, Academic Press, New York, 1979.

[9] C. C. Davis and T. A. King, 'Gaseous ion lasers', *Advances in Quantum Electronics*, **3**, 169, 1975.

[10] T. G. Webb *et al.*, 'Optical excitation in charge transfer and Penning ionisation', *J. Phys. B*, **3**, L134 (1970).

[11] W. J. Witteman, *The CO_2 Laser*, Springer Series in Optical Sciences, **53**, 1986.

[12] M. J. W. Boness and G. J. Schultz, 'Vibrational excitation of CO_2 by electron impact', *Phy. Rev. Lett.*, **21**, 1031, 1968.

[13] G. J. Schultz, 'Vibrational excitation of N_2, CO, and H_2 by electron impact', *Phy. Rev.*, **135A**, 988, 1964.

[14] P. K. Cheo, 'CO_2 Lasers', in A. K. Levine and A. DeMaria (Eds), *Lasers*, **3**, Marcel Dekker, New York, 1971.

[15] D. C. Tyte, 'Carbon dioxide lasers', in D. W. Goodwin (Ed.), *Advances in Quantum Electronics*, **1**, Academic Press, New York, 1970.

[16] 'Excimer lasers', in Ch. K. Rhodes (Ed.), *Topics in Applied Physics*, **30**, Springer-Verlag, New York, 1978.

[17] M. H. R. Hutchinson, 'Tunable lasers', in L. F. Mollenauer and J. C. White (Eds), *Topics in Applied Physics*, **59**, Springer-Verlag, New York, 1987.

[18] M. J. Weber, 'Solid state lasers', *Methods of Experimental Physics*, **15A**, 167, 1979.

[19] W. Koechner, *Solid State Laser Engineering*, 2nd edn, Springer Series in Optical Sciences, **1**, 1988.

[20] G. P. A. Malcolm and A. I. Ferguson, 'Diode pumped solid state lasers', *Contemporary Physics*, **32**, 305, 1991.

[21] F. J. Duarte and L. W. Hillman, *Dye Laser Principles*, Academic Press, New York, 1990.

[22] P. W. France (Ed.), *Optical Fibre Lasers and Amplifiers*, Blackie, Glasgow, 1991.

[23] A. L. Bloom, 'Modes of a laser resonator containing tilted birefringent plates', *J. Opt. Soc. Am.*, **64**, 447, 1974.

[24] T. F. Johnston, Jr, R. H. Brady and W. Proffitt, 'Powerful single-frequency ring dye laser spanning the visible spectrum', *Applied Optics*, **21**, 2307, 1982.

[25] K. Vahala, L. C. Chiu, S. Margalit and A. Yariv, 'On the linewidth enhancement factor α in semiconductor injection lasers', *Appl. Phys. Lett.*, **42**, 631–3, 1983.

[26] A. Bar-Lev, *Semiconductors and Electronic Devices*, 2nd edn, Prentice Hall International, Englewood Cliffs, NJ, Section 4.3.

[27] R. Guenther, *Modern Optics*, John Wiley, New York, 1990, pp. 370–1.

[28] H. A. Haus, *Waves and Fields in Optoelectronics*, Prentice Hall, Englewood Cliffs, NJ, 1984, Sections 8.1 and 8.2.

[29] R. Dingle, W. Weigman and C. H. Henry, 'Quantum states of confined carriers in very thin $Al_xGa_{1-x}As$-GaAs-$Al_xGa_{1-x}As$ heterostructures', *Phys. Rev. Lett.*, **33**, 827–30, 1974: L. Esaki, 'A bird's-eye view on the evolution of semiconductor superlattices and quantum wells', *IEEE J. Quantum Electron.*, **QE 22**, 1611–24, 1986.

[30] H. C. Casey and M. B. Panish, *Heterojunction Lasers*, Academic Press, New York, 1978, Part B, Chapters 5–7: M. J. Adams, A. G. Steventon, W. J. Devlin and I. D. Henning, *Semiconductor Lasers for long-Wavelength Optical-Fibre Communications Systems*, Peter Peregrinus, Stevenage, 1987, Chapter 5.

[31] G. C. Osbourn *et al.*, R. Dingle (Ed.), in *Semiconductors and Semimetals 24 – Applications of Multi Quantum Wells – Selective Doping and Superlattices*, Academic Press, New York, 1987, p. 459.

[32] M. A. Haas *et al.*, 'Blue-green laser diodes', *Appl. Phys Lett.*, **59**, 1272–74, 1991.

[33] E. Kapon *et al.*, 'Single quantum wire semiconductor lasers', *Appl. Phys. Lett.*, **55**, 2715–7, 1989.

8

Transient effects

Most of our discussion of lasers up to this point has assumed that the power output is constant in time. However, many interesting and useful properties of lasers arise from the fact that the output power can vary extremely rapidly with time. Sometimes this behaviour is controllable (as in '*Q*-switching'), sometimes it is spontaneous (as in 'relaxation oscillation').

8.1 Relaxation oscillation

When the first flashlamp-pumped ruby lasers were operated it was very noticeable that their outputs consisted of an irregular series of 'spikes'. Similar behaviour (though with more regularly spaced spikes) was exhibited by other solid-state lasers (such as Nd : YAG) but was rarely seen in gas lasers. In most cases the amplitude of the spikes becomes smaller as time progresses, with a stable laser output being obtained after a certain time (Figure 8.1). The spikes are typically one or two microseconds apart and decay away after a millisecond or so. The phenomenon is called *laser spiking*.

Before we delve into the mathematics it is instructive to look in a qualitative way at the genesis of the spikes. Figure 8.2 shows the variation in population inversion and photon density within the laser as pumping first takes the population inversion through the threshold condition. We assume the laser to have a population inversion threshold of N_{th} and a corresponding equilibrium photon density of n_p^0. While the laser is below threshold there will be very few photons present and most of these will have their origin in spontaneous emission. Once the population inversion

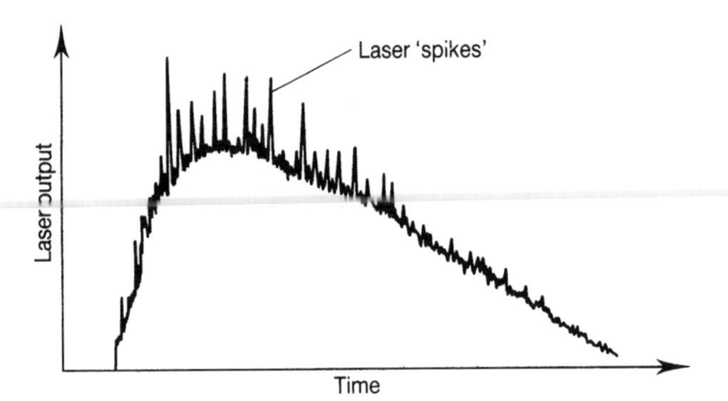

Figure 8.1 Typical output from a flashtube-pumped Nd:YAG laser, showing laser 'spiking'.

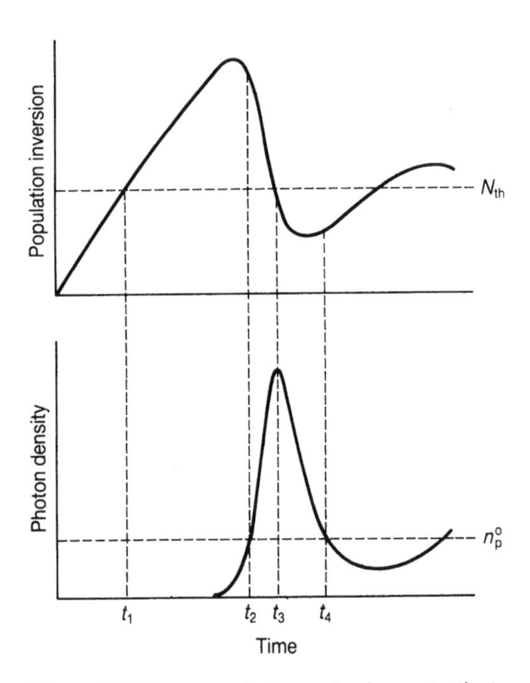

Figure 8.2 Time evolution of a laser 'spike'.

exceeds the threshold value (at time t_1) then the photon density begins to build up very rapidly through stimulated emission. This build-up is not instantaneous, however, and the population inversion has a little time in which to increase to values above the threshold value. Once the photon density has become greater than n_p^0 (i.e. at t_2) then the stimulated emission processes are numerous enough to start reducing the population inversion. The photon density then increases until the population inversion again passes through threshold (t_3). At this point the photon density reaches its maximum value. The fact that the photon density is still greater than n_p^0 ensures that stimulated emission processes dominate the situation and cause the population inversion to be pulled down to below N_{th}. Only when the photon density falls below n_p^0 (t_4) is this decline in the population inversion reversed.

The whole process is then set to repeat itself. The repetition will not be exact, however, since there will be a higher photon density present when the population inversion next exceeds N_{th}. Thus the subsequent build-up of photon density will be more rapid and the excursions in the population inversion less. Gradually the pulses will die down and an equilibrium situation reached.

We now look at the equations describing this process. To avoid unnecessary complexity we make several simplifying assumptions. Thus we assume an 'ideal' homogeneously broadened four-level system (Figure 8.3) where the population of the lower lasing level (N_1) is negligible. The losses and gains within the laser cavity are taken to be uniformly distributed with the losses being characterized by the cavity lifetime τ_c (see Eq. (5.55)). Effects such as spatial hole burning are ignored. The time-varying quantities are the upper lasing state population density (N_2) and the radiation density within the cavity. Rather than using the radiation density directly, it is convenient to deal with the photon density, n_p. Both N_2 and n_p are taken to be constant throughout the cavity at any instant in time.

We consider first the variation of N_2. N_2 will increase with time because of the presence of external pumping. We write the total pumping rate per unit volume into level 2 as R_2. Stimulated absorption from level 1 may be neglected since the population of level 1 is itself neglected. N_2 will decrease because of two factors: first, spontaneous decay to other states (characterized here by a decay lifetime τ_2) and second, because of stimulated emission (which will be proportional to the product of N_2 and n_p). Thus we may write

$$\frac{dN_2}{dt} = R_2 - \frac{N_2}{\tau_2} - CN_2n_p \tag{8.1}$$

where C is a constant associated with the stimulated emission process.

Next we turn to the behaviour of n_p. Every time a stimulated emission transition (per unit volume) takes place then n_p will increase by unity. The rate of loss of photons per unit volume due to the various cavity losses (e.g. mirror transmission,

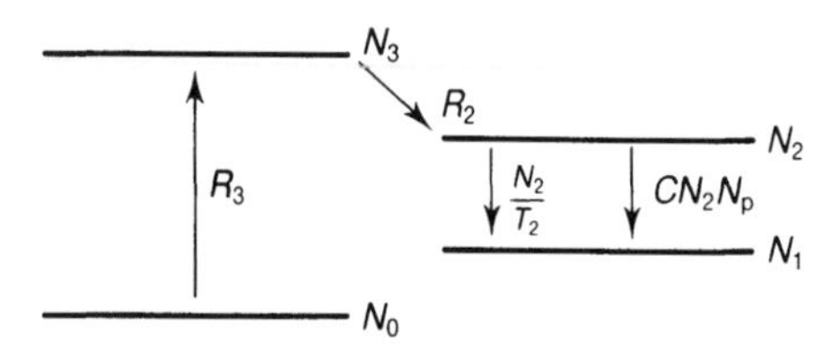

Figure 8.3 Four-level scheme assumed in the derivation of the mode-locking equations.

scattering within the cavity, etc.) is n_p/τ_c, so that the rate of change of the photon population density can be written:

$$\frac{dn_p}{dt} = CN_2n_p - \frac{n_p}{\tau_c} \tag{8.2}$$

Equations (8.1) and (8.2) form a pair of coupled differential equations (they both contain the coupling term CN_2n_p) and, despite their seeming simplicity, there are no explicit solutions which cover all circumstances.[1]

After a long enough time interval we may assume that the system will settle down to an equilibrium situation where $dn_p/dt = 0$. If the equilibrium values of n_p and N_2 are designated n_p^0 and N_2^0 respectively, then we have from Eq. (8.2) that

$$N_2^0 = \frac{1}{C\tau_c} \tag{8.3}$$

while from Eq. (8.1) it follows that

$$n_p^0 = \frac{1}{C}\left(R_2C\tau_c - \frac{1}{\tau_2}\right) \tag{8.4}$$

We see from Eq. (8.4) that we must have $R_2 > 1/(C\tau_c\tau_2)$ for there to be an equilibrium photon density in the cavity.[2] A threshold pumping rate, R_2^t, may then be defined by

$$R_2^t = \frac{1}{C\tau_c\tau_2} \tag{8.5}$$

It is convenient to write Eqs (8.1) and (8.2) in terms of the normalized variables $\mathcal{N}$, $\mathscr{n}$, R and τ where

$$\mathcal{N} = N_2/N_2^0,\quad \mathscr{n} = N_p/N_p^0,\quad R = R_2/R_2^t,\quad \tau = t/\tau_2$$

After a little algebra Eqs (8.1) and (8.2) reduce to

$$\frac{d\mathcal{N}}{d\tau} = R - \mathcal{N} - \mathcal{N}\mathscr{n}(R-1) \tag{8.6}$$

and

$$\frac{d\mathscr{n}}{d\tau} = \mathscr{n}(\mathcal{N} - 1)r_\tau \tag{8.7}$$

where $r_\tau = \tau_2/\tau_c$.

General solutions of Eqs (8.6) and (8.7) can be obtained by using numerical techniques which are relatively straightforward on modern desktop computers (see the computer project at the end of the problems section). Figure 8.4 shows a typical temporal variation in the photon density (which will be proportional to the laser output) calculated in this way. As anticipated, an initial 'spiky' oscillation is obtained whose amplitude then dies away with time.

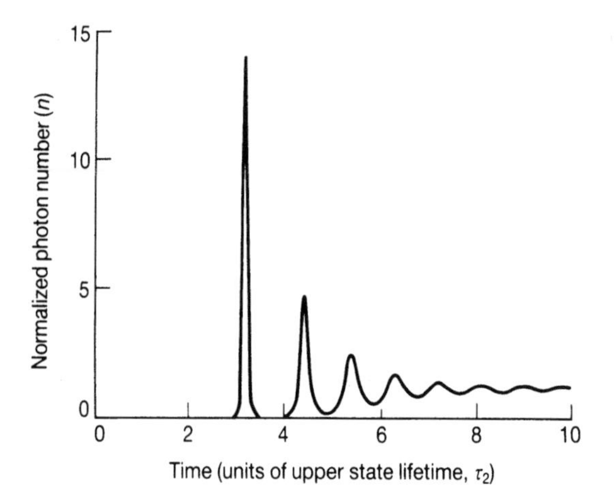

Figure 8.4 Computed variations in the photon density using the coupled Eqs (8.6) and (8.7). The pumping rate is one and a half times the threshold value (i.e. $R = 1.5$) and the ratio of the upper state lifetime (τ_2) to that of the cavity lifetime is 100 (i.e. $\tau_r = 100$). The ratio of the initial photon density to the equilibrium photon density is 10^{-4}.

It is evident from Figure 8.4 that the later stages of the oscillation resemble a damped sinusoid. The fluctuations in this region are referred to as *relaxation oscillations*. It is possible to show that the basic equations do indeed give rise to such oscillations provided departures from equilibrium are small. To do this we write

$$\mathcal{N}(t) = 1 + \delta\mathcal{N}(t)$$
$$n(t) = 1 + \delta n(t)$$

where the magnitudes of both $\delta\mathcal{N}(t)$ and $\delta n(t)$ are much less than unity. Substituting into Eqs (8.6) and (8.7) and neglecting any terms involving the product of $\delta\mathcal{N}(t)$ and $\delta n(t)$ we obtain

$$\frac{d(\delta\mathcal{N})}{d\tau} = -\delta\mathcal{N} - \delta n(R-1) \tag{8.8}$$

and

$$\frac{d(\delta n)}{d\tau} = r_\tau \delta\mathcal{N} \tag{8.9}$$

Differentiating Eq. (8.9) and then substituting for $d(\delta\mathcal{N})/d\tau$ from Eq. (8.8) then gives

$$\frac{d^2(\delta n)}{d\tau^2} + R\frac{d(\delta n)}{d\tau} + r_\tau(R-1)\,\delta n = 0 \tag{8.10}$$

To solve this equation we assume a solution of the form

$$\delta n = A \exp(p\tau) \tag{8.11}$$

whence

$$p^2 + Rp + r_\tau(R - 1) = 0 \tag{8.12}$$

Solving Eq. (8.12) for p gives

$$p = -\frac{R}{2} \pm \mathrm{j}\left(r_\tau(R-1) - \frac{R^2}{4}\right)^{1/2} \tag{8.13}$$

We may thus write

$$\delta n = A \exp(-qt) \cos(\omega_r t) \tag{8.14}$$

where (notice the change from τ to t in the time variable)

$$q = \frac{R}{2\tau_2} \tag{8.15}$$

and

$$\omega_r = \left(\frac{R-1}{\tau_2 \tau_c} - \left(\frac{R}{2\tau_2}\right)^2\right)^{1/2}$$

$$\approx \left(\frac{R-1}{\tau_2\tau_c}\right)^{1/2} \left(\text{if } r_\tau(R-1) \gg \frac{R^2}{4}\right) \tag{8.16}$$

Equation (8.14) shows that the solutions are in the form of a damped harmonic oscillation.[3] From Eqs (8.15) and (8.16) we see that both the damping rate and the oscillation frequency increase as the pumping rate (i.e. R) increases. Estimates for the values of frequencies and decay rates of relaxation oscillations in both solid-state and semiconductor lasers are given in Example 8.1.

Example 8.1 Relaxation oscillation frequencies

We consider first a YAG laser with a cavity length of 0.5 m and mirror reflectivities of 1 and 0.95. The cavity lifetime (ignoring any photon loss processes other than mirror loss) is then given by

$$\tau_c = \frac{2L}{c \log_e(1/(R_1 R_2))} = \frac{2 \times 0.5}{3 \times 10^8 \log_e(1.053)} = 6.5 \times 10^{-8}\ \text{s}$$

The upper state lifetime (τ_2) in a Nd:YAG laser is about 230 μs and so $r_\tau = \tau_2/\tau_c = 3.5 \times 10^2$. If the laser is pumped at a rate equal to twice the threshold rate, i.e. $R = 2$, then the inequality $r_\tau(R-1) \gg (R^2/4)$ holds and Eq. (8.16) then gives

$$\omega_c = \left(\frac{1}{6.5 \times 10^{-8}\ 230 \times 10^{-6}}\right)^{1/2}$$

$$= 2.6 \times 10^5 \text{ rad s}^{-1}$$

This corresponds to a linear frequency of 4×10^4 Hz (40 kHz). The decay time for the oscillation is given by $1/q$, and from Eq. (8.15) we have $1/q = 2 \times 230 \times 10^{-6}/2$ or 230 μs.

The other example we consider here is that of a semiconductor laser. Typically, such a laser based on GaAs has a cavity length of 300 μm and with mirrors that are simply the cleaved crystal ends. A flat cleaved end will give rise to a reflectivity of $((n-1)/(n+1))^2$. In GaAs $n = 3.6$ giving a reflectivity of 0.32.

Such lasers also exhibit appreciable cavity attenuation losses. Here we assume a figure of $\alpha = 2500 \text{ m}^{-1}$, so that the cavity lifetime is given by

$$\tau_c = \frac{2Ln}{c\left(2\alpha L + \log_e\left(\frac{1}{R_1 R_2}\right)\right)} = \frac{2 \times 300 \times 10^{-6} \times 3.6}{3 \times 10^8(2 \times 2500 \times 300 \times 10^{-6} + \log_e(9.76))}$$

$$= \frac{2.16 \times 10^{-3}}{3 \times 10^8(1.5 + 2.28)} = 1.9 \times 10^{-12} \text{ s}$$

The upper state lifetime can be taken to be about 3 ns, so that, assuming a value for R of 2, the inequality $r_\tau(R-1) \gg R^2/4$ again holds. The relaxation oscillation frequency according to Eq. (8.16) is then

$$\omega_c = \left(\frac{1}{3 \times 10^{-8}\ 1.9 \times 10^{-12}}\right)^{1/2}$$

$= 4 \times 10^9 \text{ rad s}^{-1}$, the corresponding linear frequency being 6.7×10^8 Hz.

Damped harmonic oscillations can also be observed when small step changes are made to the pumping rate once equilibrium has been achieved. Indeed, the relaxation oscillation frequency induced by this technique can be used to measure the quantities τ_2, τ_c or R.

It should be noted that to obtain a complex value for p we have had to assume that $r_\tau(R-1) > R^2/4$. While this is valid in most solid-state lasers it is not usually valid in gas lasers, where both τ_2 and τ_c are often of the same order of magnitude. If $r_\tau(R-1) < R^2/4$, then the solution for δn corresponds to an exponential decay back down to the equilibrium situation. When such a laser is suddenly switched on the output rises towards the equilibrium situation with little or no overshoot. Assuming that $R \gg 1$, the condition for no oscillation is that

$$R > 4r_\tau$$

so that very high pumping rates will tend to damp down any oscillations.

As mentioned at the beginning of this section, the spiking observed in ruby lasers is often rather irregular. There are several reasons for this. For example, the ruby laser is a three-level system and the above treatment is not wholly accurate. A more detailed analysis shows that the damping in such systems is less than in four-level systems so that it is vulnerable to small changes in the laser pumping rate or the temperature, etc. In addition, 'mode hopping' may be taking place. During lasing several modes may be capable of operating simultaneously and the laser energy may redistribute itself randomly between them. Spatial and temporal interference between the modes can then cause the laser output to vary in an irregular manner. In fact by restricting the laser to single-mode operation it is possible to make ruby lasers which have a much more regular spiking behaviour than usual.

8.1.1 Effects of pump modulation

We look now at the effect of a small modulation being superimposed on a steady pumping rate. If a laser exhibits a weakly damped relaxation oscillation then intuitively we might expect a rather large resonant response when the modulation frequency becomes equal to the relaxation oscillation frequency. This is borne out in practice. We start our analysis by assuming that the pumping rate can be written

$$R_2 = R_2^0 + \delta R_2 \exp(\mathrm{j}\omega_\mathrm{m} t) \tag{8.17}$$

We also assume that the photon density and population inversion exhibit corresponding oscillations about the equilibrium values, so that we may write

$$n_\mathrm{p} = n_\mathrm{p}^0 + \delta n_\mathrm{p} \exp(\mathrm{j}\omega_\mathrm{m} t) \tag{8.18}$$

and

$$N_2 = N_2^0 + \delta N_2 \exp(\mathrm{j}\omega_\mathrm{m} t) \tag{8.19}$$

If these values are substituted into Eqs (8.1) and (8.2) and any products of the (small) quantities δR_2, δn_p and δN_2 ignored, then, after a little algebra, we obtain

$$\delta n_\mathrm{p} = \frac{R_2^0 C\tau_\mathrm{c} - (1/\tau_2)}{-\omega_\mathrm{m}^2 + \mathrm{j}\omega_\mathrm{m} R_2^0 C\tau_\mathrm{c} + R_2^0 C - 1/(\tau_2\tau_\mathrm{c})}\,\delta R_2^0 \tag{8.20}$$

If we take Eq. (8.16) and express it in terms of the original variables we have

$$\omega_\mathrm{r}^2 = R_2^0 C - \frac{1}{\tau_2\tau_\mathrm{c}} - \left(\frac{R_0 C\tau_\mathrm{c}}{2}\right)^2$$

The assumption was made when obtaining Eq. (8.16) that the last term on the right-hand side of this expression could be usually neglected. Making the same assumption here, we then write

$$\omega_r^2 \approx R_2^0 C - \frac{1}{\tau_2 \tau_c} \tag{8.21}$$

Using this value in Eq. (8.20) then gives

$$\delta n_p = \frac{\tau_c \omega_r^2}{(\omega_r^2 - \omega_m^2) + j\omega_m R_2^0 C \tau_c} \delta R_2 \tag{8.22}$$

so that

$$| \delta n_p | = \frac{\tau_c \omega_r^2}{((\omega_r^2 - \omega_m^2)^2 + (\omega_m R_2^0 C \tau_c)^2)^{1/2}} \delta R_2 \tag{8.23}$$

A plot of this function is given in Figure 8.5. It reaches a peak when the denominator is at a minimum, which occurs when

$$\omega_m^2 = \omega_r^2 - 2\left(\frac{R_0 C \tau_c}{2}\right)^2$$

The last term on the right-hand side can be neglected (i.e. by using the same approximation as above in the derivation of Eq. (8.21)). The response curve then has a maximum when $\omega_m = \omega_r$. The ratio of the response at resonance ($\omega_m = \omega_r$) to that when $\omega_m \ll \omega_r$ is given by the factor

$$\frac{\omega_r}{R_2^0 C \tau_c}$$

which, in the approximation we have been using, will have a relatively large value.

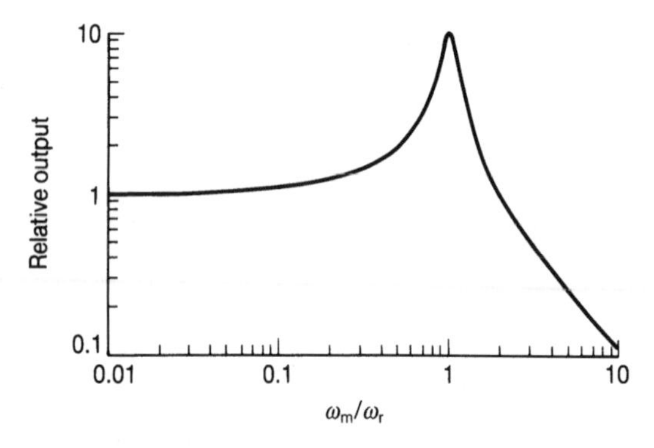

Figure 8.5 Predicted frequency response of the output of a semiconductor diode as a function of the modulation frequency ω_m as given by Eq. (8.23) and assuming a value for $\omega_r/(R_2^0 C \tau_c)$ of 10.

When $\omega_m \gg \omega_r$, then

$$|\delta n_p|/\delta R_2 = \tau_c \omega_r^2/\omega_m^2$$

This discussion has a direct relevance to the frequency response of semiconductor laser diodes when using drive current modulation. The upper operating frequency of such devices is obviously determined by ω_r since after that frequency the output will fall off as $(\omega_m)^{-2}$. It might be thought that the increase in the output near ω_r might be advantageous. However, any random noise fluctuations near this frequency are also amplified by the system, giving rise to a relatively noisy output which is generally not so desirable. The frequency response of the laser diodes may be extended by increasing ω_r. From Eq. (8.21) we see that ω_r depends approximately on $(R_2)^{1/2}$ or on the square root of the drive current. This dependence of upper operating frequency on drive current has been verified experimentally [Ref. 1].

8.2 *Q*-switching

Although the phenomenon of relaxation oscillation discussed in the previous section is more of a nuisance than a useful laser property, it does serve to verify the validity of the equations used to describe them. In this section we deal with a rather more valuable technique for influencing the output of a laser which may be described by a very similar pair of equations.

Suppose we take a laser and introduce a huge loss into the cavity (for example, we might imagine covering up one of the mirrors). This, of course, usually prevents the laser from operating at all. However, it also means that the population inversion in the laser medium can rise to well above its normal value. If then the losses are suddenly reduced to their normal values, the small signal gain is much larger than the threshold gain. The result is a very rapid increase in photon density inside the cavity, leading in turn to a rapid reduction in the population inversion through stimulated emission. A short intense pulse (sometimes called a *giant pulse*) of radiation is emitted by the laser. This process is called *Q-switching*. The name originates because of the association of cavity loss and Q value (a high loss being associated with a low Q value). The sudden change in the cavity loss implies a sudden change (or *switch*) in the Q value.

8.2.1 Theory of *Q*-switching

The mathematical approach to Q-switching is similar to that used for relaxation oscillations except that the approximations made are somewhat different. Again we assume a four-level system and effectively ignore any atoms not in levels 1 and 2 (i.e. the levels between which lasing takes place). The populations change so rapidly during the formation of the Q-switched pulse that we may ignore the effects of pumping. Similarly, we ignore the loss of atoms from levels 1 and 2 by nonradiative decay processes. Unlike the relaxation oscillation situation, however, we cannot ignore the population of level 1. The changes in population in levels 1 and 2 are now governed solely by stimulated emission, and accordingly we write

$$\frac{dN_2}{dt} = -CN_2n_p + CN_1n_p \tag{8.24}$$

$$\frac{dN_1}{dt} = -CN_1n_p + CN_2n_p \tag{8.25}$$

Subtracting Eq. (8.25) from Eq. (8.24) gives

$$\frac{d}{dt}(N_2 - N_1) = -2n_pC(N_2 - N_1)$$

Writing N for the population inversion $N_2 - N_1$, we have

$$\frac{dN}{dt} = -2n_pCN \tag{8.26}$$

The equation for the rate of change of photon density can then be written

$$\frac{dn_p}{dt} = Cn_pN - \frac{n_p}{\tau_c} \tag{8.27}$$

We see from Eq. (8.27) that a threshold inversion, N_{th}, may be defined by the value of N for which $dn_p/dt = 0$. Thus

$$N_{th} = \frac{1}{C\tau_c}$$

We now normalize both N and n_p to N_{th}, that is, we introduce parameters $\mathcal{N}$ and $\mathfrak{n}$ where[4]

$$\mathcal{N} = N/N_{th}$$

and

$$\mathfrak{n} = n_p/N_{th}$$

In terms of these parameters Eqs (8.26) and (8.27)become

$$\frac{d\mathcal{N}}{d\tau} = -2\mathfrak{n}\mathcal{N} \tag{8.28}$$

and

$$\frac{d\mathfrak{n}}{d\tau} = \mathfrak{n}(\mathcal{N} - 1) \tag{2.29}$$

where $\tau = t/\tau_c$.

Equations (8.28) and (8.29) are two coupled equations which have no direct analytical solution, but which may (as in the case of relaxation oscillations) be solved numerically.[5] To use the equations we assume that at time $t = 0$ the system is switched instantaneously from a high- to low-loss situation, so that $\mathcal{N}$ will start (at $t = 0$) from some value larger than unity. We may also note that we have to assume a nonzero starting value for n, otherwise n will always remain at zero. The result of typical computations of n (which will be proportional to the output from the laser) for various initial values of $\mathcal{N}$ are shown in Figure 8.6, while Figure 8.7 shows the variation in the population inversion during a pulse.

It is possible, however, to make a number of useful deductions directly from the equations themselves. Thus if we divide Eq. (8.29) by Eq. (8.28) we obtain

$$\frac{dn}{d\mathcal{N}} = \frac{1}{2}\left(\frac{1}{\mathcal{N}} - 1\right) \tag{8.30}$$

This equation may be integrated to give

$$n - n_i = \frac{1}{2}\left(\log_e\left(\frac{\mathcal{N}}{\mathcal{N}_i}\right) - (\mathcal{N} - \mathcal{N}_i)\right)$$

where n_i and $\mathcal{N}_i$ are the initial values of n and $\mathcal{N}$, respectively. Since the initial photon density can be neglected, we have

$$n = \frac{1}{2}\left(\log_e\left(\frac{\mathcal{N}}{\mathcal{N}_i}\right) - (\mathcal{N} - \mathcal{N}_i)\right) \tag{8.31}$$

The maximum photon density in the cavity will occur when

$$\frac{dn}{d\mathcal{N}} = 0$$

which gives $\mathcal{N} = 1$. In other words, the maximum photon density (and hence the

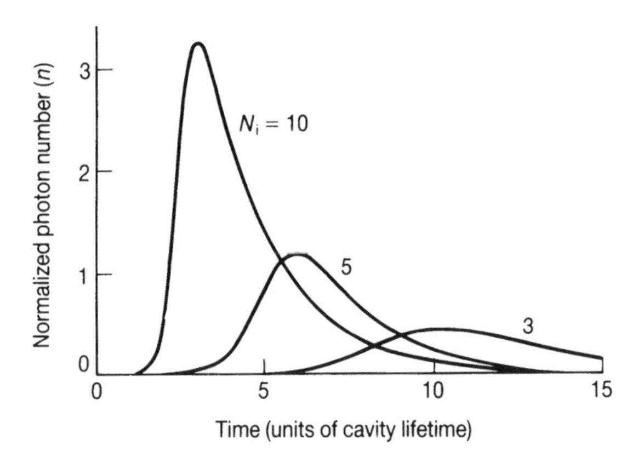

Figure 8.6 Computed solutions of the time evolution of a *Q*-switched pulse for three different initial population inversions. The indicated values of the initial population inversions (N^i) are normalized values with respect to the threshold population inversion. The ratio of the initial photon number to the equilibrium photon number is 10^{-4}.

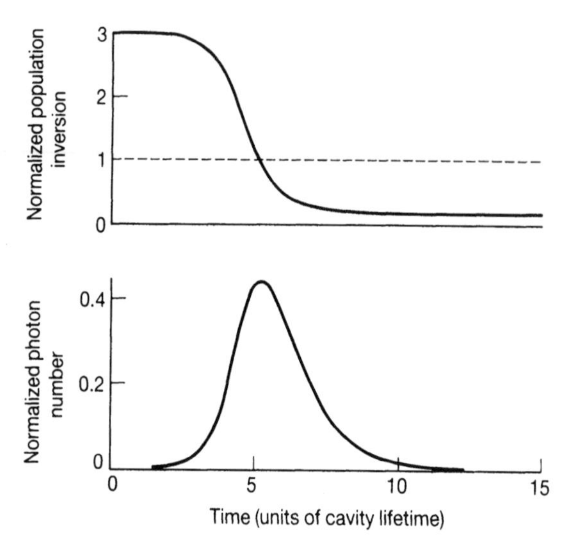

Figure 8.7 Time evolution of both the photon numbers and the population inversion during a *Q*-switched pulse.

maximum laser output) will be obtained when the population inversion falls to the threshold value (this can be clearly seen in Figure 8.7). Inserting $\mathcal{N} = 1$ in Eq. (8.31) then gives the peak normalized photon density, n_{max}, as

$$n_{max} = \frac{\mathcal{N}_i - 1 - \log_e \mathcal{N}_i}{2} \tag{8.32}$$

For situations in which $\mathcal{N}_i$ is much larger than unity we have simply that

$$n_{max} = \mathcal{N}_i/2$$

That is, the maximum photon density is just equal to half the initial population inversion density. What is happening here is that the photon population rapidly builds up, converting most of the (initially large) population inversion into photons. The photons then leak out of the cavity through the output mirror at a rate given by dividing the total number of photons in the cavity by the cavity lifetime.[6] Since the total number of photons in the cavity at the peak density can be written as $n_{max}N_v^t V$, where V is the volume of the gain medium (assuming it occupies all the space between the mirrors), the peak power output of the laser, P_{max}, can be written

$$P_{max} = n_{max} N_{th} V \hbar\omega / \tau_c \tag{8.33}$$

Another feature, evident from an inspection of Figure 8.7, is that the final population inversion left after the giant pulse is not zero. This in effect means that not all the energy in the laser has been extracted by the pulse. The total initial energy density within the laser cavity arising from population inversion will be given by

$$\left(\frac{N_2 - N_1}{2}\right)_{\text{init.}} \hbar\omega$$

or $\mathcal{N}_i N_{th} \hbar\omega$

Similarly, the total final energy density remaining when the giant pulse has passed will be given by $\mathcal{N}_f N_{th} \hbar\omega$, where $\mathcal{N}_f$ is the final value of $\mathcal{N}$. Consequently the fraction of the initial stored energy which is extracted (and which, we may assume, has gone into the the giant pulse) is given by η_p where

$$\eta_p = \frac{\mathcal{N}_i - \mathcal{N}_f}{\mathcal{N}_i} \tag{8.34}$$

We may obtain a value for $\mathcal{N}_f$ from Eq. (8.31) since at the end of the pulse the photon density will have fallen to negligible values. Putting $n = 0$ into Eq. (8.31) then gives

$$\frac{1}{2}\left(\log_e\left(\frac{\mathcal{N}_f}{\mathcal{N}_i}\right) - (\mathcal{N}_f - \mathcal{N}_i)\right) = 0$$

or

$$\mathcal{N}_f = \mathcal{N}_i \exp(\mathcal{N}_f - \mathcal{N}_i) \tag{8.35}$$

For a particular value of $\mathcal{N}_i$ Eq. (8.35) may be solved (graphically or numerically) to give $\mathcal{N}_f$ and the values then inserted in Eq. (8.34) to obtain a value for the energy-utilization efficiency. The result is shown in Figure 8.8, where we see that with a

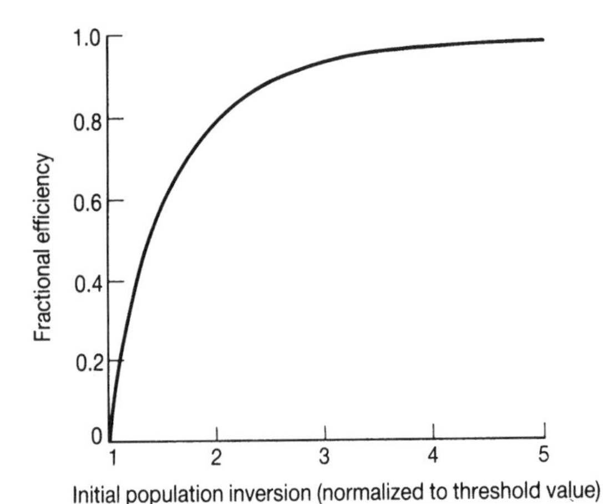

Figure 8.8 Graph of the energy extraction efficiency of a Q-switched pulse as a function of the (normalized) population inversion.

value of $\mathcal{N}_i$ greater than 5 or so, the energy-extraction efficiency is almost 100 per cent.

If all the initial normalized population $\mathcal{N}_i$ could be turned into photons and extracted from the laser, the corresponding energy as noted above would be $\mathcal{N}_i N_{th} \hbar\omega V/2$, so that the actual pulse energy, E_p, can be written

$$E_p = \eta_{pe} \mathcal{N}_i N_{th} \hbar\omega V/2 \tag{8.36}$$

We turn now to the pulse duration, $\Delta\tau_p$. This may be defined by requiring that the total energy in a pulse is given by one half of the peak power multiplied by the pulse duration. Thus

$$\Delta\tau_p = \frac{2E_p}{P_{max}}$$

Using Eqs (8.33) and (8.36),

$$\Delta\tau_p = \frac{\eta_{pe}\mathcal{N}_i}{n_{max}} \tag{8.37}$$

If $\mathcal{N}_i$ is greater than 5 or so, then $n_{max} \approx \mathcal{N}_i/2$ and $\eta_p \approx 1$, so that $\Delta\tau_p \approx 2\tau_c$.

An estimation of the total energy and peak power occurring in Q-switched pulses based on these results is given in Example 8.2.

Example 8.2 Energy and power of Q-switched pulses

We consider the energy and peak power obtainable from a Nd : YAG laser which is pumped to ten times its normal threshold inversion when the Q-switch is operated. The Nd : YAG rod is taken (for simplicity) to fill the entire cavity length of 100 mm and to have a cross-sectional area of 40 mm^2. The cavity mirrors have reflectivities of 100 per cent and 90 per cent.

Since the initial inversion is five times the threshold value, and hence $\eta_p \approx 1$ (Figure 8.8), the total energy within the Q-switched pulse is given by (Eq. 8.36)) $E_p = N_i N_v^t \hbar\omega V/2$. To evaluate this expression we need to estimate N_{th}, the threshold population inversion density of the laser. The threshold gain, γ_{th}, required to overcome the mirror losses is given by[7] (see Eq. (6.1))

$$\gamma_{th} = \frac{1}{2l}\log_e\left(\frac{1}{R_1 R_2}\right)$$

For the laser being considered here we have

$$\gamma_{th} = \frac{1}{2\times 0.1}\log_e\left(\frac{1}{0.9}\right)$$

$$= 0.527\ \text{m}^{-1}$$

The required population inversion is then obtained by dividing γ_{th} by the cross-section (σ) for the transition (see Eq. (3.112)). For the 1.06 μm Nd:YAG laser we may take σ to be 7×10^{-23} m^2 (Table 3.1). Thus the threshold population inversion density is given by

$$N_{th} = 0.527/7 \times 10^{-23} \text{ or } 7.53 \times 10^{21} \text{ m}^{3}$$

the total energy in the giant pulse is then

$$E_p = 10 \times 7.53 \times 10^{21} \hbar\omega V/2$$

where V is the volume of the gain medium. Since the energy of a photon from the laser ($= \hbar\omega$) is 1.87×10^{-19} J, we have

$$\begin{aligned} E_p &= 10 \times 7.53 \times 10^{21}\ 1.87 \times 10^{-19}\ 0.1 \times 40 \times 10^{-6}/2 \\ &= 0.028 \text{ J} \end{aligned}$$

The pulse duration can be taken to be approximately twice the cavity lifetime, $2\tau_c$ (see discussion following Eq. (8.37)). The cavity lifetime is simply the time taken for a roundtrip within the cavity ($= 2nl/c$) divided by the fractional loss per roundtrip ($= \exp(-2\alpha l)$). Thus

$$\begin{aligned} \tau_c &= \frac{2 \times 1.82 \times 0.1/(3 \times 10^8)}{\exp(-0.2 \times 0.527 \times 0.1)} \text{ s} \\ &= 1.23 \times 10^{-9} \text{ s} \end{aligned}$$

The pulse duration is thus 2.46×10^{-9} s, so that the average power during the pulse is approximately $E_p/(2\tau_c)$ or 5.7 MW.

The shape of the output pulse obtained from the laser is dependent on the value of $\mathcal{N}_i$. When $\mathcal{N}_i$ is very large, the build-up of the pulse is very rapid but once $\mathcal{N}$ has fallen significantly below unity the major change in photon density arises from photon losses, and so the falling edge of the curve is approximately an exponential decay with a time constant τ_c.

When comparing experimental curve shapes to those predicted from the above equations an initial value for the photon density is required. This may be obtained by estimating the number of spontaneously emitted photons which lie within the acceptance angle of the particular laser mode at $t = 0$.

The analysis presented so far has assumed that the Q-switching happens instantaneously. Obviously this is not possible in practice, but provided the switch can operate within a few tens of nanoseconds there is usually no problem. If the Q-switch operates much less rapidly than this then unwanted complications such as multiple pulses can occur. The origin of such multiple pulses is illustrated in Figure 8.9. The variable loss is shown as a slowly falling function of time which crosses the threshold loss curve at time t_1. A giant pulse then builds up in the usual way and reduces the population inversion, resulting in a threshold loss that is less than the

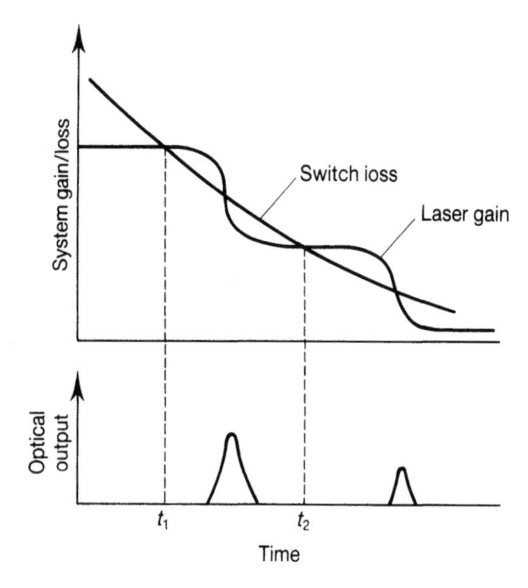

Figure 8.9 Illustration of the genesis of a secondary *Q*-switched pulse when a slow switch is present.

actual loss. However, some time later (t_2) the loss curve again crosses the threshold loss curve and a second pulse is obtained. For an analysis of pulses from slowly operating *Q*-switches see Ref. 2.

8.2.2 Techniques for *Q*-switching

8.2.2.1 Mechanical methods

The earliest method used for *Q*-switching was to introduce a rotating slit into the cavity [Ref. 3]. Although this technique produced *Q*-switched pulses the switching times were relatively long and the technique was soon abandoned in favour of rotating one of the laser end mirrors. Lasing is then only possible when the mirror was very nearly perpendicular to the laser axis. A problem with this technique is that the axis of rotation must be aligned to within a fraction of a milliradian parallel to the face of the other mirror. This difficulty can be overcome by replacing the mirror with a retroreflector type of prism.

Although such mechanical techniques are relatively simple and inexpensive they still switch relatively slowly (see Example 8.3) and hence can give rise to multiple pulses. Noise, vibration and the need for frequent maintenance are additional problems. Except in some specialized circumstances where there are difficulties with other techniques (for example, with the CO_2 laser) mechanical *Q*-switching techniques have been almost entirely supplanted by nonmechanical methods.

Example 8.3 Switching times of rotating mirror *Q*-switches

We take the mirror rotation speed to be 30 000 rpm and the required accuracy of mirror angular alignment for lasing to occur to be 1 minute of arc. The mirror will rotate through 360° or 360×60 minutes of arc in a time of 1/500 s. The time for the mirror to rotate through 1 minute of arc is then $(360 \times 60 \times 500)^{-1}$ s or 93 ns. This is appreciably longer than the 'ideal' of a few nanoseconds.

8.2.2.2 Electro-optical Q-switching

There are several nonmechanical switching techniques that may be used for *Q*-switching, and one of the most useful is based on the Pockels effect. This occurs in uniaxial crystals which develop an additional birefringence when an electric field is applied (as discussed in Section 2.1.9). In the longitudinal effect the field is applied parallel to the crystal optic axis and in the same direction as the incident light. With no field applied light of any polarization travelling along the optic axis experiences the same refractive index (i.e. n_0, see Figure 8.10(a)). When the field is switched on, however, principal axes, x' and y', are formed which are at 45° to the original x, y axes (Figure 8.10(b)). The refractive index experienced by light polarized along the x' and y' directions is given by (i.e. Eqs (2.93)and (2.94))

$$\left.\begin{aligned} n_{x'} &= n_0 + \tfrac{1}{2} n_0^3 r_{63} \mathscr{E}_z \\ n_{y'} &= n_0 - \tfrac{1}{2} n_0^3 r_{63} \mathscr{E}_z \end{aligned}\right\} \qquad (8.38)$$

where $\mathscr{E}_z$ is the electric field applied along the z direction and r_{63} is an electrooptic coefficient. For simplicity, in the following discussion we will write r_{63} as r.

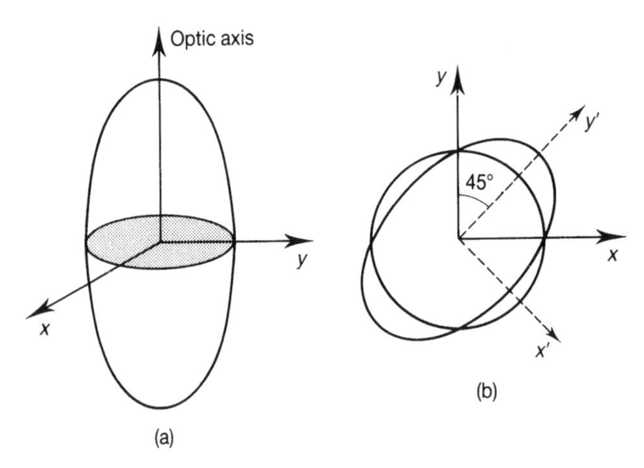

Figure 8.10 (a) The index ellipsoid for uniaxial crystals. For propagation along the optic axis the refractive index is independent of the direction of the electric field (i.e. the projection on the xy plane is a circle). (b) When an external field is applied along the optic axis, the projection changes from a circle to an ellipse whose principal axes are at 45° to the original x, y axes.

We consider now the transmission of radiation through the crystal when the radiation is initially polarized along the x axis. If the incident electric field is written $\mathscr{E} = \mathscr{E}_0 \cos(\omega t)$, then the components along the x' and y' directions will be

$$\mathscr{E}_{x'} = \mathscr{E}_{y'} = \frac{\mathscr{E}_0}{\sqrt{2}} \cos(\omega t)$$

Since the refractive indices experienced by the two components differ then they will become increasingly out of phase as they propagate through the crystal, and, in general, the emergent beam will be elliptically polarized. The phases of the emergent beams may be written (using Eq. (8.38))

$$\phi_{x'} = \phi_0 + \Delta\phi$$

and

$$\phi_{y'} = \phi_0 - \Delta\phi$$

where

$$\phi_0 = \frac{2\pi}{\lambda_0} n_0 L$$

and

$$\Delta\phi = \frac{\pi}{\lambda_0} r n_0^3 L \mathscr{E}_z$$

If the applied longitudinal voltage across the crystal is V, we have that $\mathscr{E}_z L = V$, and so

$$\Delta\phi = \frac{\pi}{\lambda_0} r n_0^3 V$$

The two emergent electric field components can now be written:

$$\left.\begin{aligned} \mathscr{E}_{x'} &= \frac{\mathscr{E}_0}{\sqrt{2}} \cos(\omega t + \phi_0 + \Delta\phi) \\ &\text{and} \\ \mathscr{E}_{y'} &= \frac{\mathscr{E}_0}{\sqrt{2}} \cos(\omega t + \phi_0 - \Delta\phi) \end{aligned}\right\} \qquad (8.39)$$

Suppose that a linearly polarizing element is placed in the path of the emergent beam with its polarization axis at right angles to the direction of polarization of the original beam (Figure 8.11). The components of field that are transmitted by the polarizer will be $\mathscr{E}_x/\sqrt{2}$ and $-\mathscr{E}_y/\sqrt{2}$, and so the transmitted field can be written

$$\mathscr{E} = \frac{\mathscr{E}_0}{2} (\cos(\omega t + \phi_0 + \Delta\phi) - \cos(\omega t + \phi_0 - \Delta\phi))$$

or

$$\mathscr{E} = -\mathscr{E}_0 \sin(\Delta\phi) \sin(\omega t + \phi_0)$$

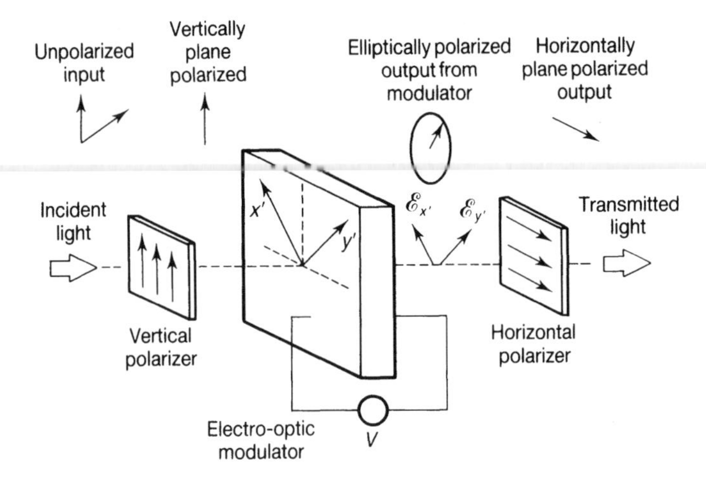

Figure 8.11 Arrangement of the components of a Pockels electro-optic switch in which the electro-optic crystal is placed between crossed polarizers.

The transmitted irradiance, I, will be proportional to the average of the square of this quantity over a time long compared with $2\pi/\omega$, so that we may write

$$I = I_0 \sin^2(\Delta\phi)$$

or

$$I = I_0 \sin^2\left(\frac{\pi}{2}\frac{V}{V_\pi}\right) \tag{8.40}$$

where

$$V_\pi = \frac{\lambda_0}{2rn_0^3} \tag{8.41}$$

The quantity V_π is called the *half-wave* voltage (a phase change of π is equivalent to a path length change of $\lambda/2$).

Thus the arrangement of a Pockels cell placed between two orthogonally orientated polarizing elements as shown in Figure 8.11 can act as a light modulator with the transfer function as given by Eq. (8.40). More importantly, as far as we are concerned, it can act as a switch. Consider, for example, the arrangement shown in Figure 8.12(b) where the above modulator is placed within a laser cavity. With no voltage applied across the cell the losses within the cavity will be very large, while by applying a voltage of V_π the losses will fall to much smaller values.

Another possible arrangement (Figure 8.12(a)) is to dispense with one of the polarizers and to switch between applied voltages of $V_\pi/2$ and zero. As can be readily verified from the above equations, the application of a voltage of $V_\pi/2$ across the Pockels cell gives rise to a phase change between the two components polarized

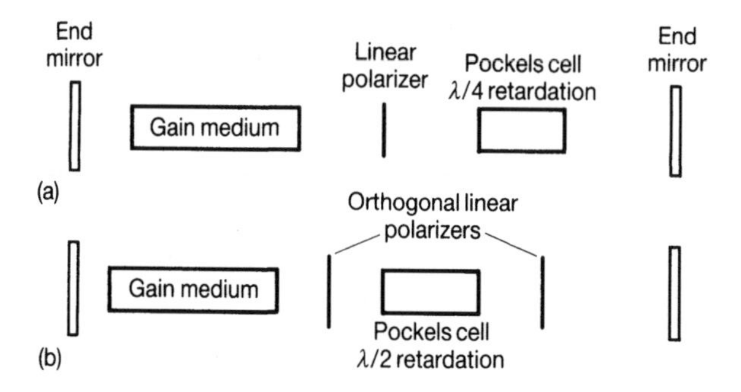

Figure 8.12 Electro-optic *Q*-switch arrangements based on a Pockels cell with (a) a λ/4 retardation and (b) a λ/2 retardation.

along the x' and y' axes of 90° (or $\pi/2$ radians). Thus an initially plane polarized beam (along the x axis) will emerge as a circularly polarized beam. After reflection from the laser end mirror the light repasses through the Pockels cell and the phase difference between the two x' and y' components is increased by a further 90° (making a total of 180°). A phase difference of 180° between the initial components gives rise again to plane polarized light but with a polarization direction which is orthogonal to that of the polarizers. The cavity losses are then very high. The cavity may be reduced to a low-loss state by reducing the applied voltage to zero.

Table 8.1 gives the experimentally determined values of n_0 and r_{63} and the resulting calculated values for the half-wave voltage (using Eq. (8.41)) for a number of materials that have been used in longitudinal modulators.

Materials such as those indicated in Table 8.1 can be readily grown from a water solution at relatively low temperatures, leading to comparatively strain-free crystals which can be easily cut and polished. They are, however, hydroscopic and must be protected from exposure to the atmosphere. Usually they are in sealed cells containing an index matching fluid with anti-reflection coatings on the outside windows. The presence of the fluid and the anti-reflection coatings help to reduce the reflection losses that would otherwise be present from the six interfaces.[8]

Table 8.1

Material	r_{63} (at 633 nm) (pm/V)	n_0	V_π (kV)
Ammonium dihydrogen phosphate $(NH_4)H_2PO_4$ or ADP	7.8	1.522	11.5
Potassium dihydrogen phosphate KH_2PO_4 or DKP	10.6	1.507	8.4
Deuterated KDP (KD*P)	24.1	1.502	3.9

One of the difficulties with the longitudinal field configuration is that of obtaining a uniform electric field across the aperture of the device. The simplest electrode structure is shown in Figure 8.13(a) where metal electrodes with circular apertures are bonded or evaporated onto the end surfaces. Obviously the field strengths will be a minimum in the centre of the aperture and will increase towards the edges. A considerable improvement in uniformity can be obtained by using cylindrical crystals with band electrodes applied along the sides of the cylinder as shown in Figure 8.13(b). In both of these cases the applied voltage required to achieve switching is somewhat greater than is implied from the calculations of Eq. (8.41). Apertures of between 10 and 20 mm are readily available.

It is also possible to design Pockels cells with transverse fields. However, this is not usually done with the phosphate crystal types used in the longitudinal design because the light is not then propagating along the optic axis, and so the ordinary and extraordinary rays travel in different directions in the crystal. However, the material lithium niobate ($LiNbO_3$) has a crystal structure such that it *can* be used in a transverse field with the light propagating along the optic axis. With light polarized along the *a* axis and an electric field also applied along the *a* axis (Figure 8.14) the half-wave voltage is given by

$$V_\pi = \frac{\lambda d}{2 r_{22} n_0^3 l} \tag{8.42}$$

where l is the length of the crystal along the light-propagation direction, d the electrode separation and r_{22} an electro-optic coefficient. As shown in Example 8.4, given appropriate values of the factor d/l the half-wave voltages for transverse field Pockels cells can be appreciably less than for longitudinal field cells.

Pockels cells are, in general, very fast-acting devices and switching times of 500 ps or less are possible.

Example 8.4 Switching voltages in transverse field modulators

For lithium niobate r_{22} has the value of 6.8 pm/V (at a wavelength of 633 nm) while n_0 is 2.286. Taking a cell with the dimensions 5 mm × 5 mm in cross-section and 30 mm long the theoretical half-wave voltage from Eq. (8.42) is given by

$$V_\pi = \frac{633 \times 10^{-9}}{2 \times 6.8 \times 10^{-12} (2.268)^3} \frac{5}{30}$$

$$= 700 \text{ V}$$

8.2.2.3 *Acousto-optic switching*

A rather slower Q-switch than those based on the Pockels effect can be obtained using the acousto-optic effect. When an acoustic wave travels through a medium

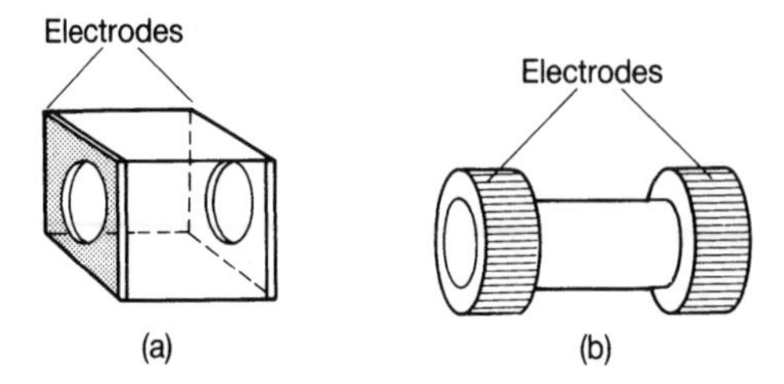

Figure 8.13 Two common electrode arrangements for producing a longitudinal field in an electro-optic switch.

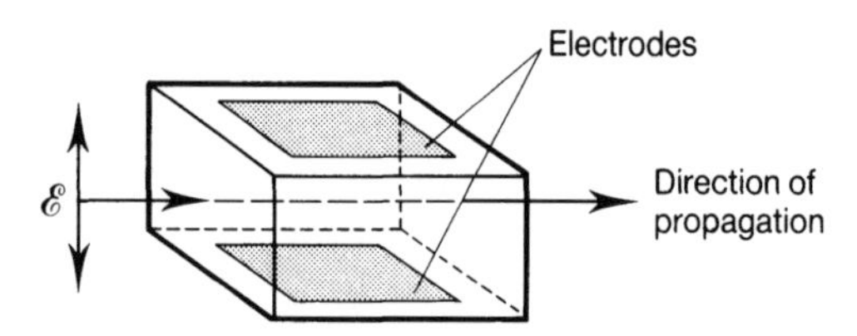

Figure 8.14 An electro-optic modulator based on lithium niobate where the applied field is transverse to the direction of propagation.

such as silica (SiO_2) it causes local changes in the material density which in turn give rise to changes in the refractive index. The resulting pattern of refractive index variation can act rather like a diffraction grating as far as a beam of light passing through the medium is concerned. In Figure 8.15 a situation is shown where an acoustic wave with a wavelength Λ and a lateral extent of L is present. The diffraction grating that is produced may be considered 'thick' provided $L\lambda \gg \Lambda^2$, in which case the diffraction produced is very similar to the Bragg type diffraction of X-rays from atomic planes in solids. Here the peaks in the refractive index act as if they were atomic planes with the light beam taking the place of the X-rays. Thus if the light beam is incident on the grating structure at the angle θ_B (see Problem 8.7) where

$$\sin\,\theta_b = \frac{\lambda}{2\Lambda} \tag{8.43}$$

then some of the incident radiation will be coherently reflected from the grating at the same angle θ_b (the 'first order' beam[9]), giving rise to a beam which is at an angle of $2\theta_B$ from the original beam direction. It should be noted that the wavelength used in Eq. (8.43) is that of the light in the medium. If the modulator is in the shape of a parallel-sided block with the acoustic wave travelling perpendicularly to two of the faces then the external ray deflection angle, θ_B', is related to the internal angle θ_B by

$$\sin\,\theta_B' = n\,\sin\,\theta_B$$

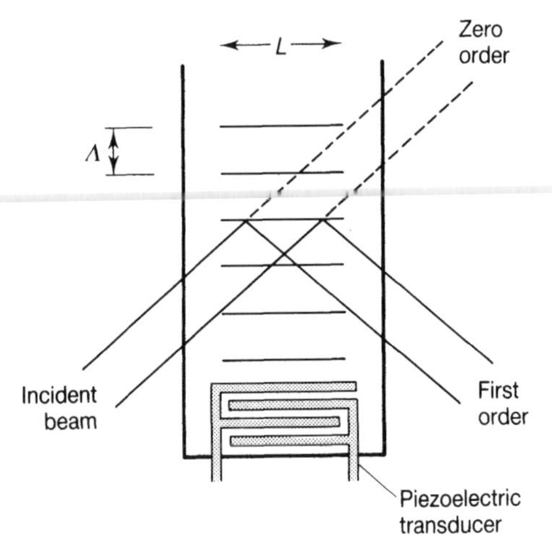

Figure 8.15 Geometry for Bragg-type acousto-optic modulation. The incident rays are scattered from successive layers of refractive index maxima.

where n is the refractive index of the medium. Thus as far as the external angles are concerned

$$\sin \theta_B' = \frac{\lambda_0}{2\Lambda} \tag{8.44}$$

where λ_0 is the vacuum wavelength of the light.

The total deflection angles (i.e. $2\theta_B'$) are usually quite small, i.e. less than 1° (see Example 8.5), but this is more than sufficient to 'spoil' a laser cavity.

Example 8.5 Beam deflection in acousto-optic modulators

We base this calculation of beam deflection on silica where the acoustic wave velocity may be taken as 3.76×10^3 m/s. With an acoustic frequency of 50 MHz the acoustic wavelength (Λ) is $3.76 \times 10^3/50 \times 10^6$ m or 7.52×10^{-5} m (i.e. 72.5 μm). If light of wavelength 1.067 μm is being used then Eq. (8.44) gives that

$$\theta_B' = \sin^{-1}\left(\frac{1.067}{2 \times 75.2}\right) \text{ or } 0.406^\circ$$

Thus the total (external) beam deflection is 0.81°.

For the Bragg scattering formula to be valid we require that $l\lambda \gg \Lambda^2$, or $l \gg \Lambda^2 n/\lambda_0$, where n is the material refractive index. Taking $n = 1.45$ the requirement in the present instance is that $l \gg 7.7$ mm. Thus a length of something like 50 mm would satisfy this requirement.

The amount of energy that is diffracted into the first-order beam depends on the amount of acoustical power present. The equations governing the efficiency of the diffraction process will not be pursued here, and the interested reader will find more details in reference 4. Although not the best material from a diffraction efficiency point of view,[10] silica, with a high damage threshold, high optical quality and low optical absorption is the usual choice for an acousto-optic *Q*-switch. Typical diffraction efficiencies of up to 40 per cent can be achieved with acoustic powers of several tens of watts.

In use the device is inserted into the laser cavity with the radiofrequency power applied to the acoustic transducer. Once the population inversion has built up to the required values the power is switched off and the *Q*-switching action can take place. The speed of operation of the device is not usually limited by the rapidity with which the power can be switched off but by the transit time of the acoustic wave across the laser beam diameter. This is given by $\tau_t = B/v_a$, where B is the laser beam diameter and v_a the acoustic wave velocity. Typically, v_a is of the order of 200 ns/mm.

8.2.3 Passive *Q*-switching

Another way of *Q*-switching a laser is to make use of so-called *saturable absorbers*. These are usually dye solutions whose transmission depends on the light irradiance. Assuming that the transition behaves like a homogeneously broadened transition, and that the theory given in Section 3.3.2 applies (although we are dealing here with absorption rather than gain) we can relate, by analogy with Eq. (3.124), the absorption coefficient α to the irradiance I by the equation:

$$\alpha = \frac{\alpha_0}{1 + I/I_s} \tag{8.45}$$

Here α_0 is the absorption coefficient at very small incident irradiances and I_0 is the irradiance at which the absorption coefficient falls by a factor two from its zero irradiance value. For simplicity, we have also assumed that the factor $\bar{g}(\omega)$ has the value unity. A typical curve of dye transmission versus incident light irradiance is shown in Figure 8.16. At high input irradiances the dye becomes completely transmitting (it is said to be 'bleached'). In operation as a *Q*-switch the dye solution is placed in a cell within the laser cavity, the best position being up against the 100 per cent reflecting mirror so that the mirror forms one wall of the cell (Figure 8.17). The thickness of the dye solution is usually only a millimetre or so.

When pumping commences the relatively low transmission of the dye cell (typically between 40 per cent and 60 per cent) means that the gain medium has to be pumped to give a high degree of population inversion before the photon population begins to build up within the cavity. To be most effective the dye solution must now rapidly bleach to give a high optical transmission. If this takes place before a significant part of the initial population inversion has been used up then the result will be the production of a giant pulse. However, it is vital that the bleaching of the dye

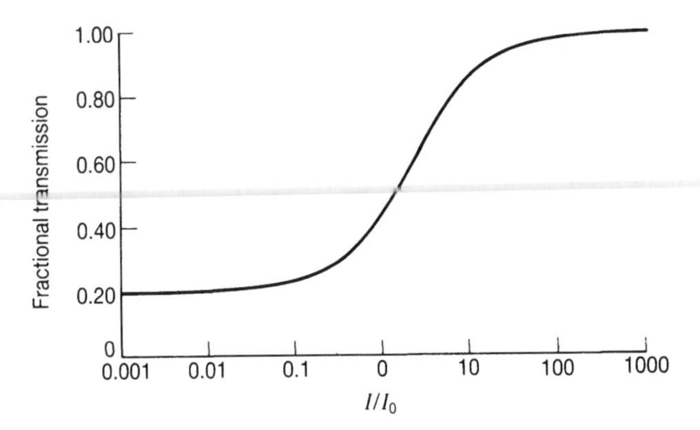

Figure 8.16 Transmission curve of a saturable dye absorber as a function of incident irradiance.

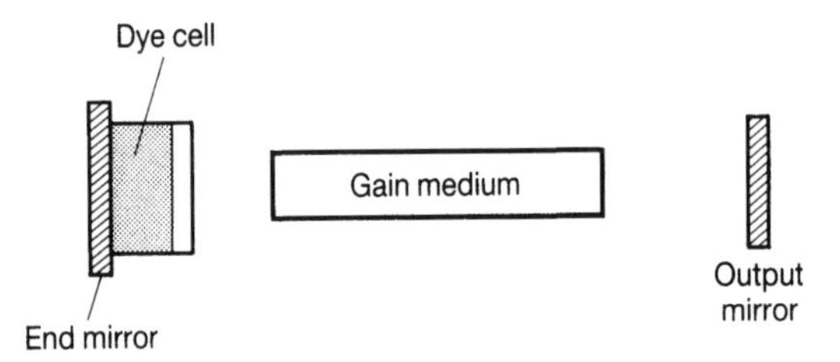

Figure 8.17 Optimum position of the dye cell for passive *Q*-switching, in contact with the 100% reflecting mirror.

occurs during the initial part of the photon build-up. This requires that the dye solution has a large photon absorption cross-section coupled with a very short upper state lifetime.

While it is possible to write down a set of three coupled equations governing the rates of change of the photon population, and the population inversions in both the amplifier and the absorber systems, it is not possible to obtain analytical solutions and they must be solved numerically. The interested reader may consult Ref. [5] for further details.

The obvious advantages of passive *Q*-switching include both simplicity and economy, but there are also a number of drawbacks. With a flashlamp-pumped laser there is always a substantial variation in the time lapse between the onset of the pumping pulse and the appearance of the giant pulse (i.e. 'jitter'). In addition, the dye solutions can degrade in use and better performance is obtained by circulating the dye using a pump and a reservoir. Finally, if the dye solution is not completely bleached during the giant pulse then a much smaller output will be obtained than with active methods of *Q*-switching.

Dye solutions are not the only type of 'saturable absorber'. In some instances certain vapours (e.g. SF_6 in conjunction with CO_2 lasers) and thin metal films can also be used.

8.2.4 Repetitive *Q*-switching

In most of the above discussion of *Q*-switching we have assumed that strong pumping is being used (e.g. flashlamp pumping in solid-state lasers) so that a large population inversion can be created prior to opening the *Q*-switch. In the case of most CW lasers the pump powers are more modest and can only achieve relatively small population inversions. The use of a *Q*-switch in these situations would lead to fairly low power pulses. Nevertheless, there are a number of instances where it is advantageous to have a train of pulses rather than a continuous output (for example, repetitively *Q*-switched Nd : YAG lasers are used for resistor trimming). Because we are dealing with relatively low population inversions the build-up of the giant pulse will not take place so rapidly (i.e. see the results of Problem 8.5) and so a relatively slow *Q*-switch such as the acousto-optic modulator can be used.

8.2.5 Gain switching and cavity dumping

Both gain switching and cavity dumping are techniques, like *Q*-switching, which enable narrow pulses of laser radiation to be generated. In gain switching the gain medium is pumped very strongly and rapidly so that a considerable population inversion is built up before the photon population has time to increase significantly. The result is the production of a laser spike, possibly followed by weaker spikes depending on the strength and duration of the pumping. The equations governing the spike production are similar to those governing relaxation oscillations.

It is obviously necessary for the photon build-up to be relatively slow so that the pumping has time to generate as high a population inversion as possible. Thus short cavities with relatively low mirror reflectivities are preferable (see Section 5.4.2). The technique has been used in a 50 μm thick dye laser to produce 5 ps wide pulses [Ref. 6]. Semiconductor lasers are also well suited to this technique because of their inherent short cavity length and low mirror reflectivity, and pulses of between 40 and 100 ps have been generated [Ref. 7]. The technique can also be applied to pulsed TEA CO_2 lasers by applying a very fast intense pumping current pulse, as was discuss in Section 7.2.6.1.

Cavity dumping is, in principle, a very simple technique which works by suddenly switching the output coupling from a low value to a high one. Thus we can imagine having a laser with high reflectivity mirrors. Then, once the cavity photon population has grown to its maximum value, one of the mirrors is suddenly made transmitting. The entire stored photon population will be released within the photon circulation time in the cavity. The pulse width will thus be approximately $2d/c$ (i.e. a cavity 0.5 m long would yield a pulse of about 3 ns duration).

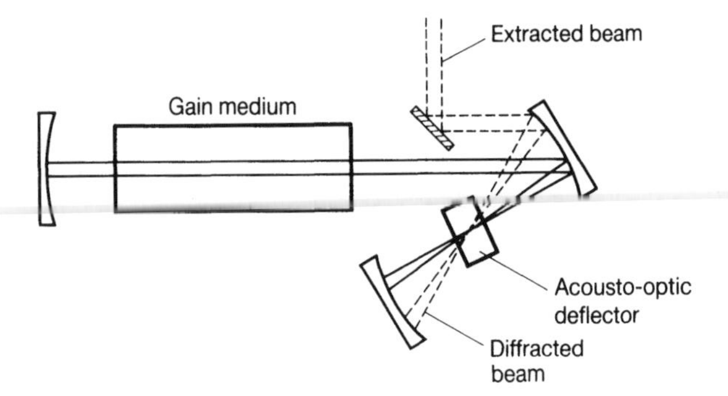

Figure 8.18 Layout for cavity dumping using an acousto-optic modulator.

In practice, it is impossible to switch the mirror reflectivities in this manner. What can be done instead is to insert a beam deflection system into the cavity. Such a system based on an acousto-optic modulator is shown in Figure 8.18. When the modulator is switched on an appreciable fraction of the beam is diffracted out of the cavity. Notice that the cavity is such that a beam waist is formed at the acousto-optic modulator. This ensures that the modulator can switch as rapidly as possible.

Cavity dumping can be carried out in both pulsed and CW lasers and offers the advantage over Q-switching that the pulse repetition rates can be much higher (up to several GHz).

8.3 Mode locking

As we have seen in Section 8.2.1, the technique of Q-switching enables single 'giant' pulses to be obtained. The duration of such pulses is governed by both the initial inversion and the rate at which the switching can be accomplished. For high initial inversion and rapid switching the pulse duration is limited (approximately) to the cavity photon lifetime. The latter is governed primarily by the length of the laser cavity and the reflectivity of the mirrors (see Section 5.4.2). A short cavity lifetime requires a short laser with low reflectivity mirrors. In lasers other than diode lasers the smallest cavity lifetime is limited to a few nanoseconds or so. Much narrower pulses than this can be obtained using the technique of *mode locking*, although such pulses are normally produced as part of a train of pulses and the power per pulse is less than in the case of Q-switched pulses.

Mode locking is possible in both homogeneous and inhomogeneously broadened laser transitions, although the two cases must be treated differently.

8.3.1 Theory of mode locking in inhomogeneously broadened lasers

From the analysis of Section 6.4.2, we know that within the gain profile of an inhomogeneously broadened laser a number of longitudinal cavity modes can oscillate independently. To simplify matters we shall assume that the cavity consists of plane mirrors of 'infinite' sideways extent; the modes are then plane wavefront standing waves. The mode frequencies will be integer multiples of the reciprocal of the roundtrip time within the cavity. Assuming therefore that the cavity contains media that have refractive indices very close to unity, the frequency of the qth mode, ν_q is given by

$$\nu_q = q\,\frac{c}{2d}$$

where d is the cavity length and q is an integer.

The difference between adjacent modes ($\delta\nu$) is $c/2d$ or $1/T$, where T is the round-trip time within the cavity. Suppose N modes are able to oscillate within the cavity. Adopting the convention that ν_1 refers to the lowest of these frequencies, a general frequency can be written ν_m where m takes on the integer values $1 \ldots N$. The field of a particular standing-wave mode is then given by

$$\mathscr{E}_m(z, t) = \mathscr{E}_m \sin\left(\frac{2\pi}{c}\,\nu_m z\right)^2 \sin(2\pi\nu_m t + \phi_m) \tag{8.46}$$

The total field ($\mathscr{E}(z, t)$) from all the N oscillating modes is obtained by summing the fields from all the modes. To simplify the summation it is convenient to make all the mode amplitudes equal ($\mathscr{E}_m = \mathscr{E}_0$). Thus

$$\mathscr{E}(z, t) = \mathscr{E}_0 \sum_{m=1}^{N} \sin\left(\frac{2\pi}{c}\,\nu_m z\right) \sin(2\pi\nu_m t + \phi_m)$$

$$\mathscr{E}(z, t) = \frac{\mathscr{E}_0}{2} \sum_{m=1}^{N} \cos\left(2\pi\nu_m\left(\frac{z}{c} - t\right) - \phi_m\right) - \cos\left(2\pi\nu_m\left(\frac{z}{c} + t\right) + \phi_m\right) \tag{8.47}$$

We now suppose that the phases ϕ_m can be *locked* together so that they have the same value ϕ_0 (the techniques used to achieve this will be described later). The first term on the right-hand side of Eq. (8.47) can be simplified by noting that

$$\sum_{m=1}^{N} \cos\left(2\pi\nu_m\left(\frac{z}{c} - t\right) - \phi_0\right) = Re \sum_{m=1}^{N} \exp \mathrm{j}\left(2\pi\nu_m\left(\frac{z}{c} - t\right) - \phi_0\right)$$

$$= Re\, \exp(-\mathrm{j}\phi_0) \sum_{m=1}^{M} \exp\left(2\pi \mathrm{j}\nu_m\left(\frac{z}{c} - t\right)\right) \tag{8.48}$$

Since $\nu_m = \nu_1 + (m - 1)\,\delta\nu$ we can write

$$\sum_{m=1}^{N} \exp\left[2\pi\mathrm{j}\nu_m\left(t - \frac{z}{c}\right)\right] = \exp\left[2\pi\mathrm{j}\nu_1\left(t - \frac{z}{c}\right)\right]\left[1 + \sum_{m=1}^{N-1} \exp\left[2\pi\mathrm{j}m\delta\nu\left(t - \frac{z}{c}\right)\right]\right]$$

To simplify the notation it is convenient to put $\Phi = 2\pi\delta\nu(t - (z/c))$. Thus

$$\sum_{m=1}^{N} \exp\left[2\pi j \nu_m \left(t - \frac{z}{c}\right)\right]$$

$$= \exp\left[2\pi j \nu_1 \left(t - \frac{z}{c}\right)\right] [1 + \exp(j\Phi) + \exp(2j\Phi) + \ldots + \exp((N-1)\, j\Phi)]$$

Now the term $1 + \exp(j\Phi) + \exp(2j\Phi) + \ldots \exp((N-1)\, j\Phi)$ on the right-hand side of this equation forms a geometrical progression, and denoting its value by S, we have

$$S = \frac{1 - \exp(Nj\Phi)}{1 - \exp(j\Phi)}$$

$$S = \frac{\exp\left(\frac{Nj\Phi}{2}\right)\left(\exp\left(\frac{Nj\Phi}{2}\right) - \exp\left(\frac{Nj\Phi}{2}\right)\right)}{\exp\left(\frac{j\Phi}{2}\right)\left(\exp\left(-\frac{j\Phi}{2}\right) - \exp\left(\frac{j\Phi}{2}\right)\right)}$$

$$S = \exp\left(\frac{(N-1)j\Phi}{2}\right) \frac{\sin\left(\frac{N\phi}{2}\right)}{\sin\left(\frac{\phi}{2}\right)}$$

So that

$$\sum_{m=1}^{N} \exp\left[2\pi j \nu_m \left(t - \frac{z}{c}\right)\right] = \exp\left(2\pi j \left(\frac{\nu_1 + \nu_N}{2}\right)\left(t - \frac{z}{c}\right)\right) \frac{\sin\left(\frac{N\pi\delta\nu}{c}(z - ct)\right)}{\sin\left(\frac{\pi\delta\nu}{c}(z - ct)\right)}$$

$$= \exp(-jk_0(z - ct)) \frac{\sin\left(\frac{N\pi\delta\nu}{c}(z - ct)\right)}{\sin\left(\frac{\pi\delta\nu}{c}(z - ct)\right)}$$

where we have put $2\pi(\nu_1 + \nu_N)/2c = k_0$. Using this result in Eq. (8.48) we obtain

$$\sum_{m=1}^{N} \cos\left(2\pi\nu_m\left(\frac{z}{c} - t\right) - \phi_0\right) = \cos(k_0(z - ct) + \phi_0) \frac{\sin\left(\frac{N\pi\delta\nu}{c}(z - ct)\right)}{\sin\left(\frac{\pi\delta\nu}{c}(z - ct)\right)} \quad (8.49)$$

Similar reasoning applied to the second term on the right-hand side of Eq. (8.47) yields

$$\sum_{m=1}^{N} \cos\left(2\pi\nu_m\left(\frac{z}{c}+t\right)+\phi_0\right) = \cos(k_0(z+ct)+\phi_0)\,\frac{\sin\left(\dfrac{N\pi\delta\nu}{c}(z-ct)\right)}{\sin\left(\dfrac{\pi\delta\nu}{c}(z-ct)\right)} \tag{8.50}$$

Thus finally using Eqs (8.49) and (8.50) in Eq. (8.47) yields

$$\begin{aligned}\mathscr{E}(z,t) &= \frac{\mathscr{E}_0}{2}\left\{\cos(k_0(z-ct)+\phi_0)\,\frac{\sin\left(\dfrac{N\pi\delta\nu}{c}(z-ct)\right)}{\sin\left(\dfrac{\pi\delta\nu}{c}(z-ct)\right)}\right.\\ &\qquad\left.-\cos(k_0(z+ct)+\phi_0)\,\frac{\sin\left(\dfrac{N\pi\delta\nu}{c}(z-ct)\right)}{\sin\left(\dfrac{\pi\delta\nu}{c}(z-ct)\right)}\right\}\\ &= \frac{\mathscr{E}_0}{2}\left\{\cos(k_0(z-ct)+\phi_0)A_{\mathrm{N}}\left(\frac{\pi\delta\nu}{c}(z+ct)\right)\right.\\ &\qquad\left.-\cos(k_0(z+ct)+\phi_0)A_N\left(\frac{\pi\delta\nu}{c}(z-ct)\right)\right\}\end{aligned} \tag{8.51}$$

where

$$A_N(x) = \frac{\sin(Nx)}{\sin(x)}$$

The resulting field can thus be thought to consist of two travelling waves which are 'modulated' by the function

$$A_N\left(\frac{\pi\delta\nu}{c}(z \pm ct)\right)$$

The function $A_N(x)$ is plotted, for various values of N, in Figure 8.19. We see that when N is fairly large the function consists principally of narrow peaks of height N separated by π. The function falls from its peak value to zero when x changes by π/N. Between the main peaks are $N-2$ subsidiary peaks whose amplitudes, relative to the main peaks, decrease as N increases.

Neglecting the subsidiary peaks, Eq. (8.51) represents two trains of pulses travelling in opposite directions. For each pulse train the pulses are separated in time by $1/\delta\nu$ (or $2L/c$) and in space by $c/\delta\nu$ (or $2L$). We can interpret this situation as arising from a single pulse which bounces backwards and forwards between the mirrors. If one of the mirrors is not 100 per cent reflecting then a train of pulses

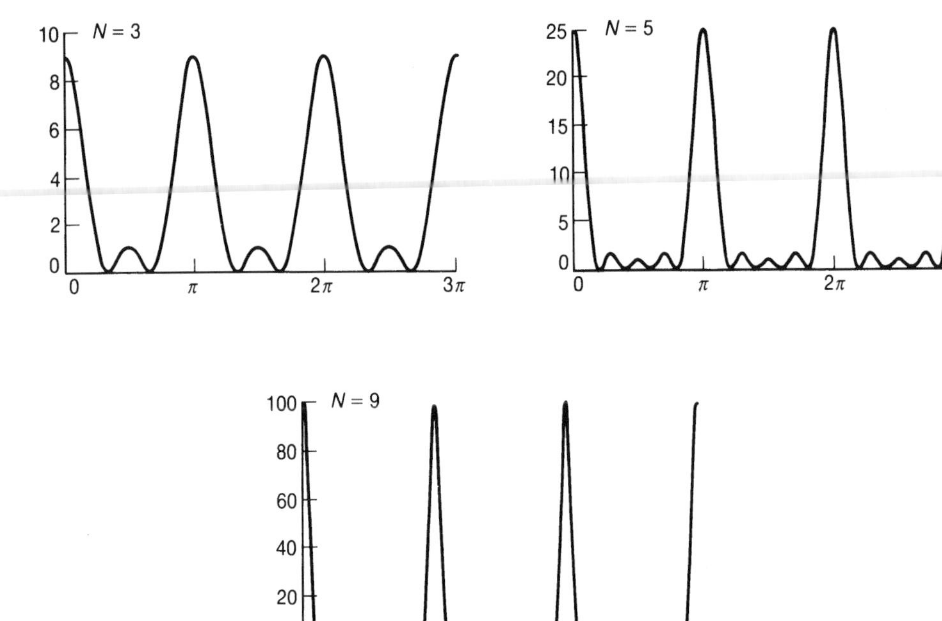

Figure 8.19 Plots of the function $\sin^2(Nx)/\sin^2(x)$ for $N = 3$, 5 and 9.

will emerge from the laser with a pulse separation of $1/\delta\nu$, each pulse having a duration of approximately $1/(N\delta\nu)$.[11] Assuming that all the modes within the full linewidth oscillate,

$$\tau_{\text{pi}} \approx \frac{1}{N\delta\nu} \approx \frac{1}{\Delta\nu_{\text{i}}} \tag{8.52}$$

where $\Delta\nu_{\text{i}}$ is the inhomogeneously broadened linewidth.

The derivation of this result has been somewhat tortuous and it is possible to obtain the same outcome rather more easily by assuming that the modes act as if they were simple harmonic oscillators. Thus writing the contribution from each mode as

$$\mathscr{E}_m \cos(2\pi\nu_m t + \phi_m))$$

the resultant of all the modes is

$$\mathscr{E}(t) = \sum_{m=1}^{N} \mathscr{E}_m \cos(2\pi\nu_m t + \phi_m) \tag{8.53}$$

Using complex notation we have

$$\mathscr{E}(t) = Re \sum_{m=1}^{N} \mathscr{E}_m \exp(\mathrm{j}(2\pi\nu_m t + \phi_m)) \tag{8.54}$$

The output irradiance, $I(t)$, will be proportional to $\mathscr{E}^*(t)\mathscr{E}(t)$. If K is the constant of proportionality then

$$I(t) = K \sum_{m=1}^{N} \mathscr{E}_m^2 + K \sum_{n \neq m}^{N} \sum_{m}^{N} \mathscr{E}_n \mathscr{E}_m \exp[2\pi\mathrm{j}(\nu_n - \nu_m) + \mathrm{j}(\phi_n - \phi_m)] \tag{8.55}$$

Provided the relative phases of the modes do not change with time, then it is easily verified (Problem 8.6(a)) that

$$I(t + 1/\delta\nu) = I(t) \tag{8.56}$$

Thus the output irradiance will repeat itself with periodicity $(\delta\nu)^{-1}$ whatever the values of the ϕ_m. The most rapid oscillations in I with time will come from the term in Eq. (8.55) where $|\nu_n - \nu_m|$ is largest, that is, at a frequency of approximately $\Delta\nu_g$. The average irradiance is simply

$$I_{\text{ave}} = K \sum_{m=1}^{N} \mathscr{E}_m^2 \tag{8.57}$$

If, however, the phases are locked together so that $\phi_m = \phi_0$, and it is assumed that the amplitudes are equal (i.e. $\mathscr{E}_m = \mathscr{E}_0$), then the expressions can be considerably simplified and it is readily shown (see Problem 8.6(b)) that

$$I(t) = K\mathscr{E}_0^2 \frac{\sin^2(N\pi\delta\nu t)}{\sin^2(\pi\delta\nu t)} \tag{8.58}$$

The temporal behaviour is thus identical to that predicted by Eq. (8.51). The advantage of this approach is that it enables us (via Eq. (8.53)) to investigate the effects of varying such parameters as the mode amplitudes and phases with relative ease.

In the derivation of Eq. (8.58), for example, we assumed the mode amplitudes to be equal. A more realistic situation, since we are assuming an inhomogenously broadened laser, would be to assume they have a Gaussian amplitude distribution. Figure 8.20(a) shows the result of such a calculation. We see that the main difference between this and the equal-amplitude situation is that the subsidiary peaks are suppressed. It may also be noted that the pulse shapes themselves are Gaussian [12] and that the pulsewidths decrease as the width of the Gaussian amplitude distribution increases. [13] A more formal analysis of the situation [Ref. 8] shows that the pulse irradiance as a function of time can be written

$$I(t) = I_0 \frac{\pi}{2\log_e 2}\left(\frac{\Delta\nu}{\delta\nu}\right)^2 \exp\left(-\left(\frac{\pi\Delta\nu}{\log_e 2}t\right)^2\right) \tag{8.59}$$

where $\Delta\nu$ is the frequency bandwidth occupied by modes that have an irradiance greater than half the peak mode irradiance.

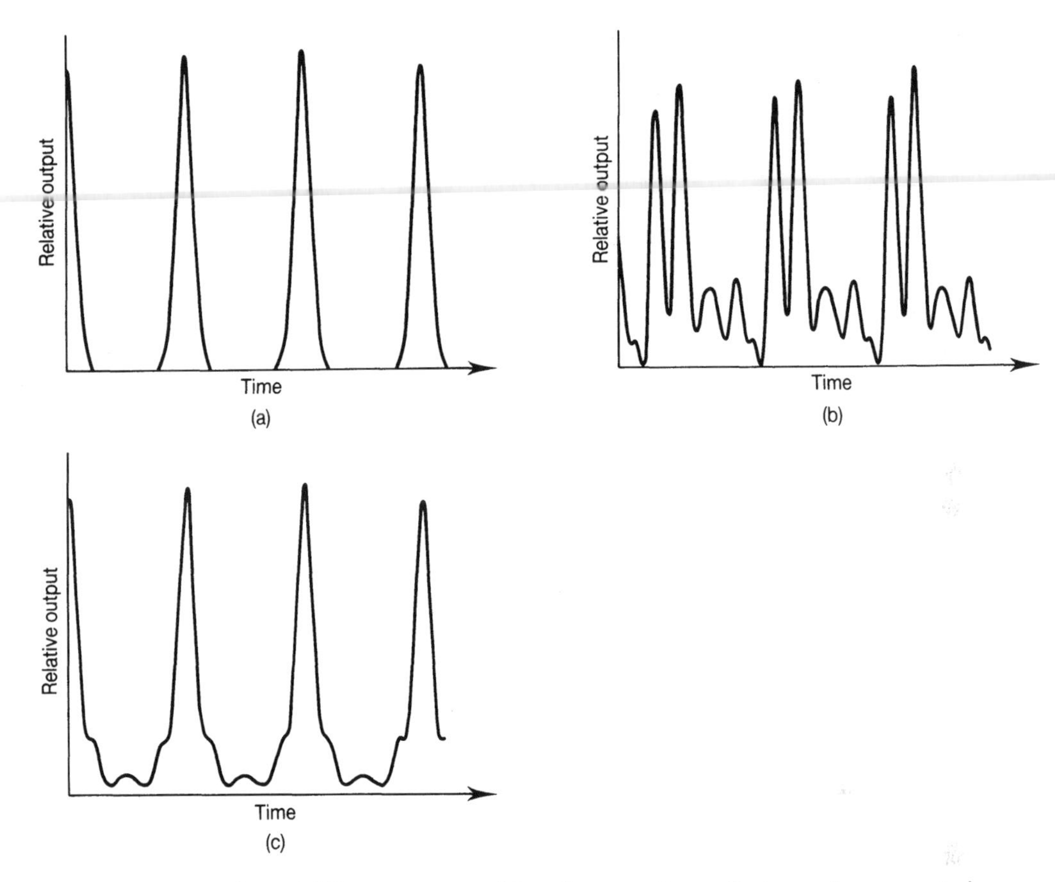

Figure 8.20 Examples of different temporal irradiance patterns that may be generated using N (here $N=7$) equally spaced frequency components with different relative amplitudes and phases. In (a) the phases are all equal and the amplitudes have a Gaussian irradiance distribution. In (b) the amplitudes are all equal but the phases random. In (c) the amplitudes are random and the phases equal.

Also illustrated in Figure 8.20 are the situations when the phases are random with equal mode amplitudes (Figure 8.20(b)), and when the phases are equal but the amplitudes are random (Figure 8.20(c)). For both of these latter two cases recalculation with different phases or amplitudes will, of course, give rise to completely different irradiance–time curves. However, the basic repetition of the irradiance pattern after a time $1/\delta\nu$, as required by Eq. (8.56), is clearly seen.

The conclusions of this theoretical discussion are that it should be possible to generate a stream of very narrow pulses from a laser if (1) we have a laser which can support the simultaneous oscillation of many different longitudinal modes and (2) we are able to 'lock' the phase of the modes together so that they are all the same.

As in the case of Q-switching there are two main types of technique used to achieve mode locking which are termed *active* and *passive*. These are dealt with in the next two sections.

8.3.2 Active mode locking (inhomogeneous lasers)

Consider a laser where a large number of longitudinal modes are oscillating simultaneously and where a very fast shutter is placed in the laser cavity. The shutter opens only for a short time every τ_{RT} seconds, where τ_{RT} is the time it takes for light to make a complete roundtrip within the cavity. This arrangement will affect drastically the circulating fields within the cavity produced by all the possible sets of values of inter-modal phase *except one*. This exception is the case where all the mode phases are equal and where the resultant field within the cavity is just a single narrow pulse passing through the modulator every τ_{RT} seconds when the shutter is open.

In practice, it is not necessary to have a shutter that closes completely. Somewhat easier to arrange is for a sinusoidally modulated transmission with the modulation frequency equal to $1/\tau_{RT}$. The transmission, T, can then be written (Figure 8.21)

$$T = (1 - \delta(1 + \cos(2\pi t/\tau_{RT})))(d \leqslant 0.5)$$

Any radiation not arriving at the time of peak transmission in the cycle will be attenuated and so the tendency will be for the modes to adjust their relative phases to give minimum overall loss. The time of arrival of the peak of the pulse will then coincide with the peak transmission of the modulator.

There will still be some loss, however, since those parts of the pulse away from the peak value will be attenuated by the modulator. This loss will depend on the pulsewidth, that is, on the number of coupled modes (i.e. see Eq. (8.52)). The higher the number of participating modes, the lower will be the loss. The tendency will thus be for as many modes as possible to couple together. However, there is a limit to this since those modes that are very far removed in frequency from the centre of the

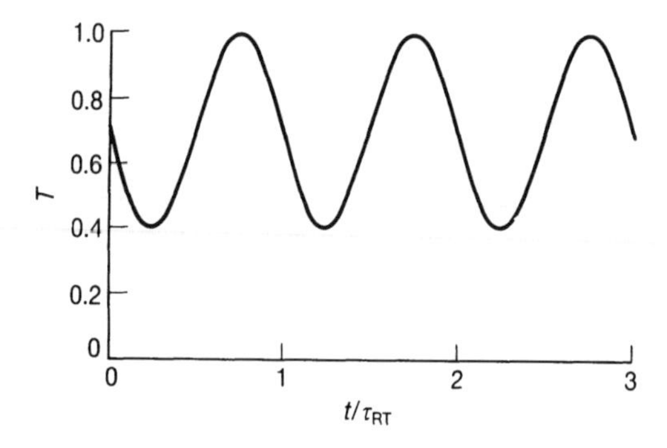

Figure 8.21 Graph of the function $T = (1 - \delta(1 + \cos(2\pi t/\tau_{RT})))$ with δ taking the value 0.3.

gain curve will be subject to absorption rather than gain. The pulsewidth will stabilize when a minimum loss within the pulsewidth in respect to the whole system (i.e. gain medium plus modulator losses plus cavity losses) has been achieved. The total number of modes involved will depend on the individual system, but we may reasonably assume that modes across the whole of the laser transition linewidth will be coupled together. The number of oscillating modes is thus given by $N_m \cong \Delta\nu_L/\delta\nu$, where $\Delta\omega$ is the halfwidth of the laser line transition and the resulting pulsewidth τ_p is given by

$$\tau_p \approx 1/\Delta\nu_L \tag{8.60}$$

Although we have discussed the effects of the mode phases being coupled together we have not as yet considered possible coupling mechanisms. Consider the situation where the amplitude modulator is operating and the laser is then turned on with the gain being turned up rather 'slowly'. There will come a point when the mode closest to the centre of the laser gain profile starts to oscillate. We take the transmission of the modulator, as far as the electric field amplitude is concerned, to be

$$T = 1 - \Delta_a(1 - \cos(\omega_c t))$$

The resultant electric field variation with respect to time can then be written:

$$\begin{aligned}\mathscr{E}(t) &= \mathscr{E}_0(1 - \Delta_a(1 - \cos(\omega_c t))\cos(\omega_0 t)) \\ &= \mathscr{E}_0\left((1 - \Delta_a)\cos(\omega_0 t) - \frac{\Delta_a}{2}\cos(\omega_0 + \omega_c)t + \cos(\omega_0 - \omega_c)t\right)\end{aligned}$$

We see that additional electric field components are created which are in phase with the initial field but which have angular frequencies of $\omega_0 + \omega_c$ and $\omega_0 - \omega_c$. These additional field components are called *sidebands*. Since ω_c is set to the intermodal frequency difference, the sidebands coincide in frequency with the cavity modes on either side of the central initial mode. At each pass through the modulator, therefore, energy is transferred from the central mode to the adjacent side modes. Once the laser gain has been increased to include gain at these side mode frequencies then it is reasonable to assume that these side modes will remain 'locked' in phase to the initial mode. These side modes will, in turn, transfer power to modes further out which will still be phase locked to the central mode.

Amplitude modulation is not the only form of modulation that can generate sidebands. Consider the situation where the phase is modulated, so that the electric field can be written:

$$\mathscr{E}(t) = \mathscr{E}_0 \cos(\omega t - \Delta_p \sin(\omega_m t))$$

We can expand the right-hand side of this equation with the help of the following identities:

$$\begin{aligned}\cos(a \sin(b)) &= J_0(a) + 2J_2(a)\sin(2b) + 2J_4(a)\sin(4b) + \text{etc.} \\ \sin(a \sin(b)) &= 2J_1(a)\sin(b) + 2J_3(a)\sin(3b) + 2J_5(a)\sin(5b) + \text{etc.}\end{aligned}$$

Here $J_n(x)$ is the Bessel function of order n. The first few such functions are shown in Figure 8.22. To some extent they can be thought of as being rather like cosine functions (when $n = 0$) or sin functions (when $n > 0$) whose amplitude decays with distance.[14] However, the initial amplitudes and the values of the argument where the functions go to zero are *not* the same as the cosine or sine functions. Using the above identities it is a simple matter to show that

$$\begin{aligned}\mathscr{E}(t)/\mathscr{E}_0 = {} & J_0(\Delta_p)\cos(\omega t) + J_1(\Delta_p)(\cos(\omega + \omega_m)t - \cos(\omega - \omega_m)t) \\ & + J_2(\Delta_p)(\cos(\omega + 2\omega_m)t + \cos(\omega - 2\omega_m)t) \\ & + J_3(\Delta_p)(\cos(\omega + 3\omega_m)t - \cos(\omega - 3\omega_m)t) \\ & + \text{etc.}\end{aligned}$$

Again we see that sidebands are created. This time, however, the spectrum is infinitely wide, although the amplitudes decrease rapidly with increasing frequency separation from ω unless Δ_p is greater than unity. The relative sideband amplitudes for two different values of the modulation parameter Δ_p are given in Table 8.2.

Figure 8.23 shows a series of active mode locked pulses obtained from a Nd : YAG laser which is simultaneously being actively Q-switched.

8.3.3 Passive mode locking

Passive mode locking is possible using somewhat similar experimental arrangements to those used for passive Q-switching. That is, a saturable absorber is inserted into the laser cavity, preferably close to one of the mirrors (Figure 8.17). We have seen that when the modes in a cavity are locked together the result is a single pulse that bounces back and forth between the mirrors. We can readily see that a saturable dye can support such a bouncing pulse provided the pulse has a sufficiently large irradiance to allow it to saturate the absorber each time it passes through. In this way, it mimics the fast shutters used for active mode locking.

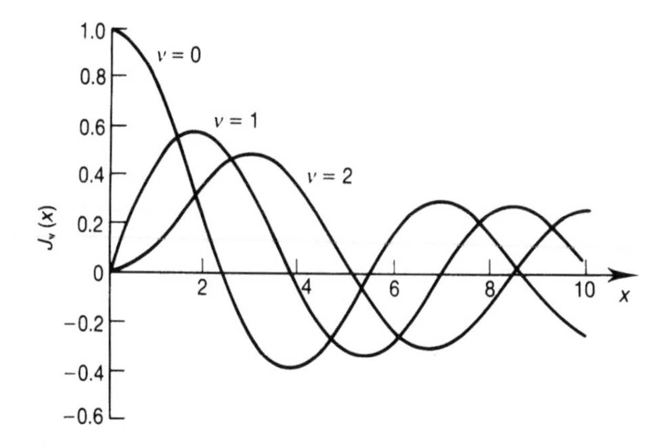

Figure 8.22 The first three Bessel functions of the first kind $J_n(x)$.

Table 8.2 Relative sideband amplitudes ($J_n(x)$) for frequency modulation

n	0	1	2	3	4
$J_n(0.5)$	0.938	0.242	0.031	0.0026	0.00016
$J_n(2)$	0.224	0.577	0.353	0.129	0.034

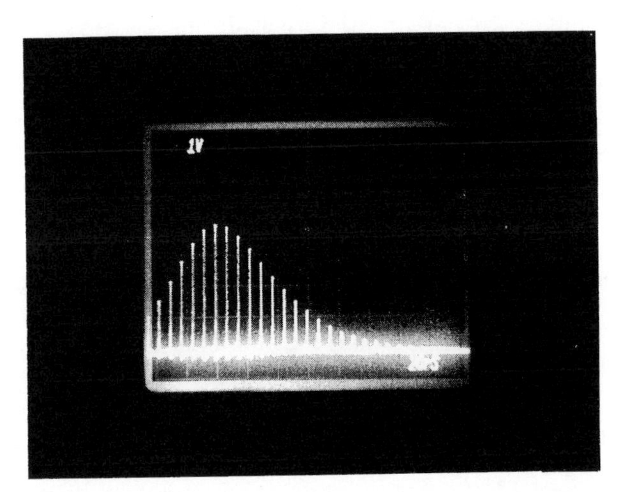

Figure 8.23 Output from a Nd:YAG laser that is being actively mode locked at the same time as being actively Q-switched. The individual pulses are some 200 ps wide. (Photograph courtesy of Mr C. Danson, Rutherford Appleton Laboratory.)

Again, as with passive Q-switching, we have the obvious requirements that the dye absorption bandwidth must overlap with the laser gain bandwidth. An additional requirement is that the recovery time must be shorter than the roundtrip time, otherwise multiple pulses could possibly form. In typical dyes used in practice the recovery time is of the order of 10 ps.

A detailed model for the way in which the single pulse within the laser cavity builds up was first given by Letokhov [Ref. 9] and experimental observations have confirmed its essential correctness. There are three main stages to the build-up process:

1. When the laser is first switched on there is insufficient gain to overcome losses in the system and the radiation that is present is essentially spontaneous emission. The bandwidth of the radiation will be approximately equal to that of the fluorescent linewidth. The cavity modes will have random, and fluctuating, phases and interference between them will lead to a rapidly and randomly fluctuating output (Figure 8.24(a)). Some time later the laser will start to exhibit

(linear) gain and the noise-like signal will be amplified. During this amplification process a certain amount of mode selection will take place since modes closest to the centre of the gain curve will be amplified most. This causes the overall bandwidth of the radiation to be reduced which in turn causes the fluctuations in the output signal to be smoothed out somewhat (Figure 8.24(b)).

2. During the second stage the laser gain remains linear but the dye absorption becomes nonlinear. The nonlinearity of the dye absorption has two main effects. First, it discriminates between the most intense pulses and the weaker ones. The latter will encounter higher losses than the former during a transit of the laser cavity and so will tend to be suppressed in favour of the more intense peaks. The second effect is that the pulsewidths are reduced. This happens because the radiation at the centre of the pulse encounters more overall gain than does the radiation in the wings (see Problem 8.9). At the end of this stage in the process the dye is completely bleached while the laser amplification is just beginning to become nonlinear (i.e. saturated). The output consists of one dominant pulse with perhaps a few subsidiary pulses (Figure 8.24(c)).
3. In the final stage the largest pulse grows rapidly at the expense of the secondary pulses. This build-up increases until the pulse grows to such a size that it extracts significant amounts of energy from the gain medium on each pass. In this final stage the total stored energy is emitted during a burst of some tens of narrow mode locked pulses which have an envelope similar to a Q-switched pulse (Figure 8.24(d)). A typical time scale for the build-up of the mode locked pulses is a microsecond or so.

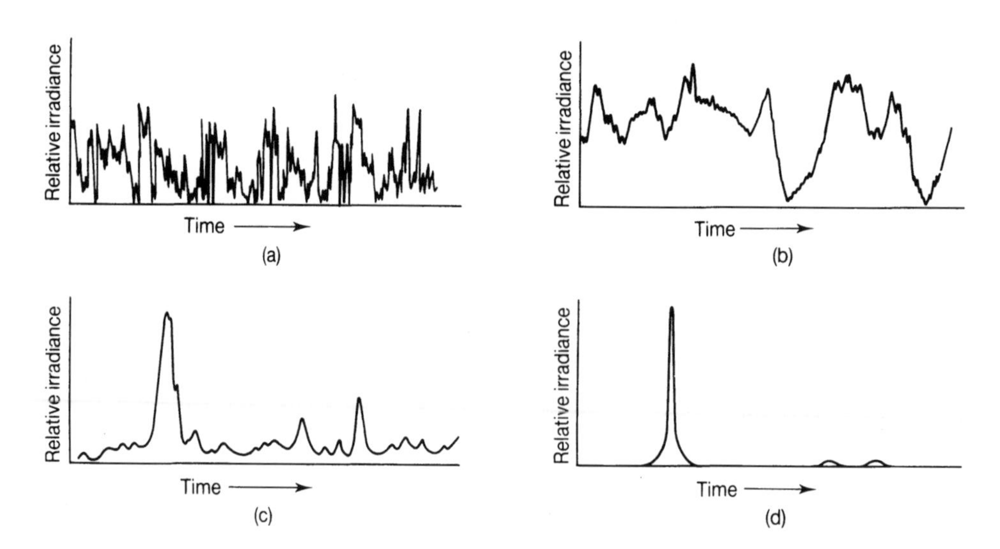

Figure 8.24 Output irradiance versus time graphs for various stages in the build-up of passive mode-locked pulses. Note that the relative amplitude scale increases rapidly from (a) to (d).

It will be evident from the above description that the final pulse is the result of a great many interactions some of them essentially random in nature and because of this it is not possible to give a simple analytical expression for the widths of the resulting pulses.

The main advantages of the passive mode locking technique are that it is relatively simple and inexpensive. There are, however, a number of drawbacks. For example, the technique is not wholly reliable. Even when extreme care is taken the probability of obtaining clean mode locked pulses is about 80 per cent. It is particularly important to avoid any reflective surfaces within the cavity (apart from the end mirrors!) since they can set up secondary cavities. The circulating pulse may then be split up into several pulses, each having different roundtrip times, the result then being an erratic output with several superimposed pulse trains. Thus within the cavity all surfaces must be either at the Brewster angle or anti-reflection coated. Care must also be taken with reflections from outside the laser. The dye cell is usually placed up against the 100 per cent reflecting mirror, with the mirror forming one of the cell walls (as in Figure 8.17). This arrangement minimizes the number of reflective surfaces within the cavity. Typical dye cells are between 1 and 10 mm in length.

Another disadvantage is that since the main pulse grows from initially random noise pulses the time delay between the switch on of the laser pump and the appearance of the mode locked pulses is variable. Any synchronization that is required has to be done directly using the pulses themselves.

8.3.4 Active mode locking in homogeneously broadened lasers

As we have discussed previously (Section 6.4.1), in homogeneous lasers only one mode can (in theory[15]) oscillate. Once the cavity mode closest to the centre of the gain curve starts to oscillate then the whole of the gain curve is reduced, thereby preventing any of the other modes achieving the condition that gains exceed loss. However, when an amplitude-modulating element is present within the cavity then, as far as this element is concerned, minimum loss may be achieved for a circulating pulse if it coincides with the maximum transmission *and* if its temporal width is as narrow as possible. From the point of view of energy loss within the entire system it may be advantageous, therefore, for several modes to couple together to produce a narrow optical pulse (and hence minimize modulator losses) even though all except one of these modes are lossy as far as the gain medium is concerned.

We start this section with a theoretical treatment of active mode locking in homogeneously broadened lasers which closely follows that given by Kuizenga and Seigman [Ref. 10] and relies on a self-consistency argument. We assume a particular form for the pulse shape and then take the pulse on a roundtrip through the cavity. Self-consistency requires that after the trip the phase change is a multiple of 2π and that the pulse shape remains unaltered.

The initial pulse shape (in the time domain) is taken to be a Gaussian which we write as

$$\mathcal{E}(t) = \mathcal{E}_0 \exp(-\Gamma t^2)\exp(j\omega_0 t) \tag{8.61}$$

where the parameter Γ essentially describes the pulsewidth. To take this pulse on a roundtrip we must investigate the effects of both the laser medium and the modulator. The lasing medium presents some problems since the amplification it gives depends on frequency and because of its finite width in the time domain the pulse will consist of a range of frequencies. To determine the frequency range we must take the Fourier transform of Eq. (8.61), the result being

$$\mathcal{E}(\omega) = \mathcal{E}_0 \exp\left(-\frac{(\omega_0 - \omega)^2}{4\Gamma}\right) \tag{8.62}$$

To investigate the effect of the laser medium we consider the results of the simple electron oscillator model of an atom given in Section 2.1.3. Although this model can give rise only to absorption and not amplification we will see how to rectify this later. When a plane wave of frequency ω passes through a medium consisting of a collection of N such oscillators per unit volume the medium behaves as if it had a complex refractive index ($n = n_r + jn_i$) and the progress of the beam is described by

$$\mathcal{E}(z) = \mathcal{E}(0)\exp\left[j\omega\left(t - \frac{n_r z}{c}\right)\right]\exp\left(-\frac{n_i\omega}{c} z\right)$$

or

$$\mathcal{E}(z) = \mathcal{E}(0)\exp\left[j\left(\omega t - \frac{\omega z}{c}(n_r - jn_i)\right)\right]$$

Thus in traversing a distance z the phase of the wave changes according to the factor

$$P(z) = \exp\left(-\frac{\omega z}{c}(jn_r + n_i)\right) \tag{8.63}$$

From Eq. (2.47) n_i is given by

$$n_i = \frac{Ne^2}{4m\varepsilon_0\omega}\left(\frac{\gamma}{(\omega_0 - \omega)^2 + \gamma^2}\right)$$

while from Eq. (2.46) we obtain:

$$n_r = 1 + \frac{Ne^2}{4m\varepsilon_0\omega}\left(\frac{\omega_0 - \omega}{(\omega_0 - \omega)^2 + \gamma^2}\right)$$

In terms of the Lorentzian function, $L(\omega - \omega_0)$, as defined in Eq. (2.48), we can then write

$$n_r = 1 + \frac{Ne^2\pi}{2m\varepsilon_0\omega}\frac{(\omega - \omega_0)}{\Delta\omega_L} L(\omega - \omega_0)$$

and

$$n_i = \frac{Ne^2\pi}{4m\varepsilon_0\omega} L(\omega - \omega_0)$$

Now the absorption coefficient, α, is related to n_i (Eq. (2.39)) by $\alpha = (2n_i\omega/c)$. The maximum absorption takes place when $\omega = \omega_0$, and we have

$$\alpha(\omega_0) = \frac{Ne^2}{m\varepsilon_0 c\ \Delta\omega_L}$$

Writing both n_r and n_i in terms of $\alpha(\omega_0)$ gives

$$n_r = 1 + \frac{c\pi}{2\omega}(\omega - \omega_0)\alpha(\omega_0)L(\omega - \omega_0)$$

$$n_i = \frac{c\pi\Delta\omega_L}{4\omega}\alpha(\omega_0)L(\omega - \omega_0)$$

Substituting these values for n_r and n_i into Eq. (8.63) gives

$$P(z) = \exp\left[-\frac{\omega z}{c}\mathrm{j} - \frac{\pi}{4}\alpha(\omega_0)L(\omega - \omega_0)\ (2(\omega_0 - \omega)\mathrm{j} + \Delta\omega_L)\right] \tag{8.64}$$

So far, we have used the expressions for the refractive indices obtained from the electron oscillator model. We do know that the resulting value for n_i gives the correct shape (with regard to frequency) for atomic electron absorption lines (i.e. Lorentzian) but fails to give the correct amplitudes. However, referencing to the peak absorption value, as we have done above, obviates this problem. We also know that when population inversion is achieved we can have amplification rather than absorption and that the shape of the gain curve is exactly that of the absorption curve. It seems reasonable, therefore, that in the presence of population inversion, the phase factor can be obtained directly from Eq. (8.64) simply by replacing $\alpha(\omega_0)$ by $-\gamma(\omega_0)$.

By using the small signal gain coefficient we are assuming that the laser transition is far from saturation. We are also assuming that the term resulting from the real part of the refractive index is correct. We have less grounds for this assumption since we have not attempted a quantum-mechanical derivation of n_r, and the reader will have to accept that it is perfectly proper to do so! Thus in a gain medium the phase factor becomes

$$P(z) = \exp\left[-\frac{\omega z}{c}\mathrm{j} + \frac{\pi}{4}\gamma(\omega_0)L(\omega - \omega_0)(2(\omega_0 - \omega)\mathrm{j} + \Delta\omega_L)\right] \tag{8.65}$$

Now the function $L(\omega - \omega_0)$ is given by

$$L(\omega - \omega_0) = \frac{\Delta\omega_L}{(\omega_0 - \omega)^2 + (\Delta\omega_L/2)^2}$$

$$L(\omega - \omega_0) = \frac{4}{\Delta\omega_L}\left(1 + \left(\frac{\omega_0 - \omega}{\Delta\omega_L/2}\right)^2\right)^{-1}$$

If we assume that the number of coupled modes is much smaller than the total possible number of modes that could oscillate under the line profile, that is, that $|\omega_0 - \omega| \ll \Delta\omega_L$, then $L(\omega - \omega_0)$ can be approximated as

$$L(\omega - \omega_0) \approx \frac{4}{\Delta\omega_L}\left(1 - \frac{4(\omega_0 - \omega)^2}{\Delta\omega_L^2}\right)$$

With this approximation the phase function now becomes

$$P(z) = \exp\left(-\frac{\omega z}{c}\,\mathrm{j} + \frac{\gamma(\omega_0)z}{2}\left(1 + \frac{2(\omega_0 - \omega)}{\Delta\omega_L}\,\mathrm{j} - \frac{4(\omega_0 - \omega)^2}{\Delta\omega_L^2}\right)\right) \tag{8.66}$$

We now consider what happens when an electromagnetic wave makes a complete traversal of the cavity. As far as the 'pure' phase terms in Eq. (8.66) (i.e. the terms in j) are concerned, after a complete traverse we expect the phase change to be a multiple of 2π. If L_c is the effective optical length of the cavity (i.e. including the effects of the refractive index of the modulator and of the lasing material at a frequency far removed from ω_0) and l is the length of the gain medium, then the requirement is that

$$\frac{2L_c\omega}{c} - 2\gamma(\omega_0)l\,\frac{(\omega_0 - \omega)}{\Delta\omega_L} = 2\pi m \tag{8.67}$$

or

$$\omega = \frac{m\pi c}{L_c} + \gamma(\omega_0)\,\frac{l}{L_c}\,\frac{(\omega_0 - \omega)}{\Delta\omega_L}\,c \tag{8.68}$$

The first term on the right-hand side of this expression is just the 'normal' condition for a longitudinal mode to oscillate. The second term provides a correction factor which takes into account the effect of the lasing transition on the real part of the refractive index in the gain medium. The ratio of the second term to the first is

$$\frac{(\omega_0 - \omega)}{\Delta\omega_L}\,\frac{\gamma(\omega_0)l}{\pi}$$

We have already made the assumption that $\omega_0 - \omega \ll \Delta\omega_L$ and in most practical situations the product $\gamma(\omega_0)l$ is less than unity, so that the relative magnitude of the correction term is small. We may now investigate the effect of the gain medium on the shape of the Gaussian pulse. After a double transit the amplitude of the frequency components is given by

$$\mathscr{E}'(\omega) = \mathscr{E}_0 \exp\left(-\frac{(\omega-\omega_0)^2}{4\Gamma}\right)\exp\left(\gamma(\omega_0)l\left(1-\frac{4(\omega_0-\omega)^2}{\Delta\omega_L^2}\right)\right) \tag{8.69}$$

The first term in the second exponential affects all frequencies equally. Since we are interested only in the pulse shape we may ignore this term. Consequently

$$\mathscr{E}'(\omega) = \mathscr{E}_0 \exp\left(-\frac{(\omega-\omega_0)^2}{4\Gamma} - \gamma(\omega_0)l\,\frac{4(\omega_0-\omega)^2}{\Delta\omega_L^2}\right)$$

$$\mathscr{E}'(\omega) = \mathscr{E}_0 \exp\left(-\frac{(\omega_0-\omega)^2}{4\Gamma'}\right) \tag{8.70}$$

where

$$\frac{1}{\Gamma'} = \frac{1}{\Gamma} + \frac{16\gamma(\omega_0)l}{\Delta\omega_L^2} \tag{8.71}$$

The result is still a Gaussian pulse but with a modified Γ parameter. Generally, the change in Γ is small and we can therefore write

$$\Gamma' - \Gamma \approx -\frac{16\gamma(\omega_0)l}{\Delta\omega_L^2}\Gamma^2 \tag{8.72}$$

In the time domain the resultant field will be given by

$$\mathscr{E}'(t) = \mathscr{E}_0 \exp(-\Gamma' t^2)$$

8.3.4.1 Amplitude modulation

We look first of all at the effect of amplitude modulation. We assume that the frequency of the modulator has been set to a frequency, ω_m, equal to the difference between axial mode frequencies (or to an integer multiple of this quantity). The transmission of the modulator as a function of time may be written as

$$T_{AM}(t) = \exp(-\delta_m(1-\cos\omega_m t)) \tag{8.73}$$

This function is plotted in Figure 8.25. Now from our previous qualitative discussion of active mode locking we know that the pulse will pass through the modulator at a time very close to its peak transmission time. When t is small we have that $1-\cos(\omega_m t) \approx \omega_m^2 t^2/2$ and so

$$T_{AM}(t) \approx \exp\left(-\frac{\delta_m\omega_m^2 t^2}{2}\right)$$

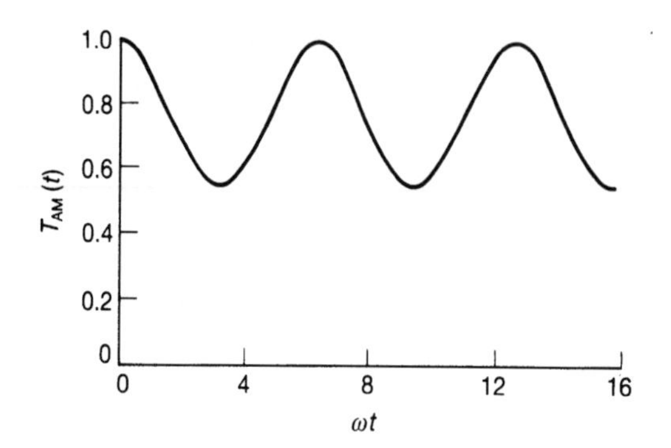

Figure 8.25 Graph of the function exp(1 – δ(1 – cos(ωt)) when δ has the value 0.3.

The field emerging from the modulator is thus

$$\mathscr{E}(t) = \mathscr{E}_0 \exp(-\Gamma' t^2)\exp\left(-\frac{\delta_m \omega_m^2 t^2}{2}\right)$$

or

$$\mathscr{E}(t) = \mathscr{E}_0 \exp(-\Gamma'' t^2)$$

where

$$\Gamma'' = \Gamma' + \frac{\delta_m \omega_m^2}{2}$$

Combining this with Eq. (8.73) gives

$$\Gamma'' - \Gamma \approx -\frac{16\gamma(\omega_0) l_g}{\Delta\omega_L^2}\Gamma^2 + \frac{\delta_m \omega_m^2}{2} \tag{8.74}$$

The requirement for self-consistency of the pulse is that its shape remains unchanged after traversing the cavity. This requirement will be met provided

$$\Gamma'' = \Gamma$$

or, using Eq. (8.74), that

$$\frac{16\gamma(\omega_0) l}{\Delta\omega_L^2}\Gamma^2 = \frac{\delta_m \omega_m^2}{2}$$

Thus the self-consistent pulse shape with amplitude modulation, Γ_{sc}^{AM}, can be written

$$\Gamma_{sc}^{AM} = \left(\frac{\delta_m}{\gamma(\omega_0) l}\right)^{1/2} \frac{\Delta\omega_L \omega_m}{4\sqrt{2}} \tag{8.75}$$

The Gaussian pulse irradiance will fall to half its maximum value when

$$t = \left(\frac{1}{2\Gamma} \log_e 2\right)^{1/2}$$

Thus the pulse halfwidth for τ_{ph} for the homogeneously broadened laser mode locked pulse is given by

$$\tau_{ph} = 2\left(\frac{1}{2\Gamma_{sc}} \log_e 2\right)^{1/2}$$

inserting the value for Γ_{sc} from Eq. (8.75) and changing from angular to linear frequency (i.e. $\omega_m = 2\pi f_m$ and $\Delta\omega = 2\pi\ \Delta\nu_h$, the subscript h being added here as a reminder that this result is for homogeneous transitions):

$$\tau_{ph} = \left[\frac{2\sqrt{2}\ \log_e 2}{\pi^2}\right]^{1/2} \left[\frac{\gamma(\omega_0)l}{\delta_m}\right]^{1/4} \left[\frac{1}{\Delta\nu_h f_m}\right]^{1/2} \tag{8.76}$$

The first, purely numerical, term on the right-hand side has the value of 0.45. In the second term, both the numerator and the denominator usually have fairly similar numerical values, say between about 0.1 and unity (a numerical calculation is given in Example 8.6). Consequently, when the ratio is taken and raised to the power of 1/4 the result will be of order unity. Thus

$$\tau_{ph} \approx \frac{1}{\sqrt{(\Delta\nu_h f_m)}} \tag{8.77}$$

If N is the maximum number of modes that could be present within the linewidth, i.e. $Nf_m = \Delta\nu_h$, then Eq. (8.77) can be written

$$\tau_{ph} \approx \frac{N^{1/2}}{\Delta\nu_h}$$

This may be compared with the situation for inhomogeneous broadening where (Eq. (8.52))

$$\tau_{pi} \approx \frac{1}{\Delta\nu_i}$$

Now N is a fairly large number (i.e. 500 for the Nd : YAG laser) so that the pulse-width is considerably broadened (by about a factor $N^{1/2}$) over the 'theoretical limit' of $1/\Delta\nu_h$. Evidently only some $N^{1/2}$ modes are effectively coupled together. In the light of the discussion at the beginning of this section indicating that only a few modes at the centre of the line profile would show gain while the remainder of those oscillating would exhibit loss, this result should not be altogether surprising. Since $f_m \approx mc/2L_c$, Eq. (8.77) implies that τ_{ph} varies (approximately) as $\sqrt{L_c}$ and hence may be reduced by reducing the cavity length.

Another way of reducing the pulsewidth is to increase the depth of modulation, δ_m. In an acousto-optic amplitude modulator δ_m is proportional to the power P_m supplied to the modulator. Thus the pulsewidth will depend on $P_m^{-1/4}$, so that, although the pulsewidth does decrease as P_m increases, it does so comparatively slowly. This behaviour can be understood in terms of a balance between the opposing tendencies of the gain medium, which favours single-mode operation, and the modulator, which would like as many modes as possible to couple together to ensure minimum losses in the modulator. When the modulator power is increased an increased amount of power is coupled into the mode system, thus enabling more modes to be coupled together and the pulsewidth to decrease.

Example 8.6 Mode locked pulsewidth in a Nd:YAG laser

We consider a Nd:YAG laser with an effective optical length of 0.5 m and a homogeneously broadened linewidth of 120 GHz (i.e. a wavelength spread of 4.5×10^{-10} m). We take the amplitude modulation index δ_m to be 0.2. To calculate the pulsewidth from Eq. (8.76) we need to know the gain coefficient at the centre of the gain profile. We may do this by taking a value for the cavity losses and then equating the cavity roundtrip gain to the roundtrip losses. Taking such losses as 10 per cent, then we have that $\exp(\gamma(\omega_0)L_c) = 1/(1-0.9)$ or $\gamma(\omega_0)L_c = 0.105$. The modulation frequency that must be applied can be taken as approximately $c/2L_c$ or 3×10^8 Hz. We now have

$$\tau_{ph} = 0.45\left(\frac{0.105}{0.2}\right)^{1/4}\left(\frac{1}{120\times 10^9\ 3\times 10^8}\right)^{1/2}$$

$$= 6.4 \times 10^{-11}\ \text{s or 64 ps}$$

8.3.4.2 Phase modulation

It is instructive to examine the effect of replacing the amplitude modulation considered above with a phase modulation. The transmission factor for the modulator can be written in a similar way to Eq. (8.73) for the AM modulator as

$$T_{FM}(t) = \exp(\mathrm{j}\delta_m \cos(\omega_m t)) \tag{8.78}$$

It can be shown that the pulse passes through the modulator when it is close to either maximum or minimum phase excursions, and, for simplicity, we consider only the former situation. Expanding the cosine function as a quadratic in $\omega_m t$ then gives

$$T_{FM}(t) \approx \exp\left(\mathrm{j}\delta_m\left(1 - \frac{\omega_m^2 t^2}{2}\right)\right)$$

The change in the value of Γ in passing through the modulator can thus be described by

$$\Gamma'' - \Gamma' \approx \mathrm{j}\,\frac{\delta_m \omega_m^2}{2}$$

while the condition for self-consistency now becomes

$$\Gamma_{\mathrm{sc}}^{\mathrm{FM}} = \sqrt{\mathrm{j}}\left(\frac{\delta_{\mathrm{m}}}{\gamma(\omega_0)L_{\mathrm{g}}}\right)^{1/2}\frac{\Delta\omega_{\mathrm{L}}\omega_{\mathrm{m}}}{4\sqrt{2}} \tag{8.79}$$

Now $\sqrt{\mathrm{j}}$ can be written as

$$\frac{1}{\sqrt{2}} - \mathrm{j}\frac{1}{\sqrt{2}}$$

so that

$$\Gamma_{\mathrm{sc}}^{\mathrm{FM}} = A(1-\mathrm{j})$$

where

$$A = \left(\frac{\delta_{\mathrm{m}}}{\gamma(\omega_0)L_{\mathrm{g}}}\right)^{1/2}\frac{\Delta\omega_{\mathrm{L}}\omega_{\mathrm{m}}}{8}$$

Before proceeding we must investigate the consequences of a complex value for Γ. Writing $\Gamma = A - \mathrm{j}B$, the expression for a Gaussian pulse becomes

$$\mathscr{E}(t) = \mathscr{E}_0 \exp(-At^2)\exp \mathrm{j}(\omega_0 t + Bt^2)$$

The instantaneous phase, $\phi(t)$, is then

$$\phi(t) = \omega_0 t + Bt^2$$

Now the angular frequency of a wave is given by the rate at which the phase changes with time, that is,

$$\omega(t) = \frac{\mathrm{d}\phi(t)}{\mathrm{d}t}$$

In the present case therefore

$$\omega(t) = \omega_0 + 2Bt$$

Within the Gaussian profile of the pulse the frequency of the electric field oscillations increases linearly with time. This is known as a *frequency chirped pulse*, and is illustrated in Figure 8.26.

We see that with phase modulation the pulsewidth is given by taking the expression as for amplitude modulation and multiplying it by the factor $2^{1/4}$ (or 1.19). The pulse will also be frequency chirped and this has the consequence of increasing the frequency bandwidth of the pulse over the unchirped value by a factor of $\sqrt{2}$ (see Problem 8.10).

To produce a pure phase modulation using $LiNbO_3$ the light, polarized along either the y or z crystal axes, is propagated along the x axis and the modulating electric field applied along the z axis. The maximum change occurs if the polarization is along the z axis, in which case the induced change in refractive index is given by[16]

$$\Delta n_z = -\tfrac{1}{2} r_{33} n_{\mathrm{e}}^3 E_z \tag{8.80}$$

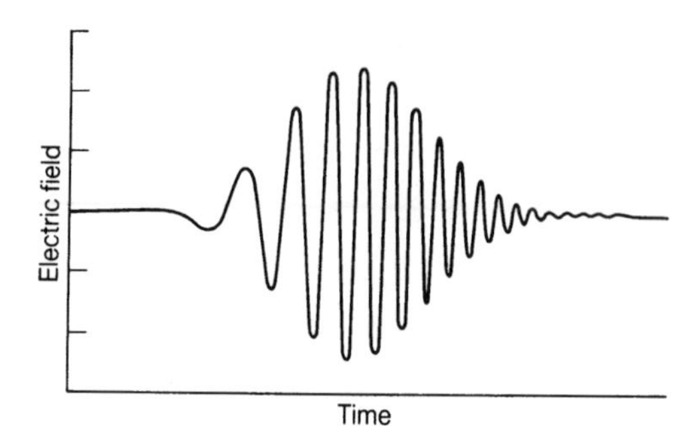

Figure 8.26 Electric field variation in a frequency chirped pulse.

If the modulating field is produced by applying an oscillating voltage with amplitude V_0 across the crystal, then the maximum phase change induced in a single pass through the crystal is given by

$$\delta_m = \frac{\pi r_{33} n_e^3 V_0 l_c}{\lambda_0 a} \tag{8.81}$$

where l_c is the length (in the x direction) and a is the width (z direction). Phase changes of the order of 1 radian are possible with voltages of a few hundred volts (see Example 8.7).

Example 8.7 Phase modulation in $LiNbO_3$
Consider a $LiNbO_3$ crystal with light of wavelength 1.06 μm polarized along the z direction propagating along the x direction. The crystal is taken to be 20 mm long and 5 mm × 5 mm in cross-section. A field of amplitude 250 V is applied along the z direction. At a wavelength of 1.06 μm the value of r_{33} for $LiNbO_3$ is 30.8×10^{-12} m/V and the value for n_e is 2.16. Inserting these values into Eq. (8.81) gives

$$\delta_m = \frac{\pi \times 30.8 \times 10^{-12} (2.16)^3 \, 250 \times 0.2}{1.06 \times 10^{-6} \, 0.05}$$

$$= 0.92 \text{ rad}$$

8.3.4.3 Approach to equilibrium
The above analysis for both amplitude and phase modulation depends on the assumption of an equilibrium state, and no indication is given of how long it takes to reach equilibrium. To obtain some idea of the build-up time we may assume that, in the case of amplitude modulation, the laser has been operating on a single mode

when the modulator is 'instantaneously' switched on. Just before the modulator is switched on the electric field amplitude within the cavity will be a constant $\mathscr{E}_0$ (say). On a first pass through the modulator the electric field amplitude will be modulated so that it becomes (see Eq. (8.73))

$$\mathscr{E}_1(t) = \mathscr{E}_0 \exp(-\delta_m(1 - \cos \omega_m t))$$

On its second pass through the modulator the modulation will be 'deepened' so that

$$\mathscr{E}_2(t) = \mathscr{E}_0 \exp(-2\delta_m(1 - \cos \omega_m t))$$

After N passes we obtain

$$\mathscr{E}_N(t) = \mathscr{E}_0 \exp(-N\delta_m(1 - \cos \omega_m t)) \tag{8.82}$$

A plot of $\mathscr{E}_{10}(t)$ is shown in Figure 8.27. The 'pulses' developed become narrower on each trip, and very quickly become quite good approximations to Gaussian pulses.[17]

We now estimate how long it takes before the pulsewidth attains its equilibrium value (i.e. when Γ becomes equal to Γ_{sc}). We have previously seen (Eq. (8.74)) that on passing through an amplitude modulator and the gain medium a Gaussian profile pulse will experience a change in Γ ($\Gamma'' - \Gamma = \Delta\Gamma$) given by

$$\Delta\Gamma \approx -\frac{16\gamma(\omega_0)l}{\Delta\omega_L^2}\Gamma^2 + \frac{\delta_m\omega_m^2}{2} \tag{8.83}$$

Using Eq. (8.75) this may be written as

$$\Delta\Gamma \approx -\frac{16\gamma(\omega_0)l}{\Delta\omega_L^2}(\Gamma^2 - \Gamma_{sc}^2) \tag{8.84}$$

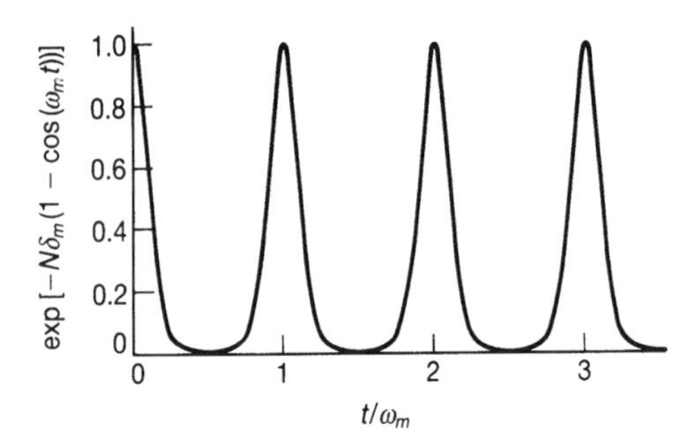

Figure 8.27 Graph of the function $\exp(-N\delta_m(1 - \cos(\omega_m t)))$ where $\delta_m = 0.3$ and $N = 10$.

This change in Γ will take place in the time τ_{RT}, the roundtrip transit time. Since the changes in Γ per roundtrip are relatively small we may write

$$\frac{d\Gamma}{dt} \approx \frac{\Delta\Gamma}{\tau_{RT}} = -\frac{16\gamma(\omega_0)l}{\Delta\omega_L^2\tau_{RT}}(\Gamma^2 - \Gamma_{sc}^2)$$

or

$$\frac{d\Gamma}{dt} \approx -\frac{\Gamma^2 - \Gamma_{sc}^2}{\Gamma_{sc}\tau_{ml}} \tag{8.85}$$

where τ_{ml} is a time constant given by

$$\tau_{ml} = \frac{\Delta\omega_L^2\tau_{RT}}{16\gamma(\omega_0)l\Gamma_{SC}} \tag{8.86}$$

We now integrate Eq. (8.85). Assuming that at $t = 0$ the Gaussian linewidth has a value much greater than any final value achieved, we may put $\Gamma(t = 0) \approx 0$, and we then have

$$\int_0^\Gamma \frac{\Gamma_{sc}}{\Gamma_{sc}^2 - \Gamma^2} d\Gamma \approx \frac{1}{\tau_{ml}} \int_0^t dt$$

The integral on the left-hand side has the value of $\tanh^{-1}(\Gamma/\Gamma_{sc})$, so that

$$\Gamma(t) \approx \Gamma_{sc} \tanh(t/\tau_{ml}) \tag{8.87}$$

Thus Γ will attain approximately 95 per cent of its final value in a time $t \approx \tau_{ml}$ $\tanh^{-1}(0.95)$ or $1.5\tau_{ml}$. We see that τ_{ml} represents the approximate time required for mode locked pulses to achieve minimum width. The number of roundtrips that this represents is given by $N_{ml} = \tau_{ml}/\tau_{RT}$. Using Eqs (8.87) and (8.75) we then have

$$N_{ml} = \frac{\Delta\omega_L^2}{16\gamma(\omega_0)l_q\Gamma_{sc}}$$

$$= \frac{1}{(8\delta_m\gamma(\omega_0)l)^{1/2}}\frac{\Delta\omega_L}{\omega_m} \tag{8.88}$$

The product $8\delta_m\gamma(\omega_0)l$ will have a value of the order of unity for reasons that have been discussed previously, so that

$$N_{ml} \approx \frac{\Delta\omega_L}{\omega_m} \tag{8.89}$$

For lasers such as Nd : YAG or Nd : glass, N_{ml} will thus be of the order of 1000 or more (i.e. representing a time of some tens of microseconds). A consequence of this is that the technique is not used in short-duration pulsed systems since there is then insufficient time for the equilibrium pulsewidth to be attained.

With phase locking the pulse build-up time is even longer. This is because there is no mechanism for generating a Gaussian pulse after a few transits of the modulator as is the case with amplitude modulation. All the phase modulator can do is generate a frequency chirp. The resulting increased frequency spread then interacts with the (frequency-dependent) gain medium to produce an amplitude-modulated pulse. The build-up of the initial Gaussian pulse thus takes much longer, so that phase modulation is even less useful than amplitude modulation for pulsed lasers.

8.4 The production of ultra-short optical pulses

We have seen in this chapter that techniques such as mode locking are capable of generating narrow pulses of radiation. With mode locking the width of the pulse is governed primarily by the width of the gain curve. Dye lasers have some of the broadest gain curves and, not surprisingly therefore, are capable of extremely narrow optical pulses (the wide gain curve is also responsible for the the other desirable feature of dye lasers, tunability over a wide wavelength range). In a typical dye laser, for example, the gain bandwidth extends over some 100 nm. The use of Eq. (8.52) would then indicate an optical pulsewidth of 10^{-14} s (i.e. 10 femtoseconds). In practice, however, it is not at all easy to achieve widths of this order. Both active and passive mode locking is possible, and with the conventional techniques described in previous sections pulses a few picoseconds long have been obtained [Ref. 11]. Sub-picosecond pulses are possible using a variety of techniques, some of which are described in the rest of this section.

As far as active mode locking is concerned some of the narrowest pulses have been obtained using a technique known as 'synchronous pumping' [Ref. 12]. Here the pump is a laser that is itself mode locked. If the effective cavity lengths of the two lasers are exactly matched[18] then the pumping pulse from the first laser will pump the second laser just as the circulating pulse is passing through the gain medium. This technique has the advantage, as far as the dye laser is concerned, of being more efficient than CW pumping since a typical upper state lifetime is of the order of a few nanoseconds, whereas the roundtrip transit time is often several times this. With continuous pumping, therefore, a considerable fraction of the energy pumped into the upper level between the arrival times of the circulating pulse is wasted. A favoured pumping system for this approach is a frequency-doubled Nd : YAG laser (i.e. $\lambda = 532$ nm). The pumping pulses from the YAG laser are typically 40 ps wide and when used to pump a rhodamine 6G laser have produced pulses of less than 1 ps [Ref. 13].

Another successful scheme has been that of the 'colliding pulse passively mode locked laser'. Although linear cavities using this scheme have been operated, the most successful arrangement is that of a ring laser. A schematic diagram of the arrangement is shown in Figure 8.28. Consider two identical pulses propagating round the ring in opposite directions. The cavity losses for both pulses will be minimized if they arrive at the saturable absorber at the same time, since they will then give rise to a higher irradiance, and hence an increased absorber saturation. In fact the system will tend to automatically 'lock-in' to this state since it gives rise to the least loss. The positioning of the gain medium is fairly critical, and best results

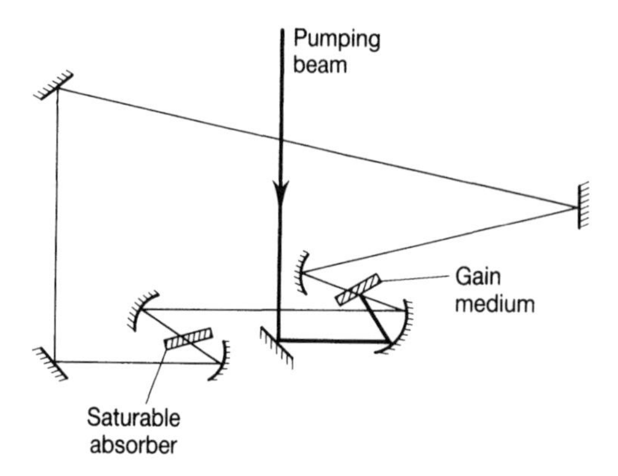

Figure 8.28 Typical layout used in a colliding pulse mode locked laser.

are obtained if it is separated from the saturable absorber by a quarter of the total ring path length (Figure 8.29). This configuration ensures (1) that the pulses do not arrive at the gain medium at the same time, and (2) that there is a constant time delay between one pulse passing through the gain medium and the other, thus ensuring that, with a constant pumping rate, both pulses encounter the same gain. If the saturable absorber is very thin, say 10 μm, then the cavity shows minimum loss if the pulses completely overlap within the absorber. Thus there will be a tendency for the pulses to take on a spatial extent as narrow as that of the absorber. A spatial width of 10 μm implies a time width of the order of $10 \times 10^{-6}/3 \times 10^{8}$ or 3×10^{-14} s (or 30 fs). In practice, widths of the order of 60 fs have been observed with this set-up [Ref. 14].

Another technique relies on the fact that when the pulses pass through the dye gain medium they can become chirped. At relatively high irradiances, for example, the chirping arises from the Kerr effect (see Section 2.1.9). The Kerr effect gives rise to a pulse where the optical frequency increases as time increases. If the pulse then

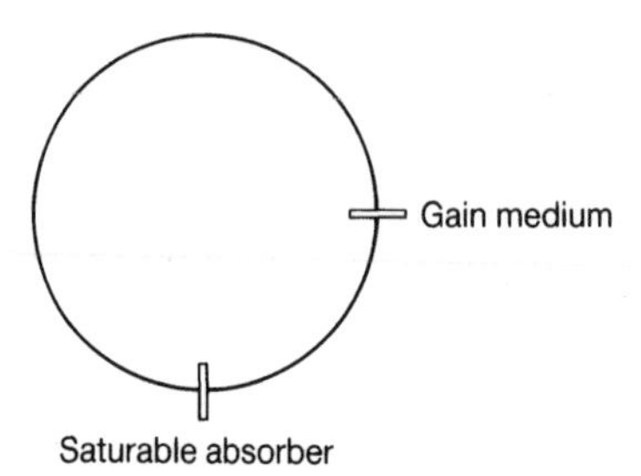

Figure 8.29 Relative topology of colliding pulse layout with absorber and gain medium separated by 1/4 of the cavity perimeter to ensure balanced gain between the counter-propagating pulses.

passes through a dispersive optical element where higher frequencies travel faster than lower frequencies the pulse will become compressed in time. Both glass prisms and optical fibers have been used as the dispersive elements. Using such techniques it has been possible to generate pulses of less than 10 fs duration [Ref. 15]. A discussion of the use of dye lasers in generating femtosecond pulses is contained in Ref. 16.

Recently laser systems with even broader oscillation bandwidths than dye lasers have become available, one of the best known being titanium-doped sapphire ($Ti:Al_2O_3$). These are capable of producing pulses as short as those from colliding pulse mode locked dye lasers [Ref. 17] and, in addition, they have a greater tuning range and greater power per pulse. ($Ti:Al_2O_3$ lasers can produce femtosecond pulses up to 1 W in power, the corresponding figure for dye lasers being tens of milliwatts.)

So far in this section we have discussed the generation of narrow optical pulses but have given no indication of how the widths may be measured. The fastest optical detectors have response times of some 10^{-11} s. Thus to measure the width of the type of pulses we are discussing here requires a different approach. The most usual technique is to use a nonlinear optical process to obtain the spatial autocorrelation trace of the pulse. A pulse duration of, say, 10^{-12} s is then 'converted' into a distance measurement of $3 \times 10^8 \times 10^{-12}$, or 0.3 mm. The most commonly used nonlinear process is that of 'frequency doubling' (or second harmonic generation, see Section 6.6.5).

A schematic diagram of the experimental layout is shown in Figure 8.30. The beam containing the mode locked pulse train is split into two by means of a beam splitter. After travelling different distances the two beams are then focused down so that they overlap within the frequency-doubling crystal. If the different paths

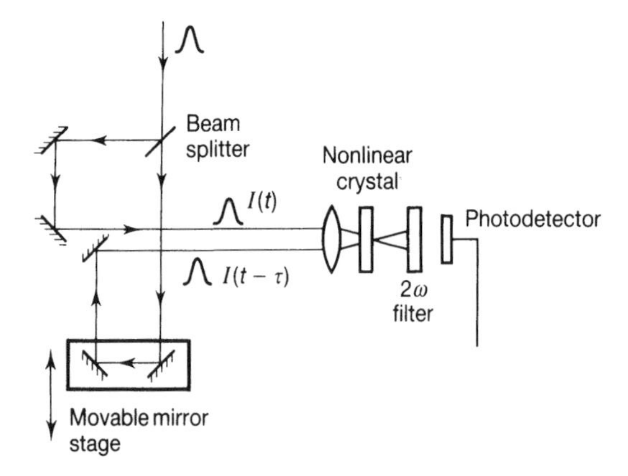

Figure 8.30 Layout used for the determination of the temporal width of a very narrow optical pulse from a measurement of its spatial autocorrelation function.

introduce a time delay, τ, then the two fields at the frequency-doubling crystal can be written $\mathscr{E}_0(t)\exp(j\omega t)$ and $\mathscr{E}_0(t-\tau)\exp(j\omega(t-\tau))$. Now the second harmonic field produced by the crystal is proportional to the square of the incident field (see discussion in Section 6.6.5), so that the output field at frequency 2ω can be written:

$$O_{2\omega}(t) = C(\mathscr{E}_0(t) + \mathscr{E}_0(t-\tau)\exp(-j\omega\tau))^2$$

where C is a constant of proportionality. Thus

$$O_{2\omega}(t) = C(\mathscr{E}_0^2(t) + \mathscr{E}_0^2(t-\tau)\exp(-2j\omega t) + 2\mathscr{E}_0(t)\mathscr{E}_0(t-\tau)\ \exp(-j\omega t)) \tag{8.90}$$

When this field is incident on the optical detector the output of the detector (which is usually in the form of a current, i_d, say) is proportional to the optical power. Thus

$$i_d = DO_{2\omega}(t)\ O_{2\omega}^*(t)$$

Substituting for $O_{2\omega}(t)$ Eq. (8.90) gives

$$\begin{aligned} i_d = DC\{&(\mathscr{E}_0(t)\mathscr{E}_0^*(t))^2 + (\mathscr{E}_0(t-\tau)\mathscr{E}_0^*(t-\tau))^2 \\ &+ 4\mathscr{E}_0(t)\mathscr{E}_0^*(t)\mathscr{E}_0(t-\tau)\mathscr{E}_0^*(t-\tau) + T(\tau)\} \end{aligned} \tag{8.91}$$

where $T(\tau)$ is composed of terms which have either a $\cos(\omega t)$ or $\cos(2\omega t)$ dependence. No detector can respond to fluctuations on the time scale of $1/\omega$, so that the term $T(t)$ will average out to zero. Again the temporal variations in $\mathscr{E}_0(t)$ will be on a picosecond time scale and the detector can only respond to the value averaged over the response time of the detector. We may put

$$\langle I(t)\rangle = \langle \mathscr{E}_0(t)\mathscr{E}_0^*(t)\rangle$$

so that

$$I_d(\tau) = DC(\langle I^2(t)\rangle + \langle I^2(t-\tau)\rangle + 4\langle I(t)I(t-\tau)\rangle) \tag{8.92}$$

If the pulse train consists of identical pulses, then

$$\langle I^2(t)\rangle = \langle I^2(t-\tau)\rangle$$

Normalizing $i_d(\tau)$ to the factor $2DC\langle I^2(t)\rangle$,

$$i_d^N(\tau) = 1 + 2G(\tau) \tag{8.93}$$

where

$$G(\tau) = \frac{\langle I(t)I(t-\tau)\rangle}{\langle I^2(t)\rangle} \tag{8.94}$$

The function $G(\tau)$ will be recognized as being similar to the mutual coherence function of Eq. (2.123), except that the present function involves the irradiance rather than the electric field.

Since the light is assumed to be coherent, the form of $i_d^N(\tau)$, as a function of τ, is a peak of relative height two units sitting on top of a background of height one unit (Figure 8.31). The exact relationship between the pulse halfwidth (τ_p) and the

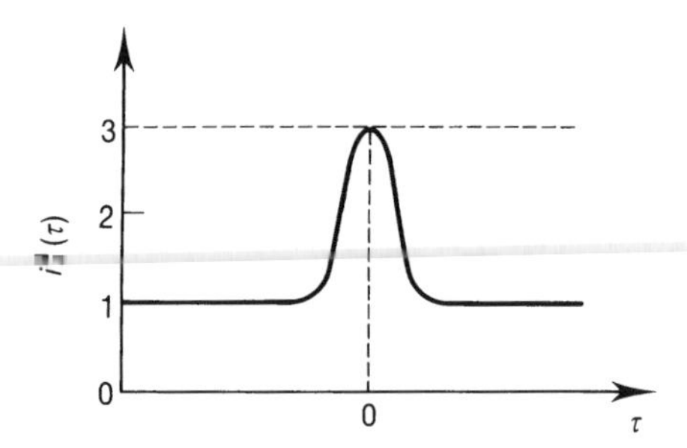

Figure 8.31 Schematic diagram of the output expected from the experimental layout shown in Figure 8.30. A 'pulse' of relative height two units is superimposed on a constant background level of relative height one unit.

halfwidth of the irradiance mutual coherence function (τ_c) depends on the shape of the pulse (see Problem 8.13). For a Gaussian pulse $\tau_p = \tau_c/\sqrt{2}$.

Problems

8.1 In the analysis of laser spiking/relaxation oscillation given in Section 8.1 the pumping rate R_2 into the upper lasing level was taken to be a constant. Given that in a 'four-level' type system the ground state population is expected to remain constant, this seems a reasonable assumption. To put this on a more quantitative basis, however, assume that the pumping rate can be written $W_p N_0$ and that, as before, N_1 and N_3 may be neglected, and show that Eq. (8.6) is now replaced with

$$\frac{d\mathcal{N}}{d\tau} = R\left(\frac{S-\mathcal{N}}{S-1}\right) - \mathcal{N} - n\mathcal{N}(R-1)$$

where $S = N_T C \tau_c$, N_T (the total density of atoms) $= N_0 + N_2$ and $R = W_p \tau_2 (S-1)$. Extend the analysis to that of a three-level system (assume that the population of the topmost level, N_3 can be neglected).

8.2 During the initial build-up of a Q-switched pulse the photon loss term N_p/τ_c in Eq. (8.20) may be neglected. Equation (8.22) can then be written

$$\frac{dn}{d\tau} = n\mathcal{N} \qquad (8.22b)$$

Show that Eqs (8.21)and (8.22b) can be solved to give

$$\mathcal{N} = \frac{\mathcal{N}_i}{1 + \exp(\mathcal{N}_i(t - t_0))}$$

and

$$n = \frac{\mathcal{N}_i}{2} \frac{\exp(\mathcal{N}_i(t - t_0))}{1 + \exp(\mathcal{N}_i(t - t_0))}$$

where $\mathcal{N}_i$ is the initial value of $\mathcal{N}$ and t_0 is a time when $\mathcal{N}$ has fallen to $\mathcal{N}_i/2$. Sketch the curves of $\mathcal{N}$ and n as a function of time, if possible superimpose them on the 'exact' computer calculations based on Eqs (8.21) and (8.22). (*Hint*: Equations (8.21) and (8.22b) can be combined to give

$$\frac{d\mathcal{N}}{d\tau} = -2\frac{dn}{d\tau}$$

Integration then yields: $\mathcal{N} - \mathcal{N}_i = -2n$. This may be used in either of the two coupled equations to give the required result.)

8.3 A ruby laser uses a laser rod 100 mm long and 10 mm in diameter, and the laser cavity itself is 500 mm long. Estimate the total energy and the peak power of a *Q*-switched pulse obtainable from this laser when it is pumped to a level which yields a gain of 5 m^{-1}. When in equilibrium (i.e. unpumped) at room temperature the crystal, which has a doping concentration of 1.6×10^{25} m^{-3}, exhibits a loss coefficient of 20 m^{-1}. The mirror reflectivities are 100 per cent and 60 per cent and the refractive index of ruby at 633 nm is 1.76. Ignore any losses apart from mirror losses.

8.4 Show that when the initial population inversion only just exceeds the threshold value, the efficiency of a *Q*-switched laser pulse, η_p, can be written

$$\eta \approx 2\frac{(\mathcal{N}_i - 1)^2}{\mathcal{N}_i^2}$$

8.5 Repetitive *Q*-switching of CW pumped solid-state lasers such as Nd : YAG is an important technique used in micromachining applications. In such a system while the shutter is 'closed' the pumping must raise the population inversion from its value of $\mathcal{N}_f$ (the normalized value at the end of the pulse) to $\mathcal{N}_i$ the value at the start. Show that this implies that

$$\mathcal{N}_i = R - (T - \mathcal{N}_f)\exp\left(-\frac{t_p}{\tau_2}\right)$$

where t_p is the time between the closing of the shutter after a pulse and the opening of the shutter for the next (t_p is approximately the time between pulses). The resulting values of $\mathcal{N}_i$ and $\mathcal{N}_f$ for a particular t_p (and R) can then be obtained by combining this equation with Eq. (8.28). Write a computer program to solve these simultaneous equations, and hence determine, for a particular pumping rate (R) how the output power efficiency varies with t_p/τ_2.

8.6 Starting from Eq. (8.53) which gives the resulting field from N oscillating modes, i.e.

$$\mathscr{E}(t) = Re\sum_{m=1}^{N} \mathscr{E}_m \exp(j(2\pi\nu_m t + \phi_m))$$

show that:

(a) The output irradiance will repeat itself with periodicity $(\delta\nu)^{-1}$, that is, $I(t + 1/\delta\nu) = I(t)$ (provided the relative phases of the modes do not change with time).

(b) If the phases are locked together so that $\phi_m = \phi_0$ and the amplitudes are equal (i.e. $\mathscr{E}_m = \mathscr{E}_0$), then

$$I(t) = K\mathscr{E}_0^2\frac{\sin^2(N\pi\delta\nu t)}{\sin^2(\pi\delta\nu t)}$$

(c) If the phases are locked so that $\phi_m = m\phi$, then the pulses are the same as in (b) above but delayed by $-\phi/(2\pi\delta\nu)$.

8.7 Show that when light of wavelength λ is incident at an angle θ onto a series of partially reflecting planes separated by a distance Λ then a strong reflection will be obtained when θ is equal to θ_B where

$$\sin \theta_B = \frac{\lambda}{2\Lambda}$$

8.8 Estimate the duration and separation of mode locked pulses obtainable from a He–Ne laser whose mirrors are 1 m apart. Take the width of the laser transition to be 1.5×10^9 Hz.

8.9 As discussed in Section 8.3.3, the linewidth of an optical pulse will be narrowed when it passes through a saturable absorber. To investigate this narrowing use may be made of the results obtained in Section 3.3.2 which deals with gain saturation. Replacing G_0 (the small signal gain) with T_0 (the small signal transmission coefficient) Eq. (3.127) becomes

$$\log_e\left(\frac{I_f}{I_i}\right) + \frac{I_f - I_i}{I_s} = \log_e T_0$$

Write a computer program to solve this equation for I_f (the transmitted irradiance) as a function of I_i (the incident irradiance). By applying this to a Gaussian pulse where the peak irradiance is $10I_s$ and where T_0 is 0.001, show that the pulse narrowing is about 0.6.

8.10 The electric field within a frequency chirped Gaussian pulse can be described by

$$\mathscr{E}(t) = \exp(-\Gamma t^2 + j\omega t)$$

where Γ is a complex quantity. If Γ is written as $a - jb$, show that the spectral halfwidth of the pulse is

$$\frac{(2 \log_e 2)^{1/2}}{\pi} \left(\frac{a^2 + b^2}{a}\right)^{1/2}$$

(*Hint*: taking Fourier transform of $\mathscr{E}(t)$ yields

$$\mathscr{E}(\omega) = \exp\left(-\frac{(\omega - \omega_0)^2}{4\Gamma}\right)$$

The power spectral density is then given by $|\mathscr{E}(\omega)|^2$). Use this result to show that the chirped pulse produced by a phase modulator will have a frequency bandwidth that is a factor $\sqrt{2}$ larger than that of an unchirped pulse of the same pulsewidth produced by an amplitude modulator.

8.11 A Nd:YAG laser is used in conjunction with a phase modulator to produce mode locked pulses. The laser cavity (total length 600 mm) uses mirrors of reflectivities 100 per cent and 90 per cent. The modulator is based on a crystal of $LiNbO_3$ which is aligned so that its x axis is along the laser axis, an oscillating voltage of amplitude 200 V is applied along the z direction while the laser beam is polarized along the y axis. Estimate the pulsewidths obtained and the frequency which must be applied to the modulator. You may assume that in $LiNbO_3$ $r_{13} = 8.6 \times 10^{-12}$ m/V and $n_0 = 2.24$.

8.12 The frequency bandwidth of a sub-picosecond pulse is relatively wide, so that any dispersive element present in the laser cavity may cause appreciable spreading (in time). A pulse of duration 0.5 ps makes a transit through a glass slide 5 mm thick. Estimate the pulse spreading caused by dispersion. You may assume that the refractive index of the glass at 656 nm is 1.5143 while at 588 nm it is 1.5168.

8.13 Show that the irradiance mutual coherence function, $G(\tau)$ (i.e. as defined by Eq. (8.94)) for the following pulse shapes ($I(t)$) are as indicated:

$I(t)$	$G(\tau)$	
Square	Square	
$\exp(-\lvert t \rvert)$	$(1 = \lvert \tau \rvert)\exp(-\lvert \tau \rvert)$	
Triangular	$(1 - \frac{3}{2}x^2 + \frac{3}{4}\lvert x \rvert^3)$	$0 < \lvert x \rvert < 1$
	$\frac{1}{4}(2 - x)^3$	$1 < \lvert x \rvert > 2$
	0	$\lvert x \rvert > 2$

where $x = \tau/\tau_p$. Hence determine the relationship between the halfwidths of the curves $I(t)$ and $G(\tau)$.

Computing project

As pointed out in the text, the analysis of relaxation oscillations and Q-switching involves coupled differential equations, which can only be solved with the use of computers. It is a very useful exercise to write computer programs which will solve these two sets of coupled equations (i.e. Eqs (8.6) + (8.7) and (8.21) + (8.22)) and display the result (in fact the graphical display aspect may cause the most difficulty!). The equations to be solved are of the form

$$\frac{dx}{dt} = f_1(x, y) \text{ and } \frac{dy}{dt} = f_2(x, y)$$

Using the Euler method, the value of x at a time $t + h$ (x_{n+1}) is related to its value at $t(x_n)$ by $x_{n+1} \approx x_n + hf_1(x_n, y_n)$, etc. for y_n. This technique will work provided a small enough interval is chosen for h. A more sophisticated procedure is to use the fourth-order Runge–Cutta approximation, where:

$$x_{n+1} \approx x_n + (c_{11} + 2c_{12} + 2c_{13} + c_{14})/6$$

$$y_{n+1} \approx y_n + (c_{21} + 2c_{22} + 2c_{23} + c_{24})/6$$

$$c_{11} = hf_1(x_n, y_n) \qquad c_{21} = hf_2(x_n, y_n)$$

$$c_{12} = hf_1(x_n + c_{11}/2, y_n + c_{21}/2) \qquad c_{22} = hf_2(x_n + c_{11}/2,\ y_n + c_{21}/2)$$

$$c_{13} = hf_1(x_n + c_{12}/2, y_n + c_{22}/2) \qquad c_{23} = hf_2(x_n + c_{12}/2, y_n + c_{22}/2)$$

$$c_{14} = hf_1(x_n + c_{13}, y_n + c_{23}) \qquad c_{24} = hf_2(x_n + c_{13}, y_n + c_{23})$$

Care needs to be taken in the choice of h (i.e. Δt). Too small an interval will give rise to very long computing times, too long an interval will lead to inaccurate values.

As well as the 'rapid' Q-switching equations, the student may like to investigate 'slow' Q-switching and Q-switching using a saturable absorber (use Eq. (8.38) for the absorber transmission).

Notes

1. The situation during the initial build-up of the pulse, when the photon population is relatively small and the cavity loss term and the stimulated emission term in Eq. (8.2) can both be ignored, has been dealt with previously (Section 6.3). The photon density in the cavity is then proportional to the factor

$$\exp\left[\left(\frac{CR_2}{2}\right)(t - t_0)^2\right]$$

2. If $R_2 < 1/(C\tau_2\tau_2)$ Eq. (8.4) predicts a negative photon density. This is, of course, a nonsense physically, and comes about because of the approximations we have made which assume a relatively high pumping rate.
3. The student may have recognized straight away that Eq. (8.10) describes damped simple harmonic motion.
4. Note that the normalized parameters $\mathcal{N}$ and n here are not the same as used earlier in the chapter for relaxation oscillations. However, this should not really cause any confusion.
5. The initial stages of the pulse evolution (when the photon population will be relatively small) may be studied by making the approximation that the photon loss term in Eq. (8.27) (i.e. n_p/τ_c) can be neglected. The coupled equations may then be solved exactly, as indicated in Problem 8.2.
6. This is only strictly true if the only source of cavity loss is because of mirror transmission.
7. For simplicity, we have ignored any losses within the cavity apart from mirror loss.
8. Assuming a typical material refractive index of 1.5, each air–material interface gives a transmission of 96 per cent, and the presence of six such interfaces gives rise to a transmission of 78 per cent. With suitable anti-reflection precautions this can be increased in practice to about 90 per cent.
9. The analogy with Bragg diffraction in crystals would lead to the supposition that there should be other orders whose diffraction angles are given by $\sin\theta = m\lambda/2\Lambda$, where m is an integer. In theory with an infinitely wide grating and because the refractive index variation is continuous as opposed to the discrete nature of the atomic plane case only the zeroth- and first-order beams can be present. In practice, the noninfinite width of the structure can give rise to such higher-order modes but these are generally much weaker than the first-order beam.
10. Lithium niobate, for example, is much more efficient.
11. We may note that we have here an example of the reciprocal relationship between pulsewidth and frequency bandwidth. The frequency bandwidth ($\Delta\nu$) is $N \times \delta\nu$ while the pulsewidth (τ_p) is $1/(N\delta\nu)$, so that $\Delta\nu \times \tau_p = 1$. The minimum possible value for the product is in fact given for Gaussian distributions where the widths are the rms values.
12. This result basically derives from the fact that the Fourier transform of a Gaussian spectrum is a Gaussian time pulse.
13. This is another instance of the inverse relationship between spectral width and temporal width as given by the Fourier transform relationship between them.

14. In fact when $x \gg 1$ then

$$J_n(x) \cong \left(\frac{2}{\pi x}\right)^{1/2} \cos(x - n\pi/2 + \pi/4)$$

15. Note, though, that spatial hole burning (Section 6.4.3) can lead to homogeneously broadened lasers oscillating on several modes simultaneously.

16. If the radiation is polarized along the y axis then the change in refractive index is given by replacing r_{33} and n_e in Eq. (8.80) by r_{13} and n_0, respectively.

17. To show this directly we may expand Eq. (8.73) about the time $t = 0$. Using the approximation $\cos \omega_m t \approx 1 - \omega^2 t^2/2(\omega_m t \ll 1)$ we have

$$\mathscr{E}_N(t) \approx \mathscr{E}_0 \exp(-(N\delta_m \omega^2/2)t^2)$$

which is a Gaussian.

18. In practical terms the effective lengths have to be matched to within a few microns.

References

[1] K. T. Lau, N. Bar-Chaim, I. Ury and A. Yariv, 'Direct amplitude modulation of semiconductor GaAs lasers up to X-band frequencies', *Appl. Phys. Lett.*, **43**, 11–13, 1983.

[2] R. B. Kay and G. S. Waldman, 'Complete solutions to the rate equations describing *Q*-spoiled and PTM laser operation', *J. Appl. Phys.*, **36**, 1319–23, 1965.

[3] R. J. Collins and P. Kisliuk, 'Control of population inversion in pulsed optical masers by feedback modulation', *J. Appl. Phys.*, **33**, 2009–11, 1962.

[4] A. Yariv, *Optical Electronics*, 4th edn, Saunders College Publishing, 1991, Section 12.3.

[5] A. Szabo and R. A. Stein, 'Theory of laser giant pulsing by a saturable absorber', *J. Appl. Phys.*, **36**, 1562–6, 1965.

[6] B. Fan and T. K. Gustafson, 'Narrow-band picosecond pulses from an ultrashort-cavity dye laser', *Appl. Phys. Lett.*, **28**, 202–4, 1976.

[7] C. Lin *et al.*, 'Simple picosecond pulse generation scheme for injection lasers', *Electron. Lett.*, **16**, 600–2, 1980. P. T. Torphammar and S. T. Eng, 'Picosecond pulse generation in semiconductor lasers using resonant oscillation', *Elect. Lett.*, **16**, 587–9, 1980.

[8] J. T. Verdeyen, *Laser Electronics*, Prentice Hall, Englewood Cliffs, NJ, 1981, Section 9.4.2.

[9] V. S. Letokhov, 'Generation of ultrashort light pulses in a laser with a nonlinear absorber', *Sov. Phys. JETP*, **28**, 562–8, 1969.

[10] D. J. Kuizenga and A. E. Siegman, 'FM and AM mode locking of the homogeneous laser Part 1: Theory. Part 2: Experimental results', *J. Quant. Elec.*, **QE-6**, 694–715, 1970.

[11] *Passive mode locking*: D. J.Bradley *et al.*, 'Picosecond pulses from mode-locked lasers', *Phys. Lett.*, **30A**, 535–6, 1969.
Active mode locking: D. J.Kuizenga, 'Mode locking of the CW Dye laser', *Appl. Phys. Lett.*, **19**, 260–263, 1971.

[12] B. H. Soffer and J. W. Linn, 'Continuously tunable picosecond-pulse organic-dye laser', *J. Appl. Phys.*, **39**, 5859–60, 1968. C. V. Shank and E. P. Ippen, 'Sub-picosecond kilowatt pulses from a mode-locked cw dye laser', *Appl. Phys. Lett.*, **24**, 373–5, 1974. C. K. Chan and S. O. Sari, 'Tunable dye laser pulse converter for production of picosecond pulses', *Appl. Phys. Lett.*, **25**, 403–6, 1974.

[13] A. M. Johnson and W. M. Simpson, 'Tunable femtosecond dye laser synchronously pumped by the compressed second harmonic of Nd:YAG', *J. Opt. Soc. Am.*, **B2** 619–25, 1985.

[14] R. L. Fork and C. V. Shank, 'Generation of optical pulses shorter than 0.1 ps by colliding pulse mode locking', *Appl. Phys. Lett.*, **38**, 671–3, 1981.

[15] R. L. Fork *et al.*, 'Compression of optical pulses to six femtoseconds by using cubic phase compensation', *Opt. Lett.*, **12**, 483–5, 1987.

[16] J.-C. Diels, 'Femtosecond dye lasers', in F. J. Duarte and L. W. Hillman (Eds), *Dye Laser Principles, with applications*, Academic Press, New York, 1990. S. L. Shapiro (Ed.), *Ultrashort Laser Pulses: Picosecond Techniques and Applications*, Topics in Applied Physics, Vol. 60, Springer-Verlag, New York, 1977.

[17] C. Huang *et al.*, 'Generation of transform limited 32 fs pulses from a self-locked Ti:sapphire laser', *Opt. Lett.*, **17**, 139–41, 1992.

Appendix 1

Quantum mechanics

It is obviously impossible to give anything like a comprehensive treatment of quantum mechanics in the space available here. All that can be attempted is a brief outline of some of the ideas relevant to the present text. We start with a discussion of some of the essential mathematical concepts.

Suppose we have a function of the N variables $q_1, q_2 \ldots q_N$ which we write as $\psi(q_1, q_2 \ldots q_N)$ or, more compactly, simply as ψ. Suppose also that we have an operator $\hat{A}$ which acts on the variables q. The operator is said to be *linear* if

$$\hat{A}(\psi_1 + \psi_2) = \hat{A}\psi_1 + \hat{A}\psi_2 \tag{A1.1}$$

for all ψ_1 and ψ_2.

We define the *scalar product* of two functions $\psi(q_1, q_2 \ldots q_N)$ and $\theta(q_1, q_2 \ldots q_N)$ as

$$\langle \psi, \theta \rangle = \int \ldots \int \psi^*(q_1, q_2 \ldots q_N)\theta(q_1, q_2 \ldots q_N)\, \mathrm{d}q_1 \ldots \mathrm{d}q_N \tag{A1.2}$$

where * indicates the complex conjugate. Thus $\langle \psi, \theta \rangle = \langle \theta, \psi \rangle^*$ and so $\langle \psi, \theta \rangle$ is real if $\langle \psi, \theta \rangle = \langle \theta, \psi \rangle$; $\langle \psi, \psi \rangle$ is *always* real.

Without loss of generality we may suppose that the functions we use are *normalized to unity*, that is, we have $\langle \psi, \psi \rangle = 1$. Two functions are said to be *orthogonal* if

$$\langle \psi, \theta \rangle = 0 \tag{A1.3}$$

The operator $\hat{A}$ is said to be *Hermitian* if, for all ψ and θ

$$\langle \psi, \hat{A}\theta \rangle = \langle \hat{A}\psi, \theta \rangle \tag{A1.4}$$

For such an operator

$$\langle \psi, \hat{A}\psi \rangle = \langle \hat{A}\psi, \psi \rangle = \langle \psi, \hat{A}\psi \rangle^* \tag{A1.5}$$

Thus $\langle \psi, \hat{A}\psi \rangle$ is always *real*.

A function $\phi(q_1, q_2 \ldots q_N)$ is said to be an *eigenfunction* of $\hat{A}$ with *eigenvalues* a if

$$\hat{A}\phi = a\phi \tag{A1.6}$$

There may, in general, be a whole set of functions ϕ_n with corresponding eigenvalues a_n where $n = 1, 2, \ldots, N$. Often for the operators used in quantum mechanics N is infinite and the eigenvalues discrete, although they may also form a continuum.

Two important results may now be deduced concerning the eigenvalues and eigenfunctions of Hermitian operators:

1. The eigenvalues of Hermitian operators are always real.
2. The eigenfunctions of a Hermitian operator corresponding to different eigenvalues are orthogonal.

To prove (1) we consider the expression $\langle \phi, \hat{A}\phi \rangle$. From Eq. (A1.5) we know that it is real. Using Eq. (A1.6) it can be written as $a\langle \phi, \phi \rangle$. Since we know that $\langle \phi, \phi \rangle$ is real, we may deduce that a must be real.

The proof of (2) is as follows. We have $\hat{A}\phi_m = a_m\phi_m$ and $\hat{A}\phi_n = a_n\phi_n$ and so

$$\langle \phi_n, \hat{A}\phi_m \rangle = a_m\langle \phi_n, \phi_m \rangle \text{ and } \langle \hat{A}\phi_n, \phi_m \rangle = a_n\langle \phi_n, \phi_m \rangle$$

Since $\hat{A}$ is Hermitian $\langle \phi_n, \hat{A}\phi_m \rangle = \langle \hat{A}\phi_n, \phi_m \rangle$, and thus $a_m\langle \phi_n, \phi_m \rangle = a_n\langle \phi_n, \phi_m \rangle$.

If $a_m \neq a_n$ then $\langle \phi_n, \phi_m \rangle = 0$. This can be expressed by

$$\langle \phi_n, \phi_m \rangle = \delta_{mn} \tag{A1.7}$$

where δ_m is the *Kronecker* δ *function*. This has the value 1 if $m = n$ and 0 otherwise.

A set of functions ϕ_n is said to be *complete* if a general function ϕ (of the same basic variables q) can always be expressed as

$$\phi = \sum_{n=1}^{N} C_n\phi_n \tag{A1.8}$$

It may be shown that the eigenfunctions of a Hermitian operator form a complete set.

To determine the values of the C_n, we take the scalar product of both sides of Eq. (A1.8) with ϕ_m:

$$\begin{aligned}\langle \phi_m, \phi \rangle &= \sum_{1}^{N} C_n\langle \phi_m, \phi_n \rangle \\ &= \sum_{1}^{N} C_n\delta_{mn} \text{ (from Eq. (A1.7))} \\ &= C_m\end{aligned}$$

Thus

$$C_m = \langle \phi_m, \phi \rangle \tag{A1.9}$$

ϕ can be regarded as an N-dimensional vector in 'ϕ_n space', with components C_n. This vector space is known as *Hilbert space*.

Two operators $\hat{A}$ and $\hat{B}$ are said to *commute* if, for all ϕ

$$(\hat{A}\hat{B} - \hat{B}\hat{A})\phi = 0 \tag{A1.10}$$

An important result is that if $\hat{A}$ and $\hat{B}$ have eigenfunctions in common, then they commute. To prove this we expand a general function ψ in terms of the eigenfunctions of $\hat{A}$ and $\hat{B}$. Thus $\psi = \Sigma\ C_n\phi_n$. Then

$$
\begin{aligned}
\hat{A}\hat{B}\psi &= \Sigma C_n \hat{A}\hat{B}\phi_n \\
&= \Sigma\ C_n b_n \hat{A}\phi_n = \Sigma C_n b_n a_n \phi_n = \Sigma C_n a_n b_n \phi_n \\
&= \Sigma C_n \hat{B}\hat{A}\phi_n \\
&= \hat{B}\hat{A}\psi
\end{aligned}
\qquad \text{(A1.11)}
$$

The converse is also true, that is commutativity implies common eigenfunctions. As a shorthand notation we write $\hat{A}\hat{B} - \hat{B}\hat{A} = [\hat{A}, \hat{B}]$.

It often seems, when first introduced, that quantum mechanics arises 'out of the blue', with little connection, if any, with classical mechanics. This is usually because many physics courses do not follow the developments of classical mechanics much beyond Newton's formulations. In the nineteenth century mathematicians such as Hamilton sought a formalism whereby Newton's equations of motion could be made independent of the coordinate system used (the expression *force* = *mass* × *acceleration* looks very different, depending on whether Cartesian or polar etc. coordinates are used). It was shown that a mechanical system can be described by *N generalized position coordinates* q_s ($1 \leqslant s \leqslant N$). (Note that these may not bear much direct relationship to x, y, z, r, etc.)

The kinetic (T) and potential energies (V) may then be expressed in terms of q_s and $\dot{q}_s$ ($\dot{q}_s = \partial q_s/\partial t$). *Generalized momenta* (p_i) may then be defined as

$$p_i = \frac{\partial}{\partial \dot{q}_i}(T - V)$$

All physical quantities may then be expressed in terms of the q_i and the p_i. Of particular interest is the total energy $T + V$, known as the *Hamiltonian H*. It can then be shown that

$$\dot{p}_i = -\frac{\partial H}{\partial q_i} \text{ and } \dot{q}_i = \frac{\partial H}{\partial p_i} \qquad \text{(A1.12)}$$

At this stage it may be useful to give a simple example. We take a particle moving in zero potential along the x axis with velocity v_x. We take $q = x$. Now

$$H = \tfrac{1}{2} m v_x^2 = \tfrac{1}{2} m \dot{x}^2$$

Thus

$$p = \frac{\partial}{\partial \dot{q}}(T - V) = \frac{\partial}{\partial \dot{x}}(\tfrac{1}{2} m \dot{x}^2) = m\dot{x} \text{ i.e. the 'normal' momentum)}$$

Since H is independent of q,

$$\dot{p} = -\frac{\partial H}{\partial q_i} = 0$$

that is, $m\dot{x} = \text{const}$, and we have 'conservation of momentum'. Similarly, we have

$$\dot{q}(=\dot{x}) = \frac{\partial H}{\partial p} = \frac{\partial}{\partial p}\left(\frac{p^2}{2m}\right) = \frac{p}{m} \text{ (i.e. } p = m\dot{x})$$

The quantities p_i and p_i are known as *canonically conjugate* variables. As we have seen in the above example, ordinary position and momentum can form such a pair.

We are now in a position to introduce some of the basic ideas of quantum mechanics. This is usually done via a series of postulates, although the precise number, ordering and indeed content of these postulates varies considerably, depending on the author in question! Here we may restrict ourselves to the following six postulates:

1. As we have seen, any classical physical quantity can be constructed from pairs of canonically conjugate variables. The corresponding quantum-mechanical quantity is an operator formed by replacing the canonical variables by their corresponding quantum-mechanical operators. The operator must be Hermitian.
2. For all pairs of canonically conjugate operators the following commutation relations hold:

$$\left.\begin{aligned} [\hat{q}_n, \hat{q}_m] &= 0 \\ [\hat{p}_n, \hat{p}_m] &= 0 \\ [\hat{p}_n, \hat{q}_m] &= \frac{h}{2\pi \mathrm{j}}\,\delta_{nm} \end{aligned}\right\} \tag{A1.13}$$

3. The state of a physical system is described by a vector in Hilbert space (the *state vector*). When a measurement is made on the system of a particular physical quantity the result will be one of the eigenvalues of the corresponding operator. Immediately after the measurement the system is described by the eigenfunction of the operator.
4. If a system is in the state ψ (not an eigenfunction of the measurement operator) then the average value of a number of repeated measurements of a quantity A whose corresponding operator is $\hat{A}$ is given by $\langle \psi, \hat{A}\psi \rangle$ (assuming that ψ is normalized).
5. The time evolution of a state is described by

$$\mathrm{j}\hbar \frac{\partial \psi}{\partial t} = \hat{H}\psi \tag{A1.14}$$

6. For a system of a number of identical particles, the only allowed state vectors are those which are either completely symmetric or completely anti-symmetric in all pairs of particle variables. (This is really a way of formulating the Pauli exclusion principle.)

From postulate 3 we know that a measurement projects the state vector into an eigenvector of the corresponding operator. It follows that if we can make simultaneous measurements of two quantities they must project the initial state into the *same* eigenstate. Two operators can do this provided they commute (see Eq. (A1.11)). For a given system there exists a maximal set of commuting operators, that is, we can perform only a limited number of compatible simultaneous measurements. Performing all these measurements at the same time gives us the maximum information that can ever be obtained about the system. Immediately after such a measurement the state vector is projected into the common eigenvector. If the eigenvalues of the measurements are a, b, c, etc. then the state is represented in Dirac's notation by

$$\psi = | a, b, c \ldots \rangle$$

However, the choice of a maximal set of observable, is not unique. The most-used set consists of the three components of the coordinate vector $\mathbf{r}$ (*coordinate representation*), but we could equally well use the three components of momentum (*momentum representation*).

In the w representation (where w stands in general for more than one variable) any arbitrary state can be expressed (see Eq. (A1.8)) as

$$| \rangle = \sum_{w'} C_{w'} | w' \rangle.$$

If w forms a continuum then this expression must be rewritten as

$$| \rangle = \int C(w') | w' \rangle$$

Once we specify the set $| w' \rangle$ the state is completely characterized by the function $C(w)$ (the *state function*). When we work in the coordinate representation then $C(w)$ is written as $\psi(\mathbf{r})$ and known as the *Schrodinger wave function*. Thus

$$| \rangle = \int \psi(\mathbf{r}') | \mathbf{r}' \rangle$$

The interpretation we place on this equation is that the function $\psi(\mathbf{r}')$ is the probability amplitude that the system is in the state $| \mathbf{r}' \rangle$ i.e. that the system is 'at' the point $\mathbf{r}'$. Since in fact $\mathbf{r}'$ forms a continuum, we say that the probability of finding the system within a small volume $\mathrm{d}\tau$ at the point $\mathbf{r}'$ is given by

$$| \psi(\mathbf{r}') |^2 \, \mathrm{d}\tau$$

From now on we will remain in the coordinate representation and also assume that, for simplicity, we are dealing with a single particle. The first thing to be decided is what our canonically conjugate pairs of variables are (postulate 1), and what mathematical form they must take to satisfy postulate 2. In the coordinate representation it turns out that position and momentum are suitable candidates and that the corresponding operators are

$$\hat{x} = x, \text{ etc. for } y, z$$

$$\hat{p}_x = \frac{\hbar}{\mathrm{j}} \frac{\partial}{\partial x} \text{ etc. for } \hat{p}_y \text{ and } \hat{p}_z$$

The energy operator (Hamiltonian) is a particularly important operator in quantum mechanics. For a single particle moving with a momentum p in a potential $V(\mathbf{r})$ the total energy, E, is given by

$$E = \frac{p^2}{2m} + V(\mathbf{r})$$

The corresponding Hamiltonian operator is then

$$\hat{H} = -\frac{\hbar^2}{2m} \nabla^2 + V(\mathbf{r}) \tag{A1.15}$$

where

$$\nabla^2 = \frac{\partial^2}{\partial x^2} + \frac{\partial^2}{\partial y^2} + \frac{\partial^2}{\partial z^2}$$

If the Hamiltonian does not depend on time, then Eq. (A1.14) may be solved by assuming that $\psi(\mathbf{r}, t) = u(\mathbf{r}) f(t))$. It is a fairly simple matter to show that

$$\psi(\mathbf{r}, t) = A \exp\left(-\frac{\mathrm{j}}{\hbar} Et\right) u(\mathbf{r}) \tag{A1.16}$$

and hence that

$$\hat{H}\psi(\mathbf{r}, t) = E\psi(\mathbf{r}, t)$$

Therefore $\psi(\mathbf{r}, t)$ is an eigenfunction of $\hat{H}$ with eigenvalue E.

Another very useful result also follows from Eq. (A1.14). The average value of the physical measurable quantity A is given by (postulate 4)

$$\langle A \rangle = \int \psi^* \hat{A} \psi \, \mathrm{d}\tau$$

and it may be deduced that

$$\frac{\partial \langle A \rangle}{\partial t} = \int \psi^* \left(\frac{\partial \langle \hat{A} \rangle}{\partial t} + \frac{1}{\mathrm{j}\hbar} [\hat{A}\hat{H} - \hat{H}\hat{A}] \right) \psi \, \mathrm{d}\tau \tag{A1.17}$$

If A does not depend explicitly on time then $\partial \langle A \rangle / \partial t = 0$ if $\hat{A}$ and $\hat{H}$ commute, that is, $\langle A \rangle$ will remain constant in time. It should be possible, therefore, to find functions that are simultaneously eigenfunctions of both $\hat{A}$ and $\hat{H}$.

With a time-independent Hamiltonian we know that the wavefunction depends on time as Eq. (A1.16), and we then seek functions $\psi_k(\mathbf{r})$, which satisfy the eigenvalue equation:

$$\hat{H}\psi_k(\mathbf{r}) = E_k\psi_k(\mathbf{r}) \qquad \text{(A1.18)}$$

Usually, however, it is not possible to decide immediately on the correct set of functions. The usual way of proceeding in these circumstances is to take a convenient complete set of functions $\phi_1(\mathbf{r})$. Because the set is complete we may then expand the $\psi_k(\mathbf{r})$ in terms of them, so that:

$$\psi_k = \sum_l C_{kl}\phi_l \qquad \text{(A1.19)}$$

It is then possible to show that the C_{kl} may then be determined from the relationship

$$\sum_l C_{kl}(H_{ml} - E_k\delta_{lm}) = 0 \qquad \text{(A1.20)}$$

where $H_{ml} = \int \phi_m \hat{H} \phi_l \, d\mathbf{r}$

The quantities H_{ml} may be regarded as elements in a matrix and Eq. (A1.20) as a (matrix) eigenvalue equation where to obtain the E_k we have to solve the equation

$$\det\,(H_{ml} - E_k\delta_{lm}) = 0 \qquad \text{(A1.21)}$$

To obtain the corresponding eigenfunctions we must diagonalize the matrix.

Perturbation theory

One of the main difficulties with quantum mechanics is that it is only possible to solve explicitly a few problems that are of physical interest. One way round this is the use of *perturbation techniques*. In essence, we assume we can solve a problem that is 'close' to the required one and then the real solutions will only require small modifications (or 'perturbations') to the solutions already obtained.

Suppose our Hamiltonian can be written

$$\hat{H} = \hat{H}_0 + \lambda\hat{H}'$$

where the eigenfunctions of $\hat{H}_0$ (which are known) are designated ψ_0, with eigenvalues E_0, and where the (required) eigenfunctions of $\hat{H}$ are ψ, with eigenvalues E. The factor λ attached to $\hat{H}'$ serves as a marker to enable the various levels of approximation to be recognized. Thus we assume that we can write

$$\psi = \psi_0 + \lambda\psi_1 + \lambda^2\psi_2 + \lambda^3\psi_3 + \ldots$$

$$E = E_0 + \lambda E_1 + \lambda^2 E_2 + \lambda^3 E_3 + \ldots$$

We must have

$$(\hat{H}_0 + \lambda\hat{H}')(\psi_0 + \lambda\psi_1 + \lambda^2\psi_2 + \ldots) =$$
$$(E_0 + \lambda E_1 + \lambda^2 E_2 + \ldots)(\psi_0 + \lambda\psi_1 + \lambda^2\psi_2 + \ldots)$$

Equating equal powers of:

$$\hat{H}_0\psi_0 = E_0\psi_0 \tag{A1.22}$$

$$\hat{H}_0\psi_1 + \hat{H}'\psi_0 = E_0\psi_0 + E_1\psi_0 \tag{A1.23}$$

$$\hat{H}_0\psi_2 + \hat{H}'\psi_1 = E_0\psi_2 + E_1\psi_1 + E_2\psi_0 \tag{A1.24}$$

For simplicity, we assume that the ψ_0 etc. are nondegenerate (i.e. have different energy eigenvalues) and form part of a set of ψ_0^n of eigenfunctions of $\hat{H}_0$, where specifically $\psi_0 = \psi_0^m$; $E_0 = E_0^m$, etc. Since the ψ^n form a complete set, we may expand ψ_1 in terms of them:

$$\psi_1 = \sum_n a_n^{(1)}\psi_0^n$$

Substituting this into Eq. (A1.19) gives

$$\sum_n a_n^{(1)}\hat{H}_0\psi_0^n + \hat{H}'\psi_0^m = E_0^m \sum_n a_n^{(1)}\psi_0^n + E_1^m\psi_0^m$$

Multiplying by $(\psi_0^k)^*$ and integrating over all space then gives

$$\sum_n a_n^{(1)}E_0^n\delta_{kn} + \int \psi_0^{k^*}\hat{H}'\psi_0^m\,\mathrm{d}\tau = E_0^m \sum_n a_n^{(1)}\delta_{nk} + E_1^m\delta_{km}$$

or putting

$$H'_{km} = \int \psi_0^{k^*}\hat{H}'\psi_0^m\,\mathrm{d}\tau$$

$$a_k^{(1)}E_0^k + H'_{km} = E_0^m a_k^{(1)} + E_1^m$$

Putting $k = m$ we obtain

$$E_1^m = H'_{mm} = \int \psi_0^*\hat{H}'\psi_0\,\mathrm{d}\tau \tag{A1.25}$$

while putting $k \neq m$ gives

$$a_k^{(1)} = \frac{H'_{km}}{E_0^m - E_0^m} \tag{A1.26}$$

To first order, therefore, the change in energy from the unperturbed eigenvalue is $\int \psi_0^*\hat{H}'\psi_0\,\mathrm{d}\tau$.

Higher-order perturbation theory can be carried out by starting from Eq. (A1.20) and proceeding similarly to the above, although the equations soon become too involved to be of much practical use.

Appendix 2

Spherical harmonics

Spherical harmonics, $Y_{l,m}(\theta, \phi)$, for $l = 0, 1, 2$

l	m	$Y_{l,m}(\theta, \phi)$
0	0	$\left(\frac{1}{4\pi}\right)^{1/2}$
1	1	$-\left(\frac{3}{8\pi}\right)^{1/2} \sin(\theta)\ e^{i\phi}$
1	0	$\left(\frac{3}{4\pi}\right)^{1/2} \cos(\theta)$
1	−1	$\left(\frac{3}{8\pi}\right)^{1/2} \sin(\theta)\ e^{-i\phi}$
2	2	$\left(\frac{15}{32\pi}\right)^{1/2} \sin^{2\theta}\ e^{2i\phi}$
2	1	$\left(\frac{15}{8\pi}\right)^{1/2} \sin(\theta)\cos(\theta)\ e^{i\phi}$
2	0	$\left(\frac{5}{16\pi}\right)^{1/2} (3\cos^2\theta - 1)$
2	−1	$\left(\frac{15}{8\pi}\right)^{1/2} \sin(\theta)\cos(\theta)\ e^{-i\phi}$
2	−2	$\left(\frac{15}{32\pi}\right)^{1/2} \sin^2\theta\ e^{-2i\phi}$

Radial functions (normalized) for the hydrogen atom

n	l	$R_{nl}(r)$
1	0	$2\left(\frac{1}{a_0}\right)^{3/2} e^{-r/a_0}$
2	0	$2\left(\frac{1}{2a_0}\right)^{3/2} e^{-r/2a_0}\left(1-\frac{r}{2a_0}\right)$
2	1	$\frac{1}{\sqrt{3}}\left(\frac{1}{2a_0}\right)^{3/2} e^{-r/2a_0} \frac{r}{a_0}$
3	0	$2\left(\frac{1}{3a_0}\right)^{3/2} e^{-r/3a_0}\left(1-\frac{2r}{3a_0}+\frac{2r^2}{27a_0^2}\right)$
3	1	$\frac{8}{9\sqrt{2}}\left(\frac{1}{3a_0}\right)^{3/2} \frac{r}{a_0}\left(1-\frac{r}{6a_0}\right)$
3	2	$\frac{4}{27\sqrt{10}}\left(\frac{1}{3a_0}\right)^{3/2} e^{-r/3a_0} \frac{r^2}{a_0^2}$

Appendix 3

Propagation in planar dielectric waveguides

We consider the propagation of a beam of light travelling down a planar symmetric waveguide in a zigzag fashion as shown in Figure 2.10. However, a beam of light has a finite width, so we must imagine a whole collection of similar parallel beams. The wavefront of such a beam can be obtained by drawing a perpendicular to any point on any of the rays. Now all points on the wavefront (if it is to be a true wavefront) must have the same phase (to within a multiple of 2π). In Figure A3.1 the line AC represents a wavefront connecting the points A and C both of which lie on the same ray trajectory. The total phase change along the ray path ABC must then be an integer multiple of 2π. Now the phase change due to the path length AB + BC is

$$(\mathrm{AB} + \mathrm{BC}) \frac{2\pi}{\lambda_0} n_1$$

where λ_0 is the vacuum wavelength of the radiation. Since $\mathrm{AB} + \mathrm{BC} = \mathrm{BC}(1 + \cos 2\theta)$ and $\mathrm{BC} = d/\cos\theta$ this phase difference can be written as

$$\frac{2d \cos(\theta) n_1 2\pi}{\lambda_0}$$

In determining the phase change between points A and C we must also include the phase change on reflection that will occur at points B and C. If this is denoted by $\phi(\theta)$[1] then the total phase shift between points A and C is

$$\frac{4\pi d n_1 \cos(\theta)}{\lambda_0} - 2\phi(\theta)$$

The term in $\phi(\theta)$ is negative since on reflection the phase increases whereas when travelling along a ray trajectory the phase decreases.

The condition for constructive interference at point C is then that:

$$2\pi m = \frac{4\pi d n_1 \cos(\theta)}{\lambda_0} - 2\phi(\theta) \tag{A3.1}$$

Equation (A3.1) may be readily solved by graphical means. Thus rewriting the equation gives

$$\frac{2\pi d n_1 \cos(\theta)}{\lambda_0} = m\pi + \phi(\theta) \tag{A3.2}$$

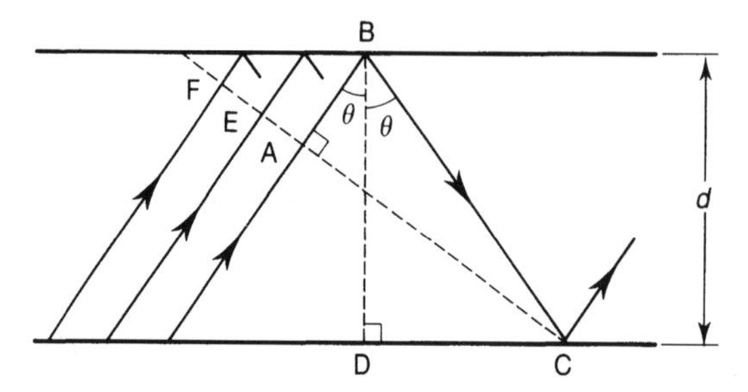

Figure A3.1 Ray paths which have the same internal angle θ within a planar waveguide. A wavefront is shown connecting the points F, E, A, C, which must therefore have the same phase.

In Figure A3.2 both the left- and right-hand sides of Eq. (A3.2) are plotted as a function of θ^2. Each different value of m will give rise to a different curve and the left-hand curve will intersect only a limited number of these. Furthermore, only when the intersection takes place between θ_c and 90° will we satisfy the total internal reflection requirement.

The resulting allowed values of θ for each value of m correspond to different modes, and are designated TE_m, or TM_m, depending on whether TE or TM radiation is involved.

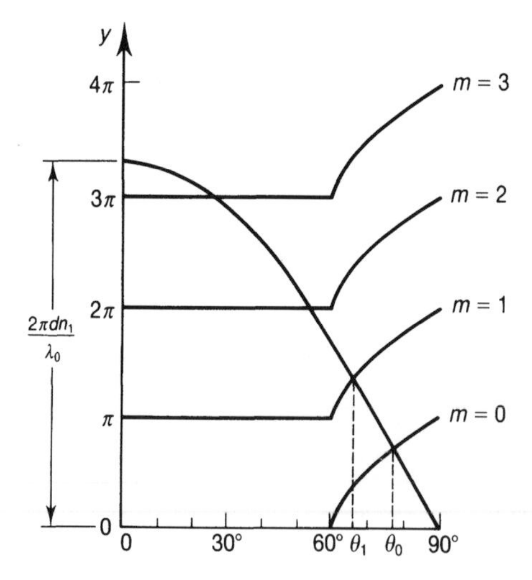

Figure A3.2 A plot of the curve $y = (2\pi d n_1 \cos(\theta))/\lambda_0$ and the set of curves $y = m\pi + \phi(\theta)$ with m taking the values 0, 1, 2 and 3 (the phase changes corresponding to TE polarization). The former intersects the $m = 0$ and $m = 1$ curves at the angle values θ_0 and θ_1, respectively. These are the only intersections at angles greater than the critical angle (here 60°). In this particular case, therefore, only two TE modes are possible, and these will correspond to internal guide angles of θ_0 and θ_1.

We can see from Figure A3.2 that as we decrease the value of $2\pi d n_1/\lambda_0$, then the number of modes will tend to decrease. However, one mode can always, in theory, propagate. The condition for mode $m = 1$ to *just* propagate is that

$$\frac{2\pi d n_1 \cos(\theta_c)}{\lambda_0} = \pi$$

or

$$\frac{2d}{\lambda_0}(n_1^2 - n_2^2)^{1/2} = 1$$

In terms of the parameter

$$V = \frac{\pi d}{\lambda_0}(n_1^2 - n_2^2)^{1/2}$$

the condition for single-mode operation is then that

$$V < \pi/2 \qquad \text{(A3.3)}$$

As we have seen, a given mode can be regarded as being made up of an 'infinite' collection of rays travelling down the guide with the same value of the internal angle. If we consider any particular point within the guide, only two rays can pass through it. One will be directed 'upwards' and the other 'downwards'. Since, in general, there will be a phase difference between these two, they will interfere and give rise to a variation in the field amplitude across the guide.

This phase difference may be determined from Figure A3.3. This shows two rays meeting at a point C, a distance y above the centre of the guide. The line AC represents a wavefront, and hence the phases at the points A and C must be the same. Thus the phase difference between the two rays meeting at C, $\Delta\phi(y)$, arises from a path difference of AB + BC together with a phase change of $\phi(\theta)$ on reflection at B.

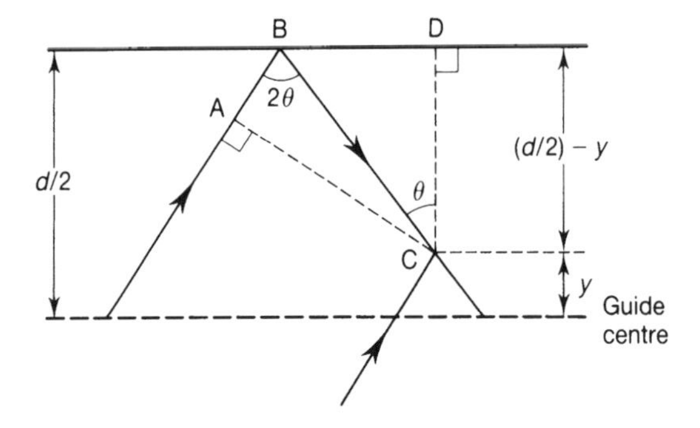

Figure A3.3 Two 'oppositely' directed rays with the same internal ray angle meeting at a point C which is at a distance y above the guide centre.

We have that $AB = BC \cos 2\theta$, so that $AB + BC = 2BC \cos^2\theta$. Since $BC = \{(d/2) - y\}/\cos\theta$, it follows that

$$AB + BC = 2\{(d/2) - y\}/\cos\theta$$

Thus we may write

$$\Delta\Phi(y) = 2\left(\frac{d}{2} - y\right)\frac{2\pi n_1 \cos\theta_m}{\lambda_0} - \phi(\theta)$$

Substituting for $\cos\theta_m$ from Eq. (A3.2) gives

$$\Delta\Phi(y) = \pi m - \frac{2y}{d}(\pi m + \phi(\theta)) \tag{A3.4}$$

The resultant of two waves with a phase difference of $\Delta\Phi(y)$ may be written

$$\mathscr{E}_0\{\cos(\omega t) + \cos(\omega t + \Delta\Phi(y))\} = 2\mathscr{E}_0 \cos\left[\omega t + \frac{\Delta\Phi(y)}{2}\right]\cos\left[\frac{\Delta\Phi(y)}{2}\right]$$

The effective amplitude of the electric field, $\mathscr{E}(y)$, is then

$$\mathscr{E}(y) = 2\mathscr{E}_0 \cos\left[\frac{\Delta\phi(y)}{2}\right]$$

or, using Eq. (A3.4),

$$\mathscr{E}(y) = 2\mathscr{E}_0 \cos\left[\frac{\pi m}{2} - \frac{y}{d}(\pi m + \phi)\right] \tag{A3.5}$$

Equation (A3.4) describes the variation of the field across the guide itself, it will be noticed that it has a finite value at the boundary of the guide. This means that there must be a field within the cladding since it is a requirement that the tangential components of the fields across the interfaces should be the same on either side of the interface. We may deduce the form of this field by examining in more detail the consequences of total internal reflection.

If θ_i is less than θ_c we have

$$n_1 \sin\theta_i = n_2 \sin\theta_t \tag{A3.6}$$

so that

$$\cos\theta_t = \left[1 - \left(\frac{n_1}{n_2}\right)^2 \sin^2\theta_i\right]^{1/2} \tag{A3.7}$$

When $\theta_i > \theta_t$, $\sin\theta_i > n_2/n_1$ and $\cos\theta_t$ becomes wholly imaginary, so that we write

$$\cos\theta_t = \pm jB \tag{A3.8}$$

where

$$B = \left[\left(\frac{n_1}{n_2}\right)^2 \sin^2\theta_i - 1\right]^{1/2}$$

Although the idea of an imaginary angle seems rather farfetched we accept it for the moment and see where it leads.

The phase factor for the transmitted wave may be written as

$$\mathscr{P} = \exp\ \mathrm{j}(\omega t - \mathbf{k}_t.\mathbf{r})$$

where $\mathbf{k}_t$ is the wave vector associated with the transmitted wave. In terms of the coordinates z (along the guide) and y, we can write r as (see Figure A3.4)

$$r = z \sin \theta_t + y \cos \theta_t$$

Hence

$$\mathscr{P} = \exp\ \mathrm{j}\left[\omega t - \frac{2\pi n_2}{\lambda_0}(z \sin \theta_t + y \cos \theta_t)\right]$$

Substituting for $\sin \theta_t$ and $\cos \theta_t$ from Eqs (A3.6) and (A3.7) gives

$$\mathscr{P} = \exp\left(\pm B \frac{2\pi n_2}{\lambda_0} y\right) \exp\ \mathrm{j}\left(\omega t - \frac{2\pi n_1 \sin \theta_i}{\lambda_0} z\right) \tag{A3.9}$$

Thus in the y direction the wave either grows or decays exponentially with distance. The former situation may be rejected as nonphysical, and so we must choose $\cos \theta_t = -\mathrm{j}B$ in Eq. (3.8).

The rate of decay with distance in the cladding is determined by the factor $F(y)$ where

$$F(y) = \exp\left\{-\frac{2\pi n_2}{\lambda_0}\left[\left(\frac{n_1}{n_2}\right)^2 \sin^2\theta_i - 1\right]^{1/2} y\right\} \tag{A3.10}$$

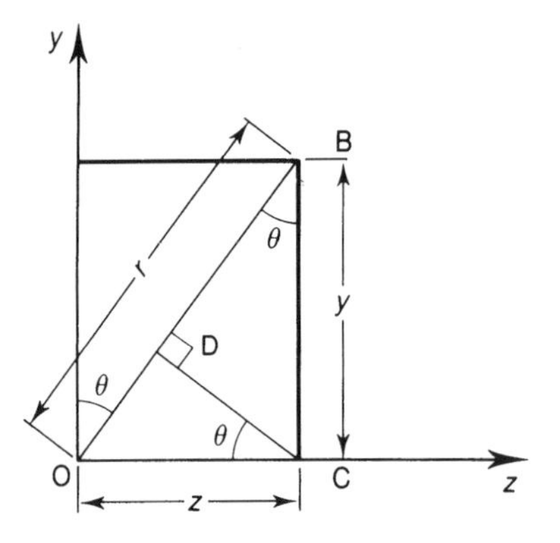

Figure A3.4 An illustration of the relationship between the rectangular coordinates y and z and the distance r measured from the origin. From the figure OB = OD + DB = OC $\sin \theta$ + BC $\cos \theta$. Thus in terms of the coordinates, $r = z \sin \theta + y \cos \theta$.

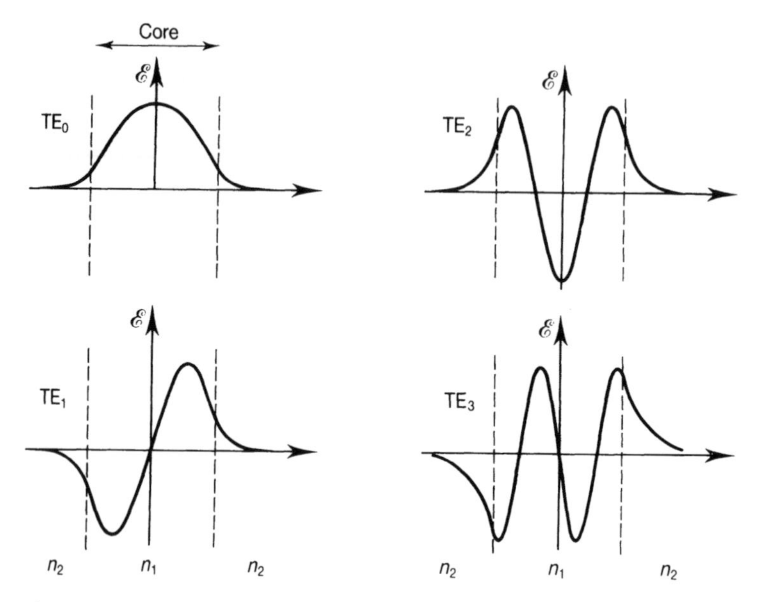

Figure A3.5 Field· distribution for TEM_m modes in a planar dielectric waveguide.

If θ_i ($=\theta$ in the main text) is not particularly close to θ_c then the term

$$2\pi n_2 \left[\left(\frac{n_1}{n_2}\right)^2 \sin^2\theta_i - 1\right]^{1/2}$$

is of order unity or greater, so that the decay takes place over a distance $\approx \lambda_0$ or less. In the context of single-mode waveguides where the layer thicknesses are $\approx \lambda_0$ in any case, then the decay factor is significant. That part of the mode that is in the cladding is often referred to as the *evanescent mode.*

Figure A3.5 shows the field variation across the core of a planar waveguide for a few of the lowest-order TE_m modes.

Notes

1. The phase change will, of course, be different for TE or TM polarizations.
2. See Eqs (2.64) and (2.65) for the functional form of $\phi(\theta)$ for TE and TM polarizations.

Appendix 4

The Wiener–Khintchine theorem

The frequency spectrum $\mathscr{E}(\omega)$ of a field $\mathscr{E}(t)$ is given by taking the Fourier transform

$$\mathscr{E}(\omega) = \frac{1}{2\pi} \int_{-\infty}^{\infty} \mathscr{E}(t)\exp(\mathrm{j}\omega t)\, \mathrm{d}t$$

Now the Fourier transform of the autocorrelation of $\mathscr{E}(t)$, $\gamma_{11}(\tau)$, is given by

$$\mathscr{T}[\Gamma_{11}(\tau)] = \frac{1}{2\pi} \int_{-\infty}^{\infty} \exp(\mathrm{j}\omega\tau)\left[\int_{-\infty}^{\infty} \mathscr{E}^*(t)\mathscr{E}(t+\tau)\, \mathrm{d}t\right] \mathrm{d}\tau$$

Changing the order of integration

$$\mathscr{T}[\Gamma_{11}(\tau)] = \frac{1}{2\pi} \int_{-\infty}^{\infty} \mathscr{E}^*(t)\left[\int_{-\infty}^{\infty} \exp(\mathrm{j}\omega\tau)\mathscr{E}(t+\tau)\, \mathrm{d}\tau\right] \mathrm{d}t$$

If we now put $t = -t'$ and $\tau - t' = T$, we have

$$\mathscr{T}[\Gamma_{11}(\tau)] = \frac{1}{2\pi} \int_{-\infty}^{\infty} \mathscr{E}^*(-t')\left[\int_{-\infty}^{\infty} \exp(\mathrm{j}\omega\tau' + \mathrm{j}\omega T)\mathscr{E}(T)\, \mathrm{d}\tau\right] \mathrm{d}t'$$

$$\mathscr{T}[\Gamma_{11}(\tau)] = \frac{1}{2\pi} \left[\int_{-\infty}^{\infty} \exp(\mathrm{j}\omega t')\mathscr{E}^*(-t')\, \mathrm{d}t'\right]\left[\int_{-\infty}^{\infty} \exp(\mathrm{j}\omega T)\mathscr{E}(T)\, \mathrm{d}T\right]$$

Now

$$\int_{-\infty}^{\infty} \exp(\mathrm{j}\omega t')\mathscr{E}^*(-t')\, \mathrm{d}t' = \int_{-\infty}^{\infty} \exp(-\mathrm{j}\omega t)\mathscr{E}^*(t)\, \mathrm{d}t$$

$$= \left[\int_{-\infty}^{\infty} \exp(\mathrm{j}\omega t)\mathscr{E}(t)\, \mathrm{d}t\right]^*$$

Thus

$$\begin{aligned}\mathscr{T}[\Gamma_{11}(\tau)] &= 2\pi(\mathscr{T}[\mathscr{E}(t)])^*\,\mathscr{T}[\mathscr{E}(t)] \\ &= 2\pi\mathscr{E}^*(\omega)\mathscr{E}(\omega) \\ &= 2\pi \mathrm{I}(\omega)\end{aligned}$$

If we now carry out the inverse Fourier transform:

$$\Gamma_{11}(\tau) = 2\pi \int_{-\infty}^{\infty} \exp(-\mathrm{j}\omega t)P(\omega)\, \mathrm{d}\omega$$

It is convenient to normalize this expression by dividing both sides by the quantity proportional to the total power in the beam. Now we have

$$\int_{-\infty}^{\infty} \mathscr{E}^*(t)\mathscr{E}(t)\,\mathrm{d}t = \int_{-\infty}^{\infty} \mathscr{E}(t)\left[\int_{-\infty}^{\infty} \exp(-\mathrm{j}\omega t)\mathscr{E}^*(\omega)\,\mathrm{d}\omega\right]\mathrm{d}t$$

$$= \int_{-\infty}^{\infty} \mathscr{E}^*(\omega)\left[\int_{-\infty}^{\infty} \exp(-\mathrm{j}\omega t)\mathscr{E}(t)\,\mathrm{d}t\right]\mathrm{d}\omega$$

$$= \int_{-\infty}^{\infty} \mathscr{E}^*(\omega)\mathscr{E}(\omega)\,\mathrm{d}\omega$$

$$= \int_{-\infty}^{\infty} P(\omega)\,\mathrm{d}\omega$$

Thus

$$\frac{\Gamma_{11}(\tau)}{\int_{-\infty}^{\infty} \mathscr{E}^*(t)\mathscr{E}(t)\,\mathrm{d}t} = \frac{2\pi \int_{-\infty}^{\infty} \exp(-\mathrm{j}\omega t)P(\omega)\,\mathrm{d}\omega}{\int_{-\infty}^{\infty} P(\omega)\,\mathrm{d}\omega}$$

or

$$\gamma_{11}(\tau) = \int_{-\infty}^{\infty} g(\omega)\exp(-\mathrm{j}\omega t)\,\mathrm{d}\omega$$

Appendix 5

The condition for direct optical transitions between conduction and valence bands

From Eq. (3.18) the (classical) interaction involves the term **A.p**. The corresponding quantum-mechanical operator will then involve (as far as spatial variables are concerned) the operator $\hat{p}$ or $-j\hbar\nabla$ (i.e. $-j\hbar(\partial/\partial x)$ in one dimension). It may be shown [Ref. 1] that appropriate (one-dimensional) wavefunctions are of the form $u(x)\exp(jkx)$, where $u(x)$ has the periodicity of the lattice. The transition probability will then be proportional to the quantity

$$M_{\rm if} = -j\hbar \int \exp(-jk_{\rm f}x)u_{\rm f}^*(x)\frac{\partial}{\partial x}\exp(jk_{\rm i}x)u_{\rm i}(x)\,{\rm d}x$$

where the integral extends over the volume of the solid. Thus

$$M_{\rm if} = -j\hbar \int \exp(jx(k_{\rm i}-k_{\rm f}))u_{\rm f}^*(x)\frac{\partial}{\partial x}u_{\rm i}(x)\,{\rm d}x$$

$$+j\hbar k_{\rm i} \int \exp(jx(k_{\rm i}-k_{\rm f}))u_{\rm f}^*(x)u_{\rm i}(x)\,{\rm d}x$$

The second integral may be neglected since over each unit cell[1] the exponential function varies slowly whereas the functions $u_{\rm f}(x)$ and $u_{\rm i}(x)$ are orthogonal.

As far as the first integral is concerned, again the exponential function varies slowly over each unit cell while the integral of the function $u_{\rm f}^*(x)\,(\partial/\partial x)\,u_{\rm i}(x)$ will be the same over each unit cell in the solid, so that we can write

$$M_{\rm if} = -j\hbar \sum_{\substack{\text{all unit}\\ \text{cells}}} \exp(jx(k_{\rm i}-k_{\rm f})) \int_{\substack{\text{unit}\\ \text{cell}}} u_{\rm f}^*(x)\frac{\partial}{\partial x}u_{\rm i}(x)\,{\rm d}x$$

If the unit cell length is l, and the value of the integral over the unit cell is $m_{\rm if}$, then we have

$$M_{\rm if} = -\frac{j\hbar}{l}\,m_{\rm if}\int \exp(jx(k_{\rm i}-k_{\rm f}))\,{\rm d}x$$

Again the integral extends over the length, L, of the solid. Now

$$\int_0^L \exp(jx(k_i - k_f))\,dx = \frac{\exp j(k_i - k_f)L - 1}{j(k_i - k_f)} \quad (k_i \neq k_f)$$

or

$$= L \quad (k_i = k_f)$$

Since both k_i and k_f are integer multiples of $2\pi/L$, this integral can only be finite if $k_i = k_f$.

Note

1. Remember, the functions $u(x)$ are periodic in the lattice spacing.

Reference

N. W. Ashcroft and N. D. Mermin, *Solid State Physics*, Holt Rhinehart and Winston, New York, 1976, pp. 132–41.

Appendix 6

Error signal using third harmonic of frequency-modulated signal

Formal derivation to show that the third derivative of a saturated absorption feature is obtained by third harmonic detection after frequency modulation of the saturating light beam.

The laser is frequency modulated over a bandwidth ν_m at an audio frequency ω and the absorption feature occurs at some frequency ν within the gain bandwidth. The laser irradiance depends on the position of the absorption in the gain bandwidth so it is described by a function $F(\nu)$. We can write for the variation of the irradiance

$$I(t) = F(\nu) + (\nu_m \sin \omega t)$$

If the modulation amplitude is small we can expand as a Taylor's series:

$$I(t) = F(\nu) + (\nu_m \sin \omega t)F'(\nu) = \left(\frac{\nu_m^2}{2} \sin^2 \omega t\right) F'(\nu)$$
$$+ \left(\frac{\nu_m^3}{6} \sin^3 \omega t\right) F'''(\nu) + \ldots$$

where the primes denote the order of derivative of F as a function of ν.

This series is of the form

$$I(t) = A \sin \omega t + B \sin 2\omega t + C \sin 3\omega t + \ldots$$

The coefficients for each of these harmonics can be shown to be

$$A = \nu_m F_2'(\nu) + \frac{\nu_m^3}{8} F'''(\nu) + \ldots$$

$$B = -\frac{\nu_m^2}{4} F''(\nu) - \frac{1}{96} F''''(\nu) + \ldots$$

$$C = -\frac{\nu_m^3}{24} F'''(\nu) - \frac{2}{2120} F'''''(\nu) + \ldots$$

Frequency modulation and phase-sensitive detection of the light intensity at the nth harmonic will produce a voltage proportional to the nth derivative with small contributions from $(n+2)$th derivative. The saturated absorption feature will have a Lorentzian form and and will appear on a background of the intensity modulation due to the gain curve. We can write the function which describes this for mathematical ease as a Lorentzian centred at frequency $\nu = 0$ (Eq. (2.53)) on a quadratic background thus:

$$F(\nu) = \frac{\gamma}{\gamma^2 + \nu^2} + a\nu^2 + b\nu + c$$

The orders of derivative then become

$$F'(\nu) = \frac{-2\nu\gamma}{(\nu^2 + \gamma^2)^2} + 2a\nu + b$$

$$F''(\nu) = \frac{6\nu^2\gamma - 2\gamma^3}{(\nu^2 + \gamma^2)^3} + 2a$$

$$F'''(\nu) = \frac{24\nu\gamma^3 - 24\nu^3\gamma}{(\nu^2 + \gamma^2)^4}$$

The first derivative will give a dispersion-shaped curve on a straight background. The second derivative is an even function and so cannot provide a zero crossing signal necessary for the servoloop. The third derivative provides the frequency discriminant on a zero background at $\nu = 0$.

Appendix 7

Optical heterodyne detection

A photon detector such as a photodiode or avalanche photodiode will produce a current, i, in the external circuit given by

$$i = e\eta_{qe}P_{abs}/h\nu$$

where P_{abs} is the power absorbed by the photodetector and η_{qe} is its quantum efficiency.[1] This is square law detector and the power absorbed is proportional to the square of the total electric field of the electromagnetic waves. Hence

$$i \propto P_{abs} \propto [\mathscr{E}_1 \cos \omega_1 t + \mathscr{E}_2 \cos \omega_2 t]^2$$
$$i \propto P_1 \cos^2\omega_1 t + 2P_1^{1/2} P_2^{1/2} \cos \omega_1 t + P_2 \cos^2\omega_2 t$$

Using the trigonometric identities:

$$P_1 \cos^2\omega_1 t = \tfrac{1}{2}P_1(1 + \cos 2\omega_1 t)$$
$$P_2 \cos^2\omega_2 t = \tfrac{1}{2}P_2(1 + \cos 2\omega_2 t)$$
$$2P_1^{1/2} P_2^{1/2} \cos(\omega_1 t)\cos(\omega_2 t) = P_1^{1/2} P_2^{1/2} [\cos(\omega_1 + \omega_2)t + \cos(\omega_1 + \omega_2)t]$$

Thus the photodetector current becomes

$$i \propto \underbrace{P_1 + P_2}_{\text{DC terms}} + \underbrace{2P_1^{1/2} P_2^{1/2} \cos(\omega_1 - \omega_2)t}_{\text{heterodyne term}}$$

$$+ \underbrace{2P_1^{1/2} P_2^{1/2} \cos(\omega_1 + \omega_2)t + P_1 \cos 2\omega_1 t + P_2 \cos 2\omega_2 t}_{\text{terms at optical frequencies}}$$

The detector and associated electronics will not respond at the optical frequencies and the resulting signal will be a DC current proportional to the total power incident on the detector and a beat signal whose amplitude is proportional to $\sqrt{P_1 P_2}$. In principle, therefore, the heterodyne signal can be increased by increasing the power of one of the sources. This principle is used in the detection of weak signals in communications where the strong signal is the local oscillator. Eventually, however, the local oscillator will generate so much shot noise that the signal-to-noise ratio of the heterodyne signal becomes too small for detection.

Note

1. The quantum efficiency is the ratio of the number of carriers generated to the number of photons absorbed.

Appendix 8

Laser safety

The question of safe and good work practices with lasers has been a subject of debate since the first laser operated in 1959. The variety of output characteristics, namely, wavelength, pulsed or CW, power or energy, duration and mode profile make the defining of safe limits exceedingly difficult, and as a result of working experience these limits are under continuous scrutiny and revision. This appendix is only a summary of the current code of practice in Europe [Ref. 1] and is not complete but will serve as an introduction the subject. The ANSI code of practice [Ref. 2] in the United States is similar to the European.

We must first distinguish between hazards due to the beam itself and other hazards associated with apparatus required to operate the laser such as electricity, chemicals and toxic and cyrogenic material. It is most unlikely that the beam could cause death, contrasting with the other nonoptical hazards which have over the years caused fatalities. The beam will cause damage to biological tissue by heating and by photochemical effects, and each is determined by the input energy and wavelength. In addition, for very short pulses thermo-acoustic shock waves cause damage. Since damage to the light-sensitive parts of the eye do not self-repair in the same way as does burning of the skin, then much more care and control need to be exercised to prevent damage to the eye. In addition, the lens of the eye focuses the light and greatly increases the irradiance by up to a factor of 10^5. The codes of practice summarized here will address mainly the hazard to the eye, but safe limits for skin burns are also specified in these codes.

An anatomical diagram of the eye is shown in Figure A8.1. Light is focused by the lens onto the retina which contains light-sensitive cones and the less sensitive colour sensors, the rods. The fovea is the most sensitive part of the retina and damage here can cause almost complete blindness while similar damage elsewhere on the retina may go completely unnoticed. In the wavelength range 400–1400 nm the greatest hazard is retinal damage since the eye is transparent at these wavelengths. Below 400 nm in the ultra-violet photochemical effects occur, causing damage to the front of the eye and above 1400 nm burning of the cornea takes place.

In Table A8.1 the *maximum permissible exposure* (MPE) levels for which the human eye will not suffer adverse effects for direct exposure to laser radiation (intra-beam viewing) are shown as functions of wavelength and exposure times. These levels have been determined by experimental data taken with animal eyes along with an added safety factor. Maximum permissible exposure for diffuse light and for exposure of the skin to laser radiation can be obtained from Ref. [1].

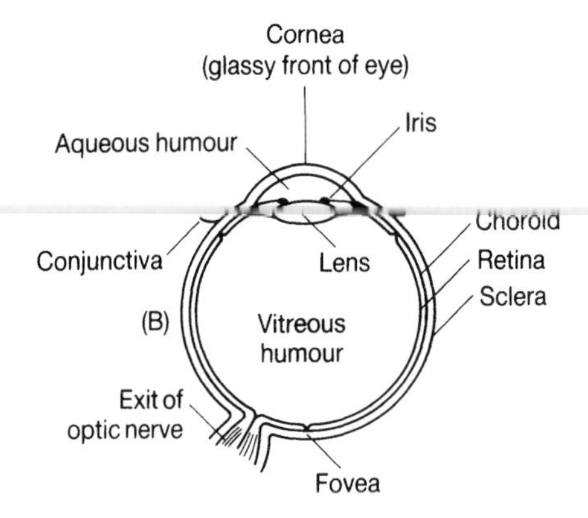

Figure A8.1 Anatomical diagram of the eye.

Lasers are given a classification depending on the hazard which they present and the control measures required to protect against these hazards will be determined by the classification of the laser. These latter control measures are known as *engineering controls*. It is the responsibility of the laser manufacturer to assess the laser classification and mark the laser head (the container from which the beam is emitted) accordingly with the markings and information determined by the code of practice. The classification of any laser is determined by its *accessible emission limit* (AEL) and the rationale for these classifications and a summary of the recommended engineering controls are summarized below. Details of the laser classifications are given in Ref. [1].

	Hazard	Control
Class 1	The output is so low that the laser is inherently safe or The laser is totally enclosed and it is safe by engineering design.	No control other than labelling
Class 2	These are visible CW and pulsed lasers where the protection can be afforded by the blink reflex of 0.25 s.	The laser head must be labelled. The beam should be terminated. No beams at eye level.
Class 3A	Class 2 is extended to exclude viewing with optical instruments.	Extra caution than class 2.
Class 3B*	This is a sub-class of 3B confined to visible lasers with less than 5 mW CW output.	Treated like 3A

Table A8.1 Maximum permissible exposure (MPE) at the cornea for direct ocular exposure to laser radiation (intra-beam viewing)

Wavelength λ(nm) \ Exposure time t(s)	$< 10^{-9}$	10^{-9} to 10^{-7}	10^{-7} to 10^{-6}	10^{-6} to 1.8×10^{-5}	1.8×10^{-5} to 5×10^{-5}	5×10^{-5} to 10	10 to 10^3	10^3 to 10^4	10^4 to 3×10^4
180–302.5		30 J m^{-2}							
302.5–315	3×10^{10} W m^{-2}	C_1 J m^{-2} $t < T_1$			C_2 J m^{-2} $t > T_1$		C_2 J m^{-2}		
315–400		C_1 J m^{-2}					10^4 J m^{-2}	10 W m^{-2}	
400–550							100 J m^{-2}		10^{-2} W m^{-2}
550–700	5×10^6 W m^{-2}	5×10^{-3} J m^{-2}			$18t^{0.75}$ J m^{-2}		$18t^{0.75}$ J m^{-2} $t < T_2$	$C_3 \times 10^2$ J m^{-2} $t > T_2$	$C_3 \times 10^{-2}$ W m^{-2}
700–1050	$5 \times C_4 \times 10^6$ W m^{-2}	$5 \times 10^{-3} \times C_4$ J m^{-2}			$18 \times C_4 t^{0.75}$ J m^{-2}			$3.2 \times C_4$ W m^{-2}	
1050–1400	5×10^7 W m^{-2}	5×10^{-2} J m^{-2}				$90 \times t^{0.75}$ J m^{-2}		16 W m^{-2}	
1400–1530		100 J m^{-2}	$5600 \times t^{0.25}$ J M^{-2}						
1530–1550	10^{11} W m^{-2}	1.0×10^4 J m^{-2}		$5600 \times t^{0.25}$ J m^{-2}			1000 W m^{-2}		
1550–10^6		100 J m^{-2}	$5600 \times t^{0.25}$ J m^{-2}						

Diameter of limiting apertures shall be for:
1 mm, $200 < \lambda < 400$ nm
7 mm, $400 < \lambda < 1400$ nm
1 mm, $1400 < \lambda < 10^5$ nm
11 mm, $10^5 < \lambda < 10^6$ nm

Table A8.2

Parameter	Spectral region
$C_1 = 5.6 \times 10^3 t^{0.25}$	302.5– 400 nm
$T_1 = 10^{0.8(\lambda - 295)} \times 10^{-15}$ g	302.5– 315 nm
$C_2 = 10^{0.2(\lambda - 295)}$	302.5– 315 nm
$T_2 = 10 \times 10^{0.02(\lambda - 550)}$ s	550 – 700 nm
$C_3 = 10^{0.015(\lambda - 550)}$	550 – 700 nm
$C_4 = 10^{(\lambda - 700)/900}$	700 –1050 nm

	Hazard	Control
Class 3B**	Eye hazard from intra-beam viewing and from specular reflections.	A designated laser area and interlocks must be used. Eye protection required for exposed beams. Beam terminated.
Class 4	Hazard from diffuse reflection and possibly to skin	As for Class 3B** Protective clothing recommended.

Having devised the engineering controls for the laser as determined by its classification the *administrative controls* must be defined. These include the appointment of a *laser safety officer* and the lines of responsibility. The laser safety officer will determine the rules for safe use of lasers in the organization and be responsible for educating users in the safe work practices.

The final protection comes from *personal control* (personal protective equipment) by using the appropriate goggles and other protective clothing. Lasers may produce light at more than one wavelength and the goggles must protect against hazardous radiation at all wavelengths emitted. If lasers are to be used out of doors they are subject to modifications of these rules and users should consult the code of practice.

The use of lasers often presents other hazards not associated with the beam some of which are potentially more dangerous. These include; electrical, water supplies, belts and rotating machinery, ultra-violet from light sources used to pump the laser, X-rays, toxic chemicals, explosion and cryogenic liquids. All these hazards must be assessed and controlled.

Example A8.1 Determination of laser classification and optical density of protective eyewear

A Nd : YAG laser operating at 1.06 μm has a peak output power in a Q-switched pulse of 20 MW of duration 100 ns having a spot size of 1 mm. Determine the classification of this laser and find the optical density of the goggles required so that the MPE is not exceeded.

$$\text{The peak irradiance of this beam} = \frac{20 \times 10^6}{\pi \times (10^{-3})^2} = 7 \times 10^{12}\ \text{W/m}^2$$

$$\begin{aligned}\text{The average fluence over the duration of the pulse} &= 7 \times 10^{12} \times 10^{-8} \\ &= 7 \times 10^4\ \text{J/m}^2\end{aligned}$$

From Ref. [1] the accessible emission limit for this type of laser is 10^5 J/m^2 if it is to be classed as a class 3B** laser.

From Table A8.1 for this laser the MPE $= 5 \times 10^{-2}$ J/m^2.

$$\text{Optical Density (OD) of goggles} = \log_{10} \frac{7 \times 10^4}{5 \times 10^{-2}} = \log_{10} 1.4 \times 10^6 \cong 6$$

This calculation refers to an output of one pulse. However, if the laser were repetitively pulsed then a correction factor must be applied to reduce the MPE. The MPE used is the most restrictive (i.e. the smallest) of the following three alternatives:

1. The exposure from any single pulse within the train shall not exceed the MPE for a single pulse (Table A8.1).
2. The average irradiance for a pulse train of duration, t, shall not exceed the MPE given in Table A8.1, where t is the time duration of the exposure. In the case of visible lasers this can be 0.25 s (i.e. the blink response) while for invisible lasers it can nominally be taken as 10 s.
3. The exposure from any single pulse within the train shall not exceed the MPE for a single pulse multiplied by $N^{-1/4}$, where N is the number of pulses in the exposure time, t.

Example A8.2 If the laser in Example A8.1 is repetitively pulsed at 50 Hz, determine the laser classification and the OD of the goggles required

Energy per pulse $= 20 \times 10^{6} \times 10^{-8} = 0.2$ J/pulse
Average power $= 0.2 \times 50 = 10$ W
This is above 0.5 W and so this is a Class 4 laser.

We will first calculate the MPE for this type of laser.

The number of pulses in a 10 s exposure = 500

(a) From above the exposure for a single pulse $= 0.05\ \mathrm{J/m^2}$.
(b) From Table A8.1 The average permitted irradiance for a $t = 10$ s exposure

$$= 90\ t^{0.75}/t$$
$$= 50\ \mathrm{W/m^2}$$

This is reduced by the duration of the pulses and becomes

$$\text{Average irradiance} = 50/50 \times 10^{-8} = 10^{8}\ \mathrm{W/m^2}$$

This is used to calculate the equivalent single pulse exposure

$$= 10^{8} \times 10^{-8} = 1\ \mathrm{J/m^2}$$

(c) MPE for a single pulse from (a) $= 0.05\ \mathrm{J/m^2}$
MPE corrected for the train of pulses $= 0.05 \times (500)^{-1/4}$
$= 0.05 \times 0.211$
$= \mathit{0.01\ J/m^2}$

The lowest of these is (c) and the MPE to be used $= \mathit{0.01\ J/m^2}$

$$\mathrm{OD} = \log_{10} \frac{7 \times 10^{4}}{0.01} = \log_{10} 7 \times 10^{6} \cong 7$$

References

[1] BS EN 60825:1992 Radiation Safety of Laser Products, Equipment and Classification, Requirements and users guide.

[2] American National Standards Institute, ANSI-Z-136, Standard for the Safe Use of Lasers.

[3] The Committee of Vice-Chancellors and Principals of the Universities of the United Kingdom: *Safety in Universities, Notes for guidance*, Part 2 : 1, *Lasers*, 3rd edition, 1992.

Appendix 9

Physical constants and properties of some common semiconductors at room temperature (300 K)

Rest mass of electron	m	$= 9.110 \times 10^{-31}$ kg $= 0.000549$ u
Rest mass of a proton	m_p	$= 1.673 \times 10^{-27}$ kg
Charge of electron	e	$= 1.602 \times 10^{-19}$ C
Electron charge/mass ratio	e/m	$= 1.759 \times 10^{11}$ C kg^{-1}
Avogadro's constant	N_A	$= 6.022 \times 10^{23}$ mol^{-1}
Planck's constant	h	$= 6.626 \times 10^{-34}$ J s
	$\hbar = h/2\pi$	$= 1.055 \times 10^{-34}$ J s
Boltzmann's constant	k	$= 1.381 \times 10^{-23}$ J K^{-1}
Speed of light (in vacuum)	c	$= 2.998 \times 10^{8}$ m s^{-1}
Permittivity of a vacuum	ε_0	$= 8.854 \times 10^{-12}$ F m^{-1}
Permeability of a vacuum	μ_0	$= 4\pi \times 10^{-7} = 1.258 \times 10^{-6}$ H m^{-1}
Stefan–Boltzmann constant	σ	$= 5.670 \times 10^{-8}$ W m^{-2} K^{-4}
Density of mercury at 20°C		$= 1.36 \times 10^{4}$ kg m^{-3}
Standard atmospheric pressure (760 mmHg)		$= 1.013 \times 10^{5}$ N m^{-2}

Property	Si	Ge	GaAs
Atomic (molecular) weight	28.09	72.60	144.6
Energy gap, E_g (eV)	1.12	0.67	1.43
Intrinsic carrier concentration n_i (m^3)	1.5×10^{16}	$24. \times 10^{19}$	1×10^{13}
Electron mobility, μ_e (m^2 V^{-1} s^{-1})	0.135	0.39	0.85
Hole mobility, μ_h (m^2 V^{-1} s^{-1})	0.048	0.19	0.045
Relative permittivity, ε_r	11.8	16.0	10.9†
Electron effective mass‡, m_e^* (×m)	0.12	0.26	0.068
Hole effective mass,‡ (×m)	0.23	0.38	0.56

† The range of values of relative permittivity of GaAs quoted in the literature is from 10.7 to 13.6. We have adopted the value 10.9.

‡ Two different 'effective masses' are defined, namely the *density of states* and *conductivity* of effective masses. The values given here are representative of those quoted for the conductivity effective mass.

Index